IUTAM SYMPOSIUM ON INTERACTION BETWEEN DYNAMICS AND
CONTROL IN ADVANCED MECHANICAL SYSTEMS

SOLID MECHANICS AND ITS APPLICATIONS
Volume 52

Series Editor: **G.M.L. GLADWELL**
Solid Mechanics Division, Faculty of Engineering
University of Waterloo
Waterloo, Ontario, Canada N2L 3G1

Aims and Scope of the Series

The fundamental questions arising in mechanics are: *Why?*, *How?*, and *How much?*
The aim of this series is to provide lucid accounts written by authoritative researchers giving vision and insight in answering these questions on the subject of
mechanics as it relates to solids.

The scope of the series covers the entire spectrum of solid mechanics. Thus it
includes the foundation of mechanics; variational formulations; computational
mechanics; statics, kinematics and dynamics of rigid and elastic bodies; vibrations
of solids and structures; dynamical systems and chaos; the theories of elasticity,
plasticity and viscoelasticity; composite materials; rods, beams, shells and
membranes; structural control and stability; soils, rocks and geomechanics;
fracture; tribology; experimental mechanics; biomechanics and machine design.

The median level of presentation is the first year graduate student. Some texts are
monographs defining the current state of the field; others are accessible to final
year undergraduates; but essentially the emphasis is on readability and clarity.

For a list of related mechanics titles, see final pages.

IUTAM Symposium on

Interaction between Dynamics and Control in Advanced Mechanical Systems

Proceedings of the IUTAM Symposium
held in Eindhoven, The Netherlands,
21–26 April 1996

Edited by

D. H. VAN CAMPEN

Department of Mechanical Engineering,
Eindhoven University of Technology,
Eindhoven, The Netherlands

SPRINGER-SCIENCE+BUSINESS MEDIA, B.V.

A C.I.P. Catalogue record for this book is available from the Library of Congress.

ISBN 978-94-010-6439-2   ISBN 978-94-011-5778-0 (eBook)
DOI 10.1007/978-94-011-5778-0

Printed on acid-free paper

CONTENTS

Contributed Papers

1. J.G. de Weger
2. K. Pottie
3. N.B.O.L. Pettit
4. R. Mettin
5. R.S. Sharp
6. M.J.G. van de Molengraft
7. M.J.M. Strik
8. L.H.G. Wcuters
9. A. Darby
10. M.F. Heertjes
11. A. Dequidt
12. J. van der Looij
13. M. de Pater
14. A.D. de Pater
15. R.H.B. Fey
16. D.L. Grozeva
17. P.M.R. Wortelboer
18. M.H.L.H. Kusters
19. H. Hu
20. S.R. Bishop
21. H. Irschik
22. D. Pogorelov
23. O.P. Salimova
24. V.V. Beletsky
25. F.E. Veldpaus
26. J.G.A.M. van Heck
27. G.H.M. van der Heijden
28. H. Nijmeijer
29. E.J. Kostelich
30. K. Schlacher
31. M. Ding
32. P. Maißer
33. E. Kreuzer
34. H. True
35. D.A. Tortorelli
36. C. Vanmarsenille
37. A.H.W.M. Kuijpers
38. K. Czolczynski
39. A. Pirotta
40. T. Kapitaniak
41. M.E. Davies
42. F. Benedettini
43. M. Radeş
44. S.A. Mikhailov
45. H. Ulbrich

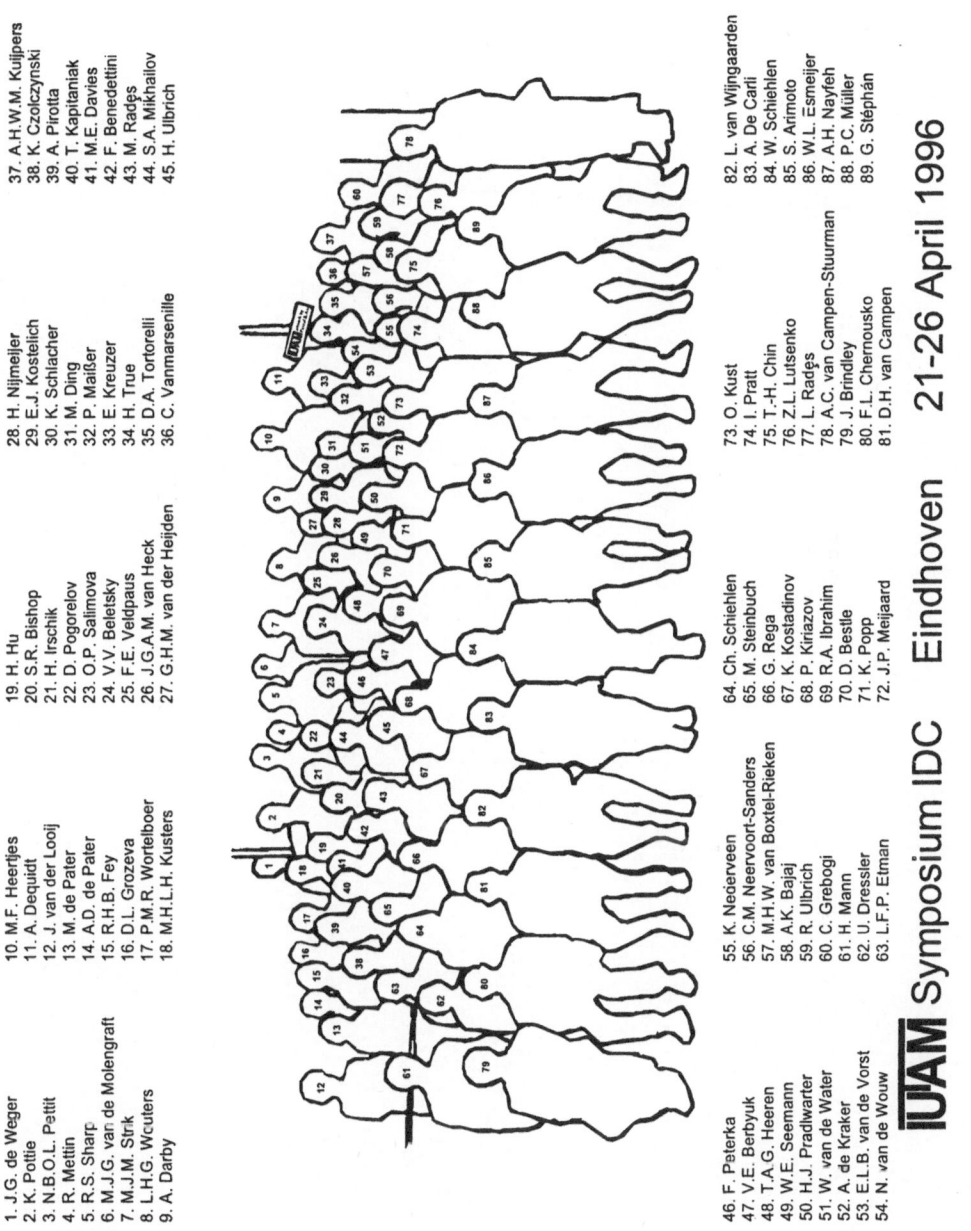

46. F. Peterka
47. V.E. Berbyuk
48. T.A.G. Heeren
49. W.E. Seemann
50. H.J. Pradlwarter
51. W. van de Water
52. A. de Kraker
53. E.L.B. van de Vorst
54. N. van de Wouw
55. K. Nederveen
56. C.M. Neervoort-Sanders
57. M.H.W. van Boxtel-Rieken
58. A.K. Bajaj
59. R. Ulbrich
60. C. Grebogi
61. H. Mann
62. U. Dressler
63. L.F.P. Etman
64. Ch. Schiehlen
65. M. Steinbuch
66. G. Rega
67. K. Kostadinov
68. P. Kiriazov
69. R.A. Ibrahim
70. D. Bestle
71. K. Popp
72. J.P. Meijaard
73. O. Kust
74. I. Pratt
75. T.-H. Chin
76. Z.L. Lutsenko
77. L. Radeş
78. A.C. van Campen-Stuurman
79. J. Brindley
80. F.L. Chernousko
81. D.H. van Campen
82. L. van Wijngaarden
83. A. De Carli
84. W. Schiehlen
85. S. Arimoto
86. W.L. Esmeijer
87. A.H. Nayfeh
88. P.C. Müller
89. G. Stéphan

IUAM Symposium IDC   Eindhoven   21-26 April 1996

PREFACE

During the last decades, applications of dynamical analysis in advanced, often nonlinear, engineering systems have been evolved in a revolutionary way. In this context one can think of applications in aerospace engineering like satellites, in naval engineering like ship motion, in mechanical engineering like rotating machinery, vehicle systems, robots and biomechanics, and in civil engineering like earthquake dynamics and offshore technology. One could continue with this list for a long time.

The application of advanced dynamics in the above fields has been possible due to the use of sophisticated computational techniques employing powerful concepts of nonlinear dynamics. These concepts have been and are being developed in mathematics, mechanics and physics. It should be remarked that careful experimental studies are vitally needed to establish the real existence and observability of the predicted dynamical phenomena.

The interaction between nonlinear dynamics and nonlinear control in advanced engineering systems is becoming of increasing importance because of several reasons.

Firstly, control strategies in nonlinear systems are used to obtain desired dynamic behaviour and improved reliability during operation, Applications include power plant rotating machinery, vehicle systems, robotics, etc. Terms like motion control, optimal control and adaptive control are used in this field of interest. Since mechanical and electronic components are often necessary to realize the desired action in practice, the engineers use the term mechatronics to indicate this field.
If the desired dynamic behaviour is achieved by changing design variables (mostly called system parameters), one can think of fields like control of chaos.

A second reason for the increasing interaction between dynamics and control is the development of parameter identification techniques for the mechanical components. These techniques are used to estimate the parameters in the mathematical model of an engineering system from experimental data and provide a solid basis for improved understanding of the appropriate system model and also for the design of nonlinear control laws.

This volume evolved from an International Symposium on Interaction between Dynamics and Control in Advanced Mechanical Systems, held in Eindhoven, The Netherlands, from 21-26 April 1996. This Symposium was initiated and sponsored by the International Union of Theoretical and Applied Mechanics (IUTAM) and was intended to bring together scientists active in different fields of dynamics and of control with the aim to exchange ideas and to stimulate the interaction between dynamics and control for advanced engineering applications.

A Scientific Committee was appointed by the Bureau of IUTAM with the following members:

D.H. van Campen	(The Netherlands, Chairman)	C. Grebogi	(U.S.A.)
S. Arimoto	(Japan)	P.C. Müller	(Germany)
J. Brindley	(U.K.)	A.H. Nayfeh	(U.S.A.)
F.L. Chernousko	(Russia)	W.O. Schiehlen	(Germany)
A. De Carli	(Italy)	G. Schweitzer	(Switzerland)

This committee selected the participants to be invited and the papers to be presented at the Symposium. As a result of this procedure, 90 active scientific participants from 18 countries followed the invitation, and 53 papers were presented in lecture and poster-discussion sessions.

The scientific presentations were devoted to the following topics:

- Control of Chaos
- Vehicle Dynamics and Control
- Motion Control
- Optimal Control
- Dynamics and Bifurcation of Nonlinear Systems
- Modelling and Dynamics of Engineering Systems
- Vibration Control
- Adaptive Control
- Optimization and Control
- Mechatronic Systems
- Modelling and Control of Engineering Systems
- System Identification
- Analysis and Control of Nonlinear Systems

Since many of the presentations are related to more than one of these topics, the papers of this volume are arranged in alphabetical order with respect to the family name of the first author.
The papers indicated a wide variety of advanced engineering applications of the interaction between dynamics and control. The presentations and discussions during the Symposium will certainly stimulate further theoretical and applied investigations with respect to the interaction between dynamics and control. The publication of the proceedings may promote this development.

The editor wishes to thank both the participants of this IUTAM Symposium and the authors of the papers for their valuable contributions to the important field of interaction between dynamics and control.

The success of the Symposium would not have been possible without the excellent work of the Local Organizing Committee. Members of that Committee were:

D.H. van Campen (Chairman), J.J. Kok (Vice-Chairman),
C.M. Neervoort-Sanders (Secretary), A.C. Stuurman (Treasurer),
M.H.W. van Boxtel-Rieken, L. Kodde, A. de Kraker, M.J.M. Strik, F.E. Veldpaus,
L.H.G. Wouters.

In addition, thanks are due to Kluwer Academic Publishers for efficient cooperation.

Eindhoven, September 1996   Dick van Campen

NONLINEAR POSITION-DEPENDENT CIRCUITS:
A LANGUAGE FOR DESCRIBING MOTIONS OF NONLINEAR
MECHANICAL SYSTEMS

SUGURU ARIMOTO
Faculty of Engineering, The University of Tokyo
Bunkyo-ku, Tokyo, 113 Japan

1. Introduction

The analogy between the dynamics of a linear lumped-parameter electric circuit consisting of resistor-inductor-capacitor (RLC) in series or in parallel and that of a linear one degree-of-freedom mechanical system of dashpot-mass-spring is well known and has been pointed out widely in the literature. This analogy was extensively made use of in robotics by Mason (1981) and Hogan (1985), who introduced new concepts of "compliance control" and "impedance control" respectively in control of physical interactions of a manipulator end-effector with objects or robot task environments. However, this impedance or compliance control concept in robotics has not yet been generalized to cope with nonlinearities in dynamics of the objective system. On the other side, generalization of electric circuits towards nonlinear circuits was attempted by Brayton and Moser (1964), who introduced a new concept called the "mixed potential function" and demonstrated a procedure to construct Lyapunov-type functions from the potential function to prove stability under certain conditions. However, this approach and successive ones such as the EL(Euler-Lagrange) formalism of an electric circuit by Meisel (1966) are limited to treatments of only "electric circuits" and do not take into consideration a large class of nonlinear dynamics related to "mechanical" motion. In other words, to express motions of mechanical systems, not only velocity vectors but also position vectors are key variables on which the potential or the kinetic energy itself depends. That is, the position q is always and explicitly paired with the velocity $\dot{q}$, and hence some nonlinear elementary blocks must be considered to be position-dependent.

More classically, there was a proposal for description of mechanical systems via a network called "bond graph" by Paynter (1960). The bond graph approach has recently been extended to cope with nonlinear mechanical systems by Rosenberg and Karnopp (1983) and in particular to bear fruit in a simulation program package for multi-body robotic systems by Felez et al.(1990). However, bond graphs do not directly express energy flows involved in systems because they are constructed on the basis of connections

1

D. H. van Campen (ed.), IUTAM Symposium on Interaction between Dynamics and Control in Advanced Mechanical Systems, 1–8.
© 1997 *Kluwer Academic Publishers*.

of physical entities. Another expression of motions of mechanical systems via a network has been proposed by Anderson (1995) on the basis of general Hamiltonian systems. The approach proposed in this paper must be within the same framework as Anderson's network, though the latter network does not describe the precise connections of energy flows conveyed by velocity vectors (corresponding to "current" vectors) through elementary blocks such as resistors, kinetic inductors(corresponding to "inductor"), displacement capacitors (generalization of "spring"), and etc..The first paper explicitly describing the idea of inducing a class of nonlinear position-dependent circuits was published in Arimoto (1995a), where direct-current analysis is applied to prove globally asymptotic stability of position control. In this paper, we first develop descriptions of motions of mechanical systems via nonlinear position-dependent circuits when both Coulomb and viscous frictions at joints are taken into account. Then, by introducing a regressor for expressing Coulomb forces and gravity forces, it is shown that insertion of a current source emanating a saturated position feedback signal and a time-varying capacitor (corresponding to the regressor) makes the closed-loop circuit dissipative. This property yields globally asymptotic stability of setpoint position control under the existence of Coulomb frictions at joints. This approach is further extended to more sophisticated dynamics in which actuator dynamics must be incorporated and/or joint flexibilities must be taken into account. It is also pointed out that extension of the approach to the case that a tool endpoint is holonomically constrained on a surface is possible in a natural may by introducing a transformer corresponding to a Jacobian transformation from Cartesian coordinates to joint coordinates (Arimoto 1995b).

2. Passivity of Nonlinear Position-dependent Circuits

Consider a block with the same n input and output physical variables $\dot{q} = (\dot{q}_1, \cdots, \dot{q}_n)^T$ like a vector of n electric currents (see Fig.1). For a mechanical system consisting of a series of n rigid links connected through n joints (see Fig.2), $\dot{q}_i$ stands for the angular velocity at joint i if the joint is rotational or for the velocity if it is prismatic. Accordingly q_i denotes the angle or linear position and hence $\dot{q}_i = dq_i/dt$. Lagrange's equation of motion for such a robot arm depicted in Fig.2 can be described by the form (see Arimoto (1995a))

$$\{H(q)\frac{d}{dt} + \frac{1}{2}\dot{H}(q)\}\dot{q} + S(q,\dot{q})\dot{q} + r(\dot{q}) + g(q) = F \qquad (1)$$

where $H(q)$ is an $n \times n$ positive definite symmetric matrix representing the inertia matrix, $S(q,\dot{q})$ a skew-symmetric matrix representing a part of Coriolis and centrifugal force terms, $g(q)$ a gravity term composed of the gradient of the potential energy $P(q)$, i.e., $g(q) = \{\partial P(q)/\partial q\}^T$, $r(\dot{q})$ a vector of frictional forces caused mainly from the motors themselves and transmission mechanisms at joints. Equation (1) can be described in a circuit like Fig.3, where F can be treated as the torque source corresponding to a voltage source in electric circuits. Applying Kirchhoff's voltage (loop) law for the circuit of Fig.3 readily leads to eqn (1). By taking an inner product between $\dot{q}$ and F in the circuit of Fig.3 or eqn (1) and referring to the skew symmetry of $S(q, \dot{q})$ it follows that

$$\int_0^t \dot{q}^T(\tau)F(\tau)d\tau = E(t) - E(0) + \int_0^t \dot{q}^T(\tau)r(\dot{q}(\tau))d\tau \qquad (2)$$

where $E(t)$ is the total energy of the system expressed as

$$E(t) = \frac{1}{2}\dot{q}^T(t)H(q(t))\dot{q}(t) + P(q(t)). \qquad (3)$$

To make the argument simple, we assume that all joints are rotational. Then we can set $min_q P(q) = 0$ because the constant of potential P can be chosen arbitrarily and P is composed of trigonometric functions of components of q. Hence $E(t)$ is always nonnegative. Next we assume that the nonlinear frictional forces $r(\dot{q})$ satisfy

$$\dot{q}^T r(\dot{q}) \geq b_0 \|\dot{q}\|^2 \qquad (4)$$

with a constant $b_0 > 0$. This assumption is reasonable in ordinary cases where frictional force at the ith joint is composed of both viscous and Coulomb frictions as described by

$$r_i(\dot{q}_i) = b_{0i}\dot{q}_i + \bar{c}_i \mathrm{sgn}(q_i) \qquad (5)$$

for $i = 1, \cdots, n$, where all coefficients b_{0i} and $\bar{c}_i$ are nonnegative and $\mathrm{sgn}(\dot{q}_i) = +1$ or -1 according to $\dot{q}_i > 0$ or $\dot{q}_i < 0$ respectively. If we define as $b_0 = \min\{b_{01}, \cdots, b_{0n}\}$, then we have eqn (4). Thus, eqn (2) is reduced to

$$\int_0^t \dot{q}^T(\tau)F(\tau)d\tau \geq -E(0) = -\gamma_0^2. \qquad (6)$$

This property is called "passivity" of the input-output pair $\{F, \dot{q}\}$ concerning the circuit of Fig.3. Evidently, the passivity can be interpreted as a generalization of impedance between the voltage (torque) source F and the induced current (velocity) $\dot{q}$.

3. Compensation for Coulomb Frictions and Gravity Forces via Regressors

The existence of Coulomb frictions in a mechanical system makes its dynamics discontinuous in the state vector, that is, the differential equation of its motion does not satisfy Lipschitz continuity. Hence, eqn (1) should be treated as a kind of VSS (Variable Structure System). To analyze the behaviour of such a VSS for setpoint position control under a typical PD servo-loop with off-line compensation for the gravity term described as

$$F = g(q_d) - A(q - q_d) - B_1\dot{q} \qquad (7)$$

4

for a given target position q_d, we consider the dynamics

$$\{H(q)\frac{\mathrm{d}}{\mathrm{dt}}+\frac{1}{2}\dot{H}(q)\}\dot{q}+\{B+S(q,\dot{q})\}\dot{q}+g(q)-g(q_d)+A\Delta q+\bar{C}\mathrm{sgn}(\dot{q})=0 \quad (8)$$

where $\Delta q = q-q_d, B = B_0+B_1, B_0 = \mathrm{diag}(b_{01},\cdots,b_{0n}), \bar{C} = \mathrm{diag}(\bar{c}_1,\cdots,\bar{c}_n),$ $\mathrm{sgn}(\dot{q}) = (\mathrm{sgn}(\dot{q}_1),\cdots,\mathrm{sgn}(\dot{q}_n))$, and $A > 0$ is a constant diagonal matrix. As is well-known (see Takegaki and Arimoto (1981)), if there is no Coulomb friction (i.e., $\bar{C} = 0$) and A is large enough to satisfy both inequalities

$$\frac{1}{2}\Delta q^T A\Delta q + P(q) - P(q_d) - \Delta q^T g(q_d) \geq \varepsilon\|\Delta q\|^2, \qquad (9)$$

$$\Delta q^T A\Delta q + \Delta q^T\{g(q) - g(q_d)\} \geq \varepsilon\|\Delta q\|^2, \qquad (10)$$

with a specified small $\varepsilon > 0$, then the equilibrium state $(q, \dot{q}) = (q_d, 0)$ of the closed-loop system (8) becomes globally asymptatically stable. However, if Coulomb friction exists at every joint, then the velocity vector $\dot{q}$ is trapped in a manifold $M = \{(q, \dot{q}) : \dot{q} = 0\}$ within a finite time $t_s(\geq 0)$ before the position $q(t)$ nears the target q_d. At this stage we assume that the magnitude c_i of static friction at ith joint is just above $\bar{c}_i$, i.e., $c_i = \bar{c}_i + \delta_i$ with a small $\delta_i > 0$. Further, we assume that each static friction at joint i stops the motion of the joint axis according to the following rule: If at instant $t = t_0$ the angular velocity $\dot{q}_i(t)$ vanishes (i.e., $\dot{q}_i(t_0) = 0$), $|f_i(t_0 - 0)| < \bar{c}_i + \varepsilon_i$ for $\varepsilon_i > 0$, and $|f_i(t_0 + 0)| \leq \bar{c}_i + \delta_i$, then $\ddot{q}_i(t_0 + 0) = 0$ and there exists a number $\varepsilon > 0$ such that, for any $t \in [t_0, t_0 + \varepsilon)$, $\dot{q}_i(t) = 0$, where $f_i(t)$ signifies the i-th component of the left hand side of eqn(8) except $h_{ii}(q)\ddot{q}_i$ and $\bar{c}_i\mathrm{sgn}(\dot{q}_i)$, $\varepsilon_i > 0$ is sufficiently small in comparison with $\delta_i > 0, f_i(t_0 - 0)$ means the limit of $f_i(t)$ as t increase towards t_0, and $\ddot{q}_i(t_0 + 0)$ has a similar meaning. Then, it is possible to show:

Theorem 1   If all $\bar{c}_i$ are positive and $A > 0$ satisfies both inequalities (9) and (10), then for any initial state $(q(0), \dot{q}(0))$ there is a finite time $t_s \geq 0$ such that for any $t(\geq t_s)$ $\dot{q}(t) = 0$ and the ith component of $\{g(q) - g(q_d) + A\Delta q\}$ is fixed in $(-c_i, c_i)$ for all i.

To avoid the trapping in the sliding manifold $\dot{q} = 0$ without attaining the target position, it is known implicitly among research workers engaged in practical experiments (see Whitcomb et al.(1993)) that the use of a regressor for Coulomb frictions is quite effective in practice. Motivated by this fact, we consider a pair of regressors $Z(q_d, \dot{q}) = (Z_0(q_d), Z_1(\dot{q}))$, where $Z_0(q_d)$ is a constant $n \times n$ matrix such that $g(q_d) = Z_0(q_d)\Theta_0$ with a unknown parameter vector $\Theta_0 = (m_1,\cdots,m_n)^T$ of link and payload masses m_i and $Z_1(\dot{q}) = \mathrm{diag}(\mathrm{sgn}(\dot{q}_1),\cdots,\mathrm{sgn}(\dot{q}_n))$ such that $\bar{C}\mathrm{sgn}(\dot{q}) = Z_1(\dot{q})\Theta_1$ with a parameter vector $\Theta_1 = (\bar{c}_1,\cdots,\bar{c}_n)^T$ of uncertain coefficients $\bar{c}_i$ of Coulomb frictions. If we denote $\Theta = (\Theta_0^T, \Theta_1^T)^T$, then

$$Z(q_d, \dot{q})\Theta = \bar{C}\mathrm{sgn}(\dot{q}) + g(q_d). \qquad (11)$$

Since Θ is unknown, we use an estimator $\hat{\Theta}(t)$ at time t which is updated by the law

$$\hat{\Theta}(t) = \hat{\Theta}(0) - \int_0^t \Gamma^{-1} Z^T(q_d, \dot{q}(\tau)) y(\tau) d\tau. \tag{12}$$

Here, y is defined as

$$y = \dot{q} + \alpha s(\Delta q) \tag{13}$$

with a constant $\alpha > 0$, and $s(\Delta q) = (s_1(\Delta q_1), \cdots, s_n(\Delta q_n))^T$ where each $s_i(\Delta q_i)$ is defined as a saturated function with a profile depicted in Fig.4. By the use of the estimator $\hat{\Theta}(t)$ for unknown parameters, we consider a simple destributed feedback control

$$F = -A \Delta q - B_1 \dot{q} - \beta y - \int_0^t \gamma y d\tau + Z(q_d, \dot{q}) \hat{\Theta} \tag{14}$$

with $\beta > 0$ and $\gamma > 0$. Substituting this into eqn (1) yields

$$\{H(q) \frac{d}{dt} + \frac{1}{2} \dot{H}(q)\} \dot{q} + \{B + S(q, \dot{q})\} \dot{q} + A \Delta q$$
$$+ g(q) - g(q_d) + \beta y + \int_0^t \gamma y d\tau + Z(q_d, \dot{q}) \Delta \Theta = 0 \tag{15}$$

where $\Delta \Theta = \Theta - \hat{\Theta}$. Note that an inner product of the last term on the left hand side of (15) with y leads to

$$y^T Z(q_d, \dot{q}) \Delta \Theta = -\{\frac{d}{dt} \Gamma \hat{\Theta}\}^T \Delta \Theta = \frac{d}{dt} \frac{1}{2} \{\Delta \Theta^T \Gamma \Delta \Theta\} \tag{16}$$

and the fifth term on the right hand side of eqn (14) can be written in the form

$$Z(q_d, \dot{q}(t)) \hat{\Theta}(t) = Z(t) \hat{\Theta}(0) - \int_0^t Z(t) \Gamma^{-1} Z^T(\tau) y(\tau) d\tau \tag{17}$$

where $Z(t) = Z(q_d, \dot{q}(t))$. Evidently, the matrix $K(t, \tau) = Z(t) \Gamma^{-1} Z^T(\tau)$ can be regarded as a kernel operator with positivity. Hence, eqn (15) can be expressed by a nonlinear position-dependent circuit as shown in Fig.5.

It should be remarked however that the above argument is valid only for the case that all angular velocities $\dot{q}_i (i = 1, \cdots, n)$ are non-zero, that is, there is no trapping of motion at any joint. If one joint axis stops moving by being trapped due to its static and Coulomb frictions, the update of parameter estimates at its corresponding component should be stopped because that vector component of eqn(15) is not valid, either. Hence, we introduce a diagonal matrix $S(\dot{q})$ defined by

$$S_{ii}(\dot{q}) = \begin{cases} 1 & \text{if } \dot{q}_i \neq 0 \\ 0 & \text{if } \dot{q}_i = 0 \end{cases} \tag{18}$$

and, instead of eqn(12), we set

$$\hat{\Theta}(t) = \hat{\Theta}(0) - \int_0^t \Gamma^{-1} Z^T(q_d, \dot{q}(T)) S(\dot{q}(\tau)) y(\tau) d\tau. \qquad (19)$$

Theorem 2 By taking α appropriately and assuming a zero gap between the magnitude of static friction and Coulomb friction at every joint (i.e., $c_i = \bar{c}_i$ for all i), the solution $(q(t), \dot{q}(t))$ to eqn (15) together with the update law (19) for any initial state $(q(0), \dot{q}(0))$ and a bounded initial guess $\hat{\Theta}(0)$ (say, $\hat{\Theta}(0) = 0$) tends to the steady state $(q_d, 0)$ asymptotically as $t \to \infty$, regardless of whether $\hat{\Theta}(t)$ converges or not to its true value.

4. Necessary Conditions for Passivity for Mechanical Systems

In Theorem 2 we cannot obtain any conclusion about whether the estimator $\hat{\Theta}(t)$ converges or not to the true value Θ as $t \to \infty$. However, it is well known (see Sadegh and Horowitz(1990)) that, after a series of various maneuver satisfying the persistent excitation, the estimator tends to the true value Θ if there is disturbance. In this section, we consider the case that the regressor term $Z_1(\dot{q})\hat{\Theta}_1$ in eqn (14) can approximately cancel the Coulomb frictions $\bar{C}\mathrm{sgn}(\dot{q})$. Thus, we assume that the closed-loop system is subject to

$$\{H(q)\frac{\mathrm{d}}{\mathrm{d}t} + \frac{1}{2}\dot{H}(q)\}\dot{q} + \{B_0 + B_1 + S(q,\dot{q})\}\dot{q} + A\Delta q + g(q) - g(q_d)$$
$$+ Z(q_d)\hat{\Theta}_0 = \Delta r(\dot{q}) + v, \qquad (20)$$

concerning eqn (20) where $\Delta r(\dot{q})$ is a structural disturbance of remaining frictional terms and v is a pure disturbance that may arise from other sources (for example, torque ripples of motors, etc.). It should be remarked that the term of structural disturbances may include other missing frictional characteristics such as a lubrication effect approximated by $-d_i\{1 - \exp(-c_i|\dot{q}_i|)\}\mathrm{sgn}(\dot{q})$ (see Tustin (1974)) and another type of negative resistance caused by time-delay of a position feedback such that $A\Delta q(t - \Delta t) \cong A\Delta q(t) - \Delta t A\Delta \dot{q}$. Hence, by taking all this into consideration, we suppose a class of structured disturbances satisfying

$$-\gamma^2 \int_0^t \Delta r^T(\dot{q}(\tau))\dot{q}(\tau)d\tau \leq \int_0^t \|\dot{q}(\tau)\|^2 d\tau \qquad (21)$$

for a fixed $\gamma^2 > 0$ and any $t > 0$. We are now able to show the following:

Theorem 3 A necessary and sufficient condition for the pair $\{v, \dot{q}\}$ linked by eqn (20) to satisfy passivity for any structural disturbance $\Delta r(\dot{q})$ satisfying eqn (21) is that

$$\gamma^{-2} \leq \lambda_m(B_0 + B_1) \qquad (22)$$

where $\lambda_m(B)$ denotes the minimum eigenvalue of B.

Based on this result, we say that the pair $\{v, \dot{q}\}$ linked by eqn (20) satisfies passivity with margin γ^{-2} if γ^2 satisfies eqn (21).

Now we derive a relation between passivity with margin γ^{-2} and H_∞-tuning in a sense of disturbance attenuation.

<u>Definition</u>   Assume that a solution of eqn (20) satisfies $q(0) = q_d, \dot{q}(0) = 0$, and $\Delta\Theta(0) = 0$. If for a fixed $\gamma^{-2} > 0$ and any time interval $[0, t]$ it holds that

$$b_0 \int_0^t \|\dot{q}(\tau)\|^2 d\tau \leq \gamma^2 \int_0^t \|w(\tau)\|^2 d\tau \tag{23}$$

where $b_0 = \lambda_m(B_0 + B_1)$ and $w(t)$ signifies the total of disturbances, i.e., $w = \Delta r(\dot{q}) + v$, then we say that the dynamics (20) establishes H_∞-tuning with level γ^{-2}.

<u>Theorem 4</u>   Assume that $q(0) = q_d, \dot{q}(0) = 0$, and $\Delta\hat{\Theta}(0) = 0$. In order that the pair $\{w, \dot{q}\}$ establishes H_∞-tuning with level γ^{-2} and the pair $\{v, \dot{q}\}$ satisfies passivity with margin γ^{-2} simultaneously, it is necessary and sufficient for γ^{-2} to satisfy the inequality $\gamma^{-2} \leq b_0(= \lambda_m(B_0 + B_1))$.

5. Conclusions

We have introduced a framework of nonlinear position-dependent circuits, that can describe dynamics of mechanical systems. Instead of Fourier and Laplace transforms, the concept of passivity becomes fundamental and can be used effectively as a basic tool for characterizing input-output properties of such nonlinear circuits.

In this paper all theorems have been presented without proofs. The details of the proofs will be provided in future papers.

References

Anderson, R. J. (1995). SMART: A modular control architecture for telerobotics, *IEEE Robotics & Automaiton Magazine*, **Sept. 1995**, 10-18.
Arimoto, S. (1995a). Stability analysis of setpoint control for robot dynamics via non-linear position-dependent circuits, *Dynamics of Continuous, Discrete and Impulsive Systems*, 1-1, 1-17.
Arimoto, S. (1995b). Nonlinear position-dependent circuits for mechanical motion control, *Proc. of IFAC Workshop 'Motion Control'*, Munich, Germany, Oct. 1995, pp.1-20.
Arimoto, S. (1996). Another language for describing motions of mechatronics systems: A nonlinear position-dependent circuit theory, *ASME/IEEE Trans. on Mechatronics*, 1-2 (to be published).
Brayton, R. K. and Moser, J. K. (1964). A theory of nonlinear networks, Parts I and II, *Quart. Appl. Math.*, **22**, 1-33 and 81-104.
Felez, J., Vera, C., San Jose, I., and Cacho, R. (1990). BOMDYM: A bond graph based simulation program for multibody systems. *J. of Dynaimic Systems, Measurement, and Control*, **112**, 717-727.
Hogan, N. (1985). Impedance control: An approach to manipulation. Parts 1, 2, and 3, *J. of Dynamic Systems, Measurement, and Control*, **107**, 1-24.
Mason, M. T. (1981). Compliance and force control for computer controlled manipulators, *IEEE Trans. on Systems, Man, and Cybernetics*, 11, 418-432. McGraw Hill, New York.

8

Meisel, J. (1966). *Principle of Electromechanical Energy Conversion*, McGraw-Hill, New York.

Paynter, H. M. (1960). *Analysis and Design of Engineering Systems*, The MIT Press, Cambridge, Massachusetts.

Rosenberg, R. and Karnopp, D. (1983). *Introduction to Physical System Dynamics*, Mc-Graw Hill, New York.

Sadegh, N. and Horowitz, R. (1990). Stability and robustness of adaptive controllers for robotic manipulators, *Int. J. of Robotics Research*, 9, 74-92.

Tustin, A. (1974). The effect of backlash and speed-dependent friction on the stability of closed-cycle control systems, *J. of IEE*, 94, No. 2A, 143-151.

Takegaki, M. and Arimoto, S. (1981). A new feedback method for dynamic control of manipulators, *Trans. ASME J. of Dynamic Systems, Measurement, and Control*, 103-2, 119-125.

Whitcomb, L. L., Rizzi, A. A., and Koditschek, D. E. (1993). Camparative experiments with a new adaptive controller for robot arms, *IEEE Trans. on Robotics and Automation*, 9-1, 59-70.

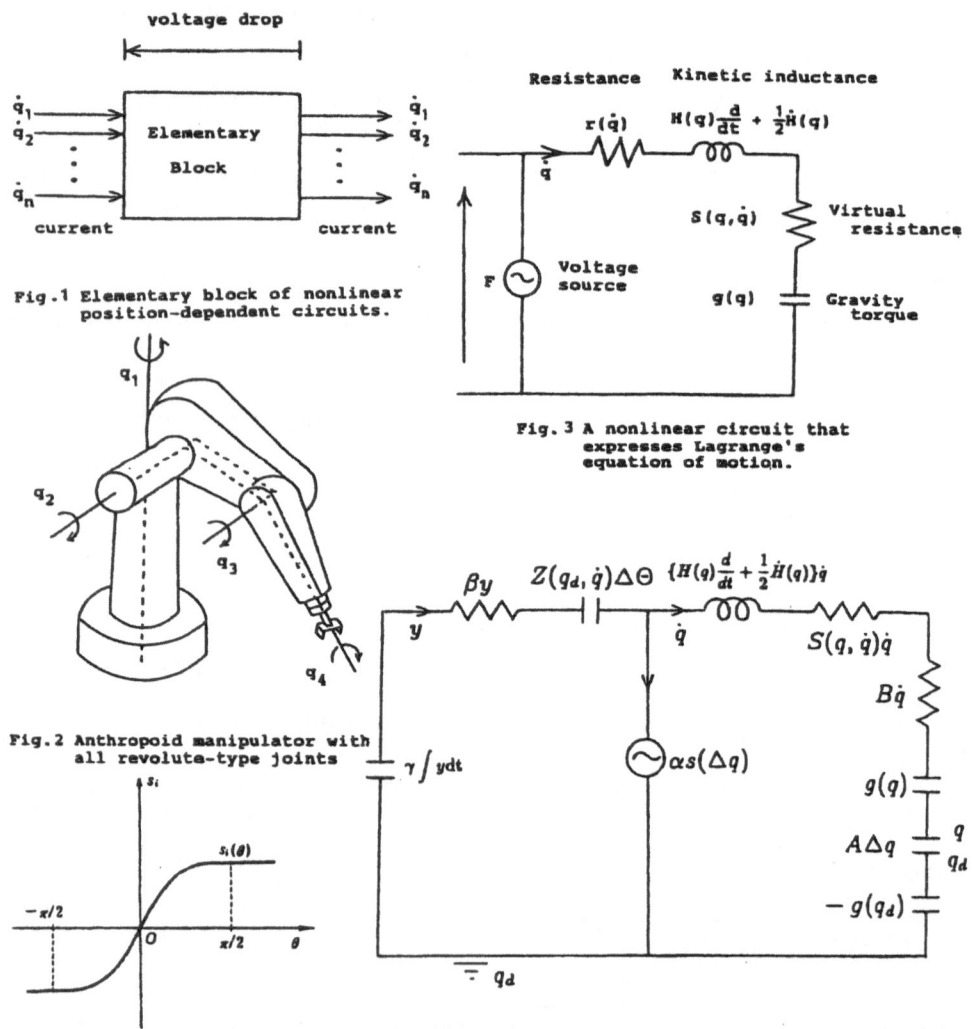

Fig.1 Elementary block of nonlinear position-dependant circuits.

Fig. 3 A nonlinear circuit that expresses Lagrange's equation of motion.

Fig.2 Anthropoid manipulator with all revoluta-type joints

Fig.4 A profile of saturated function $s_i(\theta)$.

Fig.5 The circuit expresses the dynamics in eqn(15).

DYNAMICS OF SINGULARLY PERTURBED NONLINEAR SYSTEMS WITH TWO DEGREES-OF-FREEDOM

A. K. BAJAJ
School of Mechanical Engineering
Purdue University, West Lafayette, IN 47907

I. T. GEORGIOU
Special Project for Nonlinear Science,
Naval Research Laboratory
Washington, DC 20375

M. CORLESS
School of Aeronautics & Astronautics
Purdue University, West Lafayette, IN 47907

1 Introduction

In engineering applications, complex structural systems are usually composed of simpler substructures with widely varying elasticities and damping properties. This broad diversity in flexibilities present in a complex structural system prompts us to view its motion in terms of the dynamics of interacting *stiff* and *soft* substructures. More specifically, we are here interested in *soft-stiff* structural-mechanical systems with multiple equilibrium states. A fundamental question to be asked is: *how is the dynamics of a soft-stiff structural system related to the dynamics of a simpler structure obtained in the limit as its stiff substructures become essentially rigid?* In this work, we present a systematic analytic-geometric methodology, by combining the singular perturbation theory with invariant manifolds and symbolic and numeric computation, to study the nonlinear dynamics of a soft-stiff two degrees-of-freedom system.

2 A Soft-Stiff Nonlinear Two Degrees-of-Freedom System

Figure 1 shows a linear oscillator (spring-mass-dashpot combination) of mass M, stiffness K and damping coefficient δ, coupled to a nonlinear oscillator of mass

9

D. H. van Campen (ed.), IUTAM Symposium on Interaction between Dynamics and Control in Advanced Mechanical Systems, 9–16.

M_1, linear and cubic stiffness coefficients $(-K_1)$ and K_2, and damping coefficient δ_1. In terms of the relative displacements $\eta \equiv u_2 - u_1$ and $\xi \equiv u_3 - u_2$, the motions of the two coupled oscillators are described by the following two coupled equations,

$$
\begin{aligned}
(M + M_1)\eta'' + M_1\xi'' + \delta\eta' + K\eta &= -(M + M_1)u_1'', \\
M_1\eta'' + M_1\xi'' + \delta_1\xi' - K_1\xi + K_2\xi^3 &= -M_1u_1'',
\end{aligned}
$$

(1)

where $(')$ denotes differentiation with respect to the natural time t_1. It is shown in Georgiou (1993) that the above two degrees-of-freedom system can be identified with a low order Galerkin approximation to the equations of motion of a buckled nonlinear viscoelastic beam supported at its ends by linear viscoelastic columns.

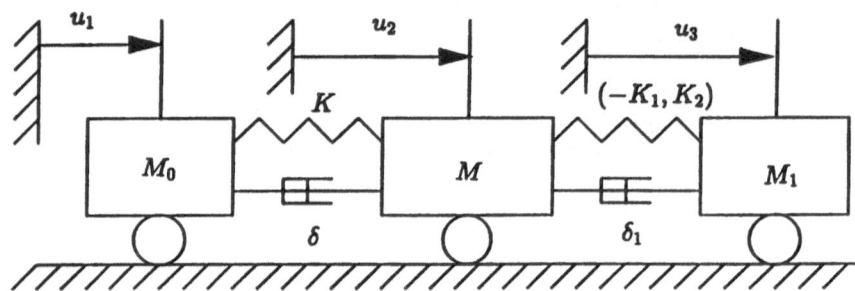

Figure 1. The two degrees-of-freedom nonlinear system.

The motivation for the study lies in the following observation: in the limiting case when the linear string stiffness $K \to \infty$, the masses M_0 and M are rigidly coupled so that $\eta = 0$ and $u_1 = u_2 = \hat{u}_1(t_1)$. The resulting single degree-of-freedom nonlinear system with multiple equilibria is well known (Moon, 1993), for certain combinations of system parameters, and under harmonic or quasiharmonic excitation $\hat{u}_1(t_1)$, to exhibit complex steady-state motions including chaotic attractors.

2.1 SINGULAR PERTURBATION FORMULATION

We wish to study the above two degree-of-freedom system when K is sufficiently large. The fundamental step in the present analysis is to cast the second-order equations of motion (1) into a set of singularly perturbed first-order equations when the natural frequency $\omega \equiv \sqrt{K/M}$ of the linear oscillator is much larger than the uncoupled linear "frequency" $\omega_1 \equiv \sqrt{K_1/M_1}$ of the nonlinear oscillator. To this end, we let $t = \omega_1 t_1$ and introduce the parameter μ defined by

$$
\mu \equiv \frac{\omega_1}{\omega} = \sqrt{\frac{\kappa}{\beta}}, \qquad \beta \equiv \frac{M_1}{M}, \qquad \kappa \equiv \frac{K_1}{K}.
$$

(2)

The linear transformations:

$$
\begin{aligned}
\xi &= x_1, & \dot{\xi} &= x_2, \\
\eta &= \mu^2 z_1, & \dot{\eta} &= \mu z_2,
\end{aligned}
$$

(3)

then cast the equations of motion (1) into the singularly perturbed form:

$$\begin{aligned}
\dot{x}_1 &= x_2, \\
\dot{x}_2 &= (1+\beta)\left(x_1 - 2\zeta_1 x_2 - \gamma x_1^3\right) + z_1 + 2\zeta z_2, \\
\dot{\psi} &= \Omega,
\end{aligned} \tag{4a}$$

$$\begin{aligned}
\mu \dot{z}_1 &= z_2, \\
\mu \dot{z}_2 &= -z_1 - 2\zeta z_2 + \beta\left(-x_1 + 2\zeta_1 x_2 + \gamma x_1^3\right) + \alpha \cos \psi,
\end{aligned} \tag{4b}$$

where $\zeta_1 = \delta_1/2\sqrt{M_1 K_1}$ and $\zeta = \delta/2\sqrt{MK}$ are linear damping factors, $\gamma = K_2/K_1$ is the strength of the normalized nonlinearity coefficient, and $\alpha \equiv \rho\Omega^2$ is the amplitude of the harmonic forcing induced by $u_1(t) = \rho \cos(\Omega t)$.

We can easily see that the unforced system ($\alpha = 0$) possesses a set of equilibria consisting of three states:

$$S \equiv (0,0,0,0), \qquad C_{-1} \equiv (-\sqrt{1/\gamma}, 0, 0, 0), \qquad C_{+1} \equiv (\sqrt{1/\gamma}, 0, 0, 0), \tag{5}$$

where S is a saddle-focus, and C_{-1} and C_{+1} are stable foci. It is transparent from equations (4) that, for small μ, the dynamics of the system depends on a slow time scale t and a fast time scale $\tau = t/\mu$.

3   Invariant Manifolds

We are interested in characterizing the global slow and fast dynamics of this system, and one way to achieve this is to use the theory of   invariant manifolds of motion (Fenichel, 1979).

Setting $\mu = 0$, the equations (4b) for the stiff oscillator degenerate to an algebraic system which is linear in z, and its unique solution $\mathbf{H}_0$ is given by

$$\begin{aligned}
z_1 &= H_{10}(x, \psi) = \beta\left(-x_1 + 2\zeta_1 x_2 + \gamma x_1^3\right) + \alpha \cos(\psi), \\
z_2 &= H_{20}(x, \psi) = 0.
\end{aligned} \tag{6}$$

Equations (6) describe a three-dimensional manifold $\mathcal{W}_0$ in the five-dimensional phase space of system (4). The manifold $\mathcal{W}_0$ is periodic in time, it passes through all equilibrium states of the unforced system ($\alpha = 0$), and is an *exact invariant manifold for the system at $\mu = 0$*. The restriction of the system, at $\mu = 0$, on $\mathcal{W}_0$, and its subsequent projection onto the space (x_1, x_2, ψ) reduces the system (4) to a three-dimensional system which is identical to the forced uncoupled soft oscillator on the slow time:

$$\begin{aligned}
\dot{x}_1 &= x_2, \\
\dot{x}_2 &= x_1 - 2\zeta_1 x_2 - \gamma x_1^3 + \alpha \cos(\psi), \\
\dot{\psi} &= \Omega.
\end{aligned} \tag{7}$$

Expressing equations (4) in the fast time scale, $\tau = t/\mu$, gives a regularly perturbed system of equations which, for $\mu = 0$, reduces to

$$
\begin{aligned}
z_1' &= z_2, \\
z_2' &= -z_1 - 2\zeta z_2 + H_{10}(x_{10}, x_{20}, \psi_0),
\end{aligned}
\tag{8}
$$

and

$$
x_1(t) = x_{10}, \qquad x_2(t) = x_{20}, \qquad \psi(t) = \psi_{10},
\tag{9}
$$

where (x_{10}, x_{20}, ψ_0) are initial conditions. Now a point on the slow invariant manifold $\mathcal{W}_0$ has coordinates $(x, \psi, z = H_0(x, \psi))$. Clearly, all motions initiated off the slow manifold $\mathcal{W}_0$ are governed by the stiff oscillator (8), and it is easy to show that they approach exponentially fast the point $(x_0, \psi_0, z_0 = H_0(x_0, \psi_0))$ of $\mathcal{W}_0$. Thus each point $(x, \psi, z = H_0(x, \psi))$ on $\mathcal{W}_0$ is an equilibrium for fast motions, the fast dynamics residing in a two-dimensional manifold attached transversely to a point of $\mathcal{W}_0$.

The above discussion for $\mu = 0$ reveals a global picture of the dynamics through the natural introduction of slow and fast invariant manifolds. We now focus on the existence of a slow invariant manifold for $\mu \neq 0$. To this end, equations (4) can be written on the fast time scale τ as the regularly perturbed equations (Georgiou $et\ al.$, 1995).

$$
\begin{aligned}
X' &= \mathbf{A}X + \mu \mathbf{F}_1(X) + \mu \mathbf{G}_1 Y, \\
\mu' &= 0, \\
Y' &= \mathbf{B}Y + \mu \mathbf{F}_2(X) + \mu \mathbf{G}_2 Y,
\end{aligned}
\tag{10}
$$

where

$$
X \equiv (x_1, x_2, x_3, x_4), \qquad Y \equiv (y_1, y_2),
\tag{11a}
$$

$$
z = y + \mathbf{H}_0(x, \psi), \qquad x_3 = \alpha \cos \psi, \qquad x_4 = \alpha \sin \psi.
\tag{11b}
$$

Note that equations (10) possess a continuum of equilibrium states given by

$$
\hat{X} = (x_1, x_2, x_3, x_4), \qquad \hat{\mu} = 0, \qquad \hat{Y} = (0, 0).
\tag{12}
$$

Three of these equilibrium states correspond to the equilibria of the unforced system, the difference being that now they are embedded in a state space augmented by the singular perturbation parameter μ and the variables x_3, x_4.

The Global Invariant Manifold Theorem (Carr, 1981) for singularly perturbed equations then guarantees the existence of a global invariant manifold $\hat{\mathcal{W}}_\mu$ passing through each of the continuum of equilibria (12). This invariant manifold is exponentially attractive and is described by the graph of a function $Y = \hat{\mathbf{H}}_\mu(x_1, x_2, x_3, x_4) = \hat{\mathbf{H}}_\mu(x_1, x_2, \psi)$. The function $\hat{\mathbf{H}}_\mu$ is determined by the $slow$ manifold condition for system (10),

$$
[\mathbf{B} + \mu \mathbf{G}_2] \hat{\mathbf{H}}_\mu(X) + \mu \mathbf{F}_2(X) - \mu D_X \hat{\mathbf{H}}_\mu(X) \left[\mathbf{F}_1(X) + \mathbf{G}_1 \hat{\mathbf{H}}_\mu(X) \right] = 0, \quad (13)
$$

whose solutions can be approximated asymptotically (Georgiou *et al.*, 1995a). Note that the invariant manifold $\hat{\mathcal{W}}_\mu$ is the $O(\mu)$ correction to the invariant manifold of the original system described by the graph of the function $\mathbf{H}_\mu(x, \psi) = \mathbf{H}_0(x, \psi) + \hat{\mathbf{H}}_\mu(x, \psi)$. Figure 2 depicts the $O(\mu^3)$ approximation to the configuration component of the slow invariant manifold for the unforced system.

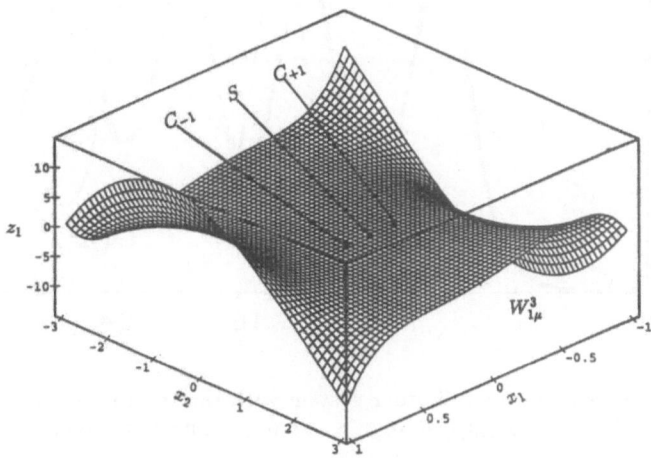

Figure 2. $O(\mu^3)$ approximation to the configuration component of the slow invariant manifold $\mathcal{W}_\mu$, $\mu = 0.1$, $\beta = 1.0$, $\gamma = 10.0$, $\alpha = 0.0$, $\zeta_1 = 0.05$, $\zeta = 0.05$.

4   Slow Dynamics

The existence of an attractive slow invariant manifold for the soft-stiff system implies that the long time dynamics are of reduced dimension. The restriction of the system on the slow manifold gives the *slow* reduced system, defined by

$$
\begin{aligned}
\dot{x}_1 &= x_2, \\
\dot{x}_2 &= x_1 - 2\zeta_1 x_2 - \gamma x_1^3 + \hat{H}_{1\mu}(x, \psi) + 2\zeta \hat{H}_{2\mu}(x, \psi), \\
\dot{\psi} &= \Omega,
\end{aligned}
\tag{14}
$$

which is a regular perturbation in μ of the uncoupled nonlinear oscillator (7). The asymptotic expansion approximates only part of the region of the slow invariant manifold. Numerical experiments indicate that the slow manifold bifurcates or undergoes a qualitative change above some critical energy state (see Georgiou *et al.*, 1995).

For all motions on the slow invariant manifold, the motion of the linear oscillator is given by $z_1 = H_{1\mu}(x, \psi)$, $z_2 = H_{2\mu}(x, \psi)$, or in terms of the original variables

$$
\eta = \mu^2 H_{1\mu}(x, \psi), \quad \dot{\eta} = \mu H_{2\mu}(x, \psi).
\tag{15}
$$

Equations (14) and (15) define the reduced-order slow system. Figure 3, for $\mu = 0.30$ (and $\beta = 1.0, \gamma = 10.0$), compares the displacement of the unforced soft nonlinear oscillator to those of the $O(\mu^0)$ and $O(\mu^3)$ approximations.

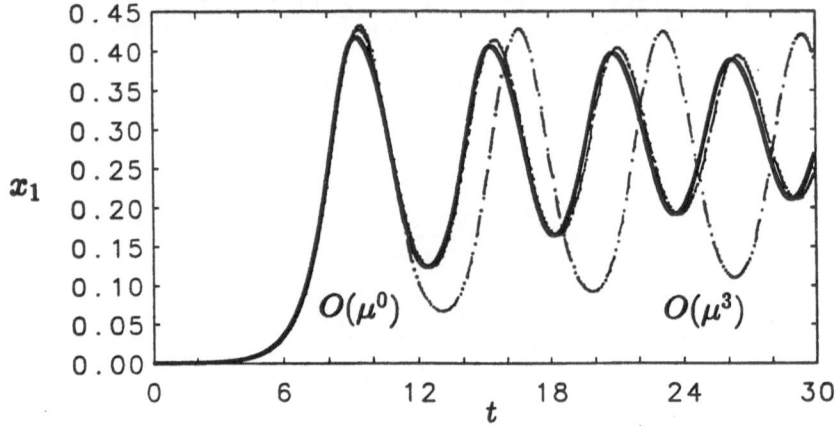

Figure 3. Comparison of the full-order system with the third-order slow oscillator, $\mu = 0.3$, $\beta = 1.0, \gamma = 10.0$, $\alpha = 0.0$, $\zeta_1 = 0.05$, $\zeta = 0.05$.

4.1 GLOBAL SLOW NORMAL MODE

As the dissipation and the forcing parameters approach zero, the slow invariant manifold approaches a two-dimensional manifold. The existence of a global invariant manifold for the conservative system cannot be shown theoretically. The Lyapunov center theory guarantees only local invariant manifolds about the three equilibria. However, a slow manifold condition can also be formulated for the conservative system. Numerical experiments show that, for small μ, the limits of the slow manifolds of the unforced and forced dissipative systems is indeed an invariant manifold for the conservative system (Georgiou et al., 1996), which can be identified with a global *slow* nonlinear normal mode (Rosenberg, 1964) of vibration.

5 Slow Chaotic Attractor

On the slow invariant manifold of the conservative system resides a homoclinic orbit to the saddle-center equilibrium S, whose location is independent of the singular parameter μ. The hyperbolic equilibrium is perturbed to a hyperbolic (or saddle type) periodic motion for the forced dissipative system. The dynamics on the slow invariant manifold of the forced system are governed by a regular perturbation of the slow oscillator for the conservative system, that is,

$$\dot{x}_1 = x_2,$$

$$\dot{x}_2 = (1+\beta)x_1\left(1-\gamma x_1^2\right) + \hat{H}_{1\mu}(x) + \epsilon g_\mu(x,\psi;\epsilon),$$
$$\dot{\psi} = \Omega, \tag{16}$$

where $\hat{H}_{1\mu}$ describes the configuration component of the slow invariant manifold of the conservative system, and the perturbation term $\epsilon g_\mu(x,\psi;\epsilon)$ is due to small damping and weak forcing. It is well known that complex dynamics such as chaotic vibrations is related directly to the interaction of the invariant manifolds of the saddle type periodic motion, which can be explored in terms of the distance,

$$M_\mu(t_0,\psi_0) = \int_{-\infty}^{\infty} h_{2\mu}(t-t_0)g_\mu\left(h_\mu(t-t_0),\Omega t + \psi_0\,;\epsilon=0\right)dt, \tag{17}$$

between the stable and unstable manifolds of S. Here, $h_\mu(t) = (h_{1\mu}(t),h_{2\mu}(t))$, $-\infty < +\infty$, denotes the homoclinic motion of the conservative slow oscillator. It is shown in Georgiou *al.* (1995a) that M_μ, the Melnikov function, possesses simple zeroes, thus implying that the invariant manifolds of the saddle periodic motion intersect transversely. Poincare sections reveal that the full-order system undergoes a cascade of period-doubling bifurcations leading to a strange attractor (Figure 4). For small μ the various orders of approximation to the forced slow oscillator predict the route to chaos observed in the full order system.

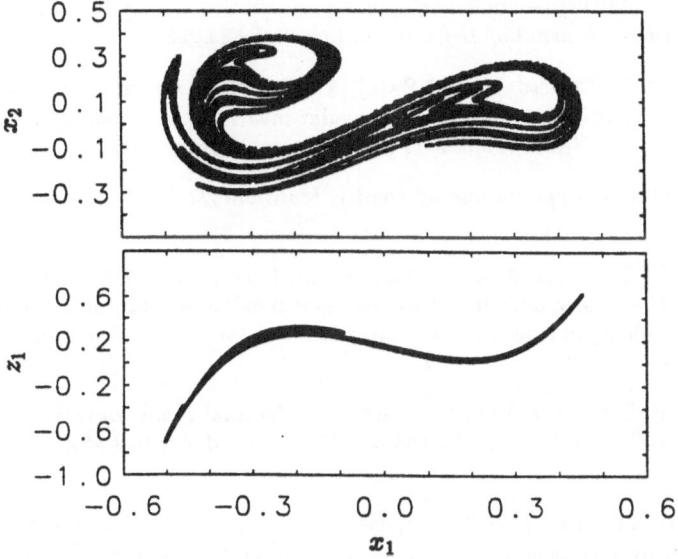

Figure 4. Projection of a strange attractor of the full-order system onto (a) the phase plane of the soft nonlinear oscillator, (b) configuration plane $x_1 - z_1$, $\mu = 0.1$, $\beta = 1.0, \gamma = 10.0, \alpha = 0.14, \Omega = 1.42, \zeta_1 = 0.05, \zeta = 0.05$.

6 Conclusions

We have studied the long time motions of a two degrees-of-freedom system consisting of a stiff linear oscillator coupled to a soft nonlinear oscillator with multiple equilibrium states. Combining the theory of singular perturbations and the theory of invariant manifolds, we have shown that the unforced dissipative system possesses a two-dimensional invariant manifold passing through all equilibrium states. Furthermore, the harmonically forced system possesses a global slow invariant manifold harmonic in time. The dynamics on the slow invariant manifold are governed by a slow nonlinear oscillator which, for small μ, is a regular perturbation of the soft nonlinear oscillator. The slow oscillator approximates well all periodic and chaotic motions on the slow invariant manifold.

7 References

1. Georgiou, I.T. (1993) Nonlinear dynamics and chaotic motions of a singularly perturbed nonlinear viscoelastic beam, Ph. D. Dissertation, Purdue University, West Lafayette.

2. Moon, F.C. (1993) *Chaotic and Fractal Dynamics*, John Wiley & Sons, New York.

3. Fenichel, N. (1979) Geometric singular perturbation theory for ordinary differential equations, *Journal of Differential Equations* **31**, 53-98.

4. Georgiou, I.T., Corless, M. and Bajaj, A.K. (1995a) Dynamics of nonlinear structures with multiple equilibria: a singular-invariant manifold approach, *ZAMP (Journal of Applied Mathematics and Physics)*, (submitted).

5. Carr, J (1981) *Applications of Centre Manifold Theory*, Springer-Verlag, New York.

6. Georgiou, I.T., Bajaj, A.K. and Corless, M. (1996) Slow and fast invariant manifolds, and normal modes in a two degree-of-freedom structural dynamical system with multiple equilibrium states, *International Journal of Non-Linear Mechanics*, (in press).

7. Rosenberg, R.M. (1964) On the existence of normal mode vibrations of nonlinear systems with two degrees of freedom, *Quarterly of Applied Mathematics* **22** (3), 217-234.

8. Georgiou, I.T., Bajaj, A.K. and Corless, M. (1995b) Invariant manifolds and chaotic vibrations in singularly perturbed nonlinear oscillators, *International Journal of Engineering Science*, (submitted).

CONTROL OF CHAOS:

IMPACT OSCILLATORS AND TARGETING

ERNEST BARRETO[1,2]
University of Maryland, College Park, MD 20742.

FERNANDO CASAS
Departament de Matemàtiques, Universitat Jaume I, 12071-Castellón, Spain.

CELSO GREBOGI[2,3,4]
University of Maryland, College Park, MD 20742.

AND

ERIC J. KOSTELICH
Department of Mathematics, Arizona State University, Tempe, AZ 85287.

1. *Department of Physics.*
2. *Institute for Plasma Research.*
3. *Department of Mathematics*
4. *Institute for Physical Science and Technology.*

Abstract. We present two applications of chaos control techniques that can be of importance in mechanical systems. First, we apply chaos control to select a desired sequence of impacts in a map that captures the universal properties of impact oscillators near grazing. Next we describe a targeting method that can significantly reduce the chaotic transients that precede stabilization when these control methods are used.

1. Introduction

Recently, the application of chaos control techniques to physical systems has commanded increasing attention. In this work we describe the application of these methods to the impact oscillator, a mechanical system of great

D. H. van Campen (ed.), IUTAM Symposium on Interaction between Dynamics and Control in Advanced Mechanical Systems, 17–26.
© 1997 *Kluwer Academic Publishers.*

importance. We also describe a targeting method that can improve the utility of control techniques when applied to higher dimensional chaotic systems.

An impact oscillator is a forced vibrating mechanical system which undergoes a sequence of contacts with motion-limiting constraints. The dynamics is therefore smooth motion, governed by a differential equation, interrupted by a series of non-smooth collisions. The collisions introduce nonlinearity into the system. Impact oscillators are used to model a variety of different systems arising in engineering (for example, moored ships colliding with fenders, forced mechanical systems with clearances such as rattling gears, and railway vehicles [1, 2])

Mathematically, impact oscillators constitute a subclass of dynamical systems that do not satisfy the usual smoothness assumptions. These discontinuities are responsible for new forms of behavior not found in smooth dynamical systems, particularly in the limit of low velocity or grazing impacts [1-8].

In engineering, systems are modelled and investigated in order to identify and avoid unacceptable responses. For impact systems, it is necessary to avoid high velocity impacts as these cause the greatest wear or damage to components. This can be accomplished by the well-known techniques of chaos control [9]. The flexibility provided by chaos allows us to select particular trajectories with a desirable sequence of impacts. This can be advantageous in many technological applications of impact oscillators.

In this work we apply the method of Ott, Grebogi and Yorke to control chaotic impacts in the Nordmark map [3, 10, 11] $(x_{n+1}, y_{n+1}) = F_\rho(x_n, y_n)$, where

$$F_\rho(x, y) = \begin{cases} (\alpha x + y + \rho, -\gamma x) & \text{for} \quad x \leq 0 \\ (-\sqrt{x} + y + \rho, -\gamma \tau^2 x) & \text{for} \quad x > 0. \end{cases} \tag{1}$$

This is a piecewise continuously differentiable map that models the behavior of a sinusoidally forced linear oscillator experiencing impacts at a hard wall. It is obtained by expanding solutions of the system in the neighborhood of a grazing orbit [3], i.e., of an orbit that just touches the wall with zero velocity. The map captures the *universal* properties of the dynamics in the regime of low velocity impacts. The equivalence with the physical system is as follows: x_n and y_n are transformed coordinates in the position-velocity space $(\xi, \dot{\xi})$ evaluated at times $t_n = 2n\pi/\omega$, where ω is the frequency of the external forcing. The quantity τ^2 is the restitution coefficient of the impacts, and ρ is related to F_0, the amplitude of the external force. The parameters α and γ depend on the intrinsic properties of the oscillator such that the limit $\gamma \to 0$ corresponds to a large coefficient of friction, and $\gamma \tau^2 = 1$ gives the opposite limit of zero dissipation. For physical systems (with positive friction) we have [10-12]

$$0 < \gamma < 1, \qquad -2\sqrt{\gamma} < \alpha < 1 + \gamma. \tag{2}$$

The top expression in (1), valid for $x \leq 0$, governs the system if there is no impact between time t_n and t_{n+1}. Otherwise, $x > 0$ and the second expression applies. Thus, the effect of impacts in the system is modelled by a square root nonlinearity.

2. Control of the Impact Oscillator

The control technique of Ott, Grebogi and Yorke has the feature that it enables one to select a predetermined time-periodic behavior embedded in a chaotic attractor by making only *small* time-dependent perturbations to a set of accessible parameters of the system. The basic idea is as follows [9]. First one chooses a desirable unstable periodic orbit embedded in the chaotic attractor according to some set of performance criteria. Second, one defines a small region around the desired periodic orbit. A trajectory starting with almost any initial condition eventually falls into this small region by ergodicity. When this occurs, one applies perturbations to available control parameters so as to move the orbit onto the stable manifold of the desired unstable orbit. The flexibility of the method allows for the stabilization of different periodic orbits for the same set of nominal values of the parameter. This is possible because a chaotic attractor typically has embedded within it a large number of different unstable periodic orbits. We choose a single control parameter, ρ. This characterizes the strength of the driving. The grazing state corresponds to $\rho = 0$, and dynamics in the neighborhood of grazing is given for $|\rho| << 1$. Bifurcations occur as the parameter ρ is increased through $\rho = 0$ with γ and α held fixed.

By applying the OGY algorithm to the Nordmark map, one can stabilize periodic trajectories with an arbitrary number and an arbitrary distribution of impacts per period. This is so even if it is not possible to get analytic expressions for the position of the physical components. Also, the necessary information needed for applying control can be extracted purely from measured data [9, 13]. Here, for simplicity, we consider only maximal periodic orbits [10], i.e., periodic trajectories for which there is exactly one impact per period.

For systems with parameters in the region $4\gamma + \frac{1}{4} < \alpha < \frac{3}{2}\gamma + \frac{2}{3}$, windows of stable maximal periodic orbits are encountered as ρ is decreased from positive values [10, 11]. In particular, a window of period p is separated from the succeeding window, of period $p + 1$, by a band of chaos. There is an infinite cascade of such windows of decreasing width in ρ and increasing period, accumulating on $\rho = 0^+$. This is illustrated by the bifurcation

20

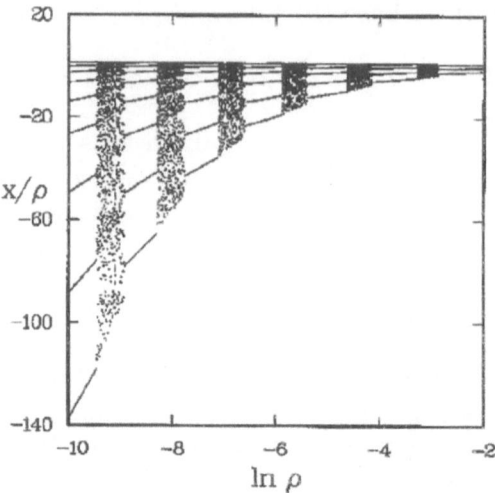

Figure 1. Bifurcation diagram for $(\gamma, \alpha) = (0.05, 0.65)$ and $\tau^2 = 1$ for small positive ρ values.

diagram of Fig. 1, obtained for $(\gamma, \alpha) = (0.05, 0.65)$ and $\tau^2 = 1$ for small positive ρ values. Here we can avoid the presence of chaotic impacts for $\rho > 0$ by applying control. As an example we take $\rho = \exp(-9.2)$, on the left band of chaos in Fig. 1. Here we have unstable maximal orbits up to period $M = 8$ embedded in the chaotic attractor.

Fig. 2 illustrates control of these periodic orbits. We plot the x-coordinate of a trajectory as a function of time. The parameter perturbations were programmed to successively control the seven different periodic orbits. Control for the $M = 2$ maximal orbit was turned on after 3000 free iterations. Each window was controlled for 500 iterations before switching to the next orbit. The figure shows that the time to achieve control is almost negligible in this case, with no apparent transients between switches. The maximum allowed parameter perturbation is $\delta = 10^{-4}$. Thus it is possible to convert chaotic impacts to controlled periodic orbits by applying only small perturbations $|\delta\rho| < 10^{-4}$ to the parameter ρ.

For parameters in the region $\frac{3}{2}\gamma + \frac{2}{3} < \alpha < 1 + \gamma$, there is an interval of ρ values occupied entirely by a chaotic attractor. As ρ increases from zero, this interval terminates at a stable maximal orbit of some period M_0 [10]. This attractor has embedded within it unstable maximal periodic orbits of increasing period as ρ approaches zero. An example of a bifurcation diagram for this case is shown in Fig. 3, obtained for $(\gamma, \alpha) = (0.15, 1)$ and $\tau^2 = 1$. Here we can control chaos by stabilizing any of the maximal orbits which are present for positive values of ρ. For $\rho = 0.05$, we have unstable

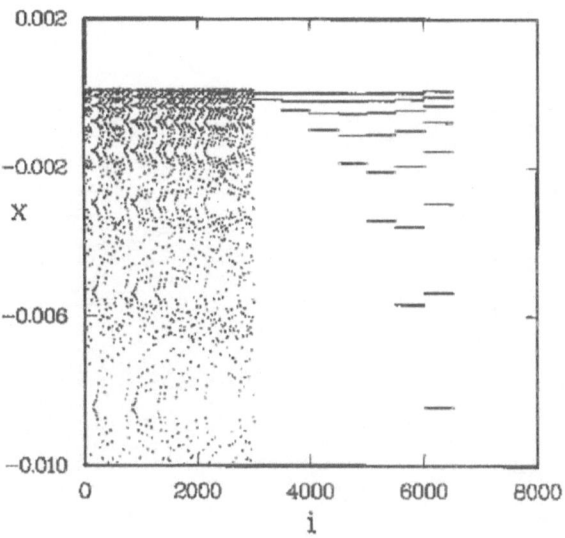

Figure 2. Successive control of unstable maximal periodic orbits for $\rho = \exp(-9.2)$, starting with period $M = 2$. The maximum parameter perturbation is $\delta = 10^{-4}$.

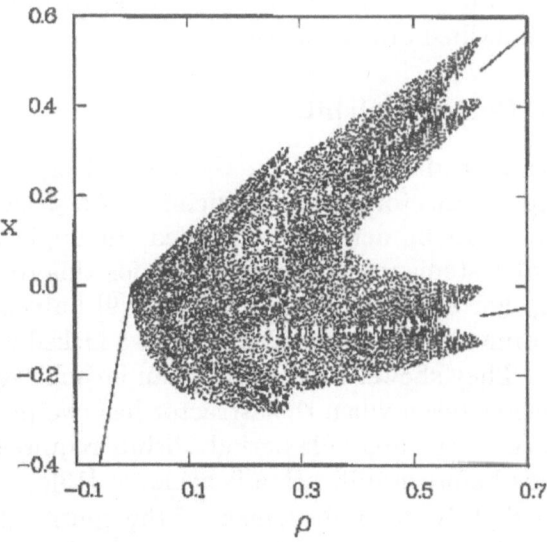

Figure 3. Bifurcation diagram for $\gamma = 0.15$, $\alpha = 1$ and $\tau^2 = 1$.

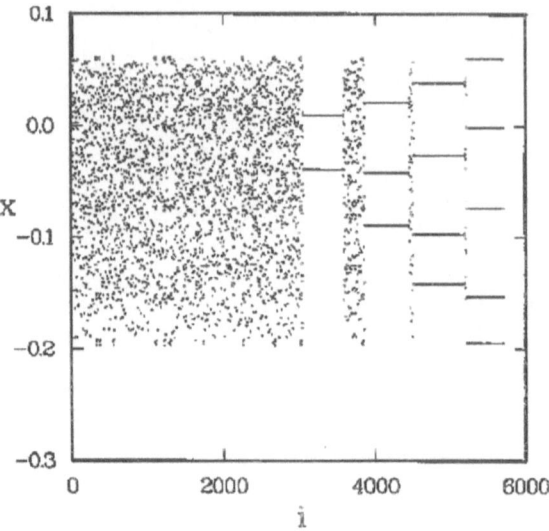

Figure 4. Successive control of the unstable maximal periodic orbits embedded in the chaotic attractor for $\rho = 0.05$, starting with period $M = 2$. The maximum parameter perturbation is $\delta = 10^{-3}$.

maximal orbits up to period $M = 5$. The control of these periodic orbits is accomplished as described above for Fig. 2.

3. Targeting of Periodic Orbits

The method described above relies on the natural ergodicity of chaotic dynamics to bring a trajectory into the vicinity of a desired unstable periodic orbit where it can be actively controlled. In applications involving higher dimensional systems, the times required for this to happen may be prohibitively long. For example, Romeiras *et al.* [9] have have applied the method to a four-dimensional map that describes a kicked double rotor [14], shown in Figure 5. They showed that control can be achieved by using only one control parameter (even when the attractor has two positive Lyapunov exponents). However, some unstable periodic orbits require several hundred thousand iterations before stabilization is achieved [15].

Targeting is a slightly different version of the control problem. We assume that we are given some initial condition on the attractor, and we wish to rapidly direct the resulting trajectory to a small region about some specified point on the attractor. Because of the inherent exponential sensitivity of chaotic time evolutions to perturbations, one expects that this can be accomplished using only small controlling adjustments of one or more

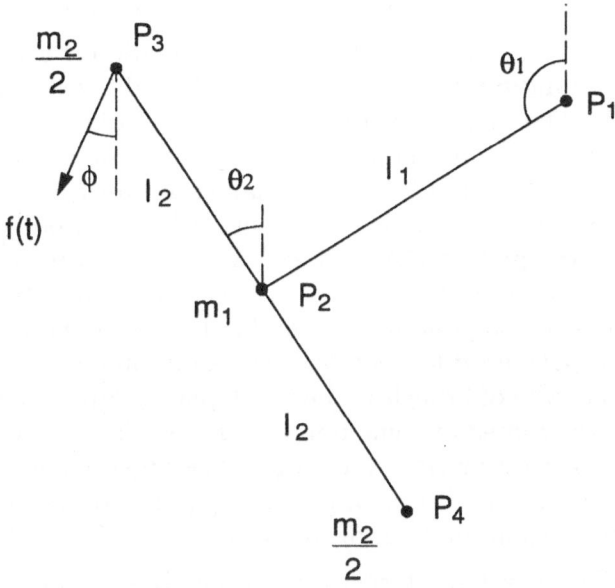

Figure 5. The Kicked Double Rotor. A massless rod of length l_1 pivots about the station-ary point P_1. A second massless rod of length $2l_2$ is mounted on pivot P_2, which in turn is mounted at the end of the first rod. Periodic impulsive kicks $f(t) = \sum_{n=0}^{\infty} \rho_n \delta(t - n)$ are applied at an angle ϕ as shown. The state of the system immediately after the $(n + 1)th$ kick is given by a four dimensional map of the form $\mathbf{X}_{n+1} = \mathbf{MY}_n + \mathbf{X}_n$ and $\mathbf{Y}_{n+1} = \mathbf{LY}_n + \mathbf{G}(\mathbf{X}_{n+1})$, where $\mathbf{X} = (\theta_1, \theta_2)^T$ are the two angular position coordinates, $\mathbf{Y} = (\dot{\theta}_1, \dot{\theta}_2)^T$ are the corresponding angular velocities, and $\mathbf{G}(\mathbf{X})$ is a nonlinear function. $\mathbf{M}$ and $\mathbf{L}$ are both constant matricies which involve the coefficients of friction at the two pivots and the moments of inertia of the rotor. Gravity is absent. Control parameters at time n are $\rho_n = 9.0 + \Delta\rho_n$ and $\phi_n = 0.0 + \Delta\phi_n$, with $|\Delta\rho|/\rho_0 \leq 0.1$ and $|\Delta\phi| \leq 0.5$. We take $l_1 = 1/\sqrt{2}$, and set all other parameters to 1. For further details, see Ref. [14].

available system parameters.

This was demonstrated theoretically and in numerical experiments for the case of a two-dimensional map by Shinbrot *et al.* [16], and also in a laboratory experiment for which the dynamics were approximated by a one dimensional map [17]. Kostelich *et al.* [18] developed an extension of the targeting procedure that can be applied to higher dimensional systems, such as the double rotor map.

Because the dimension of the double rotor attractor (for the set of pa-rameters chosen in Romeiras, *et al.*) is about 2.8, the average distance between nearest neighbors in a subset of N points on the attractor scales as $N^{-1/2.8}$. This implies that, on average, 10^{11} iterations of the map are required to come within 10^{-4} of the target without the control. Since the control procedure described in [18] can steer the initial condition to within 10^{-4} of the target in less than 10^2 steps, the method can achieve the target about 10^9 times faster than the uncontrolled chaotic process.

The method works in two steps. First, information is learned about the system by observing a very long chaotic orbit, and constructing targeting trees as follows. The map is iterated from a random initial condition while keeping in memory a short history of the iterates encountered (for example, 10 consecutive points), until the orbit lands within a suitable tolerance distance of the target. This point, together with the recorded pre-iterates, comprise the *trunk path* of the tree, and are stored in memory. The map is then iterated again, still keeping track of a brief iterate history, until the orbit lands near any one of the points already in the tree. When this happens, a new path is added as a *branch*. Continuing in this way, a tree is built with a hierarchy of branches: the trunk path is level 1; level 2 branches are those that are rooted at some point in the trunk path; level 3 branches are rooted at a level 2 branch, and so on. The objective is to build a tree with enough branches such that a typical chaotic orbit lands near a point in the tree after a small number of iterations.

Once a sufficiently large targeting tree has been built, a chaotic orbit can be steered along the tree to the target. One applies small changes to available parameters to steer the orbit to the stable manifold of a point in the tree. (The *stable manifold S* associated with a typical point x is stable in the sense that $\|F^n(x) - F^n(y)\| \to 0$ as $n \to \infty$ whenever $y \in S$.) When the method is successful, the dynamics of the system carry the orbit of the perturbed point close to an orbit that leads directly to the target. Additional details on the method are given in [18].

The targeting algorithm can be combined with the OGY control method to provide a means to rapidly switch a given chaotic process between pre-specified periodic orbits. That is, the targeting procedure can be used to steer a given initial condition on the attractor to a neighborhood of one of the periodic orbits, then the OGY control method can be used to stabilize the system near the periodic orbit. The combined method is discussed in [15], and the results of its application to the double rotor are shown in Figures 6 and 7 .

4. Conclusions

In summary, we have shown that chaotic dynamics in impact oscillators can be converted into motion on a desired periodic orbit by using only small parameter perturbations. In higher dimensional systems, it is possible to employ a targeting technique to reduce the length of the chaotic transients that precede stabilization. These results can be of importance in technological applications.

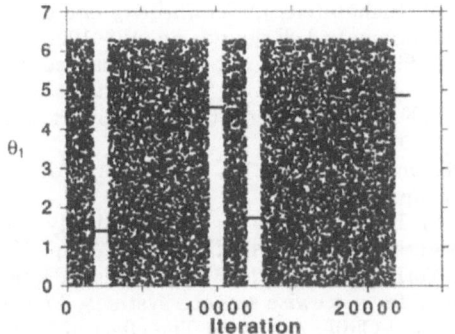

Figure 6. Graph illustrating switching between five different fixed points. The θ_1 coordinate of the state is plotted versus iteration. Here we rely on ergodicity to bring the orbit close to the desired UPO. The fifth fixed point required $153,485$ iterations to be stabilized, and is not shown.

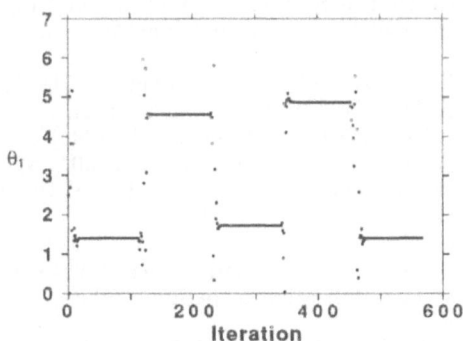

Figure 7. Improvements of up to four orders of magnitude in the switching times is evident.

ACKNOWLEDGMENTS

This work was supported by the Department of Energy (Office of Scientific Computing). E. B. was supported in part by a National Physical Science Consortium Fellowship under the sponsorship of Argonne National Laboratory. F. C. was supported by the Conselleria de Educació de la Generalitat Valenciana (Spain) and by the collaboration program UJI-Fundació Caixa Castelló 1994. E. K. is supported in part by the Department of Energy Program in Applied Mathematics, grant number DE-FG03-94ER25213.

References

1. Foale S. and Bishop, S.R. (1992) Dynamical complexities of forced impacting systems, *Phil. Trans. R. Soc. Lond.*, **A338**, pp. 547–556.
2. E. Slivsgaard and H. True (1994) Chaos in railway-vehicle dynamics, in J.M.T.

Thompson and S.R. Bishop (eds.), *Nonlinearity and Chaos in Engineering Dynamics*, John Wiley and Sons Ltd, England, pp. 183–192.

3. Nordmark, A.B. (1991) Non-periodic motion caused by grazing incidence in an impact oscillator, *J. Sound Vib.*, **145**, pp. 279–297.

4. Budd, C. and Dux, F. (1994) Intermittency in impact oscillators close to resonance, *Nonlinearity*, **7**, pp. 1191–1224.

5. Shaw, S.W. and Holmes, P.J. (1983) A periodically forced piecewise linear oscillator, *J. Sound Vib.* **90**, pp. 129–155.

6. Shaw, S.W. (1985) The dynamics of a harmonically excited system having rigid amplitude constraints, *ASME J. Appl. Mech.*, **52**, pp. 453–464.

7. Nusse, H.E. and Yorke, J.A. (1992) Border-collision bifurcations including 'period two to period three' for piecewise smooth systems, *Physica D*, **57**, pp. 39–57.

8. Budd, C., Dux, F. and Cliffe, A. (1995) The effect of frequency and clearance variations on single degree of freedom impact oscillators, *J. Sound Vib.*, **184**, pp. 475–502.

9. E. Ott, C. Grebogi and J.A. Yorke (1990) Controlling chaos, *Phys. Rev. Lett.*, **64**, pp. 1196–1199; F.J. Romeiras, C. Grebogi, E. Ott and W.P. Dayawansa (1992) Controlling chaotic dynamical systems, *Physica D* **58**, pp. 165–192; E. Barreto and C. Grebogi (1995) Multiparameter control of chaos, *Phys. Rev. E*, **52**, pp. 3553–3557 . See also E. Barreto, Y.C. Lai and C. Grebogi (1996), Controlling Chaos with applications to Mechanical Systems, in F.C. Moon (ed.), *Nonlinear Dynamics of Material Processing and Manufacturing*, John Wiley and Sons, Inc., New York, to be published.

10. Chin, W., Ott, E., Nusse, H.E. and Grebogi, C. (1994) Grazing bifurcations in impact oscillators, *Phys. Rev. E*, **50**, pp. 4427–4444.

11. Chin, W., Ott, E., Nusse, H.E. and Grebogi, C. (1995) Universal behavior of impact oscillators near grazing incidence, *Phys. Lett. A*, **201**, pp. 197–204.

12. Casas, F., Chin, W., Grebogi, G. and Ott, E. (1995) Universal grazing bifurcations in impact oscillators, *submitted for publication*.

13. U. Dressler and G. Nitsche (1992) Controlling chaos using time delay coordinates, *Phys. Rev. Lett*, **68**, pp. 1–4; P. So and E. Ott (1995) Controlling chaos using time delay coordinates via stabilization of periodic orbits, *Phys. Rev. E*, **51**, pp. 2955–2962.

14. E.J. Kostelich, C. Grebogi, E. Ott and J.A. Yorke (1987) Multi-dimensioned intertwined basin boundaries: basin structure of the kicked double rotor, *Physica D*, **25**, pp. 347–360; C. Grebogi, E.J. Kostelich, E. Ott and J.A. Yorke (1986) Multidimensioned intertwined basin boundaries and the kicked double rotor, *Phys. Lett. A*, **118**, pp. 448–452 and errata(1987), **120**, pp. 497.

15. E. Barreto, E.J. Kostelich, C. Grebogi, E. Ott and J.A. Yorke (1995) Efficient switching between controlled unstable periodic orbits in higher dimensional chaotic systems, *Phys. Rev. E*, **51**, pp 4169–4172.

16. T. Shinbrot, E. Ott, C. Grebogi and J.A. Yorke (1990) Using chaos to direct trajectory to targets, *Phys. Rev. Lett.*, **65**, pp. 3215–3218.

17. T. Shinbrot, W. Ditto, C. Grebogi, E. Ott, M. Spano and J.A. Yorke (1992) Using the sensitive dependence of chaos (the 'butterfly effect') to direct trajectories in an experimental chaotic system, *Phys. Rev. Lett.*, **68**, pp. 2863–2866.

18. E.J. Kostelich, C. Grebogi, E. Ott, and J.A. Yorke (1993) Higher-dimensional targeting, *Phys. Rev. E*, **47**, p. 305–310.

HILL'S PROBLEM AS A DYNAMIC BILLIARD

V.V.BELETSKY, O.P.SALIMOVA
The Keldysh Institute of Applied Mathematics, Russia Academy of
Sciences,Miusskaya Sq. 4, Moscow A-47, 125047, Russia,
e-mail: beletsky@applmat.msk.su, Fax: +7-(095)-9720737

Projects exist for expeditions on the surface of Mars' moons Phobos and Daimos using jumping robots. These projects require research on the dynamics of such a robot under the simultaneous actions of the gravitational fields of the planet Mars and its satellite (Phobos or Daimos) taking into account impacts of the robot on the satellite surface.

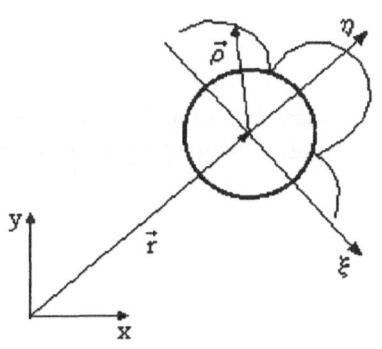

Fig. 1

In the paper a highly simplified form of this problem is investigated. The robot is taken as an elementary mass particle whose mass is negligible compared with the masses of both planet and satellite. The satellite is always pointing the same face towards the planet (like the Earth's moon does with respect to the Earth and like Phobos does with respect to Mars). The motion is investigated relative to a rotating co-ordinate system, one axis (η) of which is directed through the centres of the planet and the satellite, while another axis (ξ) is tangent to the circular satellite orbit (fig. 1). Its origin coincides with the mass centre of the moon (or satellite below). Relative to this co-ordinate system, the satellite is stationary and it is always pointing the same face towards the planet.

The satellite is considered to be a homogeneous sphere. The particle motion in the vicinity of the satellite is governed by a well-known set of Hill's equations:

$$
\begin{cases}
\xi'' + 2\eta' = -3\, \dfrac{\xi}{(\xi^2 + \eta^2)^{3/2}}, \\[4mm]
\eta'' - 2\xi' - 3\eta = -3\, \dfrac{\eta}{(\xi^2 + \eta^2)^{3/2}}.
\end{cases}
\tag{1}
$$

These equations are supplemented by equations describing impact on the satellite surface. The impacts are considered to be ideally elastic:

27

D. H. van Campen (ed.), IUTAM Symposium on Interaction between Dynamics and Control in Advanced
Mechanical Systems, 27–34.
© 1997 Kluwer Academic Publishers.

$$\begin{cases} v_\xi^+ = v_\xi^- - 2(\xi v_\xi^- + \eta v_\eta^-)\xi, \\ v_\eta^+ = v_\eta^- - 2(\xi v_\xi^- + \eta v_\eta^-)\eta. \end{cases} \qquad (2)$$

where v_ξ^-, v_η^- and v_ξ^+, v_η^+ are the velocity components of the particle before and immediately past the moment of impact accordingly.

This simplified version of the problem constitutes a kind of dynamic billiard. Since the impacts are considered to be ideally elastic, the equations of motion possess as first integral the Jacobian energy integral (3).

$$\xi'^2 + \eta'^2 - \left[3\eta^2 + \frac{6}{(\xi^2 + \eta^2)^{1/2}} \right] = -J, \quad J > 0 \qquad (3)$$

The curves below are the Hill's line or the zero-speed line for different J in the reference system. The motion is possible inside the region bounded by these curves (fig. 2). For J=10.0 this region is shown by hatching.

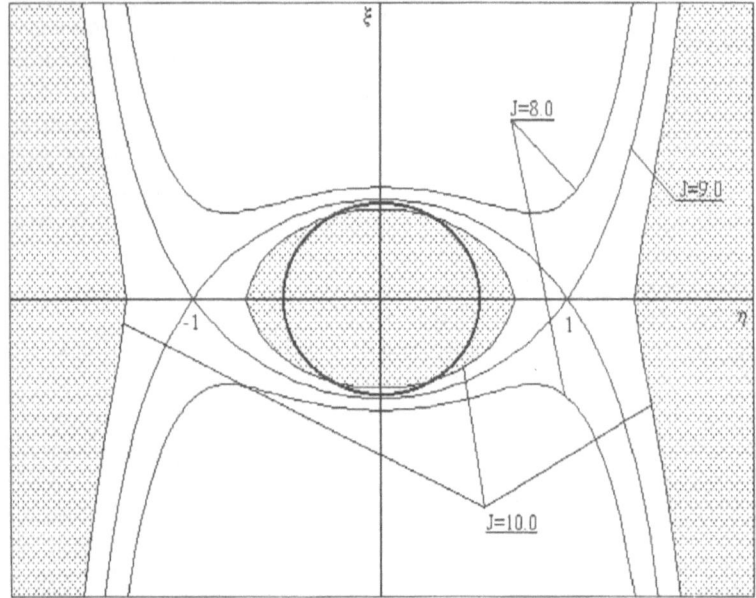

Fig. 2. The Hills' chart

The value of this constant energy (J) is one of two system parameters. The other is the radius of the spherical satellite. Thus, the problem is two-parametric and the trajectory manifold is two-parametric.

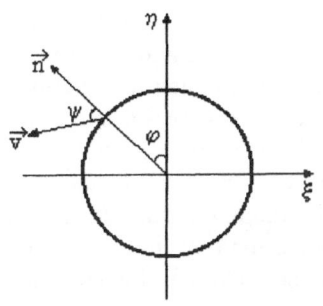

Fig.3. The Poincare map parameters.

The investigation was carried out by using numerical as well as analytical methods. To study the structure of the phase space, a numerical form of Poincare's point mapping method was used wherever possible. The phase space of the problem is four-dimensional. But at the moment of impact with the satellite surface, the distance from the origin of the co-ordinate system is known, and, in addition, the energy is constant. This means that at the moment of impact the phase space is only two-dimensional. This allows a Poincare mapping for the instants of impact in a two-dimensional space. The angle φ and the angle ψ are introduced as the Poincare map parameters. The angle φ is the angle between axis η and radius-vector of the impact point. The angle ψ is the angle between the starting velocity vector and the normal vector of planet surface in the impact point (Fig.3).

We investigate the planar trajectories in the orbit plane of the satellite. All these trajectories are symmetric about the co-ordinate origin, as the equations of motion are symmetric. Depending on the value of the two system parameters, the manifold of trajectories consists of three basic submanifolds, i.e. of three basic types of trajectories. Within each type, a more detailed classification of trajectories is made.

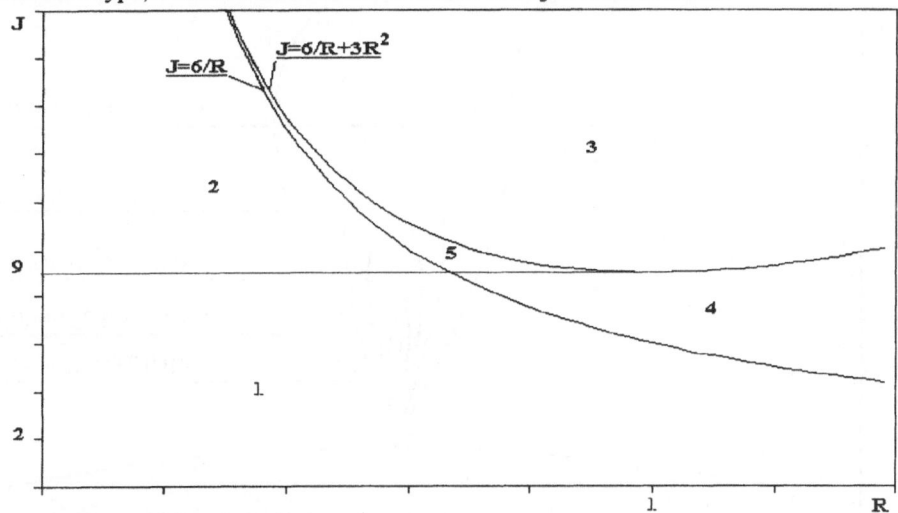

1- Unbounded trajectories cover the whole surface of the satellite.
2- Bounded trajectories cover the whole surface of the satellite.
3- The impact is impossible.
4- Unbounded polar trajectories.
5- Bounded polar trajectories.

Fig. 4. The parameter chart.

    1. *Bounded trajectories covering the whole surface of the satellite (fig.4, region 2).* These trajectories are either regular - possibly periodic - or chaotic. In typical cases the phase portrait represents a "chaotic ocean" with an archipelago of isolated "islands" of regular motions. The centres of these islands are associated with periodic motions. Periodic motions may consist of any numbers of arcs between successive impacts ranging from a single arc to multi-arc jumps. The family of single-jump trajectories is represented on fig. 5. Because it returns to its starting point, the mass particle has to jump forward in the direction of the satellite motion around the planet. This is explained by the influence of Coriolis' forces. Among the trajectories covering the whole satellite surface, there are forward or backward ones. The set of islands near the centre corresponds to trajectories covering only part of the satellite surface, so that in this case sequences of forward and backward jumps take place.

    Figures 6-9 show the maps calculated for R=0.5 and decreasing J. As we can see, the regularity of forward trajectories covering the whole surface is destroyed first of all. Then the central islands turn into chaos; backward regular trajectories covering the whole surface possess the most stability.

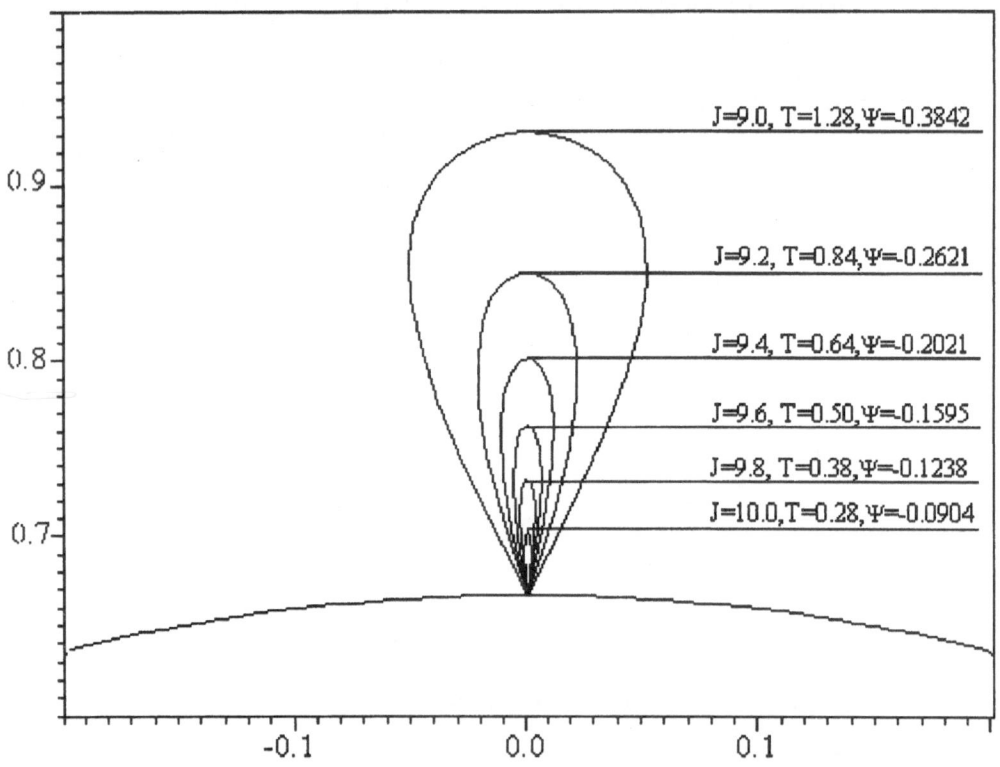

Fig. 5. The single-jump trajectory family.

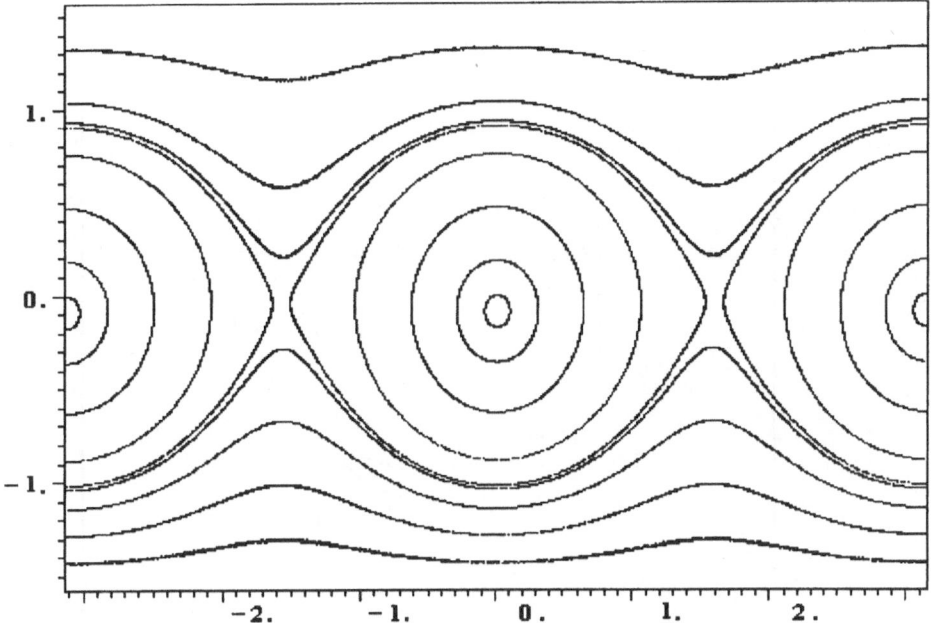

Fig . 6   R=0.5,   J=11..5

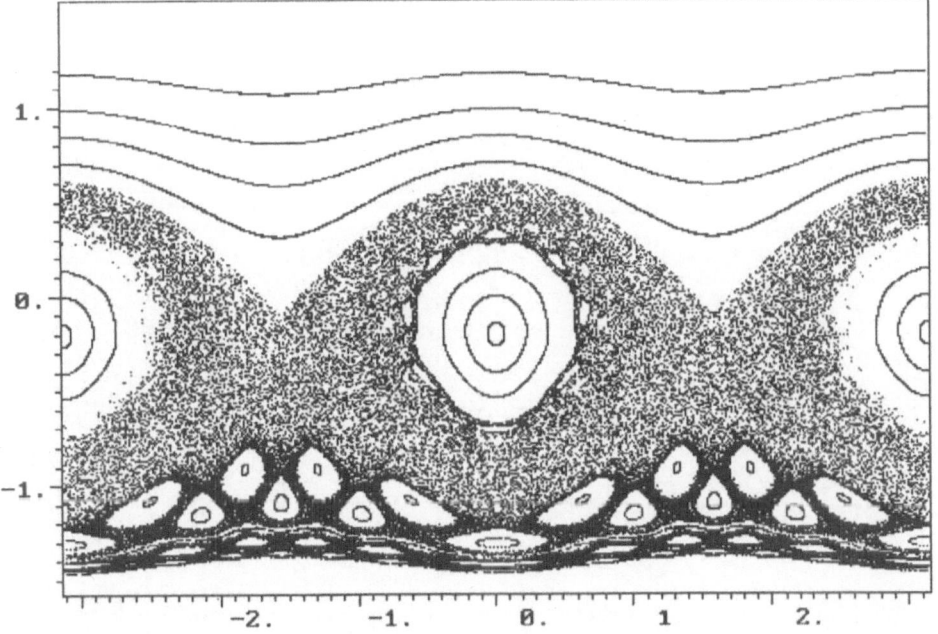

Fig.7   R=0.5,   J=9.8

32

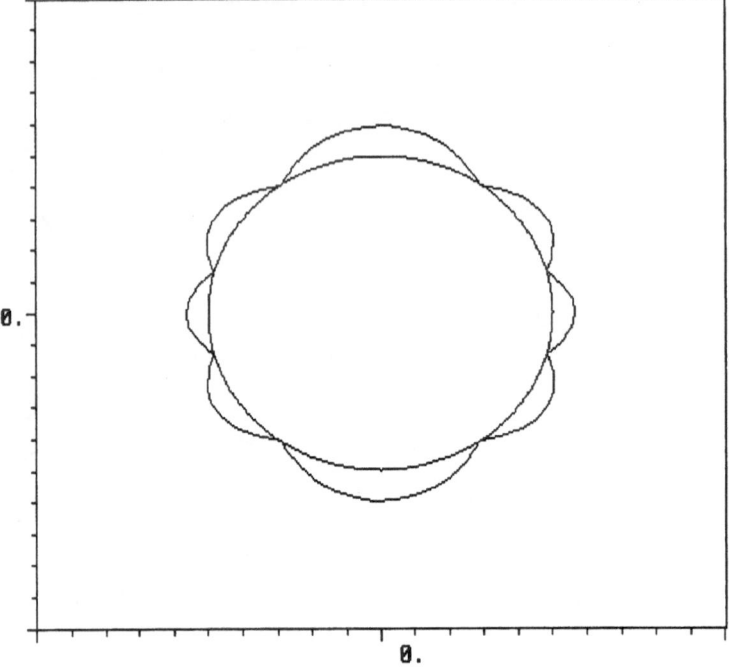

Fig. 8   Eight-jump trajectory for   R=0.5,   J=9.8.

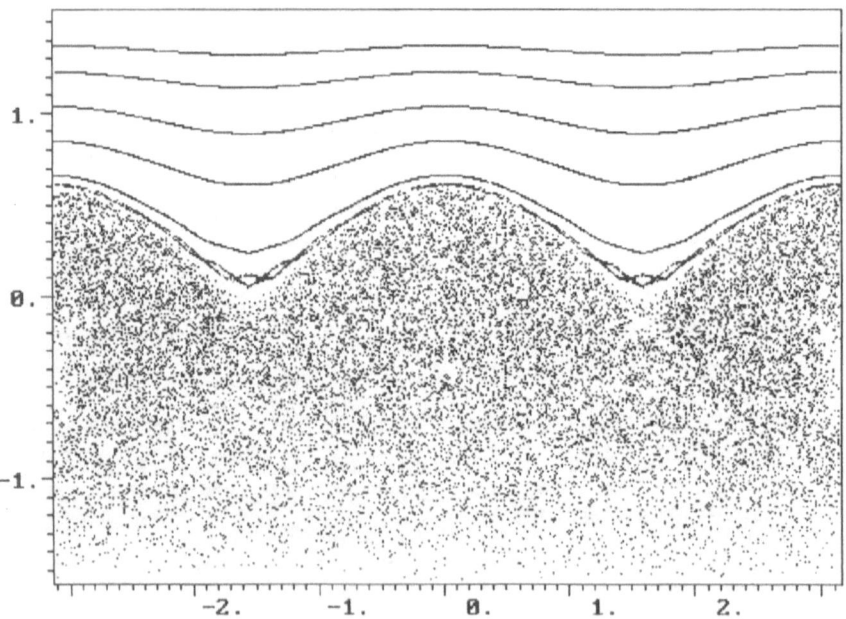

Fig. 9   R=0.5,   J=9.0.

2. *Similar, i.e. regular or periodic or chaotic trajectories which, however, do not leave a certain polar region of the satellite (fig.4, domain 5)*. Poles, in this context, are the two points of intersection of the axis pointing through the centres of the satellite and the planet, with the satellite surface. The map corresponding to this case 10 is represented in figure 10.

3. *Unbounded trajectories which after a finite number of jumps leave the satellite towards infinity (fig. 4, domain 1,4)*. An escape is possible only through "bottlenecks" of Hill's chart pointing either towards the planet or directly away from it. Before leaving the satellite, the mass particle can execute any number of jumps on the satellite surface in either a single direction or in alternating directions (fig. 11).

Among this trajectories, there are a submanifold of near-polar trajectories. In this case the jumping mass particle does not leave a certain polar region before ultimately escaping from the satellite (possible after a single jump) either towards the planet or away from it.

There are also theoretically possible motions which are characterised by the fact that finite sections of the trajectory (not only individual points) coincide with the circular satellite surface. In such cases the mass touches the satellite without impact. Such trajectories may be periodic.

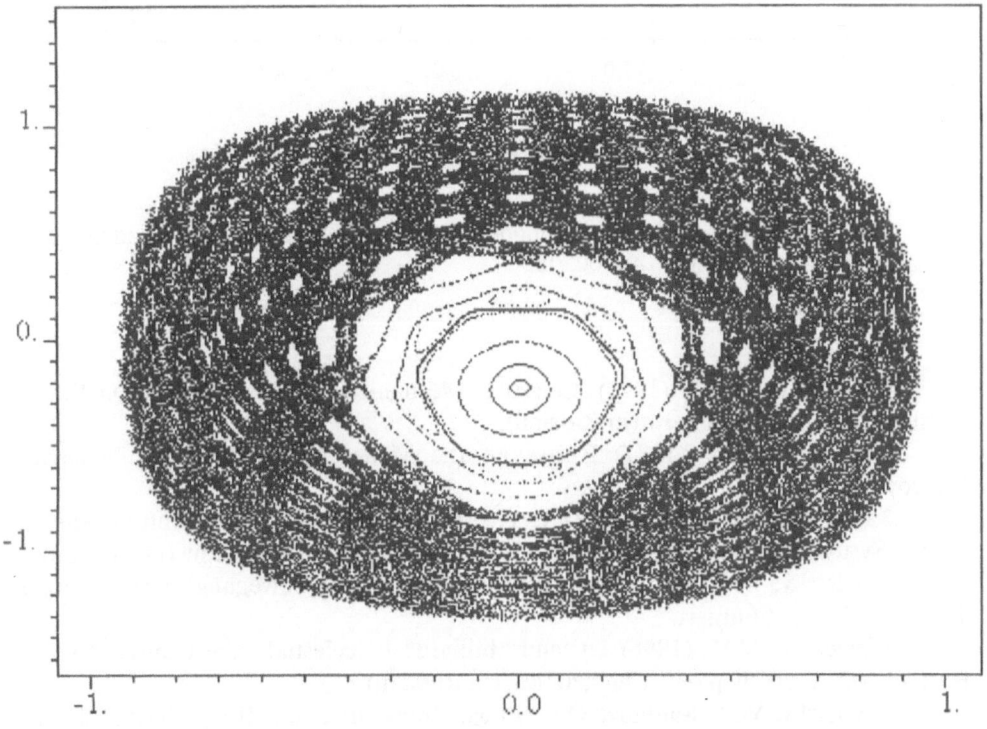

Fig. 10. The polar map. R=0.667, J=9.5.

34

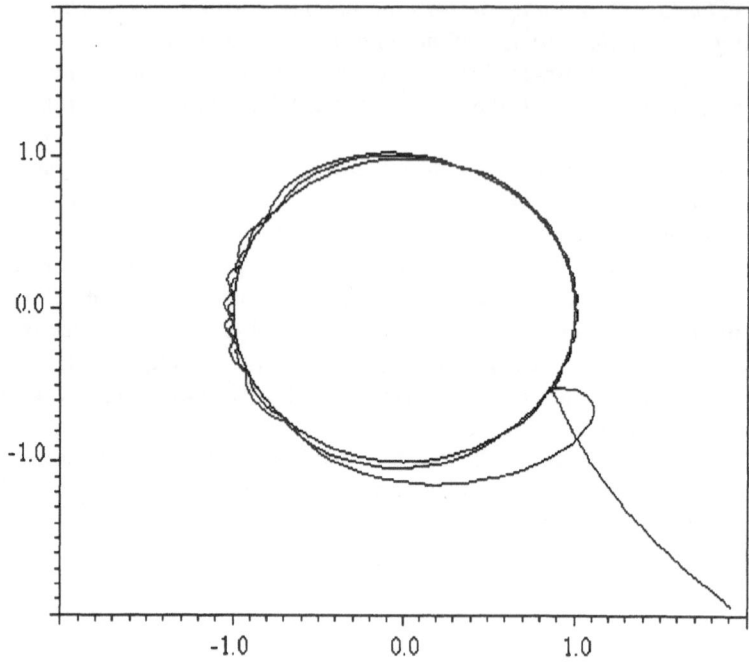

Fig. 11. The unbound trajectory escaping to the Mars. R=1.0.

This research was financially supported by the "Russian Fund for Fundamental Research" under the grant 95-01-00308.

Bibliography.

1.Duboshin, G.N. (1978) Celestial Mechanics. Analytical and Qualitative Methods, "Science", Moscow (in Russian).

2.Beletsky, V.V. (1977) Essays on the Motion of space Body, "Science", Moscow (in Russian).

3.Kozlov, V.V., Treschev, D.V. (1991) Billiards: a Genetic Introduction to Impact System Dynamics, The Moscow University Publishers, Moscow (in Russian).

4.Beletsky, V.V. (1995) Regulare und chaotische Bewegung starrer Korper, Teubner- Verlag, Stuttgart.

5.Beletsky, V.V. (1995) Dynamic billiards in celestial mechanics, Abstract Book of XIX Workshop on Cosmonautics (in Russian).

6.Beletsky, V.V., Salimova O.P., Impact trajectories in Hill's problem. Abstract Book of XIX Workshop on Cosmonautics (in Russian).

7.Beletsky, V.V. (1995) Billiards in celestial mechanics, Book of thesis, Stochastic methods and experiments in celestial mechanics (in Russian).

DYNAMICS AND OPTIMAL CONTROL PROBLEMS FOR BIOTECHNICAL SYSTEMS "MAN-PROSTHESIS"

V. BERBYUK
Pidstryhach Institute for Applied Problems of Mechanics and Mathematics of the Ukrainian National Academy of Sciences
3-B, Naukova Str., Lviv, 290601, Ukraine
e-mail:Kalyniak@IPPMM. Lviv.UA

1. Introduction

New prosthetic materials and designs have lead to many prostheses of lower limbs for amputees. As a result, it is becoming difficult for prosthetists and the physicians to choose which prosthesis is the best for the individual amputee. Presently, there is limited information about "optimal" alignment, and how the prosthesis performs dynamically in achieving optimally symmetrical gait for an amputee. Sensory feedback, better control systems, and more energy-efficient devices are strongly needed [1]. Gait studies, ambulatory physiological monitoring, mathematical modeling of a human controlled motion, and dynamic optimization techniques may be useful tools to improve and create new efficient lower limb prostheses.

Experimental data [2,3] and mathematical modeling [4-7] show that the kinematics and dynamics of a human locomotor system (HLS) are strongly sensitive to the constructive parameters of the prosthesis (massinertial, elastic, viscoelastic, etc.) and to the control parameters of a human gait (cadence, velocity, duration of the leg activity, etc.).

In this paper, a mathematical model is proposed for investigating the controlled motion of HLS with an above-knee prosthesis. To provide insight into the interaction between dynamics and control in biotechnical system *Man-Prosthesis* the energy-optimal control problem of the HLS wearing a lower limb prosthesis has been considered. The algorithm is based on special conversion of the optimal control problem for a nonlinear dynamical system which models HLS into a standard nonlinear programming problem. We solve a number of energy-optimal control problems of human locomotion with an artificial leg, and optimization problems for the constructive parameters of the prostheses under different boundary conditions and constraints.

The numerical results obtained were compared with experimental data for normal human locomotion [8]. We find energy-optimal elastic and viscoelastic characteristics of the ankle and knee joints of the prostheses.

D. H. van Campen (ed.), IUTAM Symposium on Interaction between Dynamics and Control in Advanced Mechanical Systems, 35–42.
© 1997 *Kluwer Academic Publishers.*

2. The Mathematical Model

HLS is simulated by a plane multibody system of rigid masses (Fig. 1). This system comprises an inertial body G (trunk) and two legs. Each leg consists of three elements. The two elements with mass and rotatory inertia model the thigh (link OK_i) and calf (link K_iA_i), while the third massless and inertia-free element (links $A_iH_iT_i$) models the foot.

In addition to the weights of the trunk, thighs and calves, the external forces acting on HLS include the interaction forces between the feet and the ground, which are replaced by resultant forces R_i ,($i=1,2$).

It is assumed that the control moments $q_i(t)$, $u_i(t)$, $p_i(t)$ are acting at the hip (point O) , knee (point K_i) and ankle (point A_i) joints, respectively.

The mathematical modeling of human gait with an above-knee prosthesis is based on the supposition that the force moments at the knee and ankle joints of the prosthetic leg are passive ones. The values of these moments depend not only from the gait pattern, but also on the prosthesis construction.

Henceforth the subscript 1 will refer to the prosthetic leg , 2 to the intact leg.

The set of expressions describing the dynamics of HLS are [6]:

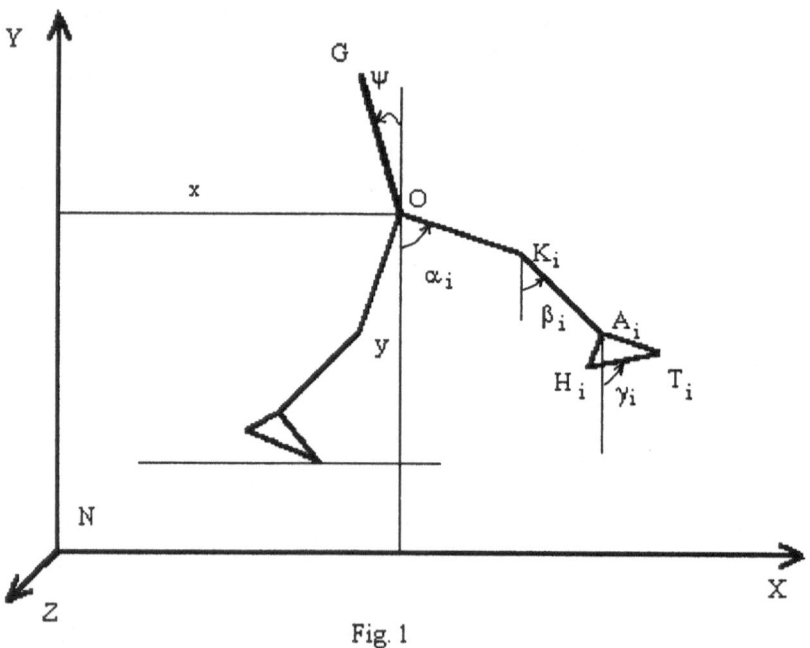

Fig. 1

$$f_1(t) - K_r(\psi''\cos\psi - \psi'^2\sin\psi) = R_{1x}(t) + R_{2x}(t)$$

$$f_2(t) - K_r(\psi''\sin\psi + \psi'^2\cos\psi) = R_{1y}(t) + R_{2y}(t)$$

$$f_{3i}(t) = q_i - u_i + a_i(R_{ix}\cos\alpha_i + R_{iy}\sin\alpha_i)$$

$$f_{4i}(t) = u_i - p_i + b_i(R_{ix}\cos\beta_i + R_{iy}\sin\beta_i)$$ (1)

$$f_5(t) = -q_1 - q_2$$

$$f_1(t) = Mx'' + \sum_{i=1}^{2}[K_{ai}(\alpha_i'\cos\alpha_i)' + K_{bi}(\beta_i'\cos\beta_i)']$$

$$f_2(t) = M(y'' + g) + \sum_{i=1}^{2}[K_{ai}(\alpha_i'\sin\alpha_i)' + K_{bi}(\beta_i'\sin\beta_i)']$$

$$f_{3i}(t) = J_i\alpha_i'' + K_{ai}(x''\cos\alpha_i + y''\sin\alpha_i) + gK_{ai}\sin\alpha_i +$$

$$a_iK_{bi}[\beta_i''\cos(\alpha_i - \beta_i) + \beta_i'^2\sin(\alpha_i - \beta_i)]$$

$$f_{4i}(t) = J_{ci}\beta_i'' + K_{bi}(x''\cos\beta_i + y''\sin\beta_i) + gK_{bi}\sin\beta_i +$$

$$a_iK_{bi}[\alpha_i''\cos(\alpha_i - \beta_i) - \alpha_i'^2\sin(\alpha_i - \beta_i)], \quad (i = 1,2)$$

$$f_5(t) = J\psi'' - gK_r\sin\psi - K_r(x''\cos\psi + y''\sin\psi)$$

$$p_i + (y_i - y_{Ri})R_{ix} + (x_{Ri} - x_i)R_{iy} = 0$$

$$M = m + m_{a1} + m_{a2} + m_{b1} + m_{b2} + m_{f1} + m_{f2}$$

$$J_i = J_{ai} + a_i^2(m_{bi} + m_{fi}), \quad J_{ci} = J_{bi} + b_i^2 m_{fi}$$

$$K_{ai} = m_{ai}r_{ai} + a_i(m_{bi} + m_{fi}), \quad K_{bi} = m_{bi}r_{bi} + b_i m_{fi}, \quad K_r = rm, \quad (i = 1,2).$$

In equations (1) : x and y are the Cartesian coordinates of the suspension point O of the legs; $\psi, \alpha_i, \beta_i, \gamma_i$ are angles that specify the position of the elements of the HLS (Fig. 1); m is the mass of the trunk; r is the distance from the suspension point of the legs to the center of mass of the trunk; J is the moment of inertia of the trunk relative to the Z axis at point O; m_{ai} is the mass of the thigh, a_i is the distance from O to the point K_i; J_{ai} is the moment of inertia of the thigh relative to the Z axis at O; r_{ai} is the distance from O to the center of mass of the thigh ; m_{bi} is the mass of the calf; b_i is the distance from the knee joint to the point A_i ; J_{bi} is the moment of inertia of the calf relative to the Z axis at the point K_i ; r_{bi} is the distance from K_i to the center of mass of calf; m_{fi} is mass of the foot located at the ankle joint A_i ; $R_{ix}(t)$, $R_{iy}(t)$ are the horizontal and vertical components of the force R_i; (x_i, y_i), (x_{Ri}, y_{Ri}) are the Cartesian coordinates of the ankle joint, and of the point of application of the vector R_i of the i-th leg, respectively; g is the acceleration due to gravity; and the prime denotes differentiation with respect to time t.

3. Statement of the Problem

It is assumed that all movement of HLS is restricted to the sagittal plane NXY of a fixed rectangular Cartesian coordinate system NXYZ (Fig 1).

Let $Z = \{x, x', y, y', \psi, \psi', \alpha_i, \alpha_i', \beta_i, \beta_i', \gamma_i, \gamma_i', \quad (i = 1,2)\}$

be a vector of the phase state, $U=\{R_{ix}, R_{iy}, q_i, u_i, p_i, \quad i=1,2\}$ be a vector of the controlling stimuli of HLS, and T be a duration of a double step, or stride [8].

There are three phases during a stride: support by the prosthetic leg alone, by both legs, and by the intact leg alone. The interaction between the dynamics of the prosthesis and the HLS is strongest during the first phase, and it is this phase of duration τ, that we will investigate.

Consider the optimization problem:

Problem A. It is required to determine the controlled process $\{Z(t), U(t)\}$, $t \in [0, \tau]$ and vector-parameters C_a, C_k which satisfy the equations of motions (1), the boundary conditions

$$x(0) = x^0, \quad x_{t2}(0) = x_{t2}^0, \quad x_{h2}(\tau) = x_{h2}^\tau, \quad y_{t2}'(0) = y_{h2}'(0) = 0, \qquad (2)$$

the following restrictions on the phase coordinates over the time $t \in [0, \tau]$

$$\alpha_i(t) \geq \beta_i(t), \quad (x-x_i)^2 + (y-y_i)^2 \leq (a_i+b_i)^2, \quad (i=1,2) \qquad (3)$$

$$x_1(t) = x_1^0, \quad y_1(t) = y_1^0, \quad y_{t2}(t) \geq 0, \quad y_{h2}(t) \geq 0, \quad y(t) \geq h \qquad (4)$$

$$\alpha_2(t) - \psi(t) = \theta_h(t), \quad \alpha_2(t) - \beta_2(t) = \theta_k(t), \quad \gamma_2(t) = \theta_a(t) + \beta_2(t) + \pi/2 \qquad (5)$$

$$R_{1y}(t) \geq 0, \quad x_{h1} \leq x_{R1}(t) \leq x_{t1} \qquad (6)$$

and the constraints on the controlling stimuli over the time $t \in [0, \tau]$

$$p_1(t) = f_a(t, C_a), \quad u_1(t) = f_k(t, C_k), \quad |q_1(t)| \leq q^0 \qquad (7)$$

and which minimize of the functional [4-6, 9,10]

$$E = \frac{1}{L} \int_0^\tau \left\{ \sum_{i=1}^{2} |q_i(\psi' - \alpha_i')| + |u_2(\alpha_2' - \beta_2')| \right\} dt \qquad (8)$$

In expressions (2)-(8): $x_{ti}(t)$, $y_{ti}(t)$, $x_{hi}(t)$, $y_{hi}(t)$ are the Cartesian coordinates of the toe and heel of the i-th leg; $f_a(t, C_a)$, $f_k(t, C_k)$ are given functions determining the dynamic characteristics of an above-knee prosthesis; C_a, C_k are the vectors defining the prosthesis structure; $\theta_k(t)$, $\theta_h(t)$, $\theta_a(t)$ are the functions given from the experimental data on human locomotion [8]; L is a stride length which is equal to the sum of two step lengths; x^0, x_{t2}^0, x_{h2}^τ, x_1^0, y_1^0, h, $q^0 \geq 0$ are given numbers.

The objective functional (8) is the integral over a stride of the sum of the absolute values of the mechanical power of all controlling stimuli acting at intact joints of the HLS.

4. Results and Discussion

Central to the approach proposed for solving problem A is the idea that any optimal control problem can be converted into a standard nonlinear programming problem by parameterizing each of the free variable functions.

The analysis of equations (1), the restrictions on the phase coordinates (3)-(5) and given constrains on the controlling stimuli shows that there is only one independently variable function in Problem A.

It is suitable to choose the function $x_2(t)$ (the abscissa of the ankle joint of the intact leg) as an independently variable function.

The function $x_2(t)$ was parameterized in the following way [11]:

$$x_2(t) = P(t) + G(t), \quad P(t) = \sum_{k=0}^{5} C_k t^k, \quad G(t) = \sum_{k=1}^{N} (a_k \cos k\omega t + b_k \sin k\omega t), \qquad (9)$$

$$\omega = 2\pi / \tau$$

The coefficients of the function $P(t)$ are determined by the following conditions:

$$P(0) + G(0) = x_2(0), \quad P'(0) + G'(0) = x_2'(0), \quad P''(0) + G''(0) = x_2''(0)$$

$$P(\tau) + G(\tau) = x_2(\tau), \quad P'(\tau) + G'(\tau) = x_2'(\tau), \quad P''(\tau) + G''(\tau) = x_2''(\tau) \qquad (10)$$

In formulae (9)-(10): $x_2(0)$, $x_2(\tau)$ are calculated using boundary conditions (2), restrictions on the phase coordinates (5) and the linear characteristics of the foot; $x_2'(0)$, $x_2'(\tau)$, $x_2''(0)$, $x_2''(\tau)$, a_k, b_k, $k=1,...N$ are the independently variable parameters.

The controlling stimuli of the above-knee prosthesis were chosen in the form:

$$p_1(t, C_a) = C_{a1}\beta_1(t) + C_{a2}\beta_1'(t), \quad u_1(t, C_k) = C_{k1}(\alpha_1 - \beta_1) + C_{k2}(\alpha_1' - \beta_1') \qquad (11)$$

where C_{ai}, C_{ki} are the parameters of the elasticity and viscoelasticity of the ankle and knee joints, respectively.

Taking into account (9)-(11) we convert Problem A into a parameter optimization problem: $E=\min Q(C)$, $g(C) \geq 0$, $f(C)=0$. Here $C=\{x_2'(0), x_2'(\tau), x_2''(0), x_2''(\tau), x'(0),$ $y'(0), C_{a1}, C_{a2}, C_{k1}, C_{k2}, a_k, b_k, k=1,..., N\}$ is a vector of the variable parameters; Q, f, g are functions determined by equations (1) and formulae (3)-(8)

Note that the procedure for calculating the objective function Q includes the Cauchy initial value problem for the fourth order system of differential equations.

To solve the above mentioned parameter optimization problem, a computional algorithm based on Rosenbrock's method [12] has been devised.

Some results for the solution of **Problem A** for the gait with natural cadence (gait with $\approx$ 105 step/min [8]) are shown in the Table (all values are in SI units) and in Fig.2-Fig.4. In all figures, centered curves (i.e. curves with asterisks) correspond to prosthetic leg, solid to intact leg.

In the model, a subject height of 1.76 m, mass of 73.2 kg, and the following parameters of the limbs have been considered: $m_{ai} = 7.08$ kg, $a_i = 0.41$ m, $J_{ai} = 0.082$ kg m^2, $m_{bi} + m_{fi} = 5.04$ kg, $b_i = 0.5$ m, $r_{ai} = 0.16$ m, $r_{bi} = 0.203$ m, $J_{bi} = 0.053$ kg m^2, $m = 46.7$ kg, $J = 7.1$ kg m^2, $r = 0.39$ m.

The following input parameters were used for numerical computation:

$T=1.1396$ s, $\tau = 0.36T$, $L=0.755$ m, $x^0 = 0.44$ m, $h=0.85$ m, $x^0_{t2} = 0.0012$ m, $x^t_{h2} = 1.492$ m, $x^0_1 = 0.755$ m, $y^0_1 = 0.0485$ m, $x_{h1} = 0.743$ m, $x_{t1} = 0.949$ m, $q^0 = 0.6$.

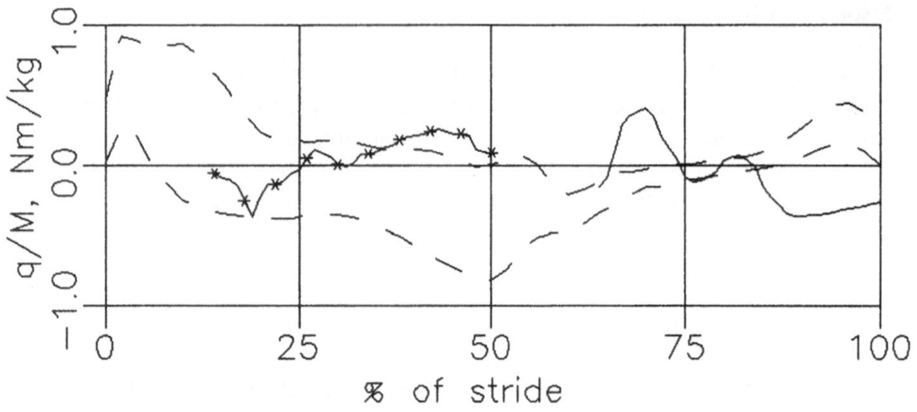

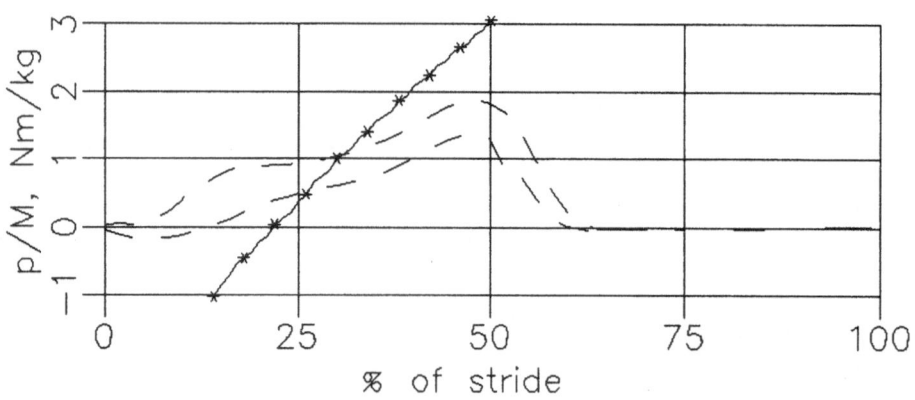

Fig.2   The torques in the hip joints

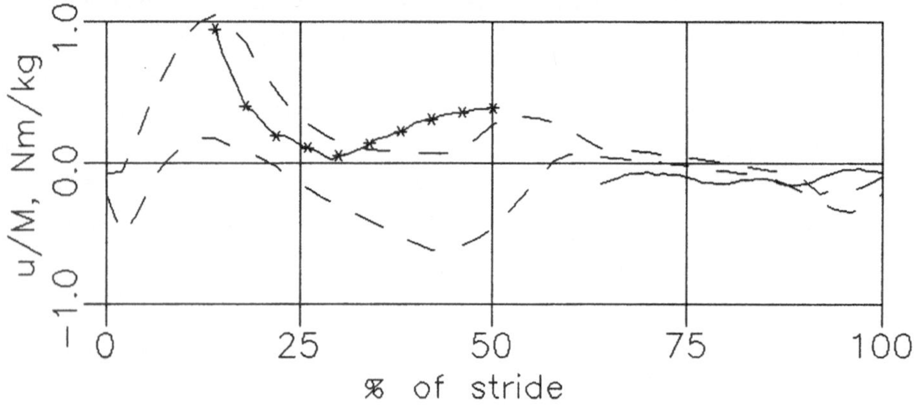

Fig.3   The torques in the knee joints

Fig.4   The torques in the ankle joints

The energetically optimal law of the ankle motion of HLS is specified by formulae (9)-(10) and by the values of the free parameters in the Table. Figures 2-4 show graphs of the control moments in the hip, knee, and ankle joints for the energy optimal control process of the biotechnical system *Man-Prosthesis*. For comparison purposes in Fig.2-4 the minimum and maximum joint moments obtained for a human normal gait with natural cadence are shown by dashed (i.e. broken) curves [8].

Comparison of these graphs with our modelling results indicates that the energy-optimal hip and knee joint moments of the biotechnical system are close to the corresponding characteristics of a human normal gait.

As numerical calculations have shown (Table), it is sufficient to use only the two first harmonics of the Fourier series (9) to determine the energy-optimal control of HLS.

TABLE

N	1	2	3	4
C_{a1}	390	390	390	390
C_{k1}	6	6	6	6
C_{a2}	0	0	0	0
C_{k2}	52	52	52	52
$x'(0)$	1.52	1.52	1.52	1.52
$y'(0)$	0.36	0.36	0.36	0.36
$x'_2(0)$	2.03	2.03	2.03	2.03
$x'_2(\tau)$	0.2	0.2	0.2	0.2
$x''_2(0)$	12.24	10.92	10.71	10.50
$x''_2(\tau)$	-11.77	-11.28	-11.01	-10.29
a_1	0.135	0.134	0.1375	0.1394
b_1	0.283	0.287	0.293	0.312
a_2	-	0.00032	0.0014	0.00163
b_2	-	-0.00011	0.00044	0.000544
a_3	-	-	-0.00037	-0.000602
b_3	-	-	-0.00026	-0.000631
a_4	-	-	-	-0.0000024
b_4	-	-	-	-0.0000028
E	20.73	20.28	19.18	19.16

5. Conclusion

In this paper which is an extension of the research [4, 6], the analysis of a controlled motion of HLS wearing an above-knee prosthesis is based on the solution of the energy-optimal control problem for a plane multibody system. The performance index used is the mechanical work spent to transfer HLS with an above-knee prosthesis from the initial phase to the final one over the given time.

To solve the nonlinear optimal control problem under the given boundary conditions, the restrictions on the phase coordinates and on the controlling stimuli, we proposed a parameter optimization approach based on the special Fourier approximation of the independently variable functions.

A key feature of the considered optimal control problem is a high level of use of the experimental data on human normal locomotion [8].

In the framework of the proposed mathematical model of HLS with an above-knee prosthesis the following conclusions have been drawn.

1. There is strong interaction between dynamics, control and essential prosthesis parameters in the biotechnical system *Man-Prosthesis*. For a given individual and cadence of the gait, there are optimal values of the stiffness and damping parameters of the ankle and knee joints of an above-knee prosthesis which give minimum energy expended per unit of distance travelled (See Table). For HLS wearing a below-knee prosthesis the same result was obtained in [6].

2. The analysis of a number of numerical simulations shows that for the energy-optimal controlled motion of HLS which was obtained, the viscoelasticity of the ankle joint of an above-knee prosthesis is negligibly small compared with the elasticity of this joint.

One of the important possible applications of the results of the present study may be the design of optimal artificial lower limbs.

6. Acknowledgements

A am grateful to N. I. Nishchenko who has joined with me in the work described in this paper. Some aspects of this work were supported by the Concern "Ukrprosthesis", Kyiv, Ukraine.

7. References

1. Michael J.W. and Bowker J.H. (1994) Prosthetics/Orthotics Research for the twenty-First Century: Summary of 1992 Conference Proceedings, *J. of Prosthetics and Orthotics*, **6**, 100-107.
2. Öberg T., Karsznia A., and Öberg K. (1994) Joint angle parameters in gait: Reference data for normal subject, 10-79 years of age, *J. Rehabilitation Research and Development*, **31**, 3, 199-213.
3. Diandelo D.J., Winter D.A. Ghista D.N, and Newcomber W.R. (1989) Perfomance assesment of the Terry-Fox jogging prosthesis for above-knee amputees, *J. Biomechanics*, **22**, 6/7, 543-558.
4. Berbyuk V.E. (1994) Modelling of human locomotion and its application in construction of prostheses. *Proc. of the IMACS Symposium on Mathematical Modelling*, February 2-4, 1994 , Technical University Vienna, Austria, Edited by I. Troch and F. Breitenecker, **2**, 352-355.
5. Berbyuk V.E. (1995) Energy-optimal control of a human leg in swing phase. *Proc. Ninth Biomechanics Seminar*, Göteborg, Sweden, **9**, 32-49.
6. Berbyuk V. (1995) Multibody System modeling and optimization problems of lower limb prostheses. In D. Bestle and W. Schielen (eds.), *IUTAM Symposium on Optimization of Mechanical Systems*, 25-32
7. Capozzo A.,Figure F., Leo T. and Macthett M. (1976) Biomechanical evaluation of above-knee prostheses. *Biomechanics* V-A, 366-372.
8. Winter D. (1991), *The Biomechanics and Motor Control of Human Gait*, University of Waterloo Press, Canada.
9. Beckett R., and Chang K. (1968) An evaluation of kinematics of the gait by minimum energy, *J. Biomechanics*. **1** , 147-159.
10. Beletskii V. V. , Berbyuk V. E. and Samsonov V. A. (1982) Parametric optimization of motions of a bipedal walking robot, *J. Mechanics of Solids*, **17**, 24-35.
11. Nagurka M., and Yen V. (1990) Fourier-based optimal control of nonlinear dynamic systems, *Trans. ASME, J. Dynamic Systems, Meas. and Contr.*, **112**, 3, 19-26.
12. Rosenbrock H.H.(1960) An automatic method for finding the greatest and least value of a function, *The Computer Journal*, **3**, 175-184.

CONTROL OF THE PARAMETRICALLY EXCITED PENDULUM

S.R. BISHOP and D.L. XU
*Centre for Nonlinear Dynamics and its Applications, Civil Engineering
Department, University College London, Gower Street,
London WC1E 6BT, UK*. email:-s.bishop@ucl.ac.uk

1. Introduction

The article by Ott *et al*. (1990), though possibly not the first on the subject, can certainly be credited with stimulating a great deal of study into the control of chaos. This novel research showed how chaos might be seen as a useful response of a dynamical system and its properties utilised to good effect. The harnessing of chaos, through tiny perturbations of parameters, is based on the principle of stabilising one of the unstable orbits which are embedded within a chaotic attractor. Since that time there have been many extensions to the original concept and various algorithms developed for implementation in different situations. Moreover, the control of chaos has been achieved in a variety of physical experiments [for instance see Roy *et al*. (1992), Schwartz & Triandaf (1992)] including mechanical systems [Hübinger *et al*. (1994), Ditto *et al*. (1990) and Starret & Tagg (1995)].

Theoretically a chaotic attractor has, embedded within it, an infinite number of unstable periodic orbits. This may be so but in practice specific unstable orbits are not easy to locate. Blind numerical searches typically only reveal the main branches. Alternative unstable fixed points corresponding to solutions which undergo complicated dynamics of high period, may be of great physical significance and so highly desirable but yet may be difficult to locate by numerical methods alone. These solutions may have associated eigenvalues which are very large, say over 100 or even 1000, so that even if they could be located, stabilization is not trivial with the slightest noise in the systems leading to a failure of the control process. We discuss here two aspects to overcome these problems and provide numerical demonstrations of the control process. In certain cases, the topological theory of dynamical systems can be used to pinpoint the location in parameter and phase space of desired orbits which can be further refined using numerical procedures. A new control method may then be applied which minimises the distance between a chaotic trajectory and the desired orbit by adjusting control parameters; typically requiring only small perturbations if the distance is small, see [Xu & Bishop (1994)]. Following the ideas in Hübinger *et al*. (1993) the method incorporates multiple sections within a recurrent time period of the solution to avoid difficulties when the divergence is rapid as a result of large eigenvalues.

D. H. van Campen (ed.), IUTAM Symposium on Interaction between Dynamics and Control in Advanced Mechanical Systems, 43–50.

2. Unstable Orbits of the Parametric Pendulum

The parametrically driven pendulum is used as a typical example of a nonlinear system exhibiting a large variety of stable periodic and chaotic motions, together with the hanging and inverted equilibrium states. These motions can be oscillatory, rotational or a combination of these. The physical system consists of a mass on a light rod vertically excited at it base or pivot point which, after suitable scaling, can be modelled by an equation of the form

$$\ddot{\theta} + c\dot{\theta} + (1 + p\cos\omega t)\sin\theta = 0 \qquad\qquad (1)$$

The asymptotic response depends crucially upon the initial conditions imparted to the system for a given frequency and amplitude of forcing, used here as parameters. The existence of a large chaotic attractor has been numerically and experimentally verified which persists for a reasonably broad range of the parameters. This chaotic solution is referred to as a *tumbling* motion since it includes rotations in both clockwise and anticlockwise directions, as well as oscillations about the hanging position.

The unstable periodic solutions embedded within the attractor may be classified according to the number of oscillations or rotations within a given number of periods of the periodic driving force. For given parameters, the position in state space of fixed points which characterise these orbits must typically be found before any control strategy applied to effect stabilisation. The major, dominant unstable points may be found simply by numerical searches using Newton's method or similar with no prior knowledge of their location. Alternatively, and in particular for experimental implementations, the method of close returns (Tuffillaro *et al.* 1992) is useful in identifying the nearby presence of an unstable orbit but the "waiting time' may be long if the period of the solution sought is of high order. More theoretically, the invariant manifolds from the global unstable, inverted, position reveals a trellis which divides the phase space. Using an equivalent topological treatment and labelling each orbit according to the portions of the phase space visited enables a close approximation to the location in the phase space of a particular orbit to be estimated. In addition to this, examining the braid formed by specific orbits enables the maximum number of solutions with a particular period to be determined. Furthermore, knowledge is also obtained about the bifurcational precedence that must occur which in turn indicates the existence of further orbits which then may be sought out. This research provides very powerful results but sadly a restriction of the theory to two dimensional maps means that extension beyond driven oscillators is not possible. Full details of the background can be found in Tuffillaro *et al.* (1991), McRobie (1992), and McRobie & Thompson (1994) while specific application of the pendulum can be found in the papers by Bishop & Clifford (1994) and Clifford & Bishop (1993, 1994a,b, 1995a,b) to which the reader is referred.

Using a combination of the above methods for fixed parameters $p=2$, $\omega=2$ over 25 unstable solutions (repellors and saddles) were located and tabled in Bishop *et al.* (to appear) whose eigenvalues may be as large as 10,000. If the system is set to operate at parameters of the driving force which cause the pendulum to undergo the chaotic tumbling motion then, using the knowledge of the location of their associated fixed points, the idea is to utilise a control process to stabilise the unstable solutions. We introduce here a method to implement control which is successful for orbits whose period is of high order with possibly large eigenvalues even in the presence of noise.

3. The Method of Control on Multiple Sections

Consider a flow whose intersection with periodic, Poincaré sections, produces the mapping

$$Z_{k+1} = F(Z_k, u), \quad k = 0, 1, 2... \tag{2}$$

for which a linear approximation may be given by

$$Z_{k+1} = F(Z^*, u^*) + D \cdot \delta Z_k + G \cdot \delta u_k \quad k = 0, 1, 2... \tag{3}$$

where $Z \in \mathfrak{R}^n$ is an n dimensional state variable, $u \in \mathfrak{R}^m$ is an m dimensional adjustable parameter vector, F defines the map, $D = \partial F(Z^*, u^*)/\partial Z$ is an $n \times n$ Jacobian matrix, $G = \partial F(Z^*, u^*)/\partial Z$ is an $n \times m$ matrix, $\delta Z_k = Z_k - Z^*$. We assume that the system behaves chaotically at the nominal value of parameters u^*, and we wish to activate small perturbations on the parameters in order to stabilize the chaotic system onto the chosen unstable periodic orbit $Z^* = F(Z^*, u^*)$. The varying parameter vector δu_k is restricted within the range

$$\|\delta u_k\| = \|u_k - u^*\| < \|\Delta u\| \tag{4}$$

where Δu is the maximal adjustable quantity of the parameters. To find an appropriate δu_k so that Z_{k+1} converges onto Z^*, the distance $\|Z_{k+1} - Z^*\|$ is minimized via the variation of parameters, which leads to the following adjustment

$$\delta u_k = -(G^T G)^{-1} G^T D \cdot (Z_k - Z^*), \quad |G^T G| \neq 0 \tag{5}$$

In the case of control onto an orbit which is divided by K Poincaré sections within a recurrent period of the orbit, at each mapping time, the perturbation should be

$$\delta u_k = -(G_i^T G_i)^{-1} G_i^T D_i \cdot (Z_k - Z^*(i)),$$

$$Z^*(i) = F^{(i-1)}(Z^*(1), u^*), \quad \left| G_i^T G_i \right| \neq 0, \tag{6}$$

where the subscript i denotes the i th fixed point mapping from the first fixed point $Z^*(1)$, $i \leq K$, D_i and G_i are evaluated at the i th fixed point. The linear approximation (3) can be extracted from experimental data by a least-square technique (as described by Ott *et al.* 1990) so that the control process need not rely on the analytical knowledge of the system.

4. Application of Chaotic Control to the Parametric Pendulum

To achieve the goal of controlling the parametric pendulum from the chaotic behaviour we choose the frequency of driving ω to be the accessible parameter for control (though equally the parameter p could have been used). The initial coefficients are set as c=0.1, p=2.0, and ω=2.0. The Jacobian matrix D and the matrix G are evaluated using small perturbations about the periodic orbit (as described in Foale & Thompson, 1991). Numerical experiments are conducted to evaluate the effects of noise considered alongside a variation of the number of sections within the period of the solution.

In figure 1 we show the time history of angular velocity $\dot{\theta}$ of an initially free-running chaotic trajectory stabilised onto a period-4 orbit in the absence of noise. The control algorithm is activated at around t=2210 and after a few periods of the driving force the system is stabilized onto the oscillating orbit whose largest amplitude of the eigenvalue is -562.3.

In the presence of noise this unstable solution can still be controlled, as shown in figure 2, where the noise level ρ is 0.03 using 12 control sections (K=12) with a restriction on the perturbations of $|\delta\omega|$<0.5. Two noise signals are added into the state variables $(\theta, \dot{\theta})$ as a series of impulses whose amplitude and frequency possess the normal property. In the phase space, the orbit is 'fuzzy' due to the effects of noise. Figure 3 demonstrates the required parameter perturbations which is renewed every control interval τ=π/3. The orbit is sampled on the 12 control sections and the mapping points are plotted in figure 4 which shows a longer time scale for the stabilization.

A relationship between the controllable noise levels ρ and the number K of control sections is shown in figure 5. The points marked by '×' joined by lines indicate the maximum controllable noise levels corresponding to the number of sections K. The result is based on the control of the above period-4 orbit. In all simulations, the initial condition of the system state is the same starting from the fixed point (-2.47421, 0.085205) and the perturbation is bounded $|\delta\omega|$<0.5 which will be set to zero if $|\delta\omega|$ exceeds this value. As can be seen, when K=1, the controllable noise level ρ is less than 0.0001, K=2, ρ=0.0005, K=3, ρ=0.004 and so on. For the orbit described, the highest noise level ρ=0.044 can be controlled with K=15 which is about 4 times of the achievable level for K=4. Selecting the proper number of control sections can greatly enhance the ability to cope with noise. Note that the controllable noise level, in general, decreases as the number of sections increases after K=15. One possible reason is that when the number of sections increases, the time for control is shortened. While, to direct a trajectory onto the desired orbit requires larger perturbations if the time interval for control is less. Due to the bound on the perturbation (which will be set to zero as it exceeds the restriction), incorrect control inputs may result in failure of the control at certain level of noise. In other numerical studies, the pattern of this relationship between ρ and K is roughly similar when the perturbation is limited to $|\delta\omega|<1.0$ but the controllable noise level is higher. Using different segments of noise time series (but with the same level) produces some differences but the relationship between ρ and K remains similar.

In figure 6, a tumbling period-5 orbit is controlled which is a repellor with two large eigenvalues 2526.8 and -146.7. The stabilization is carried out using the 15 sections (K=15) with bounded perturbation $\delta\omega<1.0$ to cope with effects of noise in the level of $\rho=0.025$. Using five control sections (K=5) one can only approach the controllable noise level being $\rho=0.003$. The ability of coping with noise is largely enhanced by using the scheme of multiple sections. This fact is ascertained in many other numerical simulations which are not shown here.

5. Remarks and Conclusions

The concept of the control of chaos is a very powerful one since it seems to offer almost limitless possibility for the dynamic outcome of a particular system at a single set of parameters using only tiny adjustments. This idea has two main prerequisites: first that one must be able to locate a desired solution, and second that a control algorithm must be able to stabilize such a solution from a free-running chaotic trajectory. The theory of symbolic dynamics can be employed to predict the existence of an orbit and locate this solution which, after further refinement, can be controlled using a suitable algorithm. Results so far indicate that although for the orbits having large eigenvalues the numerical aspects become harder, particularly in the presence of noise, in principle this goal has been achieved using the idea of multiple control sections. The method used here is efficient to stabilise highly unstable periodic orbits (including repellors which widely exist in the chaotic dynamics of the pendulum) even under the influence of relatively large noise levels.

6. References

Bishop, S. R., Xu, D.L. & Clifford (to appear) Flexible control of the parametrically excited pendulum. *Proc. Roc. Soc. Lond.* A.

Bishop, S. R. & Clifford, M. J. 1994 Non-rotating orbits in the parametrically excited pendulum. *Eur. J. Mech.* **13**, 581-587.

Clifford, M. J. & Bishop, S. R. 1993 Generic features of escape from a potential well under parametric excitation. *Phys. Lett.* A **184**, 57-63.

Clifford, M. J. & Bishop, S. R. 1994a Approximating the escape zone for the parametrically excited pendulum. *J. Sound & Vibration* **172**, 572-576.

48

Clifford, M. J. & Bishop, S. R. 1994b Bifurcational precedences for parametric escape from a symmetric potential well. *Int. J. Bif. & Chaos* 4, 623-630.

Clifford, M. J. & Bishop, S. R. 1995a Locating oscillatory orbits of the parametrically excited pendulum. *J. Aust. Math. Soc. Ser.* B, 37, 1-11.

Clifford, M. J. & Bishop, S. R. 1995b Rotating orbits of the parametrically excited pendulum. *Phys. Lett.* A. 201, 191-196.

Ditto, W. L., Rauseo, S. N. & Spano, M. L. 1990 Experimental control of chaos. *Phys. Rev. Lett.* 65, 3211-3214.

Foale, S. & Thompson, J. M. T. 1991 Geometrical concepts and computational techniques of nonlinear dynamics. *Computer Methods in Applied Mechanics and Engineering,* 89, 381-394.

Hübinger, B., Doerner, R. & Martienssen, W. 1993 Local control of chaotic motion. *Z. Phys.* B 90, 103-106.

Hübinger, B., Doerner, R., Heng, H. & Martienssen, W. 1994 Approaching nonlinear dynamics by studying the motion of a pendulum. III. predictability and control of chaotic motion. *Int. J. Bif. & Chaos* 4(4), 773-784.

McRobie, F.A. 1992 Bifurcational precedences in the braids of periodic orbits of spiral 3-shoes in driven oscillators. *Proc. R.. Soc. Lond.* A 438, 545-569.

McRobie, F.A. & Thompson, J. M. T. 1994a Driven oscillators, knots, braids and Nielsen-Thurston theory. in *Nonlinearity and Chaos in Engineering Dynamics,* (edited by Thompson & Bishop), Wiley, 317-328.

McRobie, F.A. & Thompson, J. M. T. 1994b Knot-types and bifurcation sequences of homoclinic and transient orbits of a single-degree-of-freedom driven oscillator. *Dynamics and Stability of Systems* 9, 223-251.

Ott, E., Grebogi, C. & Yorke, J. A. 1990 Controlling chaos. *Phys. Rev. Lett.* 64, 1196-1199.

Roy, R., Murphy, T. W., Maier, T. D. & Gills, Z. 1992 Dynamical control of a chaotic laser: Experimental stabilization of a globally coupled system. *Phys. Rev. Lett.* 68, 1259-1262.

Schwartz, I. B. & Triandaf, I. 1992 Tracking unstable orbits in experiments. *Phys. Rev.* A 46, 7439-7444.

Starrett, J. & Tagg, R. 1995 Control of a chaotic parametrically driven pendulum. *Phys. Rev. Lett.,* 74, 1974-1977.

Thompson, J.M.T. and Bishop, S.R. 1994 *Nonlinearity and Chaos in Engineering Dynamics.* Wiley: Chichester.

Tufillaro, N. B., Abbott, T. & Reilly, J. 1992 *An Experimental Approach to Nonlinear Dynamics and Chaos,* Addison Wesley.

Tufillaro, N. B., Holzner, R., Flepp, L., Brun, E., Finardi, M. & Badii, R. 1991 Template analysis for a chaotic NMR laser. *Phys. Rev.* A 44, 4786-4788.

Xu, D. & Bishop, S. R. 1994 Steering dynamical trajectories to target a desired state. *Chaos Solitons & Fractals,* 4, 1931-1942.

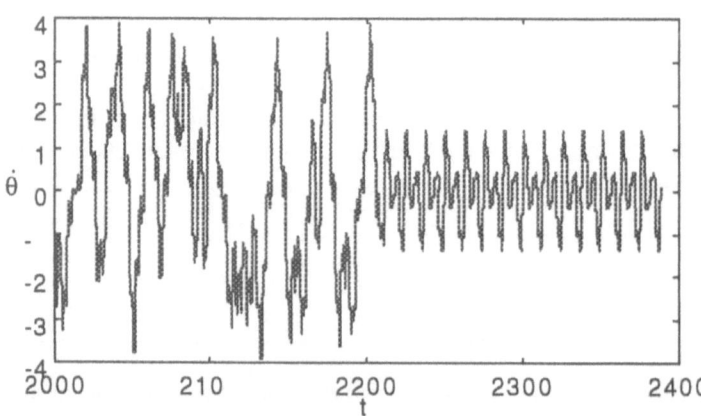

Figure 1. A chaotic trajectory of the parametrically excited pendulum is stabilised onto an oscillating period-4 orbit at t=2210 using 4 control sections per recurrent time period of the orbit in the absence of noise (K=4, ρ=0, |δω|<0.15).

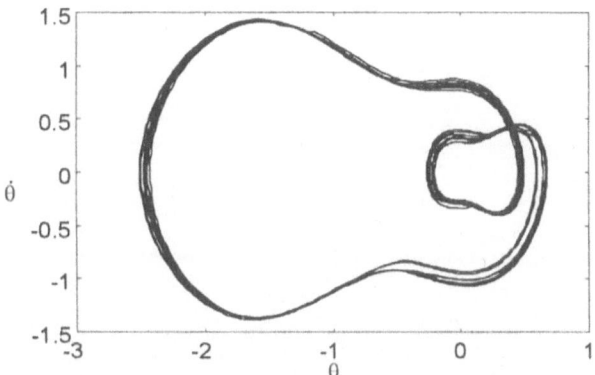

Figure 2. The phase space of the controlled period-4 orbit with the largest magnitude of eigenvalue -562.3. The orbit is 'fuzzy' due to the effects of noise with (Ḱ=12, ρ=0.03, |δω|<0.5).

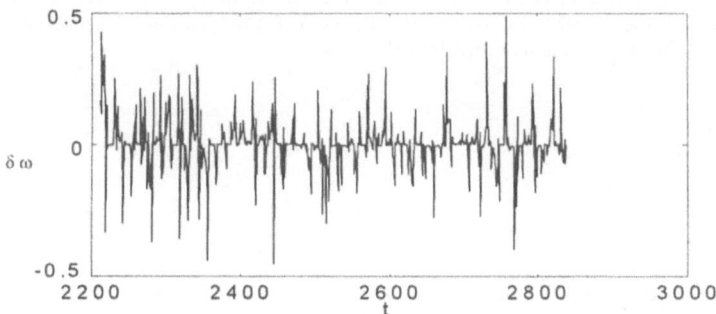

Figure 3. The required parameter perturbation to stabilize the period-4 orbit with K=12, ρ=0.03, |δω|<0.5.

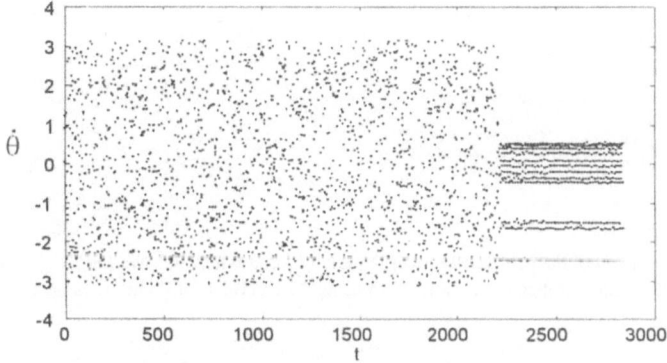

Figure 4. The stabilized period-4 orbit O(4,0) is shown by mapping points on 12 control sections, (K=12, ρ=0.03, |δω|<0.5).

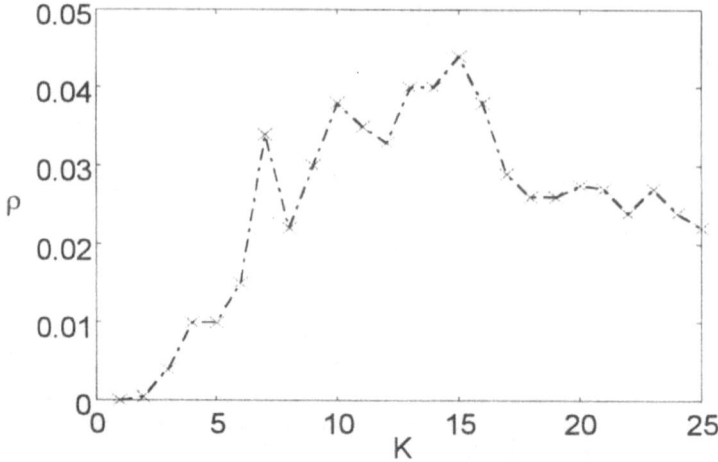

Figure 5. The relationship between ρ (the controllable noise level) and K (the number of sections) with the restriction of the parameter perturbation |δω|<0.5. This result is based on control of the period-4 orbit, where the points (marked by the symbol ×) indicate the maximum controllable noise levels.

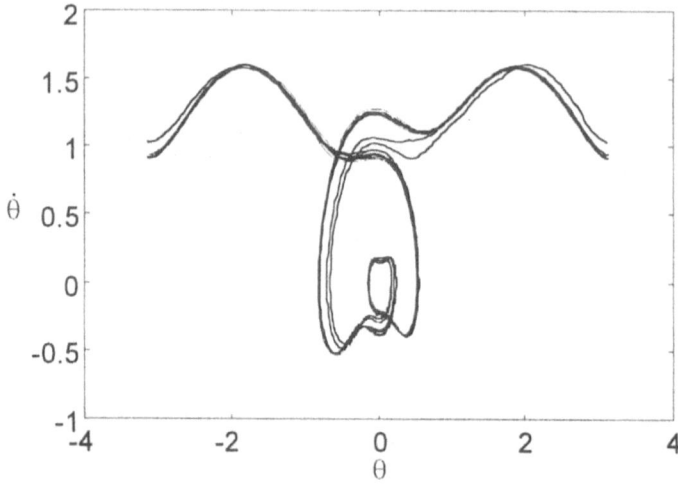

Figure 6. The phase portrait of the controlled tumbling period-5 orbit which has two large eigenvalues: 2526.8 and -146.7, and is a repellor. The stabilization is carried out using 15 control sections at the noise level ρ=0.025. (K=15, ρ=0.025, |δω|<1.0).

ON SIMULATION OF ACTIVE CONTROL OF STRUCTURES UNDER TRAVELLING INERTIAL LOADS

R. BOGACZ[*)], T. SZOLC
Institute of Fundamental Technological Research of the Polish Academy of Sciences, Warsaw, Poland
[*)] also *Cracow University of Technology, Cracow, Poland*

1. Introduction

The vibrations of structures under moving loads have been investigated for almost 130 years. Bridges and viaducts were the first objects of research in this field, as cited in Inglis (1934) and Schallenkamp (1937). The fast development of guideway systems for robotic devices and high speed vehicles needs more and more exact results of calculations for this phenomenon (Popp, 1979). Modern light-weight guideway structures are, on the one hand, more sensitive to vibrations, and on the other hand, are expected to realise precise straight-line motions for the tools or vehicles, they carry. In order to reconcile these two opposites, active control is used and applied in various ways, depending on particular cases.

In the presented paper an active control is applied to the transverse vibrations of a visco-elastic beam with a king-post truss system excited by a moving mass. The beam is used as a guideway to obtain precise horizontal motion of the mass over the whole beam length. For this purpose the 'open loop' active control approach is proposed and realised by electric actuators generating control transverse force and bending moments. The modal description of motion enables an easy implementation of the open-loop approach. Then, the control reduces to active attenuation of selected vibration modes by means of *a priori* assumed control loads eliminating the predominant modal excitations.

2. Assumptions and Formulation of the Problem

We consider a homogeneous visco-elastic beam with a king-post truss system excited to transverse vibrations by a mass m moving along it with constant speed v, as shown in *Figure 1*. The beam is supported at its ends by movable pivot bearings, and has length l, flexural stiffness EI and mass density ρA. The dynamic system is replaced by

51

D. H. van Campen (ed.), IUTAM Symposium on Interaction between Dynamics and Control in Advanced Mechanical Systems, 51–58.

52

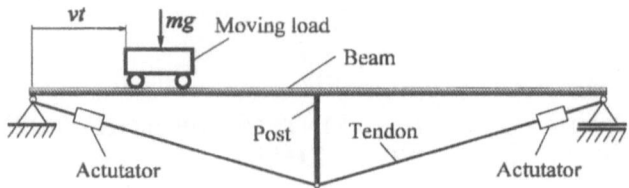

Figure 1 The investigated real system

a discrete-continuous mechanical model. In this model, for so called 'small vibrations', the interaction with the king-post truss system is realised by a post mass m_r attached to the selected cross-section of the continuous beam and supported by a mass-less spring of constant stiffness c representing an elasticity of the tendons, *Figure 2*.

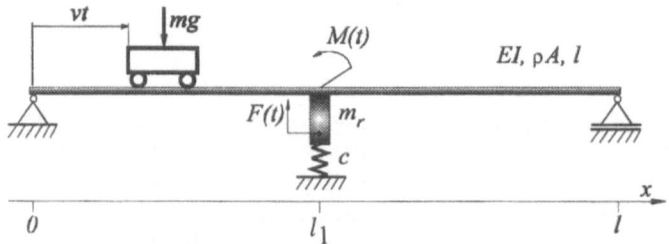

Figure 2 The mechanical model

We propose two ways of realising an active vibration control for the system. In the first, the control is performed using two electric actuators attached to the tendons and suspended from the beam supports, as shown in *Figure 1*. By tensing the tendons, these actuators generate time dependent forces which produce the control concentrated transverse force $F(t)$ and bending moment $M(t)$ acting at the beam cross-section corresponding to the post position, *Figure 2*. In the second case, the active control is realised by the bending moment $M_1(t)$ imposed on the right-hand end of the beam.

We use the Bernoulli-Euler beam theory, and represent the material passive damping by a Voigt model. Thus, the equation of motion of the beam is assumed in the following form:

$$EI\left[\frac{\partial^4 w(x,t)}{\partial x^4} + e\frac{\partial^5 w(x,t)}{\partial x^4 \partial t}\right] + \rho A\frac{\partial^2 w(x,t)}{\partial t^2} = q(x,t), \tag{1}$$

where: $q(x,t) = \delta(x - vt)\, m\left(g - \frac{d^2 w(x,t)}{dt^2}\bigg|_{x=vt}\right),$

δ denotes the Dirac function, g is the gravitational acceleration and e is the Voigt damping constant coefficient.

Equation (1) is solved for the following boundary conditions:

$$w(x,t) = 0 \quad \text{and} \quad \frac{\partial^2 w(x,t)}{\partial x^2} = 0 \quad \text{for } x = 0,$$

$$w^-(x,t) = w^+(x,t), \quad \frac{\partial w^-(x,t)}{\partial x} = \frac{\partial w^+(x,t)}{\partial x}, \tag{2}$$

$$m_r \frac{\partial^2 w^+(x,t)}{\partial t^2} + c\, w(x,t) + EI \frac{\partial^3 w^+(x,t)}{\partial x^3} - EI \frac{\partial^3 w^-(x,t)}{\partial x^3} = F(t),$$

$$M(t) + EI \frac{\partial^2 w^+(x,t)}{\partial x^2} - EI \frac{\partial^2 w^-(x,t)}{\partial x^2} + c_M\, w(x,t) = 0 \quad \text{for } x = l_1,$$

$$w(x,t) = 0 \quad \text{and} \quad M_1(t) - EI \frac{\partial^2 w(x,t)}{\partial x^2} = 0 \quad \text{for } x = l,$$

where: $c = c_s l_r^2 \left(\dfrac{1}{l_r^2 + (l - l_1)^2} + \dfrac{1}{l_r^2 + l_1^2} \right)$, $\quad c_M = c_s l_r^2 \left(\dfrac{l_1}{l_r^2 + l_1^2} - \dfrac{l - l_1}{l_r^2 + (l - l_1)^2} \right)$,

c_s denotes the tendon tensile stiffness, l_r is the length of the post, l_1 is the post position co-ordinate, and superscripts '-' and '+' denote the left and right hand side of the beam, respectively.

Using separation of variables to solve the homogeneous equation (1) with boundary conditions (2) for $F(t)=0$ and $M(t)=M_1(t)=0$, we obtain the eigenfunctions in the following form:

$$X_n(x) = B_{1n} \sin(k_n x) + D_{1n} \sinh(k_n x) \qquad \text{for } 0 \le x \le l_1, \tag{3}$$

$$X_n(x) = B_{1n} \sin(k_n x) + D_{1n} \sinh(k_n x) + B_{2n} \sin(k_n(x - l_1)) + D_{2n} \sinh(k_n(x - l_1))$$

for $l_1 \le x \le l$, where ω_n are the natural frequencies of the system, and

$$k_n = \sqrt[4]{\omega_n^2 \frac{\rho A}{EI}}, \quad n = 1, 2, \dots .$$

For the forced vibration analysis the control force $F(t)$ and moments $M(t)$, $M_1(t)$ in (2) are regarded as concentrated external excitations. The non-homogeneous equation (1) is solved using the Fourier method in the form of an infinite series of standing waves. This leads to the well known ordinary differential equations for the modal co-ordinates

$$\ddot{\zeta}_n(t) + e\omega_n^2 \dot{\zeta}_n(t) + \omega_n^2 \zeta_n(t) = \frac{1}{\rho A} Q_n(t), \quad n = 1, 2, \dots, \tag{4}$$

where:

$$Q_n(t) = \frac{1}{\gamma_n^2} \left[m \left(g - \frac{d^2 w(x,t)}{dt^2} \right) X_n(x) \bigg|_{x = vt} + F(t) X_n(l_1) + M(t) X_n'(l_1) + M_1(t) X_n'(l) \right],$$

$$\gamma_n^2 = \frac{m_r}{\rho A} X_n^2(l_1) + \int_0^l X_n^2(x)dx \quad \text{for } n = 1, 2, \dots.$$

The instantaneous acceleration of the moving mass is expressed by means of the Renaudot formula (Inglis, 1934 and Schallenkamp, 1937) which after a substitution of the Fourier solutions leads to the following series:

$$\frac{d^2 w(x,t)}{dt^2} = \sum_{n=1}^{\infty} \left[X_n(x)\ddot{\zeta}_n(t) + 2v X_n'(x)\dot{\zeta}_n(t) + v^2 X_n''(x)\zeta_n(t) \right] \quad \text{for } x = vt. \quad (5)$$

Inglis' (1934) approach for a finite number N of eigenmodes, i.e. $n=1,2,\dots,N$, and substitution of (5) into (4), leads to a system of N simultaneous ordinary differential equations with variable coefficients

$$\mathbf{M}(vt)\ddot{\mathbf{z}}(t) + \mathbf{C}(vt)\dot{\mathbf{z}}(t) + \mathbf{K}(vt)\mathbf{z}(t) = \mathbf{F}(t), \quad (6)$$

where $\mathbf{z}(t)$ is the modal co-ordinate vector. The components of the external load and control vector $\mathbf{F}(t)$ are

$$F_n(t) = mgX_n(vt) + F(t)X_n(l_1) + M(t)X_n'(l_1) + M_1(t)X_n'(l), \quad n = 1, 2, \dots, N. \quad (7)$$

The components of the $(N \times N)$ matrices $\mathbf{M}(vt)$, $\mathbf{C}(vt)$ and $\mathbf{K}(vt)$ can be found in Frischgesell *et al.* (1994). In order to obtain the dynamic system response in the form of beam deflections, equations (6) are integrated step by step using the average acceleration method.

3. Concepts of Open-Loop Active Control

The general purpose of the active control of the mechanical system is to minimise the beam deflection under the mass. We seek to motion $w(x,t) \to 0$ for $x=vt$, so that the mass moves horizontally along the beam.

The active control of the beam vibration is implemented by the bending moment $M_1(t)$ acting at the beam right-hand end, or by two actuators tensing the tendons of the king-post truss system. The actuators are able to generate the transverse control force $F(t)$ and moment $M(t)$ acting at the beam post. Using these means, we can control, at most, only the predominant first two vibration modes.

In general, the open-loop approach is usually regarded as less advanced than the closed-loop one (Meirovitch, 1989). Nevertheless, for the mechanical system with known well identified parameters and with deterministic external excitation due to the moving mass, where the vibration control process is characterised by a very short time of duration, the open-loop control can be more easily implemented in practice. On the contrary, the closed-loop approach requires a very fast processor to realise the mechanical feedback with possibly small time delays. The modal description of the continuous beam model enables an easy implementation of the open-loop approach; it reduces to active attenuation of selected vibration modes by means of *a priori* assumed control excitations. For the beam the following three kinds of active control are considered:

1. for the symmetrical king-post truss system, i.e. where the post is situated at half of the beam length, i.e. $l_1=0.5l$, the active control is realised by means of a concentrated moment $M_1(t)$ acting in the right beam end $x=l$ ($F(t)=0$ and $M(t)=0$);

2. for the non-symmetrical king-post-truss system, i.e. where the post is situated at three quarters of the beam length, i.e. $l_1=0.75l$, the active control is realised by means of a concentrated force $F(t)$ imposed on the post ($M(t)=0$ and $M_1(t)=0$);

3. for the symmetrical system, as in 1, where the active control is realised by means of a concentrated force $F(t)$ and moment $M(t)$ imposed on the post ($M_1(t)=0$).

The post position assumed in Point 2 corresponds approximately to the cross-section co-ordinate characterized by the maximum deflection of a hinged-hinged single span homogeneous uniform beam due to a force or mass moving with the critical speed in the Krylov sense (Kaliski, 1966). For the symmetrical system assumed in Points 1,3 the odd symmetrical eigenmode functions satisfy

$$X_i'(l_1) = 0, \qquad i = 1,3,5,\dots,$$

and the even anti-symmetrical ones satisfy

$$X_j(l_1) = 0, \qquad j = 2,4,6,\dots.$$

Thus, in order to control the predominant first and second vibration modes by an elimination of the respective modal excitations in (7), making $F_1(t)$, $F_2(t)=0$, it is easy to determine the *a priori* assumed control functions for the force $F(t)$ and the moment $M(t)$ in the following form:

$$F(t) = -\frac{mgX_1(vt)}{X_1(l_1)} \quad \text{and} \quad M(t) = -\frac{mgX_2(vt)}{X_2'(l_1)}. \tag{8}$$

These control functions vary with time according to the first and the second mode shape functions, respectively.

4. Numerical Results

The numerical calculations were performed for a beam of length $l=2.5$ [m], stiffness $EI=1550.2$ [Nm^2] and mass density $\rho A=3.0$ [kg/m]. The following non-dimensional parameters

$$\mu = \frac{m}{\rho Al} \quad \text{and} \quad \alpha = \frac{\pi v}{\omega_1 l}$$

are introduced: μ denotes the ratio of moving mass to beam mass, and α is the so called mass velocity parameter with respect to the first system eigenfrequency ω_1. The value 0.5 of α corresponds to the critical speed for the beam in the Krylov sense (Kaliski, 1966). The numerical calculations were performed for the following king-post truss parameters: m_r=1.77 [kg], l_r=0.40 [m], c_s=31342.2 [N/m] with an experimentally determined passive damping coefficient e=0.00019 [s].

The kind of active control assumed in Point 1 is relatively easy to apply in practice and it yields a minimisation of beam deflections for all the considered values of α, i.e. α=0.25, 0.5 and 1.0, where the deflections reach ~30% of the maximal values for the corresponding passive systems, *Figure 3*. In this case only the first vibration mode is attenuated.

The active control realised in Point 2 yields slightly better results than those obtained in Point 1. Unfortunately, for the non-symmetrical system, the eigenfunctions are neither symmetrical nor anti-symmetrical. Thus, the *a priori* assumed control force and moment are not able to eliminate the predominant first two vibration modes independently. A simultaneous action of $F(t)$ and $M(t)$ according to (8) leads to an essential increase of positive and negative beam deflections in comparison with the respective displacements of the passive system. Thus, it is more convenient to realise the active control by means of the vertical force $F(t)$ attenuating only the first vibration mode, as in Point 1. Then, one obtains a significant decrease in the beam deflections, as shown in *Figure 4*.

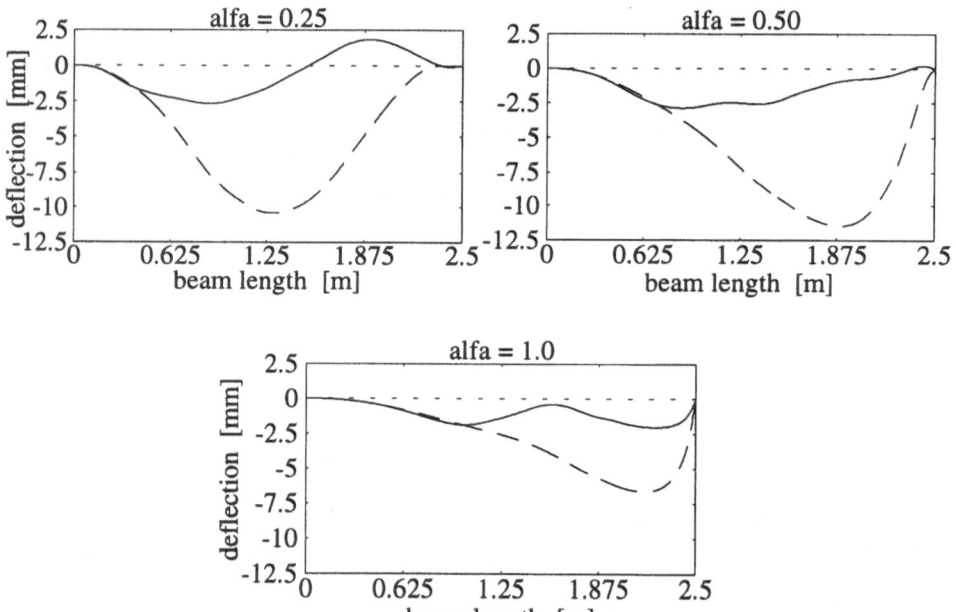

Figure 3 Deflection under the moving mass versus its current position on the beam:
- for the control moment at the beam right-hand end,
passive system (– – –), active system (———).

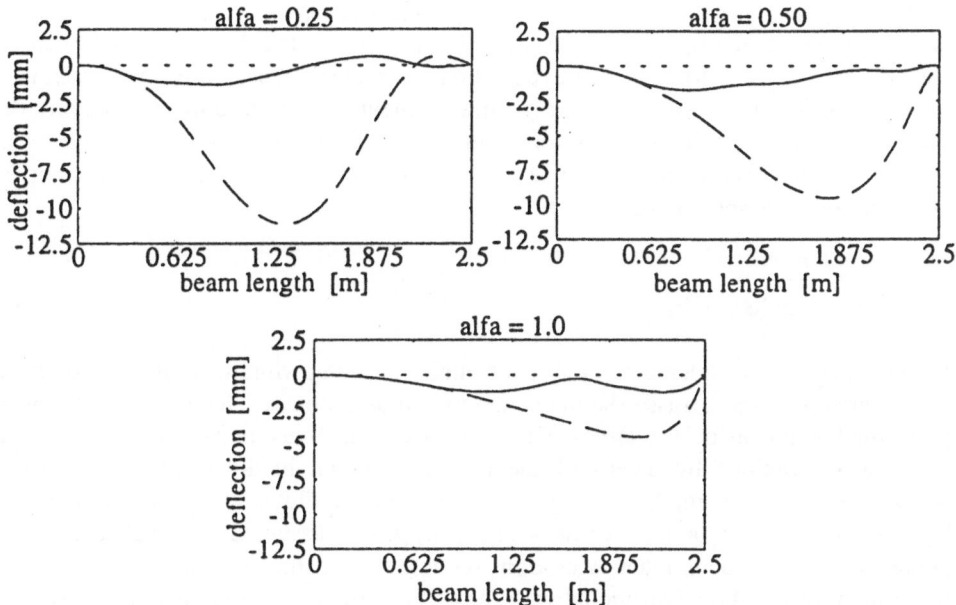

Figure 4 Deflection under the moving mass versus its current position on the beam:
- for the control force at the beam 3/4 length, passive syst. (– – –), active syst. (——).

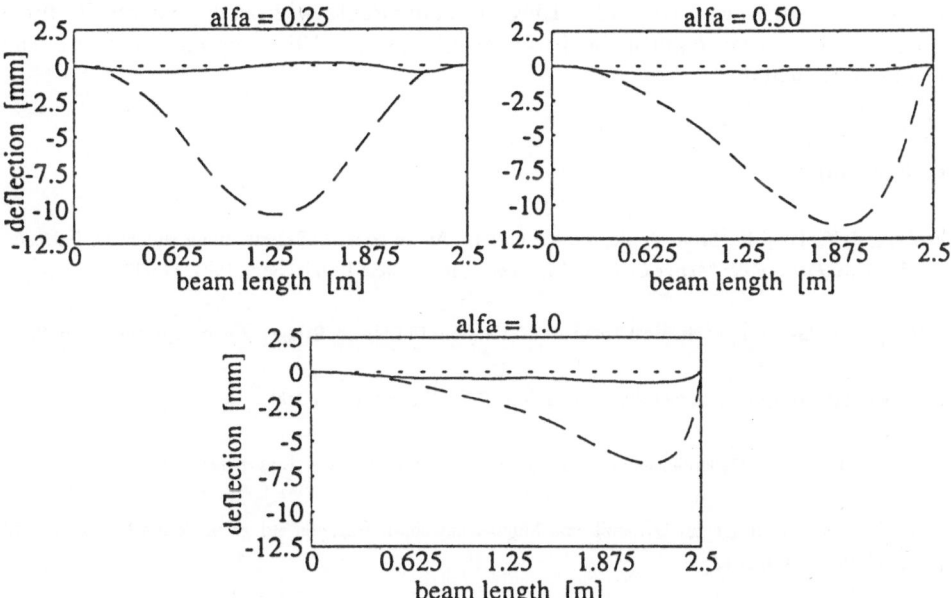

Figure 5 Deflection under the moving mass versus its current position on the beam:
- for the control force and moment at the beam 1/2 length,
passive system (– – –), active system (——).

58

The best results for the open-loop active control are achieved for the symmetrical system assumed in Point 3. In this case for the symmetrical and anti-symmetrical eigenmode functions for odd and even vibration modes, respectively, it ispossible to assume *a priori* the control force $F(t)$ and moment $M(t)$ according to relations (8). Then, one can eliminate an excitation of the first vibration mode independently of an excitation of the second mode. This leads to significant decreases in the deflections for all values of α, as shown in *Figure 5*.

5. Final Remarks

In the paper, we considered actively controlled transverse vibrations of a visco-elastic beam with a king-post truss system excited by a moving mass. The investigations were performed by means of a discrete-continuous mechanical model, in which the dynamic response is obtained by means of the modal approach in the form of a series of standing waves. The modal description of motion facilitates an introduction of the 'open-loop' active vibration control using *a priori* assumed external excitations attenuating the predominant vibration modes. From the results of calculations performed for the three considered variants of the structure and control it follows that the symmetrical structure of the beam with a king-post truss leads to symmetrical odd and anti-symmetrical even eigenmode functions, which enables an independent and simultaneous control of the two first vibration modes. For this case the best results of the active control were achieved. They are comparable with analogous results for the 'closed-loop' active control in Frischgesell *et al.* (1994) applied for the same mechanical system.

6. References

Frischgesell T., Popp K., Szolc T., Bogacz R. (1994) Active control of elastic beam structures, *I. Mech. E. Editions. Proc. of the "Active Control Conference"*, Bath, Great Britain, Sept. 1994, 115-122.

Inglis C. E. (1934) A Mathematical Treatise on Vibrations in Railway Bridges, *Cambridge University Press*.

Kaliski S. (1966) Vibrations and Waves in Solids, *PWN*, Warsaw (*in Polish*).

Meirovitch L. (1989) Dynamics and Control of Structures, *John Wiley & Sons*, New York.

Popp K. (1979) Beiträge zur Dynamik von Magnetschwebe-fahrzeugen auf geständerten Fahrwegen, *VDI-Berichte*, **35**, Reihe 12.

Schallenkamp A. (1937) Schwingungen von Trägern bei bewegten Lasten, *Ingenieur-Archiv*, Band VIII, **3**, 182-198.

ACCURATE MODELLING OF A CONTROLLED PNEUMATIC ACTUATOR WITH EXPERIMENTAL VALIDATION

R. CARACCIOLO*, E. CERESOLE*, A. GASPARETTO**,
M.GIOVAGNONI**
*DIMEG - Dept. of Innovation in Mechanics and Management -
University of Padova - Italy
**DIEGM - Dept. of Electrical, Management and Mechanical Eng. -
University of Udine - Italy

1. Abstract

This work presents an accurate model of a controlled pneumatic actuator. The work is concurrent with the development of a new control algorithm for this kind of system, based on an accurate thermo-mechanical model and oriented to perform a complete trajectory control. The actuator is composed of a proportional servovalve and a cylinder with magnetic tracking. The cylinder is supported by a pair of linear bearings. The model proposed for the valve is based on a static experimental characterization. The model of the cylinder is obtained by assuming that the thermodynamic processes in both chambers are isothermal, and that the air is a perfect gas. Numerical simulations are compared with experimental recordings for step and sine input in open chain conditions.

2. Introduction

Pneumatic actuation systems commonly use ON-OFF control valves because they are simpler than proportional flow control valves [1][4]. The behaviour of ON-OFF valves is typically non-linear; thus some non-linear terms are added in the equations of the system [2][3]. Recently, more attention has been paid to proportional valves, because nowadays they are simpler and more economical than some years ago. The problem of their sensitivity to leakages has been completely eliminated: in fact the valves are now designed to work with constant internal air leakages. Up to now these valves have been used as proportional voltage/flow devices in classical PI or PID controls, thus enabling the design of accurate positioning systems [1][5].

This work starts from several experimental tests that showed a relevant dependence of the output flow rate from the pressure in the chambers connected to the valve. This fact suggested that a more effective control could be based on a more accurate model of the valve that could exploit the dependence of the mass flow rate on the pressures.

The main contribution of this work is to propose an accurate state space model of both valve and cylinder in which all state variables are directly measured by transducers.

59

This kind of model needs no estimator for state reconstruction, so it can be easily linearized. Therefore, it is possible to explore an optimal control approach using Riccati's equation repetitively for computing the optimal gains at each linearization. Both position and velocity could be controlled with this approach.

The paper is composed of three sections.

The first section deals with the model and the experimental test of a proportional servo-valve. This type of valve differs from the classical ON-OFF valve because the mass flow rate is defined by a reference input signal and a given characteristic curve. A mathematical model and an experimental procedure oriented to validate the behaviour of the valve are presented in this work. This procedure determines the static characteristic of the valve as a function of supply pressure and of voltage reference input. Moreover, the experimental results show that the mass flow rate does not depend on the pressure in the chamber behind the valve. The dependence on the voltage input is investigated for different values of the supply pressure, and characteristic curves are obtained by interpolation.

The second section of the paper describes the mechanical model of the actuator, which takes into account non-linearities due to the compressibility of the air and to friction forces [6]. The model is valid under the following assumptions: the equation of perfect gases can be used to express the thermodynamic process inside each chamber; the thermodynamic process is isothermal.

The third section describes the whole experimental prototype, composed of a pneumatic cylinder, a proportional valve, a linear position transducer and two pressure transducers. All transducers and the valve are connected to a PC through an A/D converter. This prototype has been used to test both the developed models (i.e. the valve and the actuator models) and to demonstrate their accuracy. The standard step, pulse and sine wave functions have been input to the system and the experimental results have been compared with the numerical simulations.

3. Valve Model

The valve used to drive the pneumatic cylinder is an input voltage-output flow device (FESTO MPYE_5-1/8). In order to control the outflow section, this valve has a position feedback and operates with a dither signal (i.e. a signal having high frequency and low amplitude) superimposed on the input voltage in order to avoid stick slip phenomena, and to ensure high control frequencies.

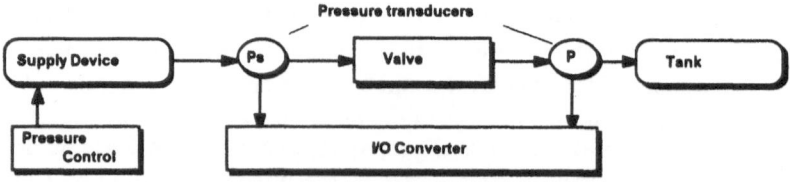

Figure 1. The apparatus for valve characterization

Nevertheless, experimental tests showed that the mass flow rate depends on the pressures measured on both sides of the valve. The proposed model for the valve is an analytical formulation obtained from a static experimental characterization. This formulation expresses the mass flow rate dependence on both the pressure and the input voltage, and represents an evolution and a generalization of the formulation proposed in [7] and [8].

A set of experimental recordings, referring to different motions obtained by imposing different constant voltages as input, is carried out using the experimental apparatus shown in figure 1. Each recording refers to the charge of a tank of known volume V. The pressure P changes from Patm to Ps in each recording. Plots of P/Ps versus time are shown in figure 2, after a triangular window filtering aimed at reducing the noise.

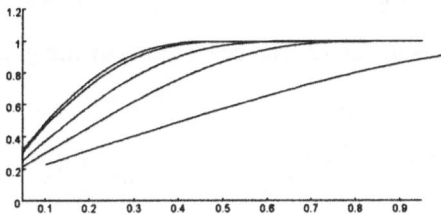

Figure 2. The experimental recordings referring to the tank charge
at different input voltages (pressure ratio versus time [s])

If the air temperature T is constant during the transformation, the ideal gas law implies:

$$\xi(t) = \dot{m}(t) = \frac{V}{RT} \dot{P}(t) \qquad (1)$$

The numeric derivator uses triangular windowing in order to minimize the influence of the noise. Plots of the mass flow rate [Kg/s] versus time for different values of the input voltage are shown in figure 3. The more traditional mass flow rate versus pressure ratio plots can be obtained by combining the plots of figures 2 and 3 (see figure 4).

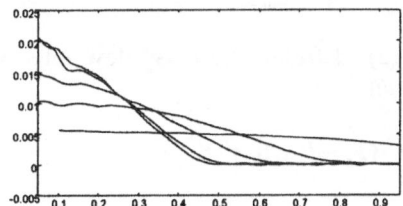 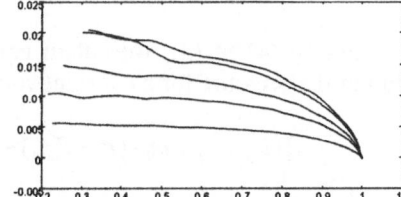

Figure 3. Mass flow rate [kg/s] vs. time [s]   *Figure 4. Mass flow rate [kg/s] vs. pressure ratio*

Now, we are seeking an analytical formulation providing a good approximation of the plots of figure 4. Heuristic considerations and the similarity of these curves with the De Saint Venant - Wantzel equation for the outflows lead to a differential equation whose solution is the combination of an exponential term and a linear term (fig. 5):

$$\dot{\xi}(P) = \alpha \cdot \xi(P) + m_1 - \alpha \cdot \left[q + m_1 \cdot \left(P - P_{atm} \right) \right] \qquad (2)$$

62

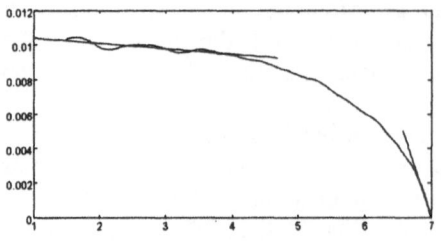
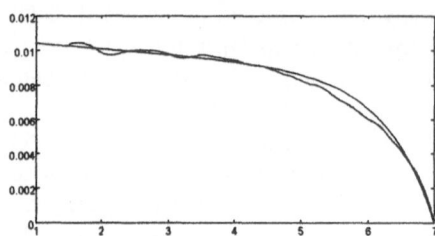

Figure 5. Boundary conditions for interpolation
(mass flow rate [kg/s] vs. pressure [bar])

Figure 6. An interpolating curve
(mass flow rate [kg/s] vs. pressure [bar])

The general integral of equation (2) can be obtained by integration:

$$\xi(P) = e^{\alpha \cdot P} \cdot \int \left[m_1 - \alpha \cdot (q + m_1 \cdot (P - P_{atm})) \right] \cdot e^{-\alpha \cdot P} dP =$$

$$= e^{\alpha \cdot P} \cdot \left\{ -\frac{m_1 - \alpha \cdot (q - m_1 \cdot P_{atm})}{\alpha} \cdot e^{-\alpha \cdot P} - \alpha \cdot m_1 \cdot \int P \, e^{-\alpha \cdot P} dP \right\} =$$

$$= -\frac{m_1}{\alpha} + q - m_1 \cdot P_{atm} - \alpha \cdot m_1 \cdot e^{\alpha \cdot P} \cdot \left[-\frac{P}{\alpha} \cdot e^{-\alpha \cdot P} - \frac{1}{\alpha^2} \cdot e^{-\alpha \cdot P} \right] +$$

$$+ C_0 \cdot e^{\alpha \cdot P} = q + m_1 \cdot (P - P_{atm}) + C_0 \cdot e^{\alpha \cdot P}$$

(3)

m_1 and α are obtained imposing the boundary conditions for $\dot{\xi}$:

In particular $m_1 = \dfrac{\partial \xi(P)}{\partial P}\bigg|_{P \to P_{atm}}$ and $\alpha = \dfrac{m_1 - m_2}{q + m_1 \cdot (P_s - P_{atm})}$ where $m_2 = \dfrac{\partial \xi(P)}{\partial P}\bigg|_{P \to Ps}$.

C_0 and q are obtained imposing the boundary conditions for ξ. Ps is the supply pressure and Patm is the atmospheric pressure:

$$\xi(P_s) = q + m_1 \cdot (P_s - P_{atm}) + C_0 \cdot e^{+\alpha \cdot Ps} = 0$$ (4)

$$\xi(P_{atm}) = q + C_0 \cdot e^{-\alpha \cdot P_{atm}} \cong q = \xi_{max.}$$

By substituting (4) in (3), we obtain equation (5), defining the mass flow rate as a function of the pressure for a constant voltage input:

$$\xi(P) = q + m_1 \cdot (P - P_{atm}) - \left[q + m_1 \cdot (P_s - P_{atm}) \right] \cdot e^{\alpha \cdot (P - P_s)}$$ (5)

Figure 6 shows the correspondence between the numerical model and the experimental recordings for a constant input voltage.
Now the parameters (α, m_1, q) are evaluated for different values of the input voltage from the experimental recordings and then are interpolated using polynomial laws (see figures 7, 8 and 9).

$$q(u) = x_{max}(u) = 1.68 \cdot 10^{-4} \cdot u^4 - 5.34 \cdot 10^{-3} \cdot u^3 + 6.11 \cdot 10^{-2} \cdot u^2 - 2.93 \cdot 10^{-1} \cdot u + 5.02 \cdot 10^{-1}$$

$$m_1(q) = -4.58 \cdot q^2 + 7.76 \cdot 10^{-3} \cdot q$$ (6)

$$a(q) = 5.16 \cdot 10^1 \cdot q + 8.16 \cdot 10^{-1}$$

The final characteristic equation is expressed in equation (7) and the numerical results are compared with the experimental ones in figure 10.

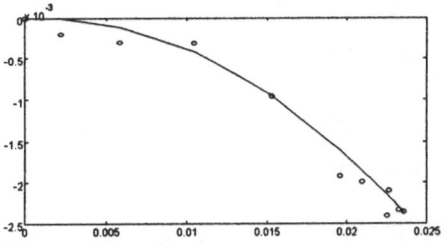

Figure 7. m_1 vs. max mass flow rate [kg/s]

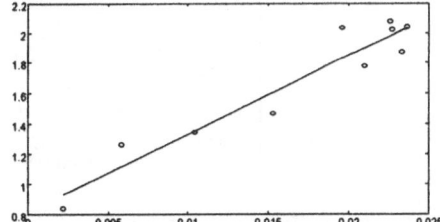

Figure 8. α vs. max mass flow rate [kg/s]

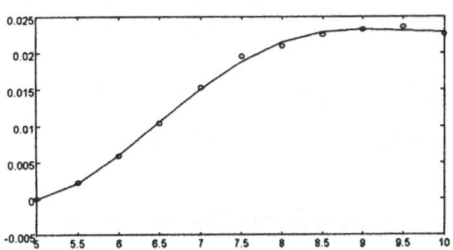

Figure 9. Max mass flow rate [kg/s] vs. voltage input [V]

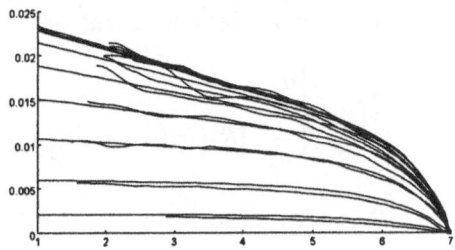

Figure 10. Mass flow rate [kg/s] vs. press. ratio

$$x_A(u,P_A,P_s) = sign(u-5) \cdot \left\{ q + m_1 \cdot (P_A - P_s - dP) - [q + m_1 \cdot dP] \cdot e^{a(P_A - P_s)} \right\} \qquad (7)$$

4. Cylinder Model

Figure 11 gives a schematic representation of the modelled cylinder.

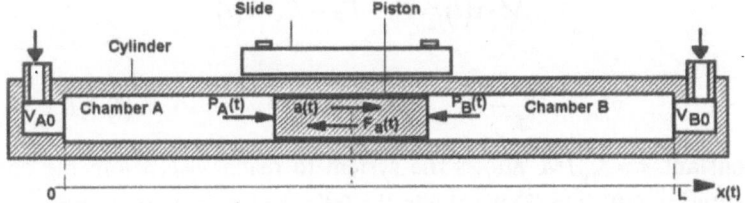

Figure 11. The modelled cylinder

The proposed model has been developed for the FESTO DGO550-40 cylinder, which has magnetic tracking and no leakages. However, it is valid for all cylinders having similar features. We assume the ideal gas law for the air and no leakages in the cylinder and isothermal thermodynamic process in both chambers. If Fa is the resultant of all friction forces, $\xi_A(t)$ and $\xi_B(t)$ the mass flow rates in the chambers, whose values are given by equation (7), the equations of the cylinder are:

$$\begin{cases} \dot{x}(t) = v(t) \\ \dot{v}(t) = \dfrac{1}{M} \cdot S \cdot [P_A(t) - P_B(t)] - \dfrac{1}{M} \cdot F_a(t) \\ \dot{P}_A(t) = \dfrac{R \cdot T}{S} \cdot \dfrac{x_A(t)}{V_{A0}/S + x(t)} - \dfrac{P_A(t) \cdot v(t)}{V_{A0}/S + x(t)} \\ \dot{P}_B(t) = \dfrac{R \cdot T}{S} \cdot \dfrac{x_B(t)}{V_{B0}/S - x(t)} + \dfrac{P_B(t) \cdot v(t)}{V_{B0}/S - x(t)} \end{cases} \qquad (8)$$

where R is the gas constant, S is the section, M is the mass of the piston, V_{A0} and V_{B0} are additional volumes accounting for tubes and connections. If we distinguish between static friction forces (named F_{stat}) and dynamic friction forces (named F_{dyn}) and express the dependence of the friction forces from the velocity sign, we find that some non-linearities are introduced and the second equation in (8) is modified as follows:

$$\dot{v}(t) = \begin{cases} 0 & \text{if} \quad |P_A(t) - P_B(t)| \le F_{stat}/S \text{ and } v(t)=0 \\ \dfrac{1}{M} \cdot \{ S \cdot [P_A(t) - P_B(t)] - \text{sign}(v) \cdot F_{dyn}(t) \} & \text{otherwise} \end{cases} \qquad (9)$$

$F_{dyn}(t)$ is assumed to be the sum of a constant and a velocity linear dependent term.

$$F_{dyn}(v) = F_0 + K_v \cdot |v| \quad \text{with} \quad F_0 = F_{dyn}(v)_{v=0} \le F_{stat} \qquad (10)$$

This assumption is supported by a set of experimental verifications. The values of the constants F_0 e K_v are determined by evaluating the speed limit of the piston after imposing a constant mass flow rate in one chamber.
In other words:

$$M \cdot a(t) = F_p - F_{dyn}(t) \quad \text{with} \quad F_{dyn}(t) = F_0 + K_v \cdot v(t) \qquad (11)$$

leads to:

$$M \cdot \dot{v}(t) = F_0 - F_P - K_v \cdot v(t) \qquad (12)$$

and:

$$v_{lim} = \dfrac{F_P - F_0}{K_v} \qquad \text{if} \qquad \dot{v}(t) = 0 \qquad (13)$$

The time constant $\tau = K_v/M$ allows the system to reach v_{lim} before the end of the course. Substituting (10) into (8) we obtain the following model, holding for $v(t) \neq 0$:

$$\begin{cases} \dot{x}(t) = v(t) \\ \dot{v}(t) = \dfrac{1}{M} \cdot \{ S \cdot [P_A(t) - P_B(t)] - F_0 \cdot \text{sign}(v) - K_v \cdot v(t) \} \\ \dot{P}_A(t) = \dfrac{R \cdot T}{S} \cdot \dfrac{x_A(u(t), P_A(t))}{V_{A0}/S + x(t)} - \dfrac{P_A(t) \cdot v(t)}{V_{A0}/S + x(t)} \\ \dot{P}_B(t) = \dfrac{R \cdot T}{S} \cdot \dfrac{x_B(u(t), P_B(t))}{V_{B0}/S - x(t)} + \dfrac{P_B(t) \cdot v(t)}{V_{B0}/S - x(t)} \end{cases} \qquad (14)$$

5. Experimental Validation

Figure 12 shows a schematic representation of the experimental apparatus used to validate the proposed models. It is composed of a proportional valve FESTO MPYE-5-1/8, a cylinder FESTO DGO 550-40, two linear bearings ROLLON TP-NRS, a linear position transducer NOVOTECHNIK Linopot TLH750, two pressure transducers RS 256-736, one for each chamber. A PC equipped with A/D-D/A converter board National Instrument AT-MIO-16H9 is used to drive the valve and to read and convert the signals coming from the transducers.

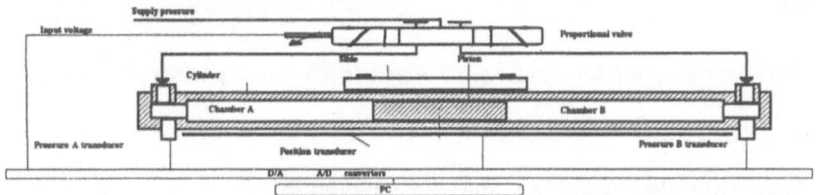

Figure 12. A schematic representation of the experimental apparatus

The system was excited by imposing the classical step and sine wave for the input voltage of the valve in open loop conditions i.e. without any feedback. Experimental results (solid lines) have been compared with numerical ones (dotted lines).

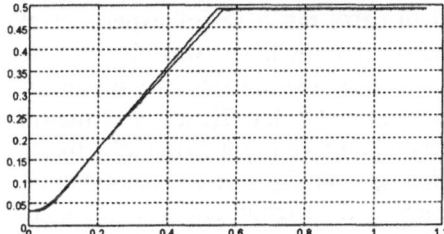

Figure 13. Slide position [m] vs. time [s], step input

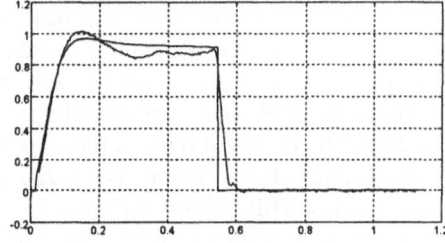

Figure 14. Slide veloc. [m/s] vs. time [s], step input

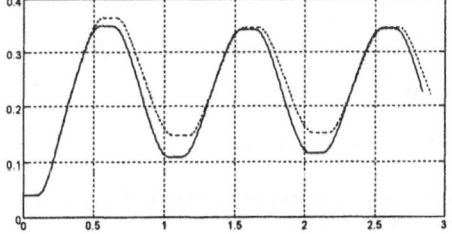

Figure 15. Slide position [m] vs. time, sine input

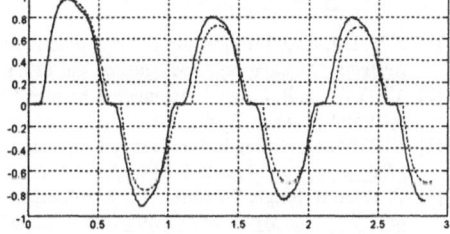

Figure 16: Slide velocity [m/s] vs. time, sine input

Plots 13 and 14 refer to the step input (2 Volts) whereas plots 15 to 18 refer to the sine wave input (2 Volts, 1 Hz). The plots show a generally good correspondence, with some discrepancies that can be attributed to the simplified model adopted for the friction forces. In particular figure 14 suggests that friction forces depend on the slide position; in fact at high speeds the slide motion plots show some irregularities. The

66

discrepancies that can be seen in figure 14 when the end of the course is reached (0.55 s) are due to the filter in the numeric derivator of the position signal.

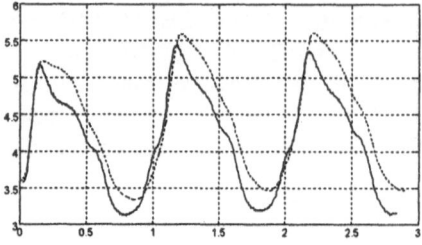

Figure 17. Chamber A pressure [bar], sine input

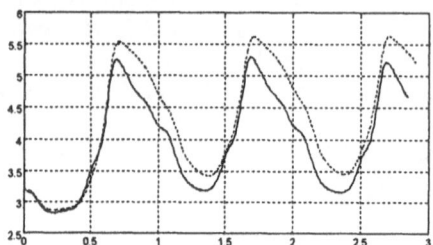

Figure 18. Chamber B pressure [bar], sine input

6. Conclusion

The paper deals with the modelling and the simulation of a pneumatic actuator controlled by means of a proportional valve. Numerical results have been compared to experimental results in order to validate the proposed models. Separate models have been proposed for the valve and the cylinder under the assumptions of isothermal thermodynamic processes and perfect gas for the air. The valve model is based on an experimental characterization and expresses the dependence of the mass flow rate on the pressures in the chambers and the supply pressure as well as on the control voltage. An experimental procedure has been developed to carry out a set of experimental recordings on which the model is based. An analytical formulation was obtained by interpolating the experimental results; the obtained polynomial functions are useful for a successive fast linearization. The cylinder model accounts for friction and inertia loading. The model of friction forces is the combination of a constant term and a linear viscous term; the coefficients of the combination are obtained from experimental verifications. This simplification and the dependence of the friction forces on the slide position can be the cause of the small discrepancy between numerical and experimental results occurring in open loop conditions.

7. References

[1] J.-Y. Lai, C.H. Menq, R.Singh, "Accurate Position Control of a Pneumatic Actuator" ASME J. of Dynamic Systems, Measurement and Control, vol 112, pp. 734-739, Dec. 1990.
[2] Y.-T. Wang, R. Singh, " Pneumatic Chamber Nonlinearities", ASME Journal of Applied Mechanics, vol 53, pp. 956-958, Dec. 1986.
[3] Y.-T. Wang, R. Singh, "Frequency Response of a Nonlinear Pneumatic System", ASME Journal of Applied Mechanics, vol 54 pp. 209-214, Mar. 1987.
[4] C. Kunt, R. Singh, "A Linear Varying Model for On-Off Valve Controlled Pneumatic Actuators", ASME J. of D. Systems, Meas. and Control, vol 112, pp 740-747, Dec. 1990.
[5] C. Kunt, R Singh, "Application of Floquet Theory to On-Off Valve Controlled Pneumatic Actuators", ASME J. of D. Systems, Meas. and Control, vol. 114, pp 299-305, Jun. 1992.
[6] Lee E. Schroeder, R. Singh, "Experimental Study of Friction in a Pneumatic Actuator at Constant Velocity", ASME J. of D. Systems, Meas. and Control, vol. 115, pp 575-577, Sept. 1993.
[7] R.Caracciolo, E. Ceresole, M. Giovagnoni, A. Rossi, Pneumatic actuators driven by proportional valves" MEMT Report in Delft University Bulletin, Delft, The Netherlands, July 1994
[8] R. Caracciolo, E. Ceresole, A. Gasparetto, M. Giovagnoni "Linearizzazione del modello di un attuatore pneumatico controllato in portata", Proc. of XII AIMETA Congress, Napoli, Italy, October 3-6, 1995

DESIGN OF CONTROL UNDER MIXED CONSTRAINTS

F.L. CHERNOUSKO

Institute for Problems in Mechanics of the Russian Academy of Sciences,
pr. Vernadskogo, 101, Moscow 117526, Russia

Abstract. Mixed constraints imposed at each time instant on both control and state variables, as well as integral constraints depending on these variables, are often to be taken into account in designing controls for dynamical systems in mechanics. It is well known that state and mixed constraints present considerable difficulties in the theory of optimal control. In this paper, Kalman's approach originally proposed for linear systems with unbounded controls is extended to the case of mixed, state, and integral constraints. As an example, a dynamical system with two degrees of freedom controlled by an electric motor is considered.

1. Introduction

Mixed constraints imposed on both state and control variables of a dynamical system, as well as integral constraints imposed on these variables, are often present in mechanical applications. For instance, if a system includes an electric drive, we are to take into account constraints imposed on the angular velocity of the rotor, on the control torques created by the motor, and also on the combination of these variables. Energy and heat restrictions are usually reduced to integral constraints. All these constraints are essential for robots controlled by electric actuators. Different mixed, state, and integral constraints arise also in various other control problems for mechanical systems. It is well known that state and mixed constraints present considerable difficulties in the frames of the optimal control theory. Solutions of optimal control problems under such constraints can be usually obtained only numerically for given sets of initial and boundary conditions. In this paper, we extend the well known method originally de-

D. H. van Campen (ed.), IUTAM Symposium on Interaction between Dynamics and Control in Advanced Mechanical Systems, 67–74.

veloped in Kalman (1960) for linear systems (see also Kalman *et al*, 1968) without control constraints to systems subject to mixed, state, and integral constraints. In earlier papers (Chernousko, 1988, 1990, Chernousko and Dobrynina, 1996), we took into account only control constraints. In Kalman's approach, the open-loop control is formed as a linear combination of the natural modes of the system. We derive sufficient controllability conditions which ensure that the obtained control satisfies all imposed constraints and brings our system to the prescribed terminal state in finite time. The terminal time is not fixed *a priori* and is to be chosen so that all constraints are fulfilled. Thus, our approach is semi-analytical: the control structure is obtained explicitly, whereas some parameters are to be calculated numerically. The proposed technique is applied to a dynamical system with two degrees of freedom which is a model for mechanical systems with oscillating elements controlled by electric drives.

2. Statement of the Problem

Consider a linear controlled system described by the equation

$$\dot{x} = A(t)x + B(t)u + f(t). \tag{1}$$

Here, $x = (x_1, \ldots, x_n)$ is the n-vector of state, $u = (u_1, \ldots, u_n)$ is the m-vector of control. The matrices A and B as well as the n-vector f are given piecewise continuous functions of time t. Let the following constraints be imposed on x and u:

$$
\begin{aligned}
|C^i(t)x(t) + D^i(t)u(t) + \int_{t_0}^{T} [G^i(t,\tau)x(\tau) + H^i(t,\tau)u(\tau)]\, d\tau \\
+ \mu^i(t)| \leq 1, \quad i = 1, \ldots, r,
\end{aligned}
\tag{2}
$$

$$
(p^j(t), x(t)) + (q^j(t), u(t)) + \int_{t_0}^{T} \{(g^j(t,\tau)x(\tau)) + (h^j(t,\tau)u(\tau))\}\, d\tau \leq 1,
$$
$$
j = 1, \ldots, s.
$$

Here and below, scalar products of vectors are denoted by $(,)$. The constraints (2) are to be fulfilled for all $t \in [t_0, T]$ where t_0 and T are the initial and terminal time instants, respectively. The instant t_0 is fixed, whereas T is not fixed. The $l \times n$-matrices C^i and G^i, $l \times m$-matrices D^i and H^i, l-vector μ^i, n-vectors p^j and g^j and m-vectors q^j and h^j are given piecewise continuous functions of $t \in [t_0, T]$ and $\tau \in [t_0, T]$. The constraints (2) imply restrictions on the absolute values and components of certain linear combinations of x, u and some integrals of these variables. These constraints include the most commonly encountered restrictions imposed on control, state, and their combinations. Let the initial and terminal states be fixed

$$x(t_0) = x^0, \qquad x(T) = x^1. \tag{3}$$

Denote by $\Phi(t)$ the fundamental matrix for the system (1) defined by

$$\dot{\Phi} = A\,\Phi, \qquad \Phi(t_0) = E \tag{4}$$

where E is the unit $n \times n$-matrix. The solution of our system (1) satisfying the initial condition (3) is given by

$$x(t) = \Phi(t) \left\{ x^0 + \int_{t_0}^{t} \Phi^{-1}(\tau)\left[B(\tau)u(\tau) + f(\tau) \right] d\tau \right\}. \tag{5}$$

Substituting (5) into the terminal condition (3), we obtain

$$\int_{t_0}^{T} \Phi^{-1}(t) B(t) u(t) dt = x^* \tag{6}$$

$$x^* = \Phi^{-1}(T) x^1 - x^0 - \int_{t_0}^{T} \Phi^{-1}(t) f(t) dt\,. \tag{7}$$

We are to find a control $u(t)$ satisfying (2) and (6).

3. Design of Control

According to the approach developed by Kalman (1960), we put

$$u(t) = Q^T(t)\,C, \quad Q(t) = \Phi^{-1}(t)\,B \tag{8}$$

where C is a constant n-vector, Q is an $n \times m$-matrix, and ^{T} denotes the transposed matrix. Substituting (8) into (6), we obtain the equation for C

$$R(T)C = x^*, \quad R(T) = \int_{t_0}^{T} Q(\tau)\,Q^T(\tau) d\tau \tag{9}$$

where $R(t)$ is a symmetric non-negative definite $n \times n$-matrix defined for $t \le t_0$. We assume that $R(t)$ is positive definite, i.e. that the system (1) is controllable. Then the first equation (9) for C has a unique solution

$$C = R^{-1}(T)\,x^*. \tag{10}$$

Using (7)–(10), we present the control and state (5) as follows

$$x(t) = \Phi(t) \left[\Phi^{-1}(T)x^1 + R_1(t,T)x^* - \int_t^T \Phi^{-1}(\tau)f(\tau)d\tau \right]$$

$$R_1(t,T) = R(t)R^{-1}(T) - E, \quad u(t) = Q^T(t)R^{-1}(T)x^*.$$

(11)

Substituting (11) into the constraints (2), we rewrite them in the form

$$|F^i(t,T)x^* + \varphi^i(t,T)| \le 1, \quad i = 1,\ldots,r,$$

$$(\psi^j(t,T), x^*) + \chi^j(t,T) \le 1, \quad j = 1,\ldots,s.$$

(12)

Here, the matrices and vectors F^i, φ^i, ψ^j, and χ^i are easily expressed through the given matrices and vectors from (2), and the defined above matrices Φ, Q, R, and R_1. The inequalities (12) which should hold for all $t \in [t_0, T]$ impose certain constraints on x^* and T. These constraints can be regarded as sufficient contrallability conditions. Let us simplify (12) assuming that

$$|\varphi^i(t,T)| \le \varphi_0^i < 1, \quad i = 1,\ldots,r,$$

$$\chi^j(t,T) \le \chi_0^j < 1, \quad j = 1,\ldots,s$$

(13)

where φ_0^i and χ_0^j are constants. Using (13) and Cauchy's inequality, we replace (12) by the conditions

$$|x^*| \le \min_i \left\{ (1 - \varphi_0^i) \left[\max_t \sum_{j=1}^l \sum_{k=1}^n \left(F_{jk}^i(t,T) \right)^2 \right]^{-1/2} \right\}$$

$$|x^*| \le \min_j \left\{ (1 - \chi_0^i) \left[\max_t |\psi^j(t,T)| \right]^{-1} \right\}$$

(14)

$$i = 1,\ldots,r, \quad j = 1,\ldots,s, \quad t \in [t_0,T].$$

According to (7), the sufficient controllability conditions (14) impose constraints on x^0, x^1, and T. If (12) or (14) hold for some T, then the control u defined by (11) satisfies all constraints and brings our system (1) from the given initial state x^0 at $t = t_0$ to the prescribed terminal state x^1 at $t = T$. Of course, the conditions (12) can be simplified also in other ways different from (14); one such possibility is used below.

4. Model of an Electromechanical System

Consider a system with two degrees of freedom described by equations

$$m_1\ddot{\xi}_1 = c(\xi_2 - \xi_1) + F, \quad m_2\ddot{\xi}_2 = c(\xi_1 - \xi_2).$$

(15)

Here, ξ_1 and ξ_2 are the coordinates of the masses m_1 and m_2, respectively, c is the stiffness of the spring connecting these masses, F is the control force (or torque) created by an electric motor and applied to the first mass. For example, the following mechanical systems driven by an electric motor are described by equations (15): two masses connected by a spring, a pendulum attached to a trolley, an elastic rotating beam where only the first mode of elastic vibrations is taken into account.

We have $F = k_1 I$, $RI + k_2 \dot{\xi}_1 = U$, where I is the electric current in the rotor circuit of the motor, U is the voltage, R is the electric resistance, and k_1 and k_2 are constants. The inductivity of the electric circuit is neglected. The constraints imposed on the angular velocity of the motor, torque, and voltage can be expressed as follows

$$|\dot{\xi}_1| \leq U_0 k_2^{-1}, \quad |F| \leq F_0, \quad |U| \leq U_0 \tag{16}$$

where U_0 and F_0 are constants. Equations (15) and constraints (16) describe various mechanical systems with two degrees of freedom controlled by electric drives. Introducing dimensionless and normalized variables and constants

$$t' = \omega t, \quad x_1 = \frac{m_1 \xi_1 + m_2 \xi_2}{(m_1 + m_2) l_0}, \quad x_2 = \frac{m_1 \dot{\xi}_1 + m_2 \dot{\xi}_2}{(m_1 + m_2) l_0 \omega},$$

$$x_3 = \frac{m_1 \xi_1 - m_2 \xi_2}{(m_1 + m_2) l_0}, \quad x_4 = \frac{m_1 \dot{\xi}_1 - m_2 \dot{\xi}_2}{(m_1 + m_2) l_0 \omega}, \tag{17}$$

$$u = \frac{F}{(m_1 + m_2) l_0 \omega^2}, \quad \omega^2 = \frac{c(m_1 + m_2)}{m_1 m_2}, \quad l_0 = \frac{F_0 m_1 m_2}{c(m_1 + m_2)^2},$$

$$p = l_0 \omega k_2 U_0^{-1}, \quad q = (m_1 + m_2) l_0 \omega^2 R k_1^{-1} U_0^{-1}, \quad \mu = m_2 m_1^{-1},$$

we reduce equations (15) and constraints (16) to the form

$$\dot{x}_1 = x_2, \quad \dot{x}_2 = u, \quad \dot{x}_3 = x_4, \quad \dot{x}_4 = -x_3 + u, \tag{18}$$

$$|u| \leq 1, \quad p|x_2 + \mu x_4| \leq 1, \quad |p(x_2 + \mu x_4) + qu| \leq 1. \tag{19}$$

Here, dots denote derivatives with respect to the dimensionless time t' which is denoted below by t. The control $u(t)$ should satisfy the mixed constraints (19) and bring the system (18) from the initial state

$$x_1(0) = x_1^0, \quad x_2(0) = x_2^0, \quad x_3(0) = x_3^0, \quad x_4(0) = x_4^0 \tag{20}$$

to the zero terminal state $x_i(T) = 0$, $i = 1, 2, 3, 4$, where T is not fixed.

5. Control for the Model

The control for our electromechanical model can be designed as proposed in Section 3. Omitting cumbersome calculations and estimates simplifying the inequalities (12), we present only final results. For simplicity, we take $T = 2\pi k$ where the integer k is to be chosen. The control (11) is given by

$$u(t) = (W(t, T), x^0)\Delta^{-1}, \quad \Delta = T(T^2 - 24),$$
$$W_1 = 12t - 6T + 24\sin t, \quad W_2 = 6tT - 4T^2 + 24 + 12T\sin t, \quad (21)$$
$$W_3 = 24t - 12T + 2T^2\sin t, \quad W_4 = -2(T^2 - 24)\cos t$$

where W_i are the components of the 4-vector W and x^0 is the initial 4-vector (see (20)). Finally, we reduce the constraints (19) to the form different from (14) and containing the absolute values $|x_i^0|$ from (20). We obtain

$$\sum_{i=1}^{4} A_i |x_i^0| \leq 1,$$
$$p\left\{D_1|x_1^0| + D_2|x_2^0| + D_3\left[(x_3^0)^2 + (x_4^0)^2\right]^{1/2}\right\} + q\sum_{i=1}^{4} A_i|x_i^0| \leq 1 \quad (22)$$

where $A_i(T)$ and $D_i(T)$ are given by

$$A_1 = 6(T + 4)\Delta^{-1}, \quad A_2 = 4(T^2 + 3T - 6)\Delta^{-1},$$
$$A_3 = 2T(T + 6)\Delta^{-1}, \quad A_4 = 2T^{-1},$$
$$D_1 = \{\max\left[3T^2/2, 24(2 + \mu)\right] + 6\mu T\}\Delta^{-1}, \quad (23)$$
$$D_2 = \left[T^3 + 4\mu T^2 + 12(2 + \mu)T + 24\mu\right]\Delta^{-1},$$
$$D_3 = \mu + \left[(4T^2 + 12\mu T + 48\mu)^2 + (2 + \mu)^2(T^2 - 24)^2\right]^{1/2}\Delta^{-1}.$$

Here, p, q, and μ are defined in (17). The inequalities (22) imposed on the initial state and the terminal time $T = 2\pi k$ are sufficient controllability conditions for our control problem (18)–(20). Note that all A_i, $i = 1, 2, 3, 4$, and D_1 tend to zero whereas $D_2 \to 1$, $D_3 \to \mu$ as $T \to \infty$. Therefore, both conditions (22) are fulfilled for sufficiently large T, iff

$$p\left\{|x_2^0| + \mu\left[(x_3^0)^2 + (x_4^0)^2\right]^{1/2}\right\} < 1. \quad (24)$$

If (24) holds, then there exists $T = 2\pi k$ such that the control (21) solves our problem. To choose T, we take $k = 1, 2, \ldots$, calculate A_i and D_i from (23), and verify (22). The smallest $T = 2\pi k$ for which both inequalities (22) hold can be taken as the terminal time. Thus, our control problem is solved by simple and explicit calculations. Computer simulation shows quite satisfactory behavior of the system under the obtained control. Though the control (21) is an open-loop one, it can also be used for a feedback control. To do that, we should, in certain time intervals, regard x_0 in (21) as a current state and recalculate the terminal time T as described above.

6. Numerical Example

Let us consider a numerical example. The control (21) was applied to the system (18) subject to the constraints (19). The dimensionless parameters (17) in the constraints (19) were taken as follows: $p = 0.2$, $q = 1$, $\mu = 0.5$. We choose the following initial conditions (20): $x_1^0 = -3$, $x_2^0 = 0$, $x_3^0 = 3$, $x_4^0 = 0$. It can be easily verified that the chosen parameters and initial data satisfy the sufficient controllability condition (24). The numerical realization of the procedure described in Section 5 provides the minimal integer k under which the both conditions (22) are satisfied. We have $k = 4$, $T = 8\pi$. After that, we obtain numerically the control $u(t)$ according to (21) and the state trajectory $x(t)$.

The projection of the obtained four-dimensional state trajectory $x(t)$ onto the planes (x_1, x_2) and (x_3, x_4) is shown in Fig. 1 by the curves 1 and 2, respectively. The both curves reach the point $(0, 0)$ at the same time $t = T = 8\pi$.

7. Conclusion

Kalman's approach is extended to the case of linear controlled systems subject to various control, state, mixed, end integral constraints. The control is presented in an explicit analytical form, and the terminal time can be obtained by a simple procedure.

The paper is based on research supported by the Russian Foundation of Basic Research, Project N 96-01-01137, and by the INTAS Project N 94-1927.

References

Kalman, R. E. (1960) On the general theory of control systems, in *Proc. 1st IFAC Congress*, Butterworth, London, pp. 481–500.

Kalman, R. E. *et al*, (1968) *Topics in Mathematical System Theory*, McGraw-Hill.

Chernousko, F. L. (1988) On the construction of a bounded control in oscillatory systems. *J. Appl. Math. and Mech. (PMM)*. **52** (4), 426–433.

Chernousko, F. L. (1990) Constrained controls in linear oscillating systems, in A. Bensoussan and J. L. Lions (eds.), *Analysis and Optimization of Systems, Proc. 9th International Conference*, Lecture Notes in Control and Information Sciences, N 144, Springer-Verlag, Berlin, pp. 580–589.

Chernousko, F. L. and Dobrynina I. S. (1996) Constrained control in a mechanical system with two degrees of freedom, in D. Bestle and W. Schiehlen (eds.), *IUTAM Symposium on Optimization of Mechanical Systems*, Kluwer Academic Publishers, Dordrecht, pp. 57–64.

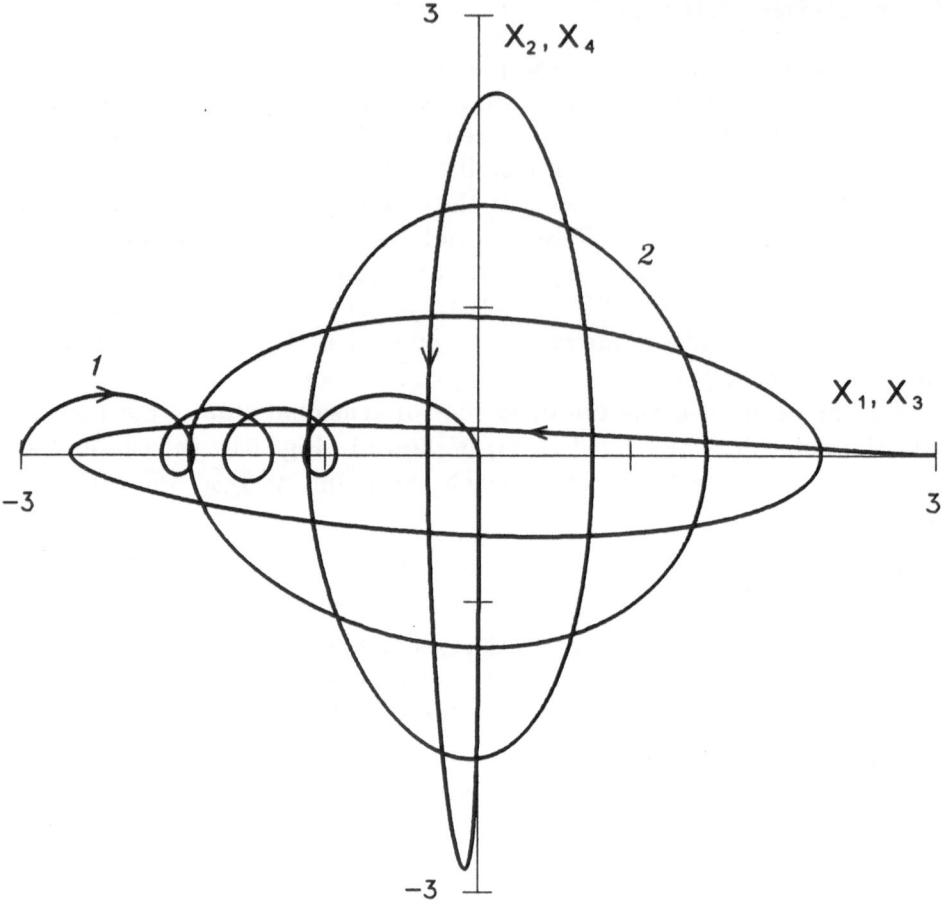

Figure 1. Projections of the state trajectory $x(t)$ onto the planes (x_1, x_2) and (x_3, x_4) (curves 1 and 2, respectively)

NEW DRIVE FOR MOTION CONTROL:

SURVEY OF THE TECHNIQUES OF VECTOR-CONTROLLED

INDUCTION MOTORS WITHOUT SPEED SENSORS

TUNG-HAI CHIN

Dept. of Electrical Electronics Engineering, Sophia Univ.
7-1 Kioi-cho, Chiyoda-ku, Tokyo 102, Japan.

1. Introduction

The DC motor has been the most popular motor for motion control, and been considered by practicing professional engineers as an ideal drive without its own dynamics and problems. However, as the DC motor possesses its congenital defect, the commutator and brushes, and due to the innovative progress of the AC drive, the DC motor has been replaced by the AC drive recently in quick tempo, especially in Japan. As well as the AC servo motors, there are PM motors (synchronous motor with permanent magnet) and induction motors. The PM motor is preferred to the induction motor for motors rating up to several kW's. However, for motors larger than these capacities, the induction motor is preferred, because the PM motor is not suitable for high power.

The characteristics of the induction motor are essentially non-linear, so it is difficult to control its torque with linearity, whereas in the case of a DC motor the induced torque is proportional to armature current autonomously. The vector control (or field oriented control) invented recently is the first method which artificially maintains the linearity of torque control to the induction motor. Today, the accompanying problem of the complexity of induction motor control has also been solved by applying micro-computers, and now the vector controlled induction motor has been established as the familiar servo drive. Moreover, the total performance of today's induction motor drive has overcome that of the DC motor, for example, by speed response, torque response, sturdiness, machine size, maintenance free, man-power saving, maximum machine capacity, maximum machine speed etc.

However, for implementation of vector control, a speed sensor is generally necessary. When the attachment of a sensor to the motor shaft is impossible or unfavorable, then its implementation is difficult. Drive engineers have dual aims: to implement vector control without speed sensor, and to make the induction motor a genuine substitute for the DC motor.

Recently, several methods with different strategies for (speed) sensor-less (vector) control of induction motors have been reported, and some of them have already been put into practice (Chin, 1992). This paper has been prepared to give a short review of sensor-less control of induction motors for motion control experts. First, the

D. H. van Campen (ed.), IUTAM Symposium on Interaction between Dynamics and Control in Advanced
Mechanical Systems, 75–82.
© 1997 *Kluwer Academic Publishers.*

fundamental principle of vector control is introduced. Then various approaches for sensor-less control are related, with emphasis on their control issues, as the induction motor itself is a complex and interesting object from the standpoint of control theory.

2. Principle of Vector Control of Induction Motors

In a DC motor, the shaft torque is induced as $T = \Psi \cdot i_a = L_m i_m \cdot i_a$. Here Ψ is the magnetic field flux interlinkage, i_m the magnetizing current (or field current) and i_a the armature current; these currents are fed through separated ports. While i_m is kept constant, the torque is proportional to the armature current.

In an induction motor, the magnetizing current $\boldsymbol{i}_m$ and the rotor current $\boldsymbol{i}_2$ are fed together from a single set of motor terminals. (Here, the bold italic characters are spatial vectors with instantaneously varying magnitude and spatial angle. They can be treated like dynamic vectors and must be distinguished from complex number vectors which are usually used for electric circuit analysis. For calculation of the instantaneous operating characteristics of an induction motor, spatial vector quantities are necessary.) The current $\boldsymbol{i}_m$ and $\boldsymbol{i}_2$ are interdependent. Consequently, the induced torque $T = L_m \boldsymbol{i}_m \times \boldsymbol{i}_2$ shows complex and non-linear behavior. The fundamental principle of vector control is to feed them in a de-coupled manner, and by keeping the amplitude of the magnetizing current $|\boldsymbol{i}_m|$ at constant value, make the induced torque $T = L_m \boldsymbol{i}_m \times \boldsymbol{i}_2$ proportional to $|\boldsymbol{i}_2|$. The principle is described as follows.

The equations of an induction motor concerning the synchronously rotating d-q frame are given by Eqs 1 and 2, with its equivalent circuit and vector diagram in Fig. 1.

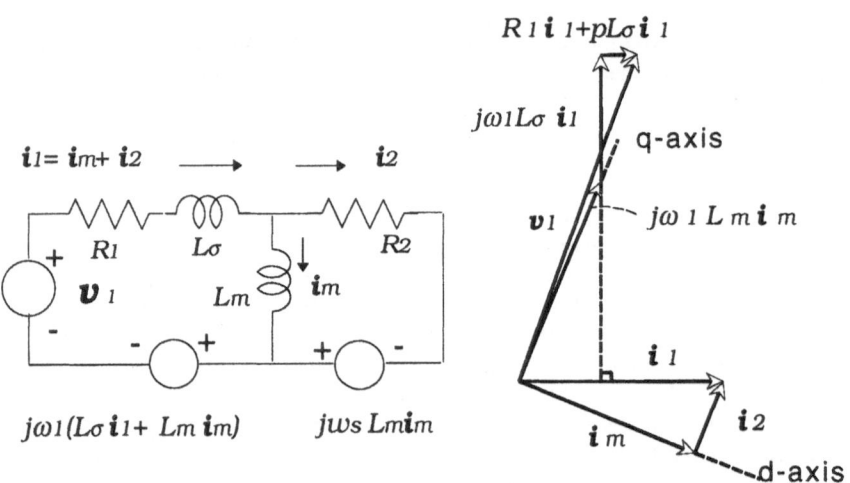

Figure 1. The equivalent circuit in the rotating flux frame and its vector diagram

Here, $\boldsymbol{i}_1 = \boldsymbol{i}_m + \boldsymbol{i}_2$, $\quad \boldsymbol{i}_2 = a \boldsymbol{i}_2^T$, $\quad \boldsymbol{i}_1 = \boldsymbol{i}_1^T$, $\quad a = L_2^T/M^T$, $\quad R_2 = R_2^T/a^2$,

$L_m = L_2^T/a^2$, $\qquad L_\sigma = L_1^T - L_2^T/a^2$, $\qquad p = d/dt$ $\qquad \omega_s = \omega_1 - \omega_r$,

$\Psi_2 = L_m \boldsymbol{i}_m = \dfrac{1}{a}\left(M \boldsymbol{i}_1^T + L_2^T \boldsymbol{i}_2^T\right)$, $\qquad T_2 = \dfrac{L_m}{R_2} = L_2^T/R_2^T$, $\qquad \omega_1 =$ source frequency, ω_r=revolving frequency of the rotor, ω_s=slip frequency of the rotor against the source frequency, and the symbols with superscript "T" show those parameters defined in the conventional T-type equivalent circuit.

$$\boldsymbol{v}_1 = R_1 \boldsymbol{i}_1 + p(L_\sigma \boldsymbol{i}_1 + L_m \boldsymbol{i}_m) + j\omega_1(L_\sigma \boldsymbol{i}_1 + L_m \boldsymbol{i}_m) \tag{1}$$

$$0 = R_2 \boldsymbol{i}_2 - j\omega_S L_m \boldsymbol{i}_m - pL_m \boldsymbol{i}_m \tag{2}$$

Eq. (2) describes the manner of current branching of the secondary current loop. The third term on its right hand side is the term induced by the amplitude variation of the magnetizing current, and it causes the interference of the currents. In vector control, the amplitude of the magnetizing current component $\boldsymbol{i}_m$ is forced to be constant, and the third term $pL_m \boldsymbol{i}_m$ equals zero. Then,

$$0 = R_2 \boldsymbol{i}_2 - j\omega_S L_m \boldsymbol{i}_m \tag{3}$$

Here, j is the symbol of perpendicular rotation, and $\boldsymbol{i}_2 \perp \boldsymbol{i}_m$. This means the currents do not interfere each other in such situation, and they can be fed in a de-coupled manner through a single set of motor terminals. Then, the induced torque $\Psi_2 \times \boldsymbol{i}_2$ can be rewritten in the simpler form $L_m \boldsymbol{i}_m \boldsymbol{i}_2$, which is just the same representation as the torque of a DC motor. As the result, the linearity of torque control is given artificially to an induction motor by forcing the amplitude of the magnetizing current to be constant.

To implement the vector control, two approaches have been proposed. One approach is to directly detect the rotating magnetic flux $\Psi_2 = L_m \boldsymbol{i}_m$ and compel it to be unchanged by feedback control. This approach has been named the direct vector control method and has rarely been put into practice, because the detection of the rotating flux is quite difficult. Another approach is to compel the remainder of Eq. 2 after excluding the term $pL_m \boldsymbol{i}_m$ to be zero as shown in Eq. (3). This can be achieved simply by feeding the stator currents with frequency $\omega_1 = \omega_r + \omega_s$ and amplitude $i_1 = \sqrt{i_m^2 + i_2^2}$; here ω_s must satisfy the requirement of Eq. (3). While Eq. (3) holds, the precondition of vector control $pL_m \boldsymbol{i}_m = 0$ will be met in an indirect manner, so the approach is named indirect vector control method. Hereafter, only the indirect method will be considered, because vector control in practical systems is almost always of the indirect type. When Eq. (3) holds, one can see that $\boldsymbol{i}_m$ is perpendicular to $\boldsymbol{i}_2$, so it can be taken as $i_{1d}=i_m$ and $i_{1q}=i_{2q}=i_2$ as shown in the vector diagram in Fig. 1. The block diagram of vector control of an induction motor is shown in Fig. 2.

3. Vector Control Implemented without Speed Sensor

As described in the previous chapter, the implementation of vector control should be accompanied by rotor speed detection. Many scientists and engineers have tried to implement vector control without a speed sensor; they get speed information by calculation instead of by mechanical detection. In this case, the problem is to correctly estimate the rotor speed by calculations, depending on the data of the terminal voltages and the input currents of the motor. Several approaches will be described in the following subsections. There are several common issues in these different approaches. From the standpoint of control theory, there are issues of integration, overlap of control

78

loops, stability and so on, whereas from the standpoint of motor proper and power electronics, there are issues of motor-parameter mismatching and voltage measurement. To suit the interests of the congress, the approaches for overcoming the control issues will be described hereafter.

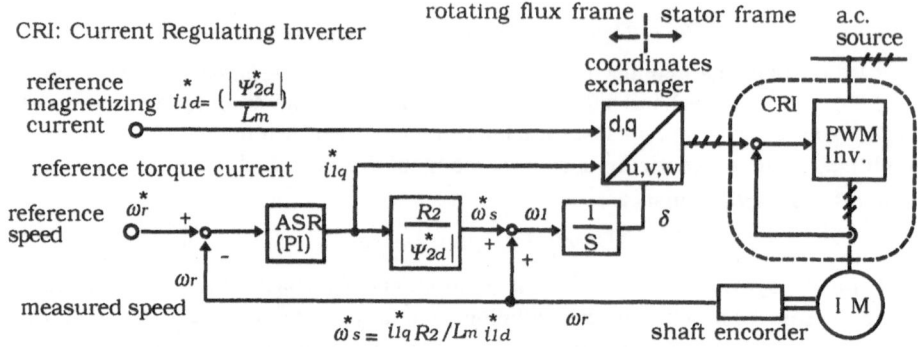

Figure 2. Block diagram of vector controlled induction motor system

3.1. DIRECT ESTIMATION USING THE TORQUE CURRENT COMPONENT

Fig. 3 shows the block diagram of the sensor-less control system. The rotor speed ω_r is estimated by the PI control circuit which removes the deviation between i_{1q}^* and i_{1q}, the command value of torque current component i_{1q} in the controller, and the calculated value from the measured motor currents, respectively. In this system,

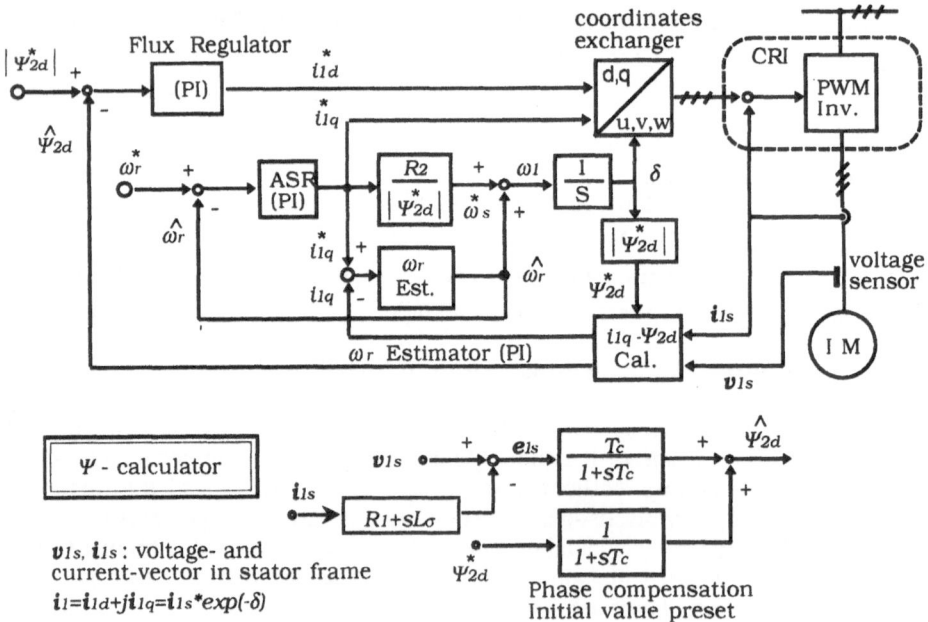

Figure 3. Sensor-less system with Current Regulated Inverter

we have to distinguish the quantities in the controller from the real quantities. An ideal CRI will bring the real input current i_1 to coincide with the command current i_1*. However, their components in d-q axes will not automatically coincide to each other, due to the deviation of the actual coordinates from that calculated in the controller. It needs some motive force to make them coincident. The PI control circuit of the ω_r Estimator forces i_{1q} to coincide with i_{1q}*, thus i_{1d} coincides with i_{1d}*. In other words, ω_s* must be equal to ω_s, because $\omega_s = (i_{1q}R_2)/(L_m i_{1d})$ under the assumption that the values of the motor parameters in the controller have been correctly preset. Thus, $\hat{\omega}_r = \omega_1$* $-\omega_s$* $= \omega_1 - \omega_s = \omega_r$, as ω_1* $= \omega_1$ is known. This means that the output of the ω_r Estimator is just the quantity which makes i_{1q} coincide with i_{1q}* through the complex control circuit, and it should simultaneously be the estimated value of the rotor speed.

The fundamental equations of the induction motor (1) and (2) can be rewritten as Eqs (4) and (5), which are the equations in the stationary frame, taking i_1 and Ψ_2 as their state variables.

$$p\Psi_2 = v_1 - R_1 i_1 - p L_\sigma i_1 \tag{4}$$

$$p\Psi_2 = \left(\frac{1}{T_2} + j\omega_r\right)\Psi_2 - R_2 i_1 \tag{5}$$

In this system, the rotor flux can be calculated by integrating Eq. (4) as follows

$$\Psi_2 = \int (v_1 - R_1 i_1 - p L_\sigma i_1) dt = \int e_m dt \tag{6}$$

Thus, the integration-issue would come to the fore. To avoid this, Ohtani et al. (1992) used first order delay instead of the integrating circuit, but added an extra circuit for compensating the phase shift, and presetting the initial value, as shown in Fig. 3. This system went well both for vector control and for speed estimation down to a lower limit of revolutions of 18 rpm (machine of 4 poles).

3.2. SENSOR-LESS SYSTEM WITH PWM VOLTAGE SOURCE INVERTER

Fig. 4 shows the block diagram of the sensor-less control system where a PWM voltage source inverter is applied. Here, the voltage sensor can be neglected. The rotor speed is estimated by PI control circuit. However, in this case the circuit has to establish two functions, one is the estimation of rotor speed and the other is to meet the condition for vector control. One can see from the figure that the control loops overlap as follows. Fig. 5 shows the block diagram concerning the ω_r estimating circuit, where the motor is also considered as a part of the circuit. Because the output of the current controller, which is also a speed estimator in the same time, is fed back to the speed controller, the response of the speed estimation is affected by the characteristics of the speed controller. Fig. 6 shows that the response of the speed controller influences the calculated response of the transfer function $\hat{\omega}_r / \omega_r$; here ω_r is the real value of the rotor speed. The response of the speed estimation decreases inversely proportional to the response of the speed controller. To avoid this, Tobari et al. (1990) suggested separating current control and speed estimation by applying the principle of the disturbance torque observer as shown in Fig. 7; this avoids the overlapping loop.

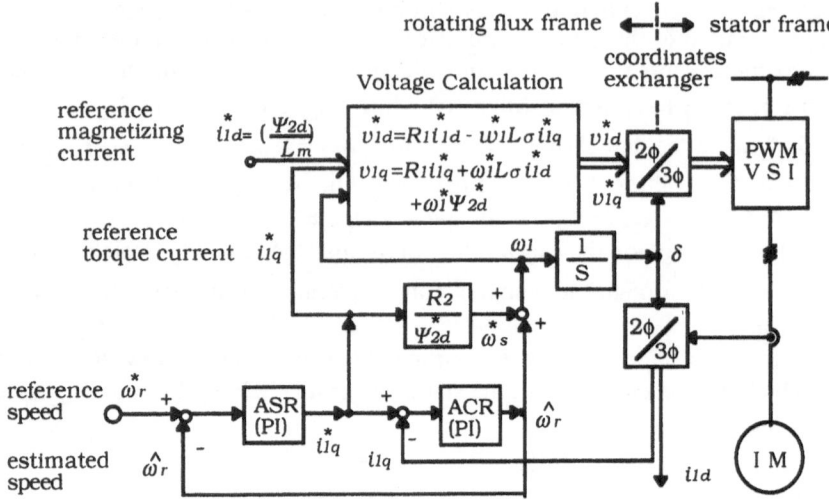

Figure 4. Sensor-less control system with PWM voltage source inverter

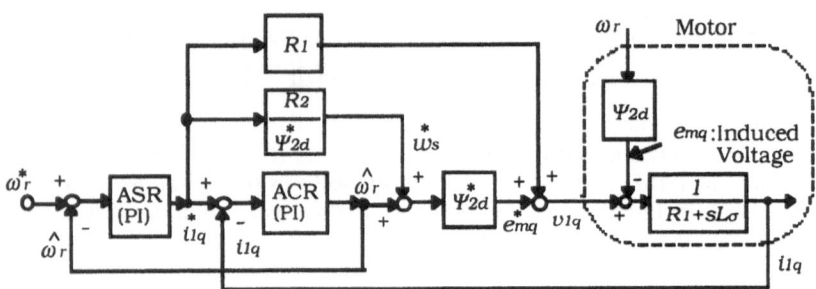

Figure 5. Block diagram concerning the loops for speed control and speed estimation

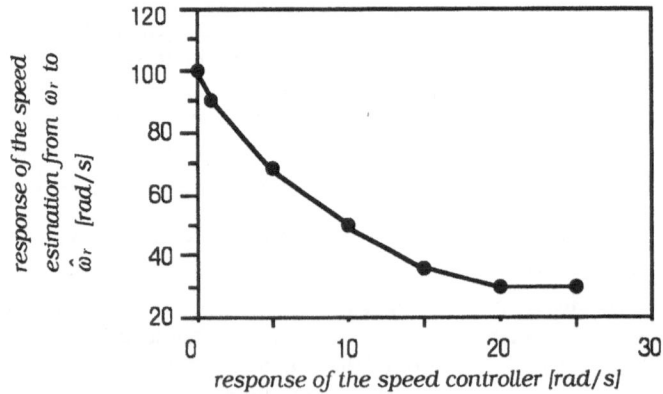

Figure 6. Interference between the response of speed estimation and the response of speed controller in the system of Fig. 5

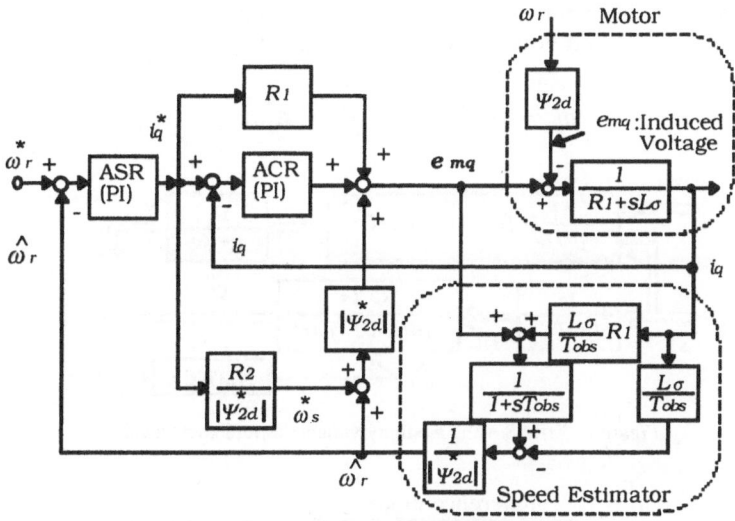

Figure 7. Proposed speed-estimating mechanism using the principle of disturbance torque observer

3.3 MRAS FOR SPEED ESTIMATION AND ITS PROGRESS

The Model Reference Adaptive System has also been used for rotor speed estimation. In this case, Eqs. (4) and (5) for the stationary frame were taken as reference model and adjustable model of the MRAS, respectively. MRAS for speed estimation can be composed by selecting an adequate reference quantity, and Fig. 8 shows the block diagram where the rotor flux inter linkage Ψ_2 is selected as the reference quantity for MRAS. However, this kind of MRAS is accompanied by the integration-issue.

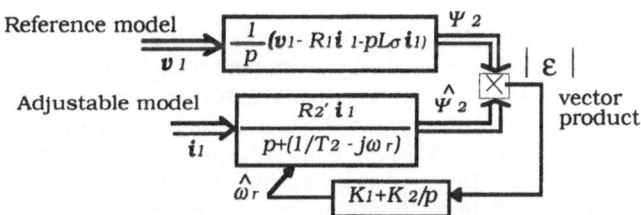

Figure 8. MRAS for speed estimation using flux Ψ_2 as reference quantity

To avoid the integration issue, Schauder (1989) introduced an auxiliary variable Ψ'_2 as reference quantity by putting in a first order delay instead of the integrator as shown in Fig. 9. The block $p/(p+K_C)$ at the output of the voltage model is an offset nullifying circuit, where K_C has a very small value, and serves to eliminate the residual offset. This system can work well around and through zero frequency. However, if the excitation dwells in the frequency range below $1/Tc$ for more than a few seconds, the integration becomes ineffective and the speed estimation becomes inaccurate.

Recently, a novel MRAS that uses induced voltage $e_m = p\Psi_2$ (see Eqs. (4) and (5)) as the reference quantity instead of the flux inter linkage has been reported by Peng et al.

82

(1993). The mechanism for tuning the adjustable model is shown in Eq. (7). As the calculation of the induced voltage does not use the integrator, the integration-issue is overcome.

$$\hat{\omega}_r = \left(K_1 + \frac{K_2}{p} \right) |\hat{\boldsymbol{e}}_m \times \boldsymbol{e}_m|$$ (7)

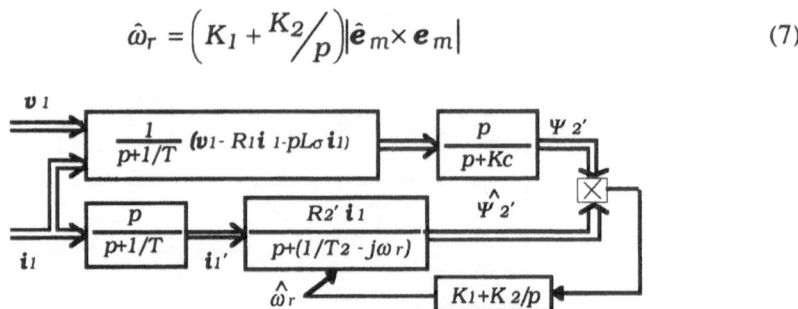

Figure 9. MRAS using auxiliary variable as reference quantity

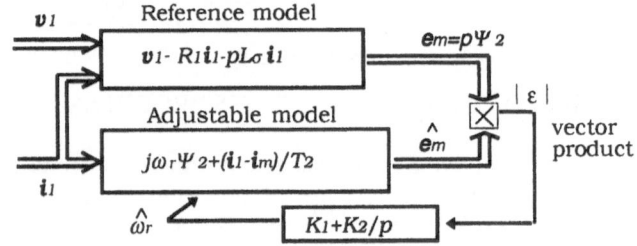

Figure 10. MRAS using induced voltage $\boldsymbol{e}_m$ as reference quantity

4. Conclusion

The approaches for sensorless vector control of induction motors and their accompanying issues have been described. Several products have already hit the market. Today's technology of sensor-less control has reached the technical level of a speed range of 1:75, speed control accuracy of ± 0.5 %, and torque control accuracy of ± 3 %, speed response of 100 rad/s, and torque control response 1000 rad/s. As sensor-less controlled induction motor drives possess various benefits over other competitive drives, it will surely spread to all fields of industry in the near future, as a general purpose high level inverter drive.

5. References

Chin, T.H. (1992) Speed Sensorless Vector Control of Induction Motor, Journal JIEE Japanese Institute of Electrical Engineers), **112**, 167-175.

Ohtani, T., Takada, N. and Tanaka, K. (1992) Vector Control of Induction Motor without Shaft Encoder, Trans. IEEE, **28-IA**, 157-165.

Peng, F.Z. and Fukao, T. (1993) Robust Speed Identification for Speed Sensorless Vector Control of Induction Motor, IEEE, IAS Annual Meeting, 419-426.

Schauder, C. (1989) Adaptive Speed Identification for Vector Control of Induction Motors without Rotational Transducers", IEEE, IAS Annual Meeting, 493-499.

Tobari, K., Okuyama, T., and Sugiyama, S. (1994) Discussion about the Response Improvement of Speed Sensorless Vector Control, Conf. JIEE '94, Tokyo, 6--26.

CONTROLLING HOPF BIFURCATION IN MECHANICAL SYSTEMS

K. CZOLCZYNSKI[1], T. KAPITANIAK[1] and J. BRINDLEY[2]
[1] Division of Dynamics
Technical University of Lodz
Stefanowskiego 1/15, 90-924 Lodz, Poland
[2] Dept. of Applied Mathematical Studies
University of Leeds
Leeds LS2 9JT, U.K.

Abstract

In this paper we describe the problem of the control of Hopf bifurcation. We illustrate this problem by the example of a rotor system supported in two gas bearings. The bearing bushes are connected to the casing by means of linear springs and viscous dampers. In the case when the stiffness coefficient of the springs is constant, at sufficiently high rotational velocity the rotor undergoes Hopf bifurcation and the bearing may be damaged by the growing amplitude of self-excited vibrations. We may avoid this danger by temporarily changing the stiffness coefficient of the springs. Our method then guarantees safe passage through the unstable zone between Hopf and reversed Hopf bifurcation points.

1. Introduction

The great majority of the vast effort put into the study of dynamical systems in the past 20 years or so has been devoted to long time behaviour of solutions, and the character of the asymptotic attractors of the system. The tools and techniques of quantitative description, such as Lyapunov exponents and various invariant measures, have naturally been driven by the need for characterisation of these attractors, and relatively little attention has been paid to the transient behaviour of trajectories during their progress from initial conditions to the appropriate attractor. Nevertheless, in a vast range of physical contexts, properly modelled by dynamical systems, the practical interest lies precisely in this behaviour; indeed the asymptotic attractor may not even be closely approached during the time interval of interest. In mechanical systems, for example those involving rotating components, it is common for the approach from an initial state to a final, often quite simple, state described by a fixed equilibrium point or limit cycle, to pass through regions of phase space quite unacceptable to the user. Thus a rotating shaft may develop large oscillations and

83

D. H. van Campen (ed.), IUTAM Symposium on Interaction between Dynamics and Control in Advanced Mechanical Systems, 83–89.
© 1997 *Kluwer Academic Publishers.*

collide with its bearing whilst slowing down to rest. In such situations the control of (perhaps chaotic) transients is required.. A number of methods by which undesirable chaotic behaviour may be controlled or eliminated have been developed (Ott *et al.*, 1990, Romeiras *et al.*, 1992, Dressler *et al.*, 1992, Ditto *et al.*, 1991, Tel 1991, Shinbrot *et al.*, 1990, Pyragas 1992, Dedieu *et al.*, 1994, Kapitaniak *et al.*, 1993, Blazejczyk *et al.*,1993). More speculatively, they indicate ways in which the existence of chaotic behaviour may be directly beneficial or exploitable. Some of these methods have been successfully applied in engineering systems (Pyragas 1992, Dedieu *et al.*, 1994, Kapitaniak *et al.*, 1993, Blazejczyk *et al.*,1993).

Chaos controlling methods, like most of the results in the theory of dynamical systems have an asymptotic character (Kapitaniak 1991, Moon 1987, Thompson *et al.*,1986) i.e. under given conditions, a dynamical system with specified parameters behaves in a particular way when $t \to \infty$. As many real systems, like the above mentioned rotor system, are characterized by some values of parameters only for finite time, these controlling methods cannot be always practically implemented, and there is a need for the development of efficient methods for control of transient nonlinear effects.

In this paper we investigate the problem of practical control of Hopf bifurcation. We consider a particular but representative example of such a situation. Let for $a = a_w$, which is a desired working condition, the dynamical system

$$\dot{x} = f(a,x) \tag{1}$$

where $x \in \mathbb{R}^n$, $a \in \mathbb{R}^m$ $(n,m = 1,2,...)$ is a vector of sytem parameters, be characterized by steady state behaviour (ie. $x = 0$ is a stable fixed point for $a = a_w$). Suppose further that to reach practically the value $a = a_w$ from $a = a_s$ (starting point), **a** has to go through the interval $[a_1,a_2]$, and that for $a \in [a_1,a_2]$, the steady state $x = 0$ of the dynamical system (1) is unstable as the system undergoes Hopf bifurcation at $a = a_1$ and reversed Hopf bifurcation at $a = a_2$. We try to answer the question: Can we reach the desired working condition $a = a_w$ safely, or do we have to restrict ourselves to the parameter range $a < a_1$?

2. Controlling Procedure

The situation described is characteristic for rotor systems, where rotor rotational velocity ω is a system parameter which has to increase continuously from its starting valve $\omega_s = 0$ to $\omega = \omega_w$ (desired working conditions), when in the interval $[\omega_1,\omega_2]$, $\omega_{1,2} < \omega_w$, the steady state $x = 0$ is unstable. Passage through this interval, as exemplified in Figure 1, is associated with an increase in amplitude of rotor vibrations as at ω_1, where the self-excited vibrations originate and are added to small oscillations due to unbalance. Practically the limit cycle of self-excited vibrations is not reached (since the time during which velocity ω increases from ω_1 to ω_2 is not sufficient to reach the attractor). At the value ω_2 the amplitude starts to decrease as the self-excited

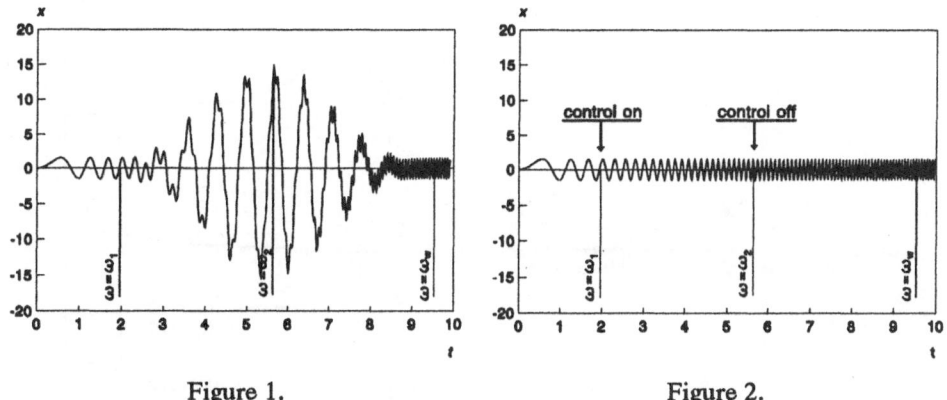

Figure 1.   Figure 2.

vibrations decay, and after a short transient the system stabilizes at the final value. In most practical systems this interval cannot be passed safely, and the self-excited oscillations can damage the system.

The transient behaviour in the dangerous zone $[a_1,a_2]$ has to be controlled in such a way that it may be safely passed. We propose a simple method which is based on the small temporal change of one of the other components of the system parameters vector a, which allows us to stabilize the fixed point $x=0$, or to significantly reduce amplitude of the self-excited oscillations for $a\in[a_1,a_2]$. When the dangerous zone is passed, vector a returns to its initial value. The passage of the controlled system through the dangerous zone is shown in Figure 2.

3. Example: Rotor Supported in Gas Bearings

We illustrate our method by the example of controlling a rotor system supported in two gas bearings. We consider the system (Figure 3) which consists of a symmetrical rigid rotor supported in two gas bearings. The joint basis of both the bearing bushes is connected with the casing by means of the linear springs K_p and the viscous dampers C_p. The force $2F_z$ is an external static load on the rotor, so F_z is the static response of the bearing to the force $2F_z$. Full dynamics of the considered system has been considered in (Czolczynski et al.,1995, Czolczynski et al.,1996, Czolczynski 1994). Here we describe only one particular case as the illustration of our controlling method.

Figure 4 shows the stability map of the system for two selected values of the stiffness coefficient K_p and various values of the damping coefficient C_p (both these coefficients are dimensionless). Λ is the dimensionless rotational velocity of the rotor; Λ_w is the working velocity. As may be seen, for $K_p=10$, the system at the desired working velocity is unstable. On the other hand, for $K_p=20$, the system at Λ_w is stable, but it cannot reach this velocity due to the unstable region below Λ_w. We have solved this problem changing the stiffness coefficient K_p from $K_p=20$ to $K_p=10$ at

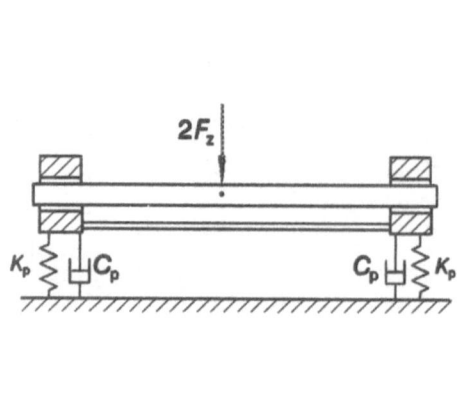

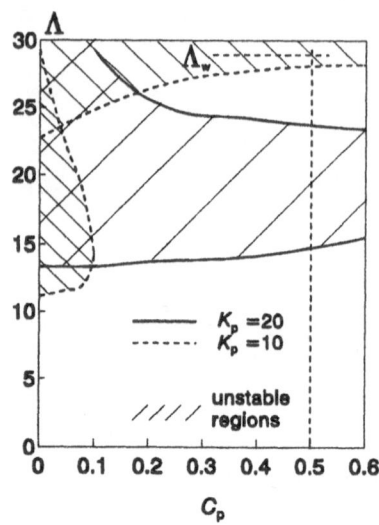

Figure 3.   Figure 4.

$\Lambda=\Lambda_1=10$, and then again from $K_p=10$ to $K_p=20$ at $\Lambda=\Lambda_2=25$. The damping coefficient C_p of the elastic support was equal to 0.5.

The elastic support of the bearing bushes, which allows us to change the stiffness coefficient K_p, may be designed as an externally pressurized air ring with the chamber feeding system, as shown in Figure 5 (1 - chamber, 2 - casing, 3 - bush, 4 - rotor, 5 - bearing gap, 6 - air ring). The air ring is fed by air through one of two orifices radius $r_{d1}=0.15\times10^{-3}$m, or $r_{d2}=1.0\times10^{-3}$m. Figure 6 shows the damping and stiffness coefficient of such an air ring as functions of the frequency ν (dimensionless) of vibration, for two cases, when the orifice r_{d1} or r_{d2} are employed. As may be seen, a decrease of the radius of the orifice causes a decrease of the stiffness coefficient. The damping coefficient remains constant, except in the range $\nu<4$, where the effect of the pneumatic hammer appears. As the possible self-excited vibrations have frequencies from the range $6<\nu<8.5$ (depending on the radius r_d and on the rotational velocity Λ), the damping coefficient C_p of the air ring may be approximately treated as $C_p\approx0.5$, and the stiffness coefficient $K_p\approx10$ for $r_{d1}=0.15\times10^{-3}$m, or $K_p=20$ for $r_{d2}=1.0\times10^{-3}$m, as on the stability map in Figure 4.

Figure 7 shows dimensionless amplitudes of the vibrations of the rotor (solid lines) and the bushes (broken lines) in the plane in which the force F_z acts, as functions of the rotational velocity Λ. The radius of the orifices is r_{d1}. Figure 4 shows that only small unbalanced vibrations appear in the wide range $\Lambda<27$. Unfortunately, at $\Lambda=27$ the system undergoes Hopf bifurcation and the self-excited vibrations grow rapidly before the working velocity is reached. At $\Lambda=\Lambda_w$ the steady state of the rotor is unstable. Contrary to this case, for r_{d2}, the steady state is stable at $\Lambda=\Lambda_w$ (see Figure 8), but below this velocity, at $\Lambda\approx14$ the rotor undergoes Hopf bifurcation (and the reversed Hopf bifurcation at $\Lambda\approx23$). The passage through the wide unstable zone

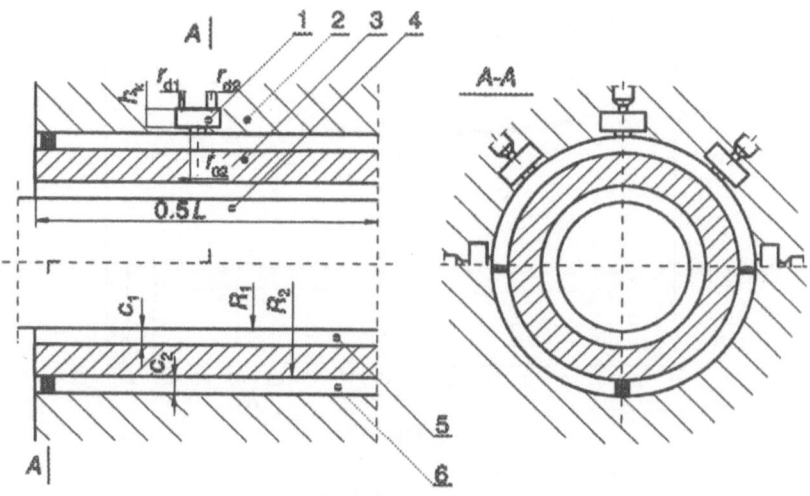

Figure 5.

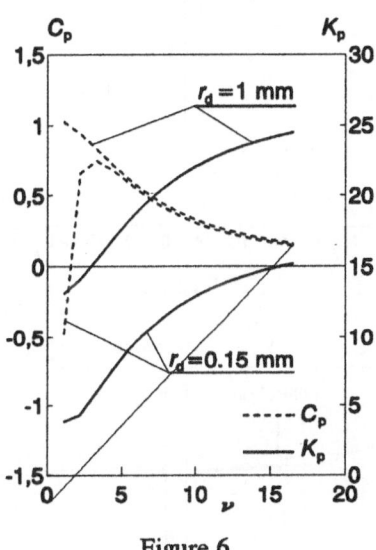

Figure 6.

$14 < \Lambda < 23$ is not possible without strong damaging contact between the journals, bushes and casing.

With the double chambers in the air ring feeding system, we may use the orifices r_{d1} for small values of Λ, then switch the feeding from r_{d1} to r_{d2} at $\Lambda_1 = 10$ (**control on** - see Figure 9), before the self excited vibrations appear, go through the dangerous zone, and finally switch again (**control off**) to r_{d1} at $\Lambda = 25$, reaching safely the working velocity Λ_w.

88

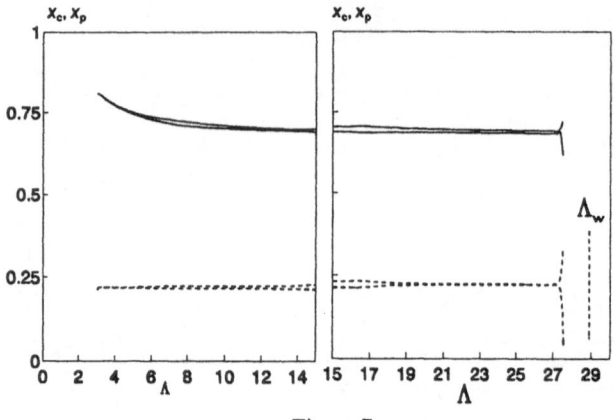

Figure 7.

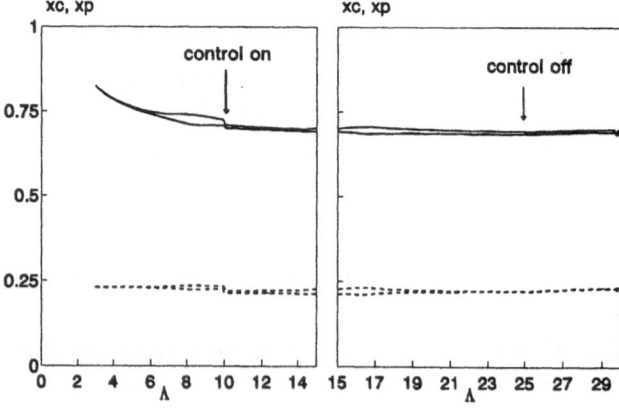

Figure 8.

Figure 9.

4. Conclusions

We have introduced a method which allows us to control Hopf bifurcation in transients. Our method guarantees passage through the unstable zone between Hopf and reversed Hopf bifurcation points. This passage is made possible by the temporal change of one of the system parameters.

Our method can have significant practical applications as it permits increase in rotational velocity of rotors of many machines, which are no longer restricted to lower velocities $\omega < \omega_1$. In many current engineering systems, a small temporal change of one of the system parameters can be easily made, and the method can stimulate the idea of "nonlinear design" i.e. the system can be designed in such a way as to exploit from nonlinear effects. We hope that the example considered in this paper will simulate further research on the transient behaviour of similar nonlinear systems.

5. References

Blazejczyk, B., Kapitaniak, T., Wojewoda. J., and Brindley, J. (1993) Controlling chaos in mechanical systems, *Appl. Mech. Rev.* **46**, 385-391.

Czołczyński, K. (1994) Stateczność i drgania samowzbudne wirnika podpartego w łożyskach gazowych (Stability and self-excited vibrations of a rotor supported in gas bearings), *Zesz. Nauk. PŁ* **694**, 2-132 (in Polish).

Czołczyński, K., and Marynowski, K. (1995) Stability of rotors supported in flexibly mounted, self-acting gas journal bearings, *Proceedings of Ninth World Congress on the Theory of Machines and Mechanisms, vol 2. Milano*, 1199-1203.

Czołczyński, K., and Marynowski, K. (1996) Stability of symmetrical rotor supported in flexibly mounted, self-acting gas journal bearings, *Wear* (to be published).

Czołczyński, K., and Marynowski, K. (1996) How to avoid self-excited vibrations in symmetrical rotors supported in gas journal bearings, *Machine Dynamics Problems*, (to be published).

Dedieu, H., and Ogorzalek, M. (1994) Controlling chaos in Chua's circuit via sampled inputs, *J. Bifurcation and Chaos* **4**, 447-456.

Ditto, W.L., Rauseo, S.W., and Spano, M.L.(1991) Experimental control of chaos, *Phys. Rev. Lett.*, **65**, 3211.

Dressler, U., and Nitsche, G. (1992) Controlling chaos using time delay coordinates, *Phys. Rev. Lett.* 1992, **68**,1

Kapitaniak, T. (1991) *Chaotic oscillations in mechanical systems*, Manchester Univ. Press, Manchester.

Kapitaniak, T., Kocarev, L., and Chua, L.O. (1993) Controlling chaos without feedback and controll signals, *Int. J. Bifurcation and Chaos* **3**, 459-468.

Moon, F.C., (1987) *Chaotic vibrations*, J. Wiley, Chichester.

Ott, E., Grebogi, C., and Yorke, Y.A. (1990) Controlling Chaos, *Phys. Rev. Lett.* **64**,1196.

Pyragas, K. (1992) Continuous control of chaos by self-controlling feedback, *Phys. Lett.* **170A**, 421.

Romeiras, F., Ott, E., Grebogi, C., and Dayawansa, W.P. (1992) Controlling chaotic dynamical systems, *Physica* **58D**, 165.

Shinbrot, T., Ott, E., Grebogi, C., and Yorke, Y.A. (1990) Using chaos to direct trajectories to targets, *Phys. Rev. Lett.* **65**, 3215.

Tel, T. (1991) Controlling transient chaos, *J. Phys. A*. **24**, L1359.

Thompson, J.M.T., and Steward, B. (1986) *Nonlinear dynamics*, J. Wiley, Chichester.

ACTIVE ALIGNMENT CONTROL OF A PAYLOAD USING NON-LINEAR, LONG STROKE ACTUATORS

A. P. DARBY AND S. PELLEGRINO

Cambridge University Engineering Department
Trumpington Street, Cambridge, CB2 1PZ, U.K.

Abstract. The paper describes a system which provides 6DOF alignment correction of a payload mounted at the tip of a flexible structure, subject to dynamic and quasi-static disturbances, using a Stewart platform arrangement. The six legs of the platform are formed by a new type of inertial slip-stick actuator. Non-linearities associated with the actuator are modelled allowing the interaction between the Stewart platform and the structure to be described. A proportional controller, capable of controlling quasi-static disturbances, and an LQG controller, for reducing dynamic disturbances, are designed. The effects of the actuator non-linearities upon the closed loop performance, are examined both experimentally and analytically.

1. Introduction

Over the last twenty years there has been a rapid increase in the number of optical structures, such as space telescopes and optical interferometers, being sent into orbit in order to perform scientific observations. These structures have been increasing in size yet requiring higher optical resolutions. A typical requirement is to maintain dimensional stability within a few microns over distances of the order of ten metres. However, disturbances in the form of quasi-static thermal distortions or initial structural deformations of spacecraft components, together with vibrations caused by momentum wheels or thermal shock, reduce the resolution that can be achieved by means of traditional, *passive* structural configurations.

Recently, a great deal of research has been carried out on actively-controlled structural systems, in order to reduce the effects of these disturbances upon space structures (Wada, 1990). The aim of the work presented in this paper is to actively control the alignment of a payload, mounted at the tip of a flexible structure, with respect to the base of the structure. Here, the flexible structure and the payload may represent, for example, respectively the main body of a telescope and the main optics, in which case the problem is to prevent defocusing and tilting of the optical plane.

It is desirable that both quasi-static distortions and vibrations of the flexible structure be controlled using a single system, but the main difficulty is that, whereas vibration control requires actuators which can change length rapidly by small amounts, quasi-static control requires actuators that can change length by several millimetres, at a slower speed, whilst maintaining a high degree of accuracy.

D. H. van Campen (ed.), IUTAM Symposium on Interaction between Dynamics and Control in Advanced Mechanical Systems, 91–100.

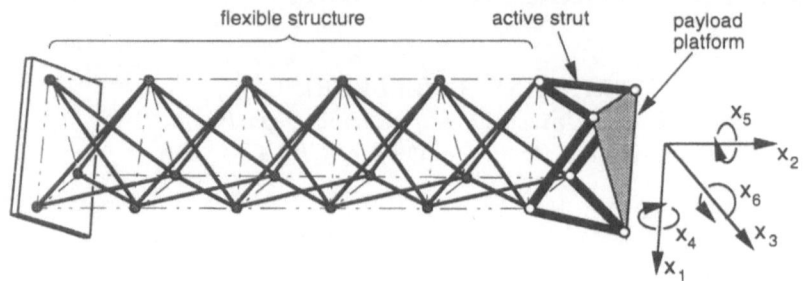

Figure 1. Flexible structure - Stewart platform arrangement.

These requirements cannot be met by most existing actuators. The solution that is presented in this paper provides 6DOF alignment correction using an active module based upon a Stewart platform arrangement. The six legs of the platform are formed by a new type of inertial slip-stick actuator that has been developed during the course of this study. A typical system, representing a pantographic deployable mast and an active module used for experimental tests, is shown diagrammatically in figure 1. The 6DOF's $x_1 \ldots x_6$ represent the position measurement outputs of the platform.

The layout of the paper is as follows. After a brief description of the long-stroke actuator, in Section 2, Section 3 describes the design of a proportional controller for quasi-static alignment control. Then, Section 4 presents a state-space model for the dynamic interaction between a 1DOF flexible structure and a 1DOF payload connected by a single actuator, and Section 5 describes the design of an LQG controller based on a state-space model of the complete structure. Section 6 concludes the paper.

2. Stick-Slip Inertial Actuator

The actuator, Figure 2, consists of two separate parts. An outer holder, connected to the right-hand-side joint fitting, and an inner sliding tube, containing the piezo-electric element and the inertial mass, connected to the left-hand-side joint fitting. The sliding tube is held within the holder purely by friction, between a set of balls.

In order to cause the sliding tube to move, a force greater than the static friction is applied to it. This force is induced by accelerating the inertial mass via a rapid change in length of the piezoelectric element. If an axial compressive force is applied to the actuator, the inertia force required to extend the actuator will be equal to the sum of the the friction force and the axial force, whereas the inertia force required to contract the actuator will be equal to the difference between these two forces.

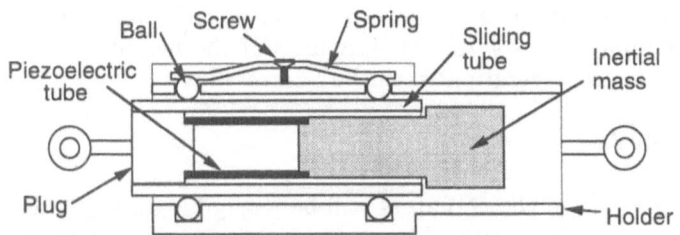

Figure 2. Schematic diagram of actuator (not to scale).

The actuator is driven by applying a suitable driving waveform, at 2kHz, to the piezoelectric element, the velocity varying with the amplitude of the waveform, u. The experimentally derived relationship between the average velocity of the actuator, $\dot{\delta}$, the input waveform amplitude and the axial force, F, is given by:

$$F = a\dot{\delta} + bu + n \qquad (1)$$

where the coefficients a, b, and n take different values depending upon the sign of the velocity. This relationship, together with the saturation limits of the actuator, define the main non-linear behaviour of the control system. The maximum velocity of the actuator with $F = 0$ is approximately 7 mm/s, and it is capable of position accuracy of less than 1 μm. Further details can be found in Darby and Pellegrino (1995).

3. Quasi-static Control

The kinematic relationship between the 6DOF position of the payload platform $y = [x_1 \ldots x_6]^T$ as defined in Figure 1, and the changes in length of the actuators, $\delta = [\delta_1 \ldots \delta_6]^T$, can be expressed by a 6×6 Jacobian matrix, J:

$$y = J\delta \qquad (2)$$

In general, the coefficients of the Jacobian matrix are configuration dependent. However, for small movements, i.e. δ_i much smaller than the length of the corresponding actuator, Equation 2 can be linearised. If y is the position error to be corrected, the solution of Equation 2 will give the corresponding changes of actuator length required to correct alignment. We choose a proportional controller which relates the velocity input $\dot{\delta}_i$ proportionally to the length error:

$$\dot{\delta}_i = c\delta_i \qquad (3)$$

where c is the controller gain, which is assumed to be the same for all actuators. By substituting Equation 2 into Equation 3, the following relationship between input and output is found:

$$\dot{\delta} = cJ^{-1}y \qquad (4)$$

The value of c should be as great as possible, in order to reject disturbances at the actuator inputs, but, since the dynamics of the flexible structure have been ignored, stability considerations must be addressed.

3.1. STABILITY

Although the forces in the actuators will be negligible during quasi-static correction, due to lack of vibration and hence inertial forces, the saturation limits of the actuators must be considered. A simple describing function analysis can be used to address the stability of the closed-loop system. The describing function for a general saturation type non-linearity is given by Cook (1994):

$$D(E) = \begin{cases} \frac{2k}{\pi} sin^{-1} \frac{m}{Ek} + \frac{2m}{E\pi} \sqrt{1 - \left(\frac{m}{Ek}\right)^2}, & \frac{Ek}{m} > 1 \\ k, & \frac{Ek}{m} \leq 1 \end{cases} \tag{5}$$

where E is the amplitude of the sinusoidal velocity input, k is the slope of the linear region of the input (here, $k = 1$), and m is the value at which saturation occurs.

Equation 1 can be linearised about $F = 0$ to give a linearised dynamic model, $G(j\omega)$, of the interaction between a modal model of the flexible structure (containing the first eight natural modes) and the actuators. To investigate the stability of the proportional controller for different gains c we substitute $k = c$ into Equation 5. Nyquist plots of the negative reciprocal of the describing function $-1/cD(E)$, with $k = c = 4$, 6 and 8, and the linearised system $G(j\omega)$ are shown in Figure 3. Where the two plots intersect an instability is likely. The two intersect at a value of $c = 6.2$ and a frequency of 20.7 rad/s which corresponds to the first bending mode of the structure. Further simulations and experimental tests show that the system does indeed become unstable with a controller gain greater that 6.2.

3.2. EXPERIMENTAL RESULTS

Based upon the stability analysis, a proportional controller with a gain of 6 was chosen to control quasi-static disturbances. Tests were carried out in order to verify the ability of the controller to counteract slowly changing disturbances. Various members of the flexible structure were heated and cooled, randomly, in order to simulate the thermal gradients that might occur in a spacecraft structure.

The open and closed-loop responses are shown in Figure 4 for each DOF. Open loop displacements, shown by the dotted lines, of greater than ± 1mm and rotations of ± 0.001 rad have been caused by the disturbance. The closed loop response, shown by the solid lines, reduces the displacements to approximately ± 0.1mm and the rotations to ± 0.00025 rad. These errors are of the same order as the accuracy of the LVDT sensors used to measure each DOF.

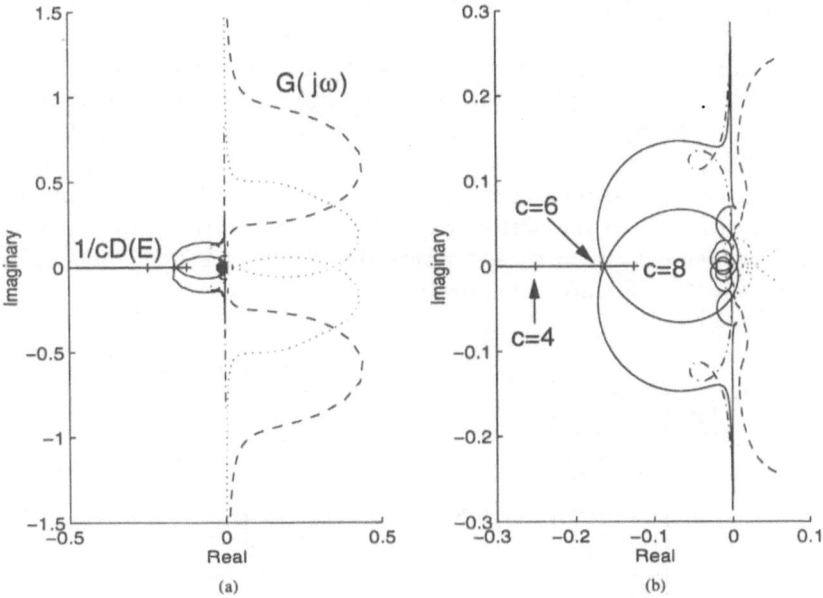

Figure 3. Nyquist plot of $G(j\omega)$ versus frequency and $-1/cD(E)$ for c=4, 6 and 8.

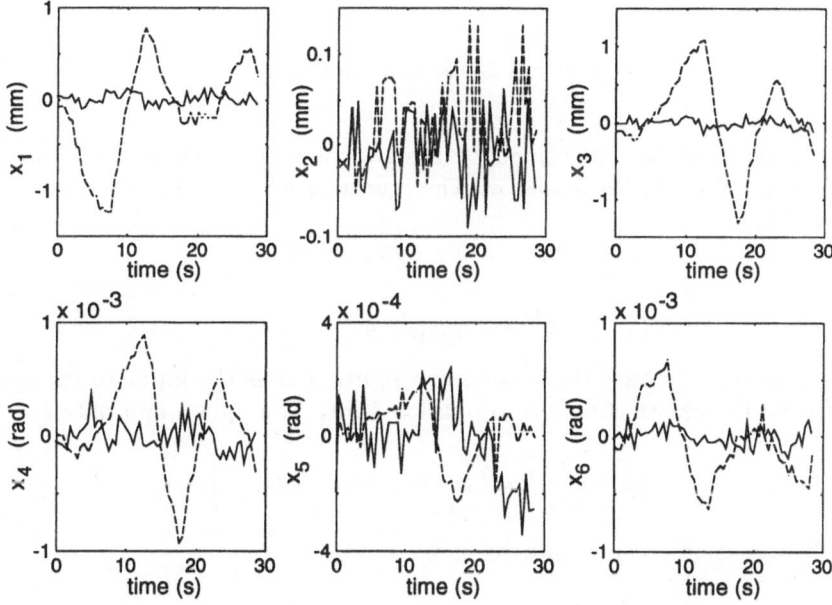

Figure 4. Measured open and closed-loop responses to a quasi-static disturbance input.

4. Dynamic Interaction in a 2DOF System

The non-linear interaction between the flexible structure and the payload will be explained through the simple 2 DOF system with a single actuator that is shown in Figure 5(a). This system represents a flexible structure with mass m_s, stiffness k, and damping d, connected to a payload of mass m_p through an actuator. The actuator extension, δ, is measured with respect to a reference configuration, and its sign is positive when the actuator gets longer. Since the actuator is assumed to be a rigid element, $x_p = x_s + \delta$ where the displacements x_p and x_s are also measured from the reference configuration.

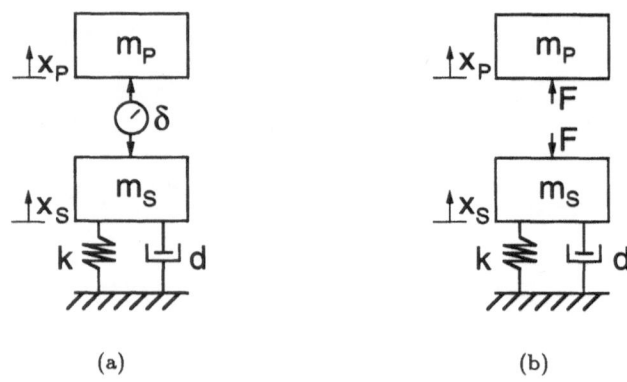

(a) (b)

Figure 5. 2-DOF dynamic system.

The system can be represented by the two subsystems shown in Figure 5(b) where F is the force in the actuator. The equations of motion for the two systems are given by:

$$m_s \ddot{x}_s + d\dot{x}_s + kx_s = -F \qquad (6)$$

and

$$F = m_p(\ddot{x}_s + \ddot{\delta}) \qquad (7)$$

However, we wish to have the waveform amplitude u as the input to the system, not force F. Hence, we substitute Equation 1 into Equation 6 and obtain:

$$\ddot{x}_s = \frac{1}{m_s}\left[-d\dot{x}_s - kx_s - a\dot{\delta} - bu - n\right] \qquad (8)$$

Similarly, Equation 7 can be rewritten as:

$$\ddot{\delta} = \frac{1}{m_p}(a\dot{\delta} + bu + n) - \ddot{x}_s \qquad (9)$$

which, substituting Equation 8, gives:

$$\ddot{\delta} = \left(\frac{1}{m_p} + \frac{1}{m_s} \right) \left[a\dot{\delta} + bu + n \right] + \frac{1}{m_s} \left[d\dot{x}_s + kx_s \right] \tag{10}$$

Equations 8 and 10 can be re-cast in a standard state-space form, by representing the system dynamics by a set of first order linear equations:

$$\dot{x} = Ax + Bu \tag{11}$$

where x is a state vector and u is an input vector. The output of the system is given by a combination of the states and the inputs:

$$y = Cx + Du \tag{12}$$

If the states are chosen as $x = [\dot{x}_s \quad x_s \quad \dot{\delta} \quad \delta]^T$ and the inputs as $y = [x_p]$, the state-space representation of the system is given by:

$$\begin{bmatrix} \ddot{x}_s \\ \dot{x}_s \\ \ddot{\delta} \\ \dot{\delta} \end{bmatrix} = \begin{bmatrix} -\frac{d}{m_s} & -\frac{k}{m_s} & -\frac{a}{m_s} & 0 \\ 1 & 0 & 0 & 0 \\ \frac{d}{m_s} & \frac{k}{m_s} & \frac{a}{m_p} + \frac{a}{m_s} & 0 \\ 0 & 0 & 1 & 0 \end{bmatrix} \begin{bmatrix} \dot{x}_s \\ x_s \\ \dot{\delta} \\ \delta \end{bmatrix}$$

$$+ \begin{bmatrix} -\frac{b}{m_s} \\ 0 \\ \frac{b}{m_p} + \frac{b}{m_s} \\ 0 \end{bmatrix} u + \begin{bmatrix} -\frac{n}{m_s} \\ 0 \\ \frac{n}{m_p} + \frac{n}{m_s} \\ 0 \end{bmatrix} \tag{13}$$

and

$$x_p = \begin{bmatrix} 0 & 1 & 0 & 1 \end{bmatrix} \begin{bmatrix} \dot{x}_s \\ x_s \\ \dot{\delta} \\ \delta \end{bmatrix} + [0] u \tag{14}$$

This approach can be extended to the MDOF Stewart platform/flexible mast system, in order to simulate its non-linear motion. A linear numerical model of the flexible mast is used to derive a modal representation of Equation 6.

5. Vibration control

The Linear Quadratic Gaussian (LQG) design method (Kwakernaak and Sivan, 1972) has been used to design a controller that, unlike the proportional controller, takes full advantage of the knowledge of the dynamics of the system. The basic aim of the LQG controller design method is to produce a state gain feedback matrix, K_c, giving a linear feedback control law:

$$\dot{\delta} = -K_c x \tag{15}$$

where x are the states of the open loop system. The criterion for designing the gain matrix is the minimisation of a quadratic cost function:

$$\int_0^\infty \left(x^T Q x + u^T R u \right) dt \tag{16}$$

where Q and R are weighting matrices. The solution to the optimal controller gain problem involves the solution of an algebraic Riccati equation.

Since in our experiments we do not measure the states, but only $x_1, \ldots x_6$, a state estimator must be formed. The estimator gain matrix is given by:

$$K_f = L C^T V^{-1} \tag{17}$$

in which L satisfies another algebraic Riccati equation involving input noise covariance, W, and measurement noise covariance, V. The noise represents uncertainty in the system rather than physical noise and disturbances. Therefore, the estimator makes the controller robust to system uncertainties.

5.1. SIMULATION

A controller using the LQG method was designed, based upon a linearised state-space model of the test structure. The controller was then incorporated into a full non-linear simulation of the system, using the dynamic interaction model described earlier, in order to check stability.

A sinusoidal input disturbance was applied in order to excite the first natural frequency of the system. The controller was switched on after 2 seconds. Figure 6 shows the output response of the closed-loop system for each DOF. A reduction in amplitude by a factor of approximately 2.5 is evident for the DOF's most affected by the disturbance.

5.2. EXPERIMENTAL RESULTS

The same controller was examined experimentally. The structure was excited sinusoidally using an electro-magnetic shaker. The closed-loop time responses for each DOF are shown in Figure 7.

The performance of the controller is similar to the simulation and the system remains stable. The measured responses compare well to the non-linear simulation estimates, suggesting that the interaction between the flexible structure and the inertial actuators is modelled sufficiently well to represent the main non-linearities in the system.

6. Discussion and Conclusions

The most significant non-linearities of the system that has been investigated stem from the actuators. Hence, incorporating a non-linear model of the actuators,

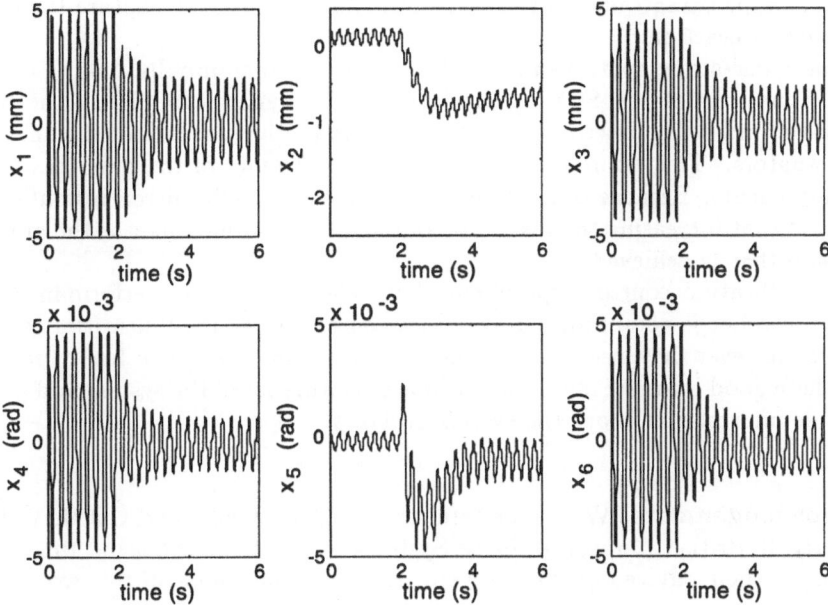

Figure 6. Simulated response to sinusoidal disturbance; closed-loop after 2 seconds.

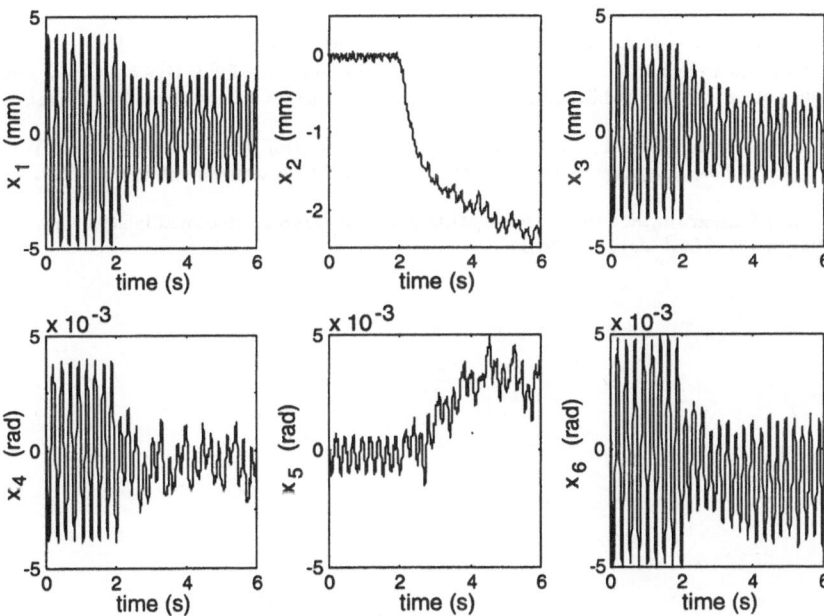

Figure 7. Measured response to sinusoidal disturbance; closed-loop after 2 seconds.

albeit an approximate one, in the system model is essential to capture the control-structure interaction.

The quasi-static control problem has been solved simply by linking the kinematic equations of the Stewart platform to the dynamic model of the flexible structure, and using describing functions to model the saturation non-linearity of the actuators. This approach provides a good estimate of the gain at which instability occurs. Because our actuators are capable of sub-micron resolution, it is expected that increasing the resolution of the sensors would increase the alignment precision that is achieved.

The vibration control experiments have shown that the performance of the LQG controller is compromised by saturation and force non-linearities of the actuators, however the closed-loop system remains stable. The non-linear simulation provides a good estimate of the closed-loop performance of the system and provides valuable information about the system stability.

Acknowledgements. We are grateful to Drs G.W. Game and C.J.H. Williams, formerly at British Aerospace Space Systems, and to our colleague Professor K. Glover for their advice and encouragement during the course of this work.

References

Cook, P.A. (1994) *Nonlinear Dynamical Systems*, Prentice Hall, UK.

Darby, A.P. and Pellegrino, S. (1995) Inertial stick-slip actuator for active control of shape and vibration, submitted for publication in *International Journal of Intelligent Material, Systems and Structures*.

Kwakernaak, H. and Sivan, R. (1972) *Linear Optimal Control Systems*, Wiley, New York.

Wada, B. K. (1990) Adaptive structures: an overview, *AIAA Journal of Spacecraft*, **27**, 4, 330-337.

You, Z. and Pellegrino, S. (1996) Cable-stiffened pantographic deployable structures. Part 1: triangular mast, *AIAA Journal*, **34**,4.

A BENCHMARK EXAMPLE TO QUALIFY A CONTROL STRATEGY FOR MOTION CONTROL

A. DE CARLI, L. ONOFRI
University of Rome "La Sapienza",
Department of Computer and System Sciences,
Via Eudossiana 18, I 00184 Roma, Italy,
Fax ++39 6 4458 5367

1. Introduction

The control of mechanical motion has importance in industrial automation. In discrete event production systems, the quality of the motion determines plant productivity, product quality, energy saving and implementation efficiency. In continuous production systems, the control of mechanical motion improves the performances of pumps, compressors, valves etc. Control of mechanical motion is applied to isolated devices or interacting mechanisms. The rated power ranges between less then one watt to many kW.

Mechanical motion is conventionally achieved by connecting a drive to the load by means of a coupling device. The drive is generally made up by connecting the motor, the supply converter, the motor speed or position measuring device, and the drive controller, to form a feedback system.

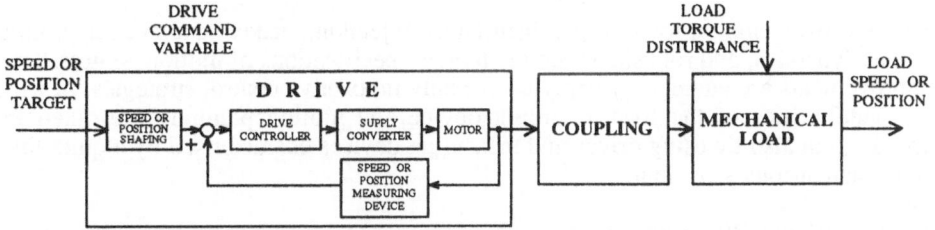

Figure 1. Conventional implementation for the control of the mechanical motion.

The desired motion is planned, in general, by taking into a limited account the behaviour of the coupling device and the interactions between the motor and load behaviour. A suitable shaping of the target speed or position allows to limit the undesired phenomena by reducing the rate of the motor torque variation, and consequently by slowing the dynamics of the drive. The conventional implementation, shown in Fig. 1, does not allow the compensation of the load torque disturbance. Some improvement in dynamics and in rejection of load torque disturbances are achieved by overrating the drive and the mechanical structure.

D. H. van Campen (ed.), IUTAM Symposium on Interaction between Dynamics and Control in Advanced Mechanical Systems, 101–108.
© 1997 *Kluwer Academic Publishers.*

Innovative approaches require the following higher quality performance specifications:

- robust stability with respect to load parameter variations;
- high accuracy at the steady-state;
- high rapidity and accuracy in dynamics;
- performance robustness in the whole range of operating conditions.

The design of a suitable motion controller is therefore a highly complex problem which can however be implemented by a low cost digital device. It is obtained by implementing the control of mechanical motion by a closed loop system including the drive, the load, the measuring device of the load speed or position and the motion controller. The speed or position shaping is necessary only to avoid the attainment of the maximum instantaneous torque of the drive motor. Innovative implementation should be designed to obtain the substantial improvements of the performance specifications. Problems solved in this framework are indicated as the new topic: *motion control*.

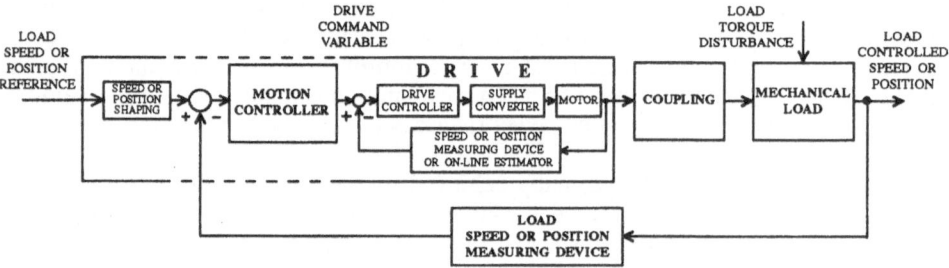

Figure 2. Innovative implementation for the control of mechanical motion.

The improvements on load torque disturbance rejection, steady-state accuracy, input-output dynamics, and robustness are the targets specifications of motion control. These improvements are attained by applying not only innovative control strategies, in which the peculiarities of the load, coupling device and motion.planning are taken into account, but also by using drives and measuring devices characterised by a quite linear and instantaneous behaviour.

A conventional PID regulator does not allow us to attain all the above mentioned performance specifications. The desired steady-state accuracy and robustness are in fact achieved by derating the input-output dynamics and consequently the productivity of the plant. Innovative control strategies on the contrary allow us to attain higher target specifications without overrating the drive and the mechanical structure. Mechanical loads require a rotary and/or linear motion; since the latter is generally obtained by suitable mechanisms, only rotary motion will herewith be taken into account.

To design the innovative control strategies, implemented into the motion controller, the peculiarities of mechanical loads can be classified as follows:

- the load inertia is practically constant and the dissipative load, due to load disturbances, viscous and friction, is variable over a wide range;

103

- the load torque and inertia variations can be characterised by their nominal values and the variations from the nominal values;
- load torque oscillations due to the elastic coupling between the load and motor shaft can be taken into account;
- uncertainties of the instantaneous load torque are due to nonlinearities and unforeseen behaviour and/or unmodelled dynamics.

Innovative control strategies are characterised by one or more of the following peculiarities:

- the on-line estimation of the load torque or the static friction, and their direct compensation by acting on the command variable of the drive;
- the robustness of the controlled system behaviour in spite of load uncertainties and unmodelled dynamics;
- the simultaneous fulfilment of different target specifications involving the input-output dynamics and the load torque disturbance rejection.

In order to illustrate the advantages obtained by applying an innovative control strategy, the behaviour of the motion controlled system will be compared with the behaviour obtained by a conventional PID, under the same operating conditions. The simulation approach will be used in order to outline the relationship between the peculiarities of the load and the innovative control strategy.

The simulated mechanical system consists of two masses connected by an elastic shaft. The motor and the load are respectively represented by the first and second mass.

In Figure 3 the simulated load and the block diagram representing the load dynamics are shown together with the value of the parameters.

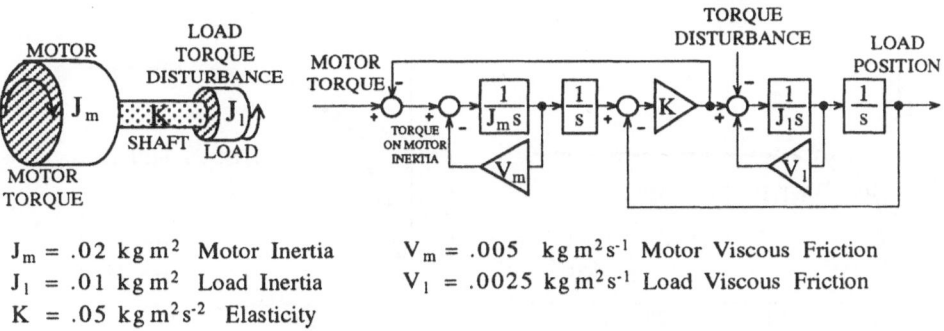

$J_m = .02$ kg m^2 Motor Inertia $V_m = .005$ kg m^{2}s^{-1} Motor Viscous Friction
$J_1 = .01$ kg m^2 Load Inertia $V_1 = .0025$ kg m^{2}s^{-1} Load Viscous Friction
$K = .05$ kg m^{2}s^{-2} Elasticity

Figure 3. Mechanical load and its dynamic model.

A suitable model of the static friction is included in the simulation of the mechanical load. By varying the nominal value of some parameters, a particular behaviour of the mechanical system can be simulated in order to evidence improvements obtainable by applying an innovative control strategy.

104

The selection of the most suitable innovative control strategy should be done by first analysing why the system is performing poorly, and then by designing the control strategy to improve it. This is carried out in two steps: to compensating for the effect of the load and friction torque, and to improving the input-output dynamics and the robustness. Both should be implemented into the motion controller.

The poor performance is due to friction, stiction, and load torque disturbance. The effect of friction and stiction can be reduced by applying an appropriate compensating torque. The compensation for the load torque can be carried out by the on-line estimation of its value, once a feasible dynamic model of the load is available.

Different control strategies can be used to improve the input-output dynamics and to achieve robustness. In this paper, only constant structure and constant parameter control strategies will be considered. They are is classified as follows:

- PID with robust tuning of parameters;
- zero-pole compensation with parameter optimisation for robustness (H_{inf});
- zero-pole compensation with parameter tuning for independent fulfilment of disturbance rejection and input-output dynamics.

Some peculiarities of these control strategies have already been outlined in [1]. In this paper the simulation results will be presented only with the aim to offer the results of a benchmark example.

2. Friction and stiction compensation

In open loop operation, friction, stiction, shaft oscillations, and load parameter variations decrease the accuracy of the mechanical motion, as shown in Figure 4.

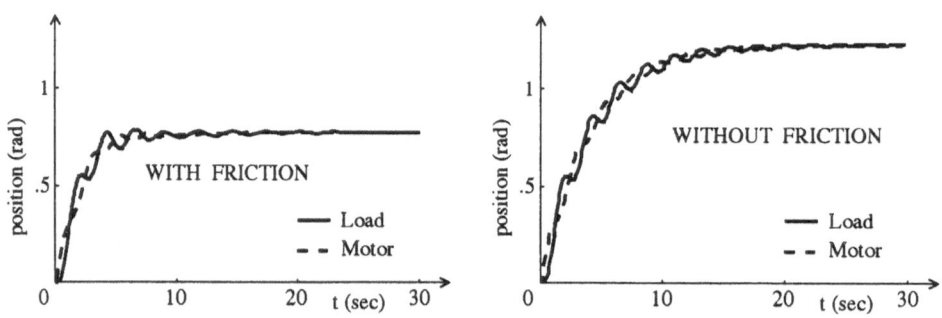

Figure 4. Load position corresponding to the application of a pulse-wise motor torque with and without friction effects.

Friction and stiction cause nonlinearity in the mechanical load behaviour. Restoring the linearity of the mechanical load is very convenient to attain the target specifications by means of a simple structured control strategy. To this aim, the most convenient way is to compensate the friction and stiction effects by acting directly on the command variable of the drive. The indirect compensation via the feedback loop does not produce comparable improvements in the behaviour of the mechanical load without an overrating of the drive.

Friction and stiction compensation is effected by using a suitable model of these phenomena. A feasible model is given by the sequential one illustrated in [2,3,4]. On the basis of this model, the compensation has been carried out according to the block diagram shown in Figure 5.

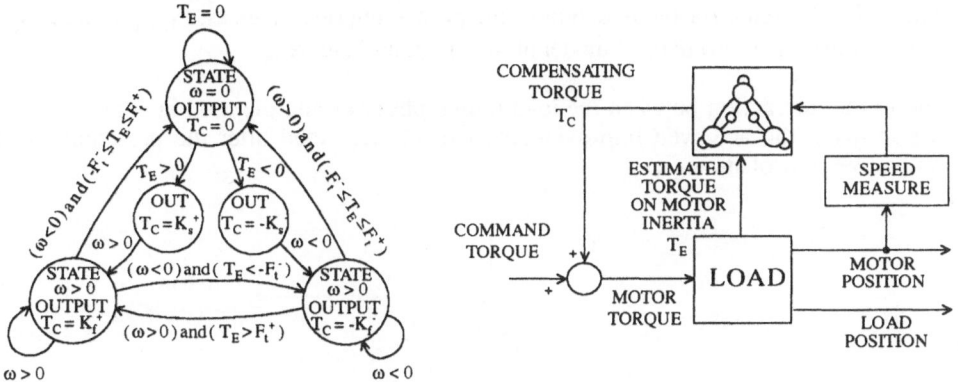

$$K_s = .0011 \ Kg \, m^2 s^{-2} \quad K_f = .001 \ Kg \, m^2 s^{-2} \quad F_t = .001 \ Kg \, m^2 s^{-2}$$

Figure 5. Sequential model for the friction and stiction compensation and block diagram of the compensated mechanical load.

The effect of the stiction and friction compensation is more evident in Figure 6, in which the step response of the controlled system is shown for different loads when a conventional PID, is applied with and without the stiction and friction compensation.

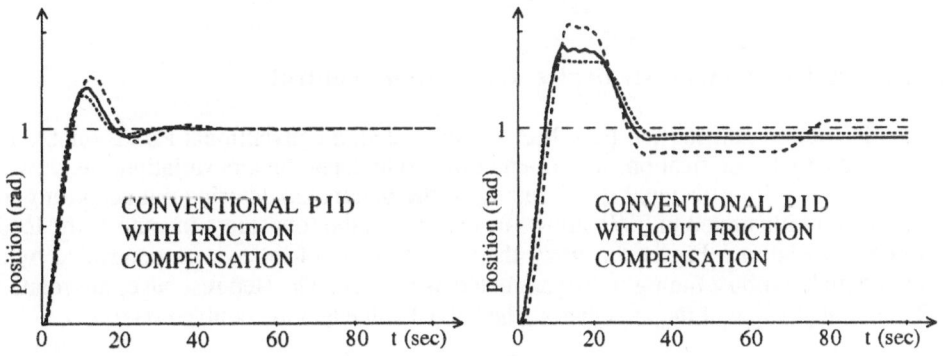

Figure 6. Step response of the controlled load with and without stiction and friction compensation.

3. Load torque observer

The most effective way to design a load torque observer is the on-line processing of the instantaneous measured values of the load and motor position (if available). The accuracy of the estimated load torque is strictly related to the feasibility of the load model. The knowledge of the motor and load inertia is fundamental for an approximate estimation. This approach has been widely used; detailed information is given in [5].

The load torque observer has been implemented in terms of an accurate linearized model of the motor and load behaviour. Figure 7 shows the related block diagram; the low-pass filter is fundamental for the rejection of the processing disturbances. P_{motor} and P_{load} represent the linear mechanical model of the motor and the load.

Due to the interaction between the load torque observer and the motion controller, the test showing the achieved improvements will be described after the presentation of innovative control strategies.

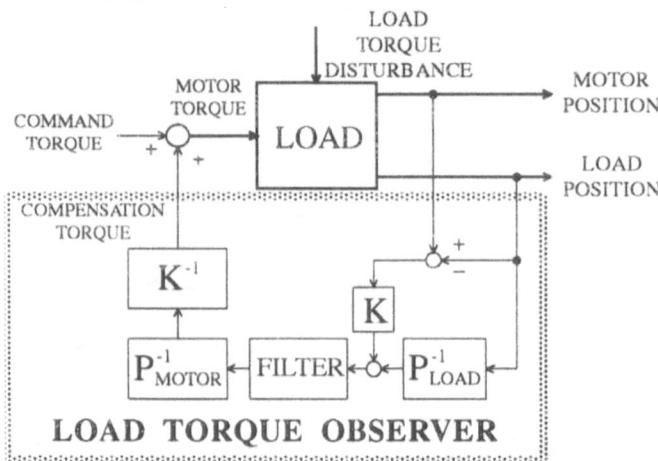

Figure 7. Block diagram of the load torque observer.

4. Innovative control strategies for motion control

Once the friction and stiction have been compensated, a conventional PID is unable to satisfy the target specifications on dynamics due to load parameters variations, as shown in Figure 6. An improvement in the tuning of the parameters, carried out according to the procedure illustrated in [6,7], allows us to increase the robustness in spite of the load parameter variations. Figure 8a shows the step response of the system controlled by a PID in which a robust tuning of its parameters is applied. The step responses are related to the nominal value of the load inertia, the halved value and the doubled one.

When uncertainties of the instantaneous load torque value, due to unforeseen behaviour and/or unmodelled dynamics, have a relevant effect, and an improvement of target specifications is required, the parameter of the control strategy should be worked out according to a procedure which allows for the optimisation of the robustness. This procedure is generally indicated as H_{inf}. Many papers illustrate the application of this procedure; some of these are listed in the References [8]. Figure 8b shows the step responses of the system controlled by a H_{inf} controller with the variations of the load inertia.

When the performance specification related to the disturbance rejection has peculiarities different from the input-output ones, the control strategy should be worked out according to a two-degree-of-freedom procedure. Many papers illustrate the application of this procedure; some of these are listed in the References [9]. Figure 8c shows the step responses of the system controlled by a two-degree-of-freedom controller having the variations of the load inertia.

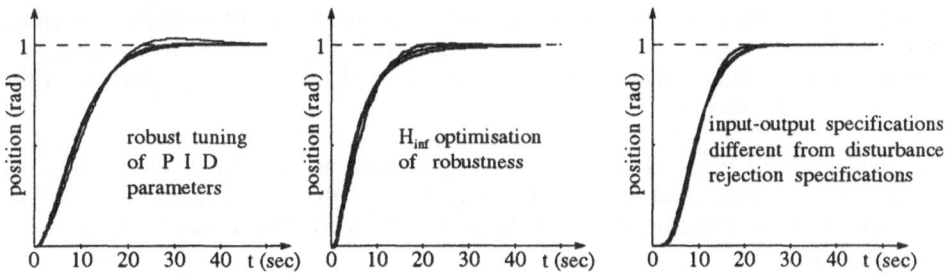

Figure 8. Step responses of the controlled system by applying innovative control strategies.

In connection with the previously presented control strategy, designed by taking into consideration mainly the input-output performance specifications, the effects of an pulse-wise load torque disturbance have been taken into account. To illustrate the effects of the load torque observer, the step responses, related to the nominal inertia, are compared with the responses obtained without the application of this device. Figure 9 shows the waveforms related to this control strategy, with and without the application of the load torque observer.

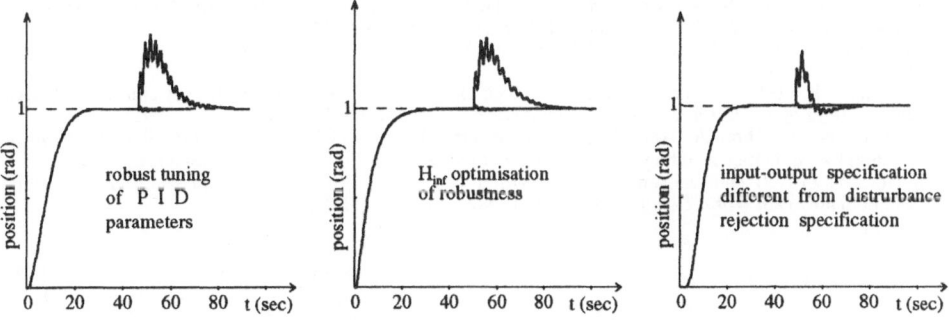

Figure 9. Input-output and load torque responses related to the innovative control strategies, with and without the application of the load torque observer.

108

5. Conclusions

In motion control, higher quality performance specifications can be achieved by applying innovative control strategies without overrating the drive and the mechanical structure. However the drive and the measuring device should behave almost linearly and rapidly.

The innovative control strategies, proposed in this paper, can be considered as direct extensions of conventional ones, because they are based on a fixed parameter controller and on a continuous shaping of the command variable of the drive. The main difficulty for the application consists in the background necessary for their design. Dedicated Matlab toolboxes give the necessary support to carry out the design problem.

The higher cost of the controller is mainly due to the design of an innovative control strategy, since the hardware for the digital implementation of a fixed parameter controller has a very low cost. The convenience of applying innovative approaches depends on many factors that should be accurately evaluated. The improvement of the efficiency is the main one.

The application of innovative control strategy should be carried out jointly with the design of the mechanical system, the selection of the instrumentation, and the realisation of the computing device. It should therefore be carried out by a team rather than by one person.

6. References

[1] De Carli A. (1995) Control Strategies for Advanced Motion Control, *Proceedings of the IFAC Workshop on Motion Control*, pp 31-40, Munich.
[2] Colombi S. (1995) Comparison of Different Strategies and Friction Compensation Algorithms in Position and Speed Controls, *Proceedings of the IFAC Workshop on Motion Control*, pp 173-180, Munich.
[3] Cristadoro G., De Carli A., Onofri L. (1996) A Robust Control Strategy for an Unstable Mechanical System, *Proceedings of the 4th International Workshop on Advanced Motion Control*, vol 2, pp 488-493, Mie University, Japan.
[4 Iwasaki M., Matsui N. (1996) Observer-Based Nonlinear Friction Compensation in Servo Drive system, *Proceedings of the 4th International Workshop on Advanced Motion Control*, vol 1, pp 344-348, Mie University, Japan.
[5] Hori Y. (1995) Vibration Suppression and Disturbance Rejection Control on Torsional Systems, *Proceedings of the IFAC Workshop on Motion Control*, pp 41-50, Munich.
[6] Ackerman J. (1993) *Robust Control Systems with Uncertain Physical Parameters*, Springer-Verlag, Germany.
[7] Isermann R. (1995) Mechatronic systems -A challenge for the design of intelligent control systems-, *Proceedings of 3rd European Control Conference*, vol 3, pp 2708-2713, Rome.
[8] Shimakage M., Moran A., Nagai M., Hayase M. (1995) Design of Vibration Suppression H_{inf} Control System for High-Rise Buildings, *Proceedings of the IPEC*, vol 1, pp 464-469, Yokohama.
[9] Hori Y. (1993) Generalised Robust Position and Force Controller Design Method for Robot Manipulator, *12th World Congress of the IFAC*, vol 9, pp 439-442, Sydney.

CONTROLLING CHAOS IN A TEMPORALLY IRREGULAR ENVIRONMENT AND ITS APPLICATION TO ENGINEERING SYSTEMS

M. DING

Center for Complex Systems and Department of Mathematics
Florida Atlantic University, Boca Raton, FL 33431, USA

1. Introduction

Responses of many man-made systems (e.g., ships or oil-drilling platforms), when subject to irregularly time varying environments, can be described by irregularly driven dynamical systems. Consequently, failures of such systems (e.g., capsize of a ship or collapse of a platform), under increasingly severe environmental conditions, come about when the system state escapes from a destroyed chaotic attractor located in some favorable region of the phase space. In this paper we review a control strategy [see Ding et al. (1994) for more details], based on a previous method of chaos control, which can prevent such failures from taking place. The key feature of the new method is the incorporation of prediction of the evolution of the environment. This makes effective operation of the control possible even when the temporal behavior of the environment has substantial irregularity. We illustrate the ideas using ship capsizing as an example. We then apply the technique to a nonlinear oscillator model of a ship-borne crane. The purpose here is to eliminate uncertainties associated with sudden changes (crises) in attractor structures as a result of environmental drift.

2. Controlling Chaos to Prevent Ship Capsizing

Imagine a ship rolling in lateral ocean waves. Its dynamics can be modeled by the following nonlinear oscillator,

$$\ddot{x} + \nu\dot{x} + \omega^2(x - \alpha x^3) = W(t), \tag{1}$$

D. H. van Campen (ed.), IUTAM Symposium on Interaction between Dynamics and Control in Advanced Mechanical Systems, 109–117.
© 1997 *Kluwer Academic Publishers.*

where x is a variable characterizing the angle from the ship's mast to the vertical direction, ν is the friction coefficient, ω is the eigenfrequency of small vibrations around the origin, α denotes the strength of nonlinearity, and $W(t)$ represents the effect of the ocean waves impinging on the ship. In the absence of waves, i.e., $W(t) = 0$, for a small displacement in x, the subsequent motion damps out and the ship reverts to its upright position. For large displacements in x, gravity overcomes buoyancy, and x tends to the attractor at $|x| = \infty$. When this occurs we say the ship has capsized. Models of similar types have been used in other studies of ship dynamics (Virgin, 1987; Nayfeh and Sanchez, 1990; Thompson et al., 1990). We assume that the irregular ocean waves have the following form,

$$W(t) = f(t)[1 + \epsilon_a g(t)]\sin\phi(t) \equiv F(t)\sin\phi(t), \qquad (2)$$

where $F(t)$ is the amplitude and the phase $\phi(t)$ is determined by

$$\dot{\phi}(t) = \Omega + \epsilon_p h(t). \qquad (3)$$

Here $g(t)$ and $h(t)$ are irregular functions of time. In what follows, we model the temporally irregular functions $g(t)$ and $h(t)$ as outputs of well known chaotic systems. Specifically, we use

$$g(t) = [y(t)- < y(t) >]/\sqrt{< [y(t)- < y(t) >]^2 >},$$

where $y(t)$ is a solution of the following driven Duffing system,

$$\ddot{y} + 0.05\dot{y} + y^3 = 7.5\cos t,$$

and

$$h(t) = [z_1(t)- < z_1(t) >]/\sqrt{< [z_1(t)- < z_1(t) >]^2 >},$$

with $z_1(t)$ taken from the Rössler system

$$\begin{aligned}
\dot{z}_1 &= -z_2 - z_3, \\
\dot{z}_2 &= z_1 + 0.398z_2, \\
\dot{z}_3 &= 2 + z_3(z_1 - 4).
\end{aligned}$$

The symbol $<>$ denotes temporal average. Since $g(t)$ and $h(t)$ are normalized, ϵ_a and ϵ_p measure the relative strength of amplitude irregularity and that of phase irregularity, respectively.

For $\epsilon_a = \epsilon_p = 0$ and $f(t) = f_0$, where f_0 is a constant, we have purely sinusoidal waves, and the dynamics in this case will be used as the basis for control later when irregularities are introduced into the system. The evolution of Eq. (1) with increasing wave amplitude f_0 and $\epsilon_a = \epsilon_p = 0$ is shown in the bifurcation diagram in Fig. 1. (We use $\nu = 0.5$, $\omega = \Omega = 1.0$ and $\alpha = 1.0$ for numerical computations.) The *surface of section* here is taken every time $W(t)$ crosses zero with $dW(t)/dt > 0$ (i.e., $\Omega t_n = 2n\pi$). For $0 < f_0 < 0.7$ the oscillation of the ship is periodic, with the period equal to that of the wave. We henceforth call this orbit the *period one* orbit. Following a period doubling cascade, the ship's response becomes chaotic. At $f_0 \approx 0.726$ a crisis takes place, in which the bounded chaotic attractor is destroyed by colliding with its basin boundary. As the wave amplitude increases past this crisis value, since no bounded attractor exists in the system, almost all initial conditions tend to the attractor at $|x| = \infty$ (i.e., the ship capsizes). Also shown in Fig. 1, as a dashed curve, is the *period one* orbit after it loses stability. Now suppose that $f(t)$ is a gradually increasing function of time starting from $f(t) = 0$. We anticipate that in the absence of control the ship will capsize some time after $f(t)$ exceeds the crisis value. To prevent this, one starts to apply control to stabilize the period one orbit in Fig. 1 when the wave is still small, and continues to do so as the wave increases. As a result, we show that the ship survives waves whose amplitude significantly exceeds the crisis value. More importantly, by incorporating a prediction feature to be detailed below, and suitably modifying the control procedure in Ott et al. (1990) and Shinbrot et al. (1993), we are able to prevent capsizing in the presence of substantial wave irregularities.

The equation of motion for the variable x, when control is applied, is again Eq. (1), but with the right hand side replaced by $W(t) + C(t)$, where $C(t)$ is the control. We take $C(t)$ to be a constant between successive crossings of the *surface of section*. In practical terms, $C(t)$ can be thought of as the balancing force provided by temporally shifting ballast on the ship, and the value of $C(t)$ is assumed to be bounded between $-C_0$ and C_0, i.e., $-C_0 \leq C(t) \leq C_0$. In fact, we are interested in the case where C_0 is small compared to the force exerted by the waves. We take the *surface of section* to specify the system state at times $t = t_n$ such that $W(t) = 0$ and $dW(t)/dt > 0$. If $f(t) > 0$, and the right hand side of Eq. (3) is positive, this corresponds to $\phi(t_n) = 2n\pi$ [see Eq. (2)]. Figure 2 illustrates the prediction and control method. A fixed point of the Poincaré map and its stable and unstable directions are constructed for $\epsilon_a = \epsilon_p = 0$ and $f(t) = f(t_n)$. Here we imagine that $f(t)$ can be treated as a constant in the interval $t_n < t < t_{n+1}$. (Note that due to the slow variation of $f(t)$, the fixed point location at $t = t_n$ changes[1] with n.) By letting $\epsilon_a \neq 0$

---

[1] An algorithm for tracking and control of unstable periodic orbits is developed by I. Schwartz and I. Triandaf, Phys. Rev. A46, 7439(1992), and implemented in a laser experiment by Z. Gills et al., Phys. Rev. Lett. **64**, 3169(1992), and in a circuit experiment by T. Carroll, I. Triandaf, I. Schwartz, and L. Pecora, Phys. Rev. A46, 6189(1992).

and $\epsilon_p \neq 0$, we introduce irregularity into the wave. The task now is to stabilize the motion around the fixed point. Suppose that at $t = t_n$, the system state is denoted by $\mathbf{z}_n$, where $\mathbf{z} = (x, y = \dot{x})$. From observation of waves propagating toward the ship, we assume that we can make an accurate prediction of $W(t)$ for $t_n \leq t \leq t_{n+1}$. Integrating Eq. (1) with this predicted $W(t)$, and with various values of $C(t) = \hat{C}$, where $\hat{C}$ is a constant in the interval $[-C_0, C_0]$, we obtain images of the system state on the next *surface of section* at $t = t_{n+1}$, which form a curve as shown in Fig. 2. The value $\hat{C} = C_n$ of the control that we actually apply to the ship is chosen so that the system state falls on the stable direction of the fixed point.

Figure 3(a) shows part of the wave for $\epsilon_a = 0.15$, $\epsilon_p = 0.1$, and $f(t)$ a linearly increasing function of t. As shown in Fig. 3(b), in the absence of control, the ship capsizes after several cycles of ocean waves. Using control with prediction we prevent the capsizing in Fig. 3(b). A more extended controlled orbit is shown in Fig. 3(c). Figure 3(d) shows $C(t)$ as a function of t. Comparing Fig. 3(d) with Fig. 3(a) we see that $C(t)$ is much smaller than the wave amplitude $W(t)$.

3. Controlling Chaos to Eliminate Uncertainties in Nonlinear Oscillators

Consider the double pendulum shown in Fig. 4. This model is motivated by the dynamics of a ship-borne crane. Here we assume that the variable x obeys Eq. (1) and the variable y does not affect the dynamics of x. The equations of motion for the system are thus

$$\ddot{y} + \mu(\dot{x} + \dot{y}) + R\cos(x+y)\ddot{x} - R\sin(x+y)\dot{x}^2 = -\eta^2 \sin(y), \quad (4)$$
$$\ddot{x} + \nu\dot{x} + \omega^2(x - \alpha x^3) = W(t), \quad (5)$$

where μ is the friction coefficient, L is the length of the lower arm, l is the length of the upper arm, $R = L/l$, $\eta = \sqrt{g/l}$ and $W(t)$ is described by (2) and (3). In what follows we fix $R = 1.5$, $\eta = \sqrt{3}$, and $\mu = 0.5$.

Our main focus is the response dynamics of the mass m. We restrict our attention to small waves ($f_0 < 0.5$) so that the ship motion is stable. Figure 5 shows the bifurcation diagram for y, which is similar to that in Fig. 1 ($\epsilon_a = 0$ and $\epsilon_p = 0$). When the waves are very small, the cargo piece oscillates with the same frequency as the wave. As the wave amplitude increases, the response goes through a bifurcation sequence leading to chaotic motion. The important aspect to note is that there are sudden changes in the attractor structure as f_0 varies. This leads to uncertainties in the response dynamics. In Fig. 6 we demonstrate that, by stabilizing the response around a period one unstable orbit using control, we can eliminate the uncertainties associated with chaotic motion and with the sudden changes in the attractor, even in the presence of substantial irregularities ($\epsilon_a = 0.1$ and $\epsilon_p = 0.05$). Specifically, Fig. 5(a) shows the increasing wave. Figure 5(b) shows the response dynamics without control. Figure 5(c) shows the controlled time series. The control is achieved by varying the length of l slightly, according to the method introduced earlier. Clearly, the large unpredictable swings in the uncontrolled case have been converted to a regular periodic orbit.

4. Conclusion

To conclude we remark that, although ship capsizing and uncertainty elimination in nonlinear oscillators have been used here to illustrate the ideas of the control strategy, other situations are also candidates for the approach in this paper. In particular, our method is potentially important in applications where a momentary loss of control due to environmental irregularity can lead to a catastrophic failure of the system.

References

M. Ding, E. Ott, and C. Grebogi, *Phys. Rev. E* **50**, 4228(1994); *ibid.*, *Physica* **D 74**, 386(1994).

A. H. Nayfeh and N. E. Sanchez, *Int. Shipbuild. Progr.* **37**, 331(1990).

E. Ott, C. Grebogi, and J. A. Yorke, *Phys. Rev. Lett.* **64**, 1196(1990).

T. Shinbrot, C. Grebogi, E. Ott, and J. A. Yorke, *Nature* **363**, 411(1993).

J. M. T. Thompson, R. C. Rainey, and M. S. Soliman, *Phil. Trans. R. Soc. Lond.* **A 332**, 149(1990).

L. N. Virgin, *Appl. Ocean Res.* **9**, 89(1987).

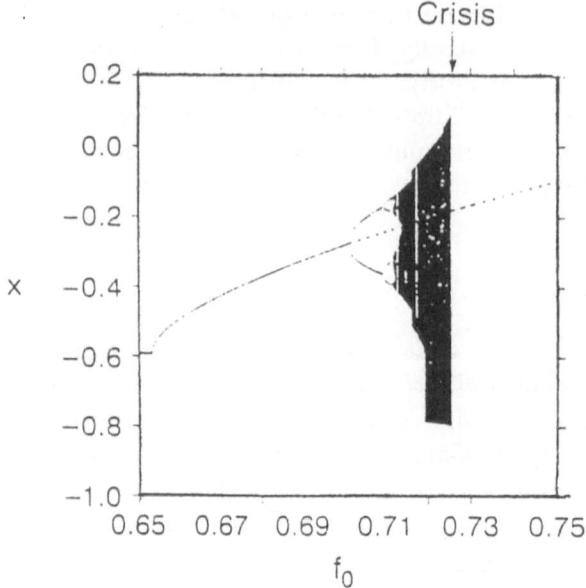

FIG. 1. Bifurcation diagram of Eq. (1). The dashed line indicates the unstable period one orbit after the period doubling bifurcation.

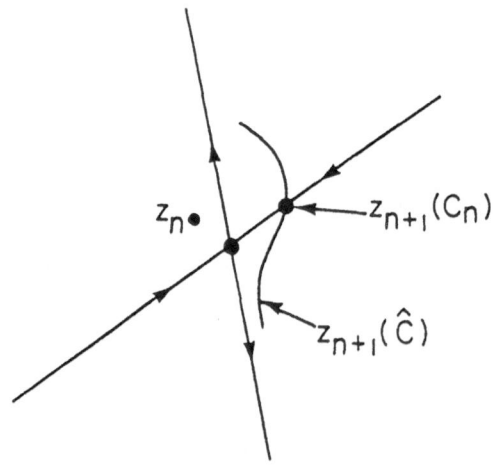

FIG. 2. Schematic of the control method.

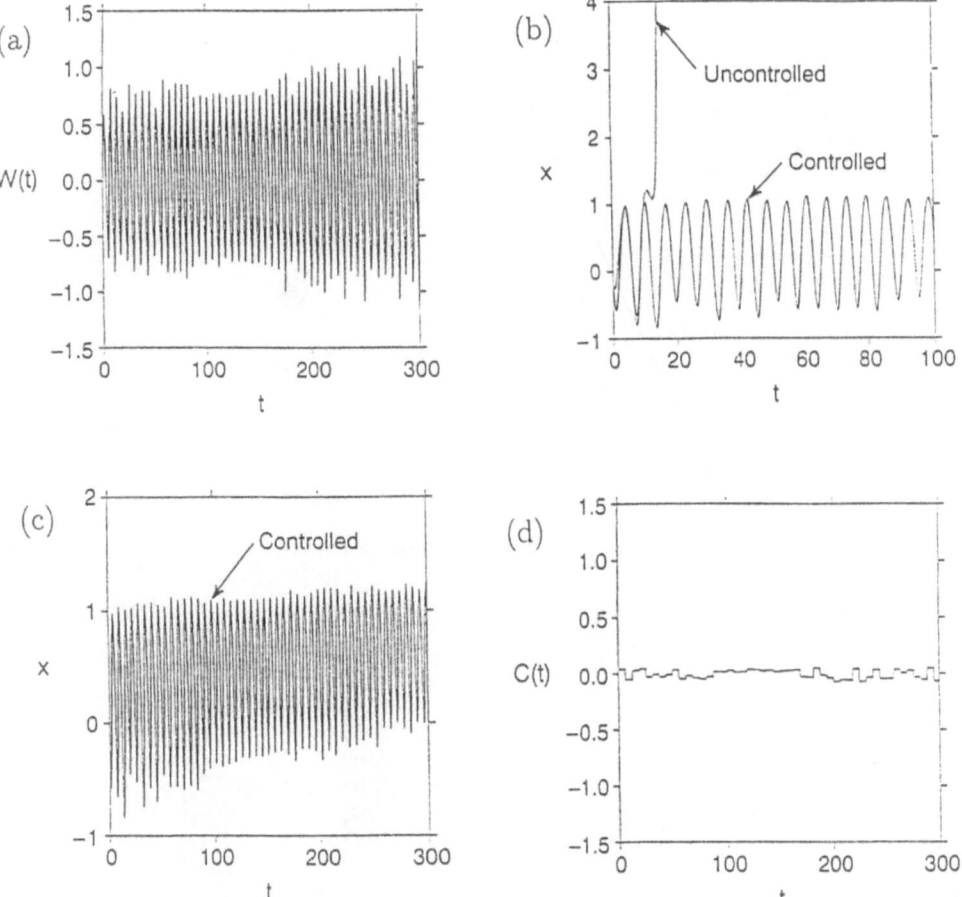

FIG. 3. (a) Segment of the irregular increasing wave with $\epsilon_a = 0.15$, $\epsilon_p = 0.1$ and $f(t)$ a linearly increasing function of t, (b) controlled together with un-controlled orbit, (c) more extended plot of controlled orbit and (d) $C(t)$ versus t. The following parameter values are chosen for numerical computations: $\nu = 0.5$, $\omega = \Omega = 1.0$, $\alpha = 1.0$, and integration step size $= 2\pi/200$.

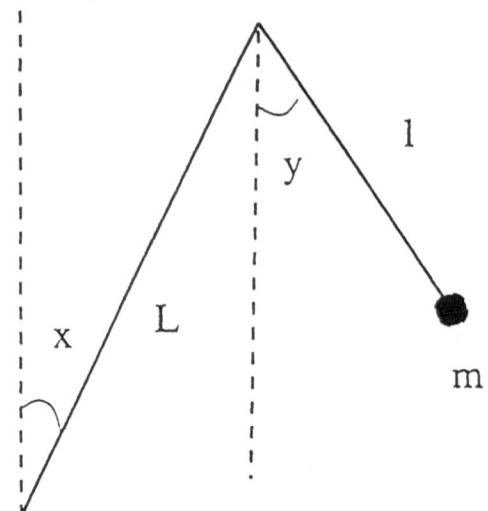

FIG. 4   A double pendulum

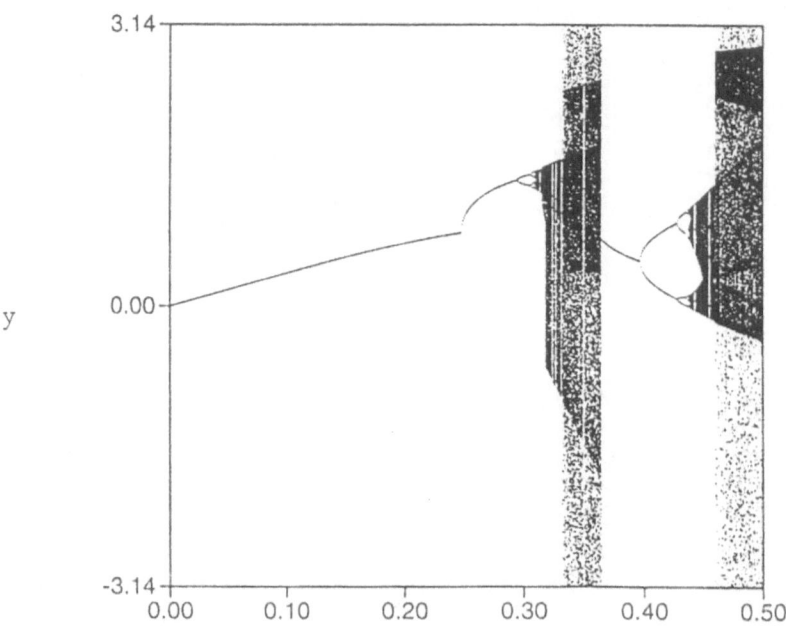

FIG. 5   Bifurcation diagram for Eq. (4) and (5)

117

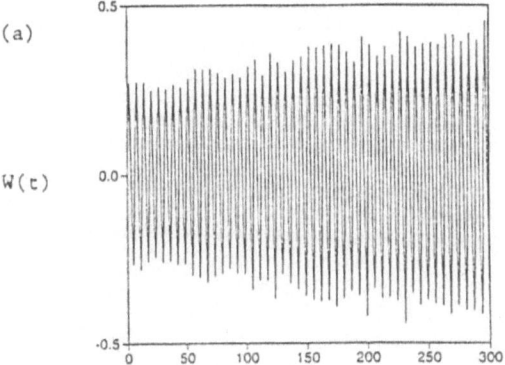

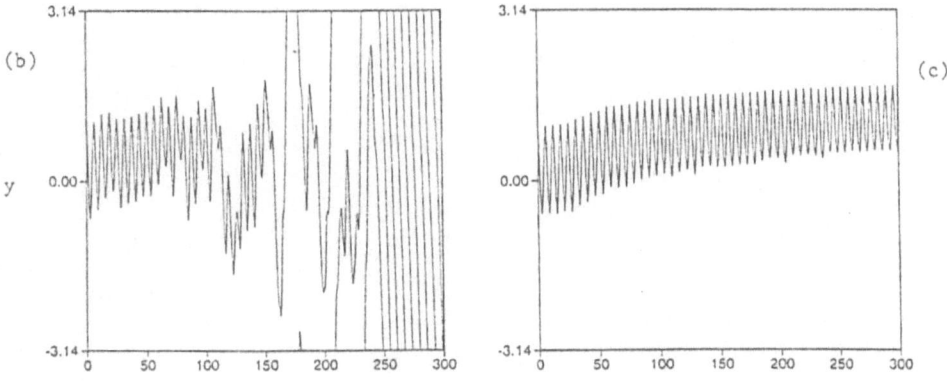

time

FIG. 6   (a) Increasing wave, (b) Uncontrolled time series, and
(c) controlled time series.

MOTION PLANNING STRATEGIES TO IMPROVE THE DYNAMIC
BEHAVIOUR OF CONTROLLED MECHANICAL SYSTEMS

R. FAGLIA
Università degli Studi di Brescia, Mech. Engineering Dept.
Via Branze, 38 - 25123 Brescia ITALY

Abstract

The aim of the work is to give practical suggestions for the "motion planner" of a servosystem in all those cases where the choice of the movement becomes difficult due to the so called "parasitic effects" which prevent the theoretical motion of the moving load as programmed in the machine controller. Together with a brief description of some important motion planning techniques, the paper suggests two original classes of parametric definition of motion which can be useful for the solution of the problem described above.

1. Introduction

In most actual servosystems, from the simplest mono-axis positioning machine to the most complex robot, it is possible to observe that the real movement of the load is quite different from the programmed or theoretical one. This is due to a series of so called "parasitic effects" (Coiffet, 1981) such as the noise of electric circuitry, the limitations of the driver-motor group (current limit, velocity gain limit, bandwidth limit, etc.) the controller sampling of the motion profile, the compliance of the transmission in the motor-load coupling, the friction and the hysteresis of the joints, etc. As an example, an experimental verification of the previous sentence is given in Figure 1.

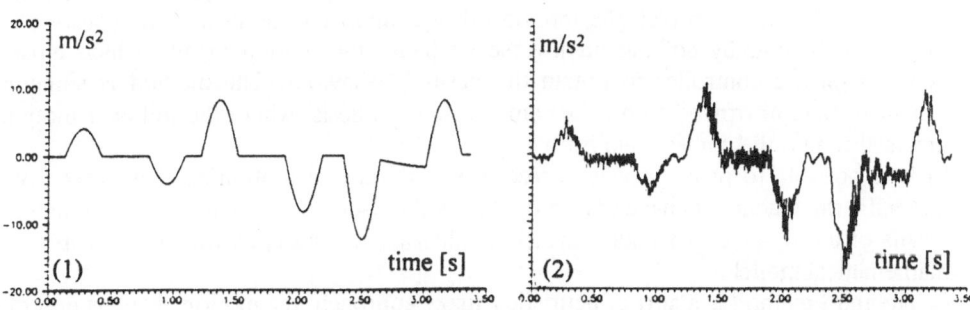

Figure 1. Comparison between the diagrams of theoretical (1) and experimental (2) acceleration vs. time of a SCARA robot gripper.

119

D. H. van Campen (ed.), IUTAM Symposium on Interaction between Dynamics and Control in Advanced Mechanical Systems, 119–126.
© 1997 *Kluwer Academic Publishers.*

The picture represents the dynamic behaviour of a SCARA robot and compares the theoretical acceleration profile obtained by means of a simulation of a "perfect" robot with the actual acceleration measured on the gripper. It is clear that the actual behaviour of the system is quite different from the one imposed. In this case the difference is mainly due to the Harmonic Drive transmission irregularity and to the compliance of the robot joints (see Legnani and Faglia, 1992).

As a further example, Figure 2 refers to the behaviour of a mono-axis system and compares the velocity profile imposed on the driver with the actual movement obtained on the load: it can be seen how the actual behaviour of the system is quite different from the one imposed (this particularly bad situation is due to the driver current limit; note also the asymmetry of the behaviour of the driver in performing a positive and a negative acceleration).

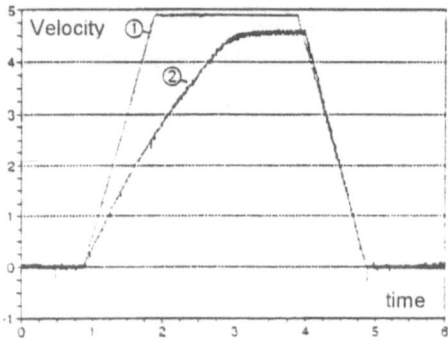

Figure 2. An actuator velocity profile: comparison between theoretical (1) and experimental (2) data.

There are some procedures, based on a correct or "optimised" choice of the motion profile, which reduce the influence of the parasitic effects or, in some situations, overcome them.

a) The first procedure is known as "pre-shaping" technique. If the model of the system is known, it is possible to plan the motion off-line, imposing the desired trajectory of the load, and finding, by solving an inverse problem, the motion profile which must be imposed on the controller to obtain the desired behaviour. The method is something similar to (and borrowed from) the cams profile synthesis, where the follower motion is imposed to find the correct cam profile.

b) It is possible to progressively correct the input law of motion so that, after several attempts, the machine behaves more or less as the user wishes. This can be obtained by means of an iterative procedure directly applied on the system without the help of any mathematical model.

c) The third method is a sort of heuristic / fuzzy approach: the motion shape is described by one or more simple and friendly parameters; when a parameter changes, the user knows a-priori the effect of its variation on the motion profile. In this way, if a qualitative correlation between the motion shape and the parasitic effects is known, it is possible to qualitatively decrease (or delete) the influence of the latter, modifying the motion shape by means of the above described parameters.

This paper briefly shows these techniques (other interesting motion planning strategies can be found in the literature, i.e. in Bayo and Paden (1987), Heimann and Kruger (1995), Potkonjak (1988), Tu and Rastegar (1993), Wei Min Yun *et al.* (1994)) and stresses the goodness of the third method in the very common case when we need to obtain i) a pre-fixed displacement S of the device in a pre-fixed time T and ii) to start and end the movement with null velocity.

2. The pre-shaping technique

For the purpose of showing how this form of trajectory planning strategy operates, we will use a very simple servosystem model, considering the following characteristics:
- the theoretical imposed profile is not modified by the controller;
- the motor is not endowed with inertia;
- the sensors perfectly detect the position and velocity of the motor shaft;
- the transmission does not induce velocity irregularity;
- the driver-motor group has no limits (neither current/torque limit, nor finite bandwidth).

The only parasitic effect is that the linkage between the motor and the loading mass is not rigid, the compliance being modelled by a linear elastic damped element.

A simple scheme of the servosystem studied is given in Figure 3.

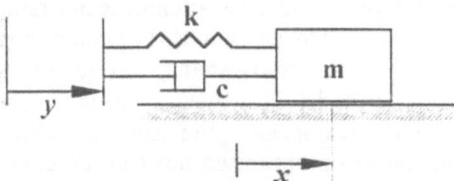

Figure 3: The scheme of a system endowed with elastic linkage.

Let x be the coordinate representing the mass displacement, y the coordinate describing the motor shaft displacement (which in the limit free motor-driver hypothesis coincides with the theoretical input profile), m the load mass, k the stiffness constant and c the damping of the transmission.

The linear relation which links x to y has the following well-known differential form:

$$m\ddot{x} + c\dot{x} + kx = c\dot{y} + ky. \tag{1}$$

The pre-shaping technique imposes a-priori our desired x(t), to solve the differential equation (1) from right to left, obtaining y(t), and so getting the movement of the motor shaft which takes account of the transmission compliance.

The difficulties of this method lie in the model definition and, if the model itself is particularly accurate and complex, in the mathematical procedures for solving the inverse problem. On the contrary, if the model is simple but too rough and does not consider such parasitic effects as driver limits, then it is possible to obtain motions y(t) which cannot be reproduced by the actual system.

122

3. The autotuning technique

Consider again the system depicted in Figure 3. In practical cases (we refer in particular to motion planning in the field of robotics), the elasticity of the transmission is neglected, so that the motion is planned on y. That is, y is imposed by the user who hopes that x is as near as possible to y. The most sophisticated planners, after imposing y, measure (or determine) x and try to modify y to obtain a better x. In comparison to the pre-shaping approach, this method has the advantage of knowing a-priori the characteristics of the motion to be imposed on the motor, so that, for example, it is possible to remain far from the limitations of the driver-motor group.

The author has proposed, in other papers (Faglia and Incerti, 1992; Faglia, 1992), a type of automation of the procedure directly achieved on an actual servosystem (also named "autotuning" approach), here briefly summed up by steps:

1) The user defines the motion shape by parameters.

2) The user imposes a motion profile for a first attempt and starts the machine working.

3) A sensor (an accelerometer, for example) placed on the system, detects the behaviour of the machine.

4) If the performance is good (according to the planner purpose), the right motion profile has been found. If not, the feature of the system to be improved must be defined (for example: the overshooting is too big, so there is the necessity to decrease it).

5) Re-start from 1) with a new motion profile optimising the goal defined at step 3) (e.g. minimising the overshooting). The new attempt is not randomly performed, but the parameters of the new motion are chosen by means of an optimisation algorithm.

This procedure, tested on actual servosystems, has given good results and can be developed without the aid of a model. However, it needs the presence of a sensor which is able to detect the system behaviour, an optimisation algorithm together with a precise and well established function to be optimised must be defined a-priori, and the motion shape must be defined by means of parameters, the variations of which must change the motion shape without modifying the boundary conditions (for example, displacement S in time T with null velocity at the ends).

Figure 4 shows the behaviour of the moving load of a controlled mono-axis system, endowed with an elastic transmission, before and after the application of the optimisation method, in which the goal was to minimise the maximum positive acceleration.

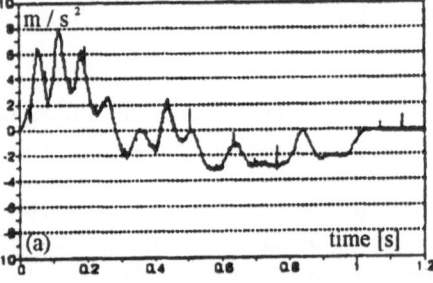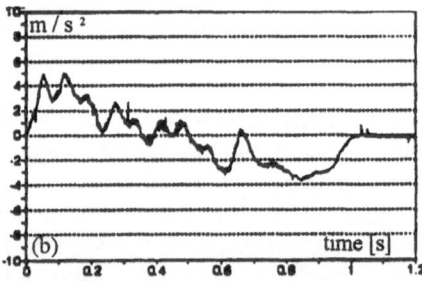

Figure 4. The result of the application of the autotuning method on a mono-axis servosystem at the beginning (a) and at the end (b) of a procedure aiming to minimise the maximum positive acceleration.

4. A qualitative motion planning strategy

With the same steps as the previous method, but instead of using an optimisation algorithm acting on a well defined functional, it is possible to utilise the qualitative knowledge of the planner to improve the system behaviour. By experience, it is possible to build such tables as Table 1, the first two columns of which relate the planner's whishes to the action he needs to perform on the motion profile. Now, if we are able to define a motion profile whose shape is easily controllable by friendly parameters, it is possible to complete Table 1 with other columns which indicate the way to modify the parameters themselves to improve the system behaviour. Our proposal is to use, as parametric motions, the two kinds of profiles hereafter briefly presented.

TABLE 1. Correlation between the planner actions and the features of the motion profiles

PLANNER ACTION	MOTION PROFILE FEATURES	TRIGONOM. PROFILE	POLYN. PROFILE
Decrease overshooting	Smooth profile at the end of motion	$\uparrow p_1$	$\uparrow n$
Decrease the kinetic energy	Low maximum velocity	$p_1 \rightarrow 0$	$\downarrow n$
Decrease inertia actions (Decrease inertia couple on the motor)	Low maximum acceleration	$p_1 \rightarrow 1/8$	$\downarrow n$
Overcome the current limits	Low maximum positive acceleration	$p_1 \rightarrow -1/8$	$\downarrow n$
Decrease vibration at the starting motion	Smooth profile at the beginning	$p_1 \rightarrow -1/4$	$\uparrow n$
Avoid finite bandwidth problems	Null acceleration at the beginning	$p_1 = -1/4$	Always
Avoid finite bandwidth problems	Null acceleration at the end	$p_1 = 1/4$	Always
Avoid resonance effects due to the transmission irregularity	Decrease maximum velocity under the critical value	$p_1 \rightarrow 0$	$\downarrow n$

4.1. THE TRIGONOMETRIC PROFILE

Parameter "p_1" of Table 1, refers to the following parameterized law of motion, formed by a combination of harmonic functions (Schmitt *et al.*, 1985):

$$x(t) = \frac{1}{2}S\left(1 - \cos\pi\frac{t}{T}\right) + \sum_{k=1}^{n}\frac{1}{2}S\,p_k\left[1 - \cos\left(2k\pi\frac{t}{T}\right)\right] \qquad (2)$$

where:
n is the number of the harmonic functions, p_k is the amplitude (or weight) of the k-th harmonic; t is the time variable, S is the total requested displacement, performed in total time T.
Function x(t) satisfies the following constraints for any combination of the weights p_k :

$$x(0) = 0; \qquad x(T) = S; \qquad \dot{x}(0) = \dot{x}(T) = 0. \qquad (3)$$

Consider the case in which only the first weight can be manipulated, and all the other p_k (with k>1) are null. It is easy to show that if p_1=0 too, we obtain the typical harmonic profile, which is very commonly used, for its simplicity and for the continuity of the acceleration in the interval $[0,T]$. Furthermore, for every value of p_1 external to the interval [-1/4, 1/4], a non monotonous x(t) function is obtained (or better, there are some points in which the velocity inverts), and this is clearly no good for the motion planning of an actuator. Finally, for all the values of p_1 included in the interval [-1/4, 1/4], we observe a progressive decrease of the final acceleration and the increase of the initial one. Furthermore, in this interval, the effect of changing the parameter upon the minimum and maximum acceleration and on the maximum velocity is very clear, as shown in Figure 5.

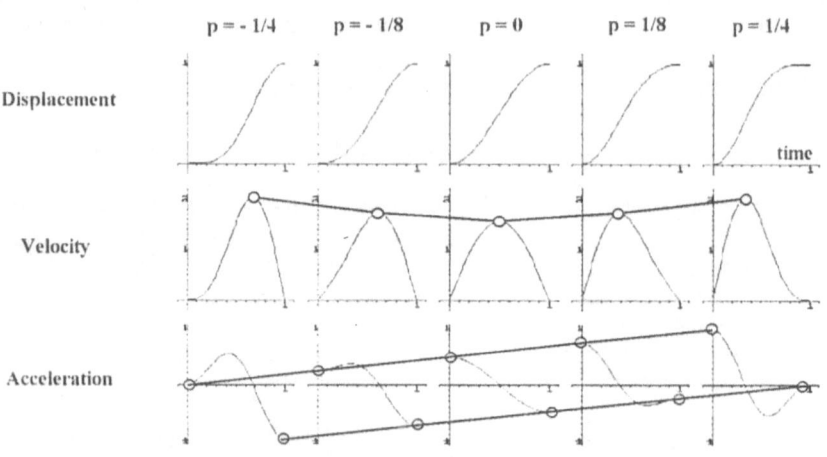

Figure 5. Trigonometric motion shapes for different values of the parameter.

4.2. THE POLYNOMIAL PROFILE

The polynomial motion profiles are defined by the following expression:

$$x(t) = C_0 + C_1 t + C_2 t^2 + ... + C_n t^n \qquad (4)$$

The coefficients C_0, C_1, C_2, ... ,C_n can be determined by imposing the boundary conditions. Among the wide variety of motions contained in expression (4), it is possible to select classes which can be recognised by a couple of numbers. We define as a polynomial function of class (i,j) the expression containing "i" consecutive terms starting from the j-th term. For example, class (3,3) is represented by the following expression:

$$x(t) = C_3 t^3 + C_4 t^4 + C_5 t^5 \qquad (5)$$

In this example, coefficients C_0, C_1 and C_2 are null. This means that, automatically, expression (5) gives:

$$x(0) = 0; \quad \dot{x}(0) = 0; \quad \ddot{x}(0) = 0 \qquad (6)$$

In general, it is possible to say that value (j-1) represents the degree of null derivatives at the initial point of the motion. The other "i" coefficients can be found by imposing other conditions, solving the consequent system of equations.

Let's now consider the class having i = j, with i = j≥3: this class gives null derivatives at the ends of the motion up to degree (j-1): class (3,3) gives null initial and final acceleration, class (4,4) gives null initial and final jerk, class (5,5) gives null initial and final quirk, etc.

Consider the group (k,k) with k≥3: the "k" parameter can be used as a parameter to change the motion profile acting on the initial and final value of the acceleration and its derivatives, without varying the basic boundary conditions of null velocity at the ends and displacement S at time T.

From Figure 6 it is possible to see that if k increases, the acceleration graph shows a growing smoothing at the end points and increasing maximum velocity and acceleration.

By manipulating k, in this class of motion, we can easily control: a) the "smoothness" of the profile, choosing thereby the one which does not introduce frequencies which cannot be reached by the driver-motor group; b) the peak acceleration value, which could not be fulfilled due to the current limit; c) the maximum velocity value.

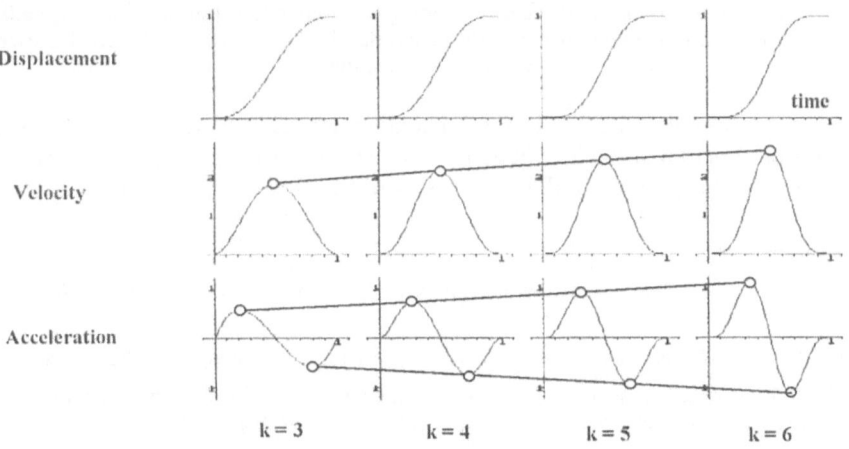

Figure 6. Polynomial motion shapes for different values of the parameter.

5. Conclusions

After a brief discussion of procedures for servosystem trajectory planning two kinds of parametric definition of the motion have been analysed: 1) the profile based on a linear combination of harmonic functions which allows easy control of the amplitude of the acceleration discontinuity, the acceleration peaks value and the maximum velocity; 2) a particular class of polynomial functions which permits one parameter to control the "smoothness" of the motion, the acceleration peaks and the maximum velocity.

It seems to the author that of all the numerous parametric laws of motion that can be defined, these two kinds of parametric functions best control the most important parameters for a good trajectory planning, as given in Table 1.

Figure 7 shows, as an example, how evident is the changing of the overshooting in a servosystem with elastic transmission, when the motor is commanded by different

126

trigonometric motion profiles (parameter p_1 is -0.25 in the left picture and 0.125 in the right one).

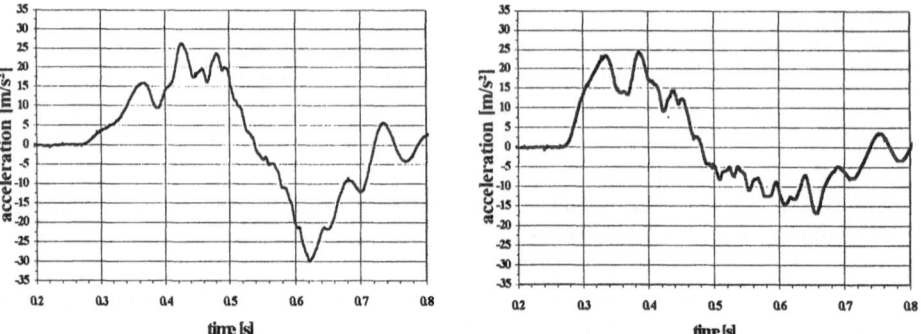

Figure 7. Acceleration vs. time of the load in a servosystem with elastic transmission, when different trigonometric profiles have been imposed on the motor: the change of the motion profile has decreased the final overshooting.

As a future development, thanks to the qualitative approach to the problem, the author thinks it is possible to link the described planning strategy to fuzzy procedures so that the choice of profile is supported both by the ability of the planner and by a logic structure.

6. References

Bayo E., Paden B. (1987) On trajectory generation for flexible robots. *Journal of Robotic Systems*, Vol.4 No.2.

Coiffet P. (1981) *Les Robots. Modelisation et commande.* Hermes Publishing, Neuilly, France.

Faglia R. (1992) Trajectory shaping for flexible manipulators: an optimisation algorithm and an application to a SCARA robot. *Proc. 3rd Int.ARK*, Ferrara, Italy, 7-9 Sept. 1992.

Faglia R., Incerti G. (1992) Optimisation of the dynamic behaviour of a controlled system having an elastic transmission. *Proc. XI National Congress, AIMETA 92*, Trento, Italy, 28 Sept. 2 Oct., 1992.

Heimann B., Kruger M. (1995) Optimal path-planning of robot-manipulators in the presence of obstacles. *Proc. IX IFToMM*, Milan, 29 Aug. - 2 Sept. 1995, Vol. 4.

Legnani G., Faglia R. (1992) Harmonic drive transmissions: the effects of their elasticity, clearance and irregularity an the dynamic behaviour of an actual SCARA robot. *Robotica*, Vol.10, 369-375.

Potkonjak V. (1988) Contribution to the dynamics and control of robots having elastic transmissions. *Robotica*, Vol.6.

Schmitt, Soni, Srinivasan, Naganathan, (1985) Optimal Motion Programming of Robot Manipulators. *ASME Trans. Journal of Mechanisms Transmissions and Automation in design*, Vol.107/293, June, 1985.

Tu Q., Rastegar J. (1993) Robot manipulator trajectory synthesis for minimal high frequency component of the actuating torques. *Proc. 3rd AMR, Vol.II*, 7-10 Nov. 1993, Cincinnati, Ohio, USA.

Wei Min Yun, Yu Geng Xi (1994) Genetic Optimum Motion Planning in Joint Space with Dynamic and Control Constraints. *Proc. ICARCV'94*, Vol.1, Singapore 9-11 Nov., 1994.

MULTIVARIABLE IDENTIFICATION OF ACTIVE MAGNETIC BEARING SYSTEMS

*C. Gähler**, M. Mohler**, R. Herzog****

*Internat. Center for Magnetic Bearings, ETH-Technopark,
Technoparkstr. 1, 8005 Zürich, Switzerland.
e-mail: gaehler@ifr.mavt.ethz.ch

**Mecos Traxler AG, Gutstr. 36, 8400 Winterthur, Switzerland.
e-mail: rmbmecos@dial.eunet.ch

Abstract

Magnetic bearing systems are unstable MIMO[1] plants. In addition, there may be poorly damped resonances of the levitated rotor. Advanced control is a crucial issue. System identification is therefore an important prerequisite for fast and reliable commissioning. Reported algorithms have difficulties to estimate the real unstable poles in magnetic bearing systems. Thus, a novel identification algorithm has been developed. Experimental results are included.

1.   Introduction

ACTIVE MAGNETIC BEARINGS

Active Magnetic Bearings (AMB's) [1] allow contactless levitation. They do not require lubrication, allow high circumferential velocities at high loads, do not have friction nor wear, and no maintainance is needed. In the domain of rotating machinery, they are used in an increasing number of applications, including high vacuum pumps, pipeline compressors/expanders, tool machines, and others.

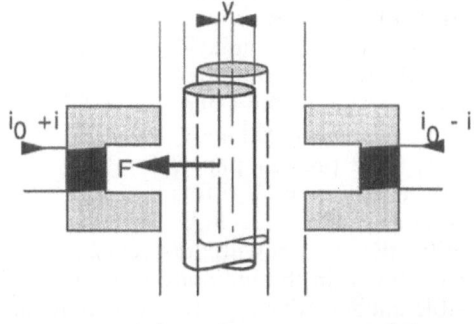

Figure 1.1: Basic principle of AMB

Figure 1.1 shows the principle of a magnetic bearing. The bearing force depends on current and displacement in a non-linear way. This relation can be linearised to

$$F = k_i \cdot i + k_s \cdot y \tag{1.1}$$

The positive coefficient k_s reflects a *negative stiffness* of the bearing.

The multivariate plant considered in this paper consists of the (flexible) non-rotating rotor suspended in two magnetic bearings. We will consider only one plane.

---

[1] MIMO = Multiple Input, Multiple Output

D. H. van Campen (ed.), IUTAM Symposium on Interaction between Dynamics and Control in Advanced Mechanical Systems, 127–134.
© 1997 *Kluwer Academic Publishers.*

128

1.3. MOTIVATION AND GOAL

AMB systems are unstable without control. A position controller is needed to stabilise the system and to provide a sufficient stiffness w.r.t. disturbance forces, often over a large frequency range. This results then in a large controller - and system - bandwidth. There are often eigenfrequencies of the rotor within the system bandwidth. It is therefore important for reliable and fast controller design to have an accurate plant model for a large frequency range (sometimes 1 .. 3000 Hz).

A dynamic model can also be obtained from theory (FE modelling of the rotor), modal analysis of the rotor, and static force measurements of the bearings. However, many effects cannot be assessed with this approach. Dynamic identification can provide a more accurate plant model, and reduce the time required for modelling. It is thus an important prerequisite for fast and reliable commissioning of AMB systems.

1.4. WHY YET ANOTHER IDENTIFICATION ALGORITHM?

Why is it necessary to reconsider the identification problem anew especially for AMB systems?

The answer can be found looking at a typical pole/zero configuration. The poles and zeros can be grouped in two sets: A set close to the imaginary axis (flexible modes) and a set on the positive and negative real axis (rigid body modes). The present paper shows that especially for multivariable AMB

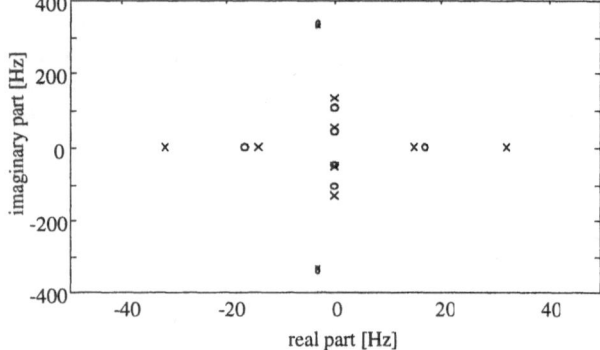

Figure 1.2: Typical pole/zero configuration

identification, the set with the 4 real poles is very hard to estimate. None of the methods that we found in the literature was capable of estimating both rigid body modes and flexible modes robustly. In particular, an application of standard modal analysis methods is not successful. We therefore developed a novel multivariable identification method capable of coping with the above-mentionned plant characteristics.

2. Plant Description and Problem Formulation

2.1. MODAL AMB SYSTEM DESCRIPTION: THE THEORETICAL MODEL

The plant (AMB System in one plane) has the set control currents, $i = \begin{bmatrix} i_A & i_B \end{bmatrix}^T$, at bearings A and B as inputs and the measured displacements, $y = \begin{bmatrix} y_A & y_B \end{bmatrix}^T$, as outputs. The plant's FRF (= frequency response function) in the Laplace domain is defined by

$$y(s) = H(s) \cdot i(s), \text{ or } \begin{bmatrix} y_A(s) \\ y_B(s) \end{bmatrix} = \begin{bmatrix} h_{11}(s) & h_{12}(s) \\ h_{21}(s) & h_{22}(s) \end{bmatrix} \cdot \begin{bmatrix} i_A(s) \\ i_B(s) \end{bmatrix} \tag{2.1}$$

The (flexible) rotor is a mass-stiffness-damper system (MKD system). The magnetic bearings act as a negative stiffness onto the rotor. This makes the rigid-body modes unstable and moves the corresponding poles from the origin onto the positive and negative real axis in the Laplace plane (see figure 1.2).

Still, our plant can be described as a MKD system with modal damping. We can therefore use the following modal description:

$$H(s) = \Phi\left(Ms^2 + Ds + K\right)^{-1}\Psi^T \tag{2.2}$$

Let us consider m modes of the rotor, where the first two modes are the rigid-body modes and the higher modes are the flexible modes; i.e., mode 3 is the first flexible mode. Then,

$$\Phi = \begin{bmatrix} | & | & | \\ \varphi_1 & \cdots & \varphi_m \\ | & | & | \end{bmatrix}; \quad \Psi = \begin{bmatrix} | & | & | \\ \psi_1 & \cdots & \psi_m \\ | & | & | \end{bmatrix} \tag{2.3}$$

M, K and D are the diagonal stiffness, damping, and mass matrices of the plant. Using mass-normalised co-ordinates, we can set $M = I$ (identity matrix).

2.2. PLANT PARAMETERISATIONS

Various plant parameterisations will be used in the proposed algorithm. They are presented in this section, and relations between them are discussed.

Since M, K and D are diagonal, equation (2.3) can be written as *sum of second-order systems*:

$$H(s) = \sum_{r=1}^{2} \frac{\varphi_r \cdot \psi_r^T}{s^2 + d_r s - p_r^2} + \sum_{r=3}^{m} \frac{\varphi_r \cdot \psi_r^T}{s^2 + 2\delta_r \omega_{0r} s + \omega_{0r}^2} \tag{2.4}$$

with

$$\begin{aligned} K &= diag\left(\begin{bmatrix} -p_1^2 & -p_2^2 & \omega_{03}^2 & \cdots & \omega_{0m}^2 \end{bmatrix}\right) \\ D &= 2 \cdot diag\left(\begin{bmatrix} d_1 & d_2 & \delta_3 \omega_{03} & \cdots & \delta_m \omega_{0m} \end{bmatrix}\right) \end{aligned} \tag{2.5}$$

d_1 and d_2 are small, such that the real-valued rigid-body poles are almost symmetrical to the imaginary axis (cf. fig. 1.2).

Equation (2.4) can be re-written in the form

$$H(s) = \sum_{r=1}^{2} \frac{R_r}{s^2 + d_r s - p_r^2} + \sum_{r=3}^{m} \frac{R_r}{s^2 + 2\delta_r \omega_{0r} s + \omega_{0r}^2} \tag{2.6}$$

Thereby, the dyadic products

$$R_r = \varphi_r \cdot \psi_r^T \tag{2.7}$$

are called *residual matrices*.

Further, (2.6) can be transformed to

$$H(s) = \frac{N(s)}{d(s)} = \frac{\begin{bmatrix} n_{11}(s) & n_{12}(s) \\ n_{21}(s) & n_{22}(s) \end{bmatrix}}{d(s)} \tag{2.8}$$

with a *common denominator polynomial* $d(s)$ of order $2m$, and nominator polynomials of order $2m$-2.

Last but not least, (2.2) and (2.4) can also be formulated as

$$H(s) = C \cdot (sI - A)^{-1} \cdot B \qquad (2.9)$$

with the *state space description*

$$\left. \begin{array}{l} s \cdot x(s) = A \cdot x(s) + B \cdot i(s) \\ y(s) = C \cdot x(s) \end{array} \right\} \qquad (2.10)$$

It is straight-forward to construct A, B and C from (2.2) or (2.4):

$$A = \begin{bmatrix} 0 & I \\ -K & -D \end{bmatrix}; \quad B = \begin{bmatrix} 0 \\ \Psi^T \end{bmatrix}; \quad C = \begin{bmatrix} \Phi & 0 \end{bmatrix} \qquad (2.11)$$

3.   The Identification Problem

3.1.   THE GOAL: A STATE-SPACE MODEL OF ORDER $2m$

Multivariate controller design can best be done in state space. A state space model of the plant is therefore needed.

If this model is allowed to have a higher order than necessary, the identification algorithm will produce a model with a poorly observable and/or controllable part. If this part is unstable, the whole model becomes unstabilisable, although the true plant can be stabilised. To reduce the identified unstable model to the desired degree is then not at all a trivial problem. It is therefore of paramount importance to control the model order during the identification process.

3.2.   THE RANK 1 CONDITION

We use the terms "model order" and "system order" in the sense of *the minimal order of a model's state space representation*.

With SISO systems, the system order is equal to to the degree of $d(s)$ in (2.8). Unfortunately, this is not true with MIMO systems [6].

From (2.7) it follows for the residual matrices R_r in representation (2.6) of our plant that the rank condition

$$rank(R_r) = 1 \qquad \forall r \qquad (3.1)$$

holds. Conversely, a term

$$H_r(s) = \frac{R_r}{s^2 + 2\delta_r \omega_{0r} s + \omega_{0r}^2}$$

is a second-order system if and only if (3.1) is satisfied, but it is a 4th order system if the rank of the residue matrix is 2.

It is obvious that the rank condition (3.1) has a counterpart in representation (2.8). A horrible non-linear relation between the coefficients of all polynomials results.

3.3.   THE IDENTIFICATION CRITERION

Let $\tilde{H}(s)$ be the FRF measured at a number of discrete frequencies $s = j \cdot \omega_k$. FRFs computed by evaluation of some parametric model (*e.g.*, 2.2 or 2.9) will be denoted by $\hat{H}(s)$.

The identification problem can be stated as follows:

Find a state-space model of order $2m$ such that (2.9) evaluated at the measurement frequencies $s = j \cdot \omega_k$ fits the measured FRF data $\tilde{H}(j\omega_k)$ in an optimal way.

The identification performance criterion that is to be minimised can be defined in different ways. We have chosen the following relative criterion:

$$J = \sqrt{\sum_k \left(\sum_i \sum_j \left(e_{ij}(j\omega_k) \right)^2 \right)}, \tag{3.2}$$

where

$$e_{ij}(j\omega_k) = w_{ij}(j\omega_k) \cdot \frac{\hat{h}_{ij}(j\omega_k) - \tilde{h}_{ij}(j\omega_k)}{\tilde{h}_{ij}(j\omega_k)} \tag{3.3}$$

$W(s)$ is a weighting function that can be used for tuning the algorithm.

3.4   SOME STANDARD IDENTIFICATION ALGORITHMS

3.4.1.   Non-linear parameter optimisation in (2.4)
Model (2.4) satisfies the rank 1 condition and therefore can be directly converted to a state-space model of correct order.
However, the optimisation problem is strongly non-linear in the parameters. It converges very slowly and only if the starting values for the parameters are good.

3.4.2.   Modal analysis
In modal analysis, the system poles are estimated with good accuracy based on the resonances of the FRF data. Because the poles are accurate, rank 1 residual matrices (dyadic products of eigenvectors) can be estimated after that. This procedure breaks up the large non-linear problem into several smaller and more tractable ones.
However, this approach cannot cope with the real rigid-body poles of an AMB system.

3.4.3.   Identification of the polynomial parameters in (2.8)
Although this is a non-linear problem as well, it can be solved using an iterative linear Least Squares procedure proposed by Sanathanan and Koerner [3]. This algorithm has already been successfully applied to AMB systems [2].
The problem is that the rank 1 condition (or, respectively, its counterpart) appears as a crude non-linear constraint which cannot be included into the problem formulation in a tractable way. Therefore, the identified system will in general have order $4m$ if the denominator polynomial was assigned the correct order $2m$. The FRF data can then be matched well with completely wrong rigid-body poles. Because the system is unstable, model reduction is not easily possible.

4.   The Novel MIMO Identification Algorithm

The new algorithm should

- yield a minimal state-space model
- cope with particular pole/zero configuration of AMB systems (real unstable poles)
- convert the large non-linear problem into a sequence of tractable problems.

To achieve this, again the poles are estimated in a first step, but from the *determinant* of the FRF. In a second step, the residual matrices (2.6) are estimated. Since the poles are estimated well, the resulting residuals are close to rank 1 matrices. Their approximation by rank 1 matrices (which is necessary to reduce the system order to $2m$) does therefore not deteriorate the model accuracy by much.

4.1.   IDENTIFICATION OF THE POLES FROM THE DETERMINANT OF THE FRF

We make use of the following fact:

$$\det\left(C(sI - sA)^{-1} B\right) = \det(H(s)) = \det\left(\frac{N(s)}{d(s)}\right) = \frac{\det(N(s))}{(d(s))^q} = \frac{n_{det}(s)}{d(s)} \qquad (4.1)$$

In words: Let us consider a MIMO system (C, A, B) with q inputs and q outputs (in our case: $q=2$). The determinant of the FRF of this system is a SISO polynomial fraction. The denominator polynomial is $d(s)$ in power one *(not power q!)*. This can be proved from system theoretic considerations.

The system poles can therefore be computed using the following algorithm:

- Compute the determinant $\det(\tilde{H}(s))$ from the modified measured FRF data.
- Estimate the coefficients of $d(s)$ from $\det(\tilde{H}(s))$. Apply the iterative linear Least Squares estimation scheme [2,3].
- Compute the poles as the roots of $d(s)$.

Consider the following AMB system with rigid rotor to illustrate the idea:

$$H(s) = \frac{\begin{bmatrix} 1 & 0 \\ 0 & 0 \end{bmatrix}}{s^2 - p^2} + \frac{\begin{bmatrix} 0 & 0 \\ 0 & 1 \end{bmatrix}}{s^2 - p^2} = \frac{\begin{bmatrix} s^2 - p^2 & 0 \\ 0 & s^2 - p^2 \end{bmatrix}}{\left(s^2 - p^2\right)^2} = \frac{\begin{bmatrix} 1 & 0 \\ 0 & 1 \end{bmatrix}}{s^2 - p^2}$$

It is not visible from either of the individual transfer functions that $H(s)$ is a 4th order system. However, it is clearly visible from the determinant:

$$\det(H(s)) = \frac{1}{\left(s^2 - p^2\right)^2}$$

4.2.   IDENTIFICATION OF RANK 1 MATRICES USING SVD

In a first step, the rank 1 constraint is neglected. Estimation of the elements of the rank 2 residual matrices $R_r^{(2)}$ from (2.6) is a linear Least Squares problem. Every $R_r^{(2)}$ can then be approximated by a rank 1 matrix using Singular-value decomposition [4, 5]:

- Decompose $R_r^{(2)}$ as $R_r^{(2)} = U \cdot \Sigma \cdot V^T$ [5]. U and V are orthonormal matrices with columns u_i, v_i. Σ is diagonal and contains the singular values σ_i with decending magnitudes.
- The best possible rank 1 approximation of $R_r^{(2)}$ is then

$$R_r^{(2)} \approx R_r = u_1 \cdot \sigma_1 \cdot v_1^T = \varphi_r \cdot \psi_r^T \qquad (4.2)$$

- Construct a state-space model of order $2m$ using (2.3), (2.5) and (2.11).

4.3. INSERTION OF PROPORTIONAL FEEDBACK FOR IDENTIFICATION

Sections 4.1-4.2 already describe an algorithm superior to those of section 3. However, the following procedure further improves its performance w.r.t. the real poles:

- Modify the measured FRF data $\tilde{H}(s)$ with a proportional feedback matrix K. K should be chosen such that it over-compensates the negative magnetic bearing stiffness.
- Identify a model $\hat{H}_K(s)$ from the modified data $\tilde{H}_K(s)$
- Remove the proportional feedback from the model $\hat{H}_K(s)$ to get the model $\hat{H}(s)$ of the true plant.

There are two advantages of this procedure:

- The real poles are moved to near to the imaginary axis. Their effect on the FRF is then more clearly visible, and therefore they can be identified more precisely.
- Every controller will contain a proportional feedback part to overcompensate the negative AMB stiffnes. With the described procedure, the identified model becomes more accurate in the vicinity of potential closed-loop poles and therefore more relevant for predicting the closed loop performance of the controlled plant.

Note that inserting or removing proportional feedback does not affect the system order.

4.4. SUMMARY OF THE ALGORITHM

To summarise, the proposed algorithm consists of the following steps:

1) Modify the measured multivariable FRF matrix with proportional feedback.
2) Compute the determinant $\det\!\left(\tilde{H}_K(s)\right)$ from the modified measured FRF data. Compute the system poles using the iterative lin. LS algorithm [2, 3].
3) Estimate the rank 2 residue matrices associated with each mode (lin. LS).
4) Approximate these by rank 1 matrices using SVD Decomposition.
5) Construct a minimal state space plant description.
6) Remove the proportional feedback to get back to a model of the original system.

The proposed identification algorithm solves the strongly non-linear identification problem with a sequence of iterative linear Least Squares, ordinary Linear Least Squares, and Singular-value decomposition steps.

5. Results with Experimental Data

The algorithm has been tested with FRF measurement data from an AMB system. The model includes the two rigid-body modes and four flexible modes. Figure 5.1 shows that magnitude and phase of all four transfer functions agree very well with the measured data. This holds for both the rigid body modes (low frequency range) and for the flexible modes, for both the resonances and the antiresonances of the individual transfer functions, and for both diagonal and off-diagonal transfer functions.

134

6. References

[1] Schweitzer, G., Bleuler, H., Traxler, A.: Active Magnetic Bearings. Verlag der Fachvereine, Zurich, Switzerland, 1994
[2] Gähler, C., Herzog, R.:Identification of Magnetic Bearing Systems. Mathematical Modelling of Systems, Vol. 1, No.1, 1995
[3] Sanathanan, C. K., Koerner, J.: Transfer Function Synthesis as a Ratio of Two Complex Polynomials. IEEE Trans. Automatic Control, Vol. 8, 1963
[4] Balmès, Etienne: New Results on the Identification of Normal Modes from Experimental Complex Modes. 12th Int. Modal Analysis Conf., Honolulu, 1994
[5] Golub, C., van Loan, G.: Matrix Computations. John Hopkins Univ. press, 1990
[6] Kailath, T.: Linear Systems. Prentice Hall, 1980

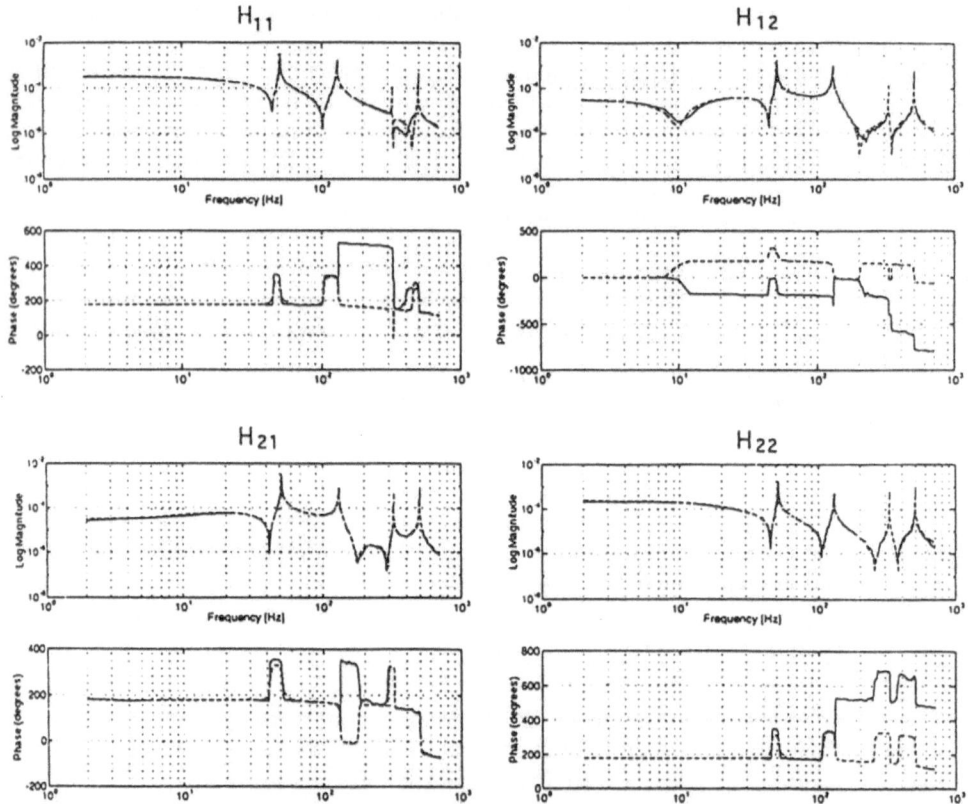

Figure 5.1: Identification results. Solid: Measured FRF; dashed: Identified model. Note that the larger differences in the phase are multiples of 360 degrees.

VIBRATION CONTROL OF A NONLINEAR BEAM SYSTEM

M.F. HEERTJES, M.J.G. VAN DE MOLENGRAFT AND J.J. KOK
Department of Mechanical Engineering
Eindhoven University of Technology
P.O. Box 513 Eindhoven, The Netherlands

AND

R.H.B. FEY AND E.L.B. VAN DE VORST
TNO-Building and Construction Research
P.O. Box 49 Delft, The Netherlands

1. Introduction

In many engineering applications high amplitude vibrations are undesirable because they may cause wear and damage and may lead to high levels of noise. In nonlinear dynamic systems the steady-state response often exhibits certain frequency ranges where two or more solutions of the system equations coexist. Our objective is to reduce the amplitude of the response by controlling the system into its natural solution with lower amplitude.

Research in this field has been done by Ott *et al.* (1990), who succeeded in converting a chaotic attractor to a time-periodic motion. An overview of various methods for controlling chaos is given by Chen and Dong (1993). All of this research focuses on single Degree Of Freedom (DOF) systems. This paper, however, deals with multi DOF systems with local nonlinearities (e.g. backlash) regardless the absence of chaos (Van de Vorst *et al.*, 1996).

Vibration reduction will be accomplished for a beam with a one-sided spring attached to the middle. A Sliding Computed Torque Controller (SCTC) forces the response of the beam from the stable $\frac{1}{2}$ subharmonic solution of high amplitude towards the coexisting unstable harmonic solution of low amplitude. Firstly, a 3-DOF model of the beam system will be described. Secondly, the controller will be described and simulation results will be discussed. Thirdly, a description will be given of the experimental implementation, followed by a discussion of the experimental results. This paper will end with conclusions and suggestions for further research.

D. H. van Campen (ed.), IUTAM Symposium on Interaction between Dynamics and Control in Advanced Mechanical Systems, 135–142.
© 1997 *Kluwer Academic Publishers.*

2. Beam system

The beam system consists of a steel beam supported at both ends by two leaf springs which exhibit great stiffness in transversal and low stiffness in longitudinal direction, see figure 1a. The one-sided spring is constructed by a clamped steel beam which adds stiffness to the main beam for positive displacements $q_m(t)$ of the middle of the main beam. Harmonic excitation is realized by means of a rotating mass unbalance attached to the middle of the main beam which is driven by a tacho-controlled motor via a flexible shaft. The shaft has practically no stiffness in transversal direction which leads to a free motion of the main beam in this direction. The shaft, however, has great stiffness in rotating direction to avoid phase lag between the desired and the realized excitation force. The control force is applied to the main beam by means of an actuator which is placed halfway the excitator and one of the supporting leaf springs, see figure 1a.

A model of the beam system is obtained with the finite element package DIANA (1996). The linear components, i.e. the main beam, the leaf springs and lumped masses are modelled with 86 finite elements. The model is reduced with the use of a component mode synthesis method. With this method, a multi DOF model is obtained where the number of DOFs depends on the number of kept free-interface eigenmodes of the linear system and the number of flexibility modes corresponding to the number of interface DOFs. Damping is added by means of modal damping which is chosen to be equal for every mode ($\zeta = 0.05$). In our case the model has three DOFs, i.e. one free-interface eigenmode and two flexibility modes corresponding with the interface DOFs for the actuator force and the forces originating from the one-sided spring and the excitator. The model equation is given by:

$$M\ddot{q}(t) + B\dot{q}(t) + Kq(t) + H_1 F_{nl}(q,t) = H_2 F_e(t) + H_3 u(t), \qquad (1)$$

with mass matrix M, damping matrix B and stiffness matrix K of size 3×3. The one-sided spring is modelled as a local nonlinearity:

$$F_{nl}(q,t) = \varepsilon[q_m(t)]k_{nl}q(t), \;\; \varepsilon[q_m(t)] = \begin{cases} 1, & \text{if } q_m(t) \geq 0, \\ 0, & \text{if } q_m(t) < 0, \end{cases} \qquad (2)$$

where k_{nl} represents the stiffness of the one-sided spring ($k_{nl} = 1.65\ 10^5 \frac{N}{m}$) as a weighted sum of a clamped and a supported beam. $F_e(t)$ represents the excitation force, $u(t)$ represents the control force and H_1, H_2 and H_3 are the so-called 1×3 transition columns. The column with DOFs $q(t)$ is given by: $q^T(t) = [q_a(t)\ q_m(t)\ \xi(t)]$, with $q_a(t)$ the controlled DOF, $q_m(t)$ the DOF at the middle of the main beam and $\xi(t)$ the modal DOF corresponding with the first kept free-interface eigenmode of the reduced linear system.

The usefulness of the model can be assessed by comparing simulation with experiment, figure 1b. Simulation results were acquired using a path

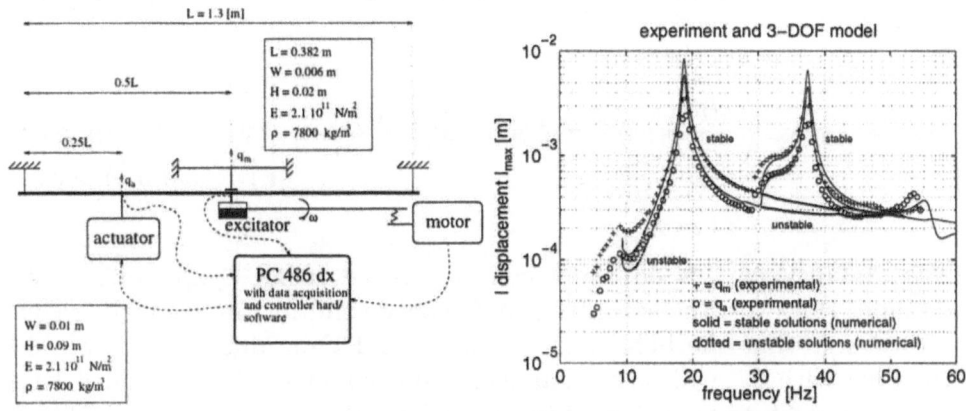

(a) Representation of the beam system (b) Behaviour of the beam system

Figure 1. Beam system

following technique in combination with a finite difference method to solve the two-point boundary value problems. The good agreement between simulation and experiment justifies our choice to control the system on the basis of a 3-DOF model. The difference in amplitude between the stable $\frac{1}{2}$ subharmonic and the unstable harmonic solutions in the frequency range between 30 and 50 Hz clearly illustrates our objective to reduce vibration amplitudes by stabilizing the harmonic solution using control, especially near 37 Hz.

3. Controller

A SCTC (Slotine and Li, 1991) is used to achieve the desired vibration reduction. The SCTC is based on controlling one DOF, in this case $q_a(t)$, by means of feedback of the difference between a realized and a desired trajectory. The tracking error is defined as:

$$e_a(t) = q_a(t) - q_{a_d}(t).\qquad(3)$$

The desired displacement $q_{a_d}(t)$ represents the unstable harmonic solution which is calculated beforehand with DIANA. The SCTC exhibits two features. Firstly, a sliding function $s(t)$ is defined by:

$$s(t) = \dot{e}_a(t) + \lambda e_a(t), \ \lambda \in \Re^+.\qquad(4)$$

When the system enters sliding motion, $s(t)$ becomes zero and a stable differential equation remains with the solution $e_a(t) = 0$ for $t \to \infty$. Secondly, an input has to be defined that puts the system in sliding motion. For this purpose the sliding function is subjected to the following differential equation:

$$\dot{s}(t) = -\eta \, \text{sat}\left(\frac{s(t)}{\sigma}\right), \quad \eta \wedge \sigma \in \Re^+, \quad \text{sat}\left(\frac{s(t)}{\sigma}\right) = \begin{cases} -1, & \text{if } s(t) < -\sigma, \\ \frac{s(t)}{\sigma}, & \text{if } |s(t)| \leq \sigma, \\ 1, & \text{if } s(t) > \sigma. \end{cases}$$

$$(5)$$

The stability of this equation can be proved using Lyapunov's second method. However, the stability proof is only valid for the controlled DOF. We assume that the behaviour of the other DOFs, due to their physical coupling with q_a, will be similar to the behaviour of the controlled DOF. With the definitions of the sliding function (4) and (5), the error (3) and the 3-DOF model (1), the control force $u(t)$ can be written as:

$$\begin{aligned} u(t) &= \frac{1}{M_{11}^{-1}}\left[\ddot{q}_{a_d}(t) - \lambda\dot{e}_a(t) - \eta \, \text{sat}\left(\frac{s(t)}{\sigma}\right) \right. \\ &\quad + \left. \left(M^{-1}(Kq(t) + B\dot{q}(t) + H_1 F_{nl}(q,t) - H_2 F_e(t))\right)_1 \right]. \end{aligned}$$

$$(6)$$

For implementation of this controller, knowledge of the state at any time is required. Because only part of the state is measured ($q_a(t)$, $\ddot{q}_a(t), q_m(t)$ and $\ddot{q}_m(t)$), a simple state reconstruction is carried out. The velocities $\dot{q}_a(t)$ and $\dot{q}_m(t)$ are approximated using a second order implicit derivation scheme:

$$\dot{q}_i(t_n) \approx \frac{q_i(t_n) - q_i(t_{n-1})}{t_n - t_{n-1}} + \frac{\ddot{q}_i(t_n)(t_n - t_{n-1})}{2}, \quad i \in \{a, m\}, \ n \in \aleph^+. \quad (7)$$

In this way the two equations of motion for $q_a(t)$ and $q_m(t)$ can be written as algebraic equations which can be solved to give values for $\xi(t)$ and $\dot{\xi}(t)$.

Simulations, as displayed in figure 2a, were carried out with the controller as outlined above ($\lambda = 100 \, \frac{1}{s}$, $\eta = 100 \, \frac{m}{s^2}$, $\sigma = 0.5 \, \frac{s}{m}$). The controller becomes active at $t = 0.2$ s. At this point the system has nearly reached the steady-state $\frac{1}{2}$ subharmonic response starting from rest. It can be seen that by controlling DOF $q_a(t)$ to its desired trajectory, the other DOFs $q_m(t)$ and $\xi(t)$ also resemble their desired (harmonic) trajectories. Thus, in this case it is sufficient to control one DOF in order to force the entire multi DOF system in its harmonic solution. It can also be seen that due to the fact that the harmonic solution is indeed a natural solution of the beam

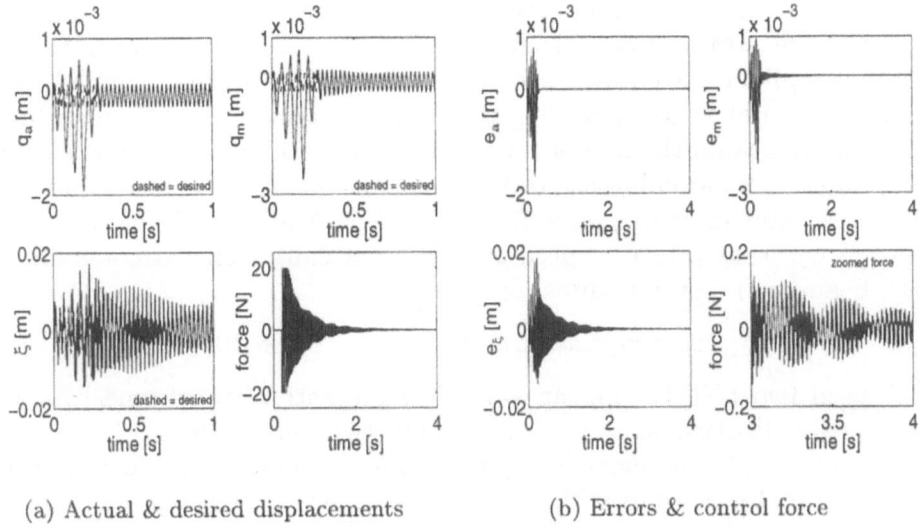

(a) Actual & desired displacements   (b) Errors & control force

Figure 2.   SCTC numerical results at 37 Hz

system, the resulting input force decreases when the error $e(t)$ vanishes. It should be noted that the input force is bounded at $|20|$ N. In figure 2b the same results are shown, but now with respect to the tracking errors.

4. Implementation

The implementation of the outlined theory in an experimental setup has been divided into a couple of stages which mainly relate to: the data acquisition and the control software environment TCE (Tools for Control Experiments (Banens, 1995)), the actuator, the excitator and the sensors. The TCE software environment uses a personal computer (type 486 dx) with a 12 bit board installed for taking in the measurements and sending out the control force. The control force is calculated between two sample moments in C^{++}-written code. The sample frequency is restricted by the required calculation time, and is set to 500 Hz. The actuator is modelled as:

$$F_a(t) = -m_{act}\ddot{q}_a(t) - b_{act}\dot{q}_a(t) - k_{act}q_a(t) - Bli(t), \qquad (8)$$

where $F_a(t)$ represents the force on the beam, m_{act} the core mass, b_{act} the damping of the actuator caused by the electromotive force due to the displacement of the core in the actuator, k_{act} the stiffness of the actuator, B the magnetic induction of the actuator and l the wire length from the coil in the magnetic field ($m_{act} = 0.15$ Kg, $b_{act} = 95$ $\frac{Ns}{m}$, $k_{act} = 1200$ $\frac{N}{m}$ and

$Bl = 20.3\ Tm$). The current $i(t)$ must be chosen in such a way that $F_a(t)$ corresponds to the desired control action. This current is generated by a linear amplifier driven by a set point voltage. The force computed by the controller will be corrected using the inverse dynamics of the actuator model in order to obtain the proper input voltage for the amplifier. The excitator generates the harmonic excitation force by means of a mass unbalance. When a point mass m_e, on a distance r_e from the axis of rotation, rotates with a constant rotation speed $\omega = 2\pi f_e$, it causes an harmonic force on the beam in transversal direction:

$$F_{exc}(t) = m_e r_e \omega^2 \cos(\omega t), \quad m_e r_e = 0.986\ 10^{-3}\ kgm. \qquad (9)$$

We used two LVDTs (Linear Variable Differential Transformer) for measuring the displacements $q_a(t)$ and $q_m(t)$, two piezoelectric accelerometers for measuring the accelerations $\ddot{q}_a(t)$ and $\ddot{q}_m(t)$, and two piezoelectric force transducers for measuring the realized forces acting on the beam due to the actuator and the excitator.

5. Experiments

Experiments have been carried out, resulting in the behaviour as shown in figure 3. It shows that we are able to force the system into the harmonic

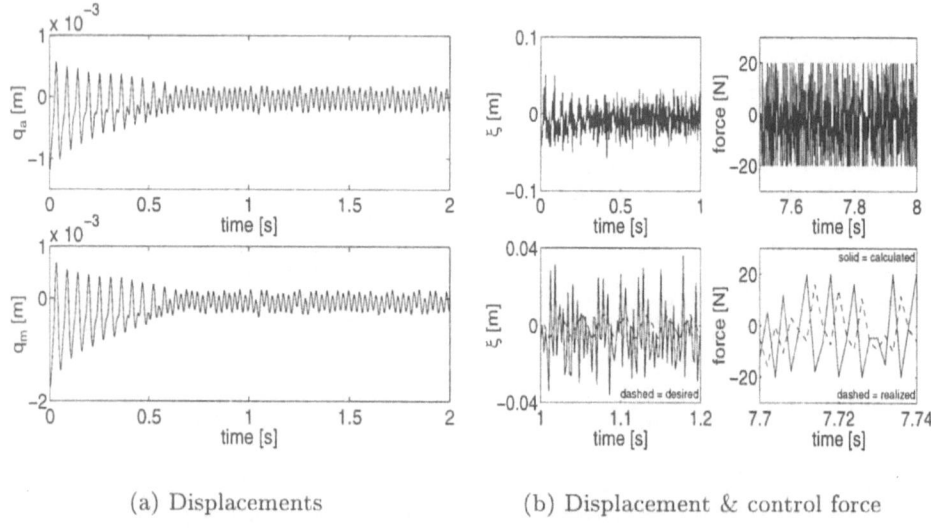

(a) Displacements   (b) Displacement & control force

Figure 3. SCTC experimental results at 37 Hz

motion, giving an enormous reduction of the displacement amplitudes and

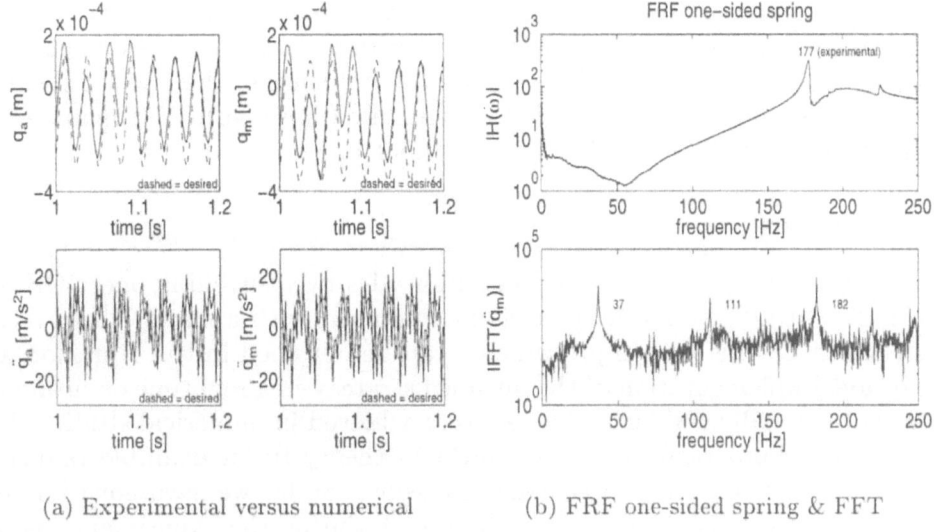

(a) Experimental versus numerical   (b) FRF one-sided spring & FFT

Figure 4.   Experimental & numerical results at 37 Hz

doubling the base frequency of the response. By controlling only DOF $q_a(t)$ the DOFs $q_m(t)$ and $\xi(t)$ were also forced in the harmonic motion. Figure 3b shows that the control force does not tend to zero during the experiment. This is caused by a number of error sources ranging from unmodelled dynamics of the beam, poorly modelled actuator dynamics to a delay in the control. The control delay sometimes even leads to a desired and realized control force that are each other opposites. Therefore, we conclude that the sample frequency is not high enough. It can also be seen in figure 3b that next to the control delay a poor resemblance occurs which is caused by the limitations of the actuator model. It should be noted that the actuator is subjected to saturation, which results in a bounded actuator force.

In figure 4a the measured and desired displacements based on a 3-DOF model are displayed. The small differences for $q_a(t)$ and $q_m(t)$ can be ascribed to an offset occurring in the displacement signals, as can be seen in the difference between the measured and the calculated accelerations, see figure 4a. Accelerometers exhibit practically no offset. The differences between the measured and the desired displacement of the modal DOF $\xi(t)$ (figure 3b) are due to the limitations of the state reconstruction, which is sensitive to variations in the displacement and acceleration signals, mainly because of the calculation of the velocities. Another difference between experiment and simulation becomes clear from the acceleration signals. Higher order harmonics can be displayed more clearly in acceleration signals then in displacement signals. Thus, differences between experiment and simulation due to model simplification have greater impact on acceleration signals, see figure 4a. A closer look at the acceleration signal $\ddot{q}_m(t)$, by means of a Fast Fourier Transform (FFT), shows the influence of the one-sided spring besides the influence of unmodelled dynamics (figure 4b). We conclude that mass effects of the one-sided spring cannot be neglected.

6. Conclusions

It is possible to reduce vibration amplitudes by controlling one DOF and using a 3-DOF model of the system combined with a simple state reconstruction. In this way improvements can be obtained in the vibration amplitude. Implementation of the outlined strategy in a real time environment leads to globally similar results as were achieved in numerical studies. This favours the practical use of the outlined theory in for example reduction of vibrations occurring in gear transmissions. So far we have not been able to show a decrease in the controlling force during the experiments as was obtained in simulations. The reason is attributed to a number of causes ranging from hardware limitations, poorly state reconstruction to model errors. Further research will focus on the use of other controllers than the SCTC and on the optimal placement of actuators and sensors in interaction with multi DOF dynamical systems with local nonlinearities.

References

Banens, J. (1995). Documentation on TCE modules. Technical Report, Department of Mechanical Engineering, Eindhoven University of Technology, WFW-Report 94.050.

Chen, G. and Dong, X. (1993). From chaos to order-Perspectives and methodologies in controlling chaotic nonlinear dynamical systems. *International Journal of Bifurcation and Chaos*, **Vol. 3**, pp. 1363–1409.

DIANA 6.0 (1996). TNO Building and Construction Research, Delft, The Netherlands.

Ott, E., Grebogi, C. and Yorke, J.A. (1990). Controlling chaos. *Physical Review Letters*, **Vol. 64**, pp. 1196–1199.

Slotine, J. and Li, W. (1991). *Applied nonlinear control*. Prentice-Hall, New York.

Van de Vorst, E. L. B., Van Campen, D.H., Fey, R.H.B., De Kraker, A. and Kok, J.J. (1995). Vibration control of periodically excited nonlinear dynamic multi-DOF systems, *Journal of Vibration and Control*, **Vol. 1**, pp. 75–92.

HOMOCLINIC BIFURCATION AND LOCALISED TORSIONAL BUCKLING OF ELASTIC RODS

G.H.M. VAN DER HEIJDEN*, A.R. CHAMPNEYS[†], J.M.T. THOMPSON*

*Centre for Nonlinear Dynamics and Its Applications,
 University College London, Gower Street, London WC1E 6BT, UK.

[†] Department of Engineering Mathematics,
 University of Bristol, University Walk, Bristol BS8 1TR, UK.

In this paper we study localised buckling in elastic rods with non-symmetric cross-section, which has applications in undersea cable and drillstring buckling as well as in DNA engineering [10]. Strong evidence will be given of the existence of an infinity of localised buckling modes. In addition, we show how these solutions can be computed numerically, and how some structure can be found in their multitude.

1. The Mathematical Model

Consider the classical problem of determining the equilibrium configuration of a long, thin, elastic rod subject to end moment and tension. Suppose, as an idealisation for long rods, that the rod is inifinitely long. We shall further assume that the rod is naturally straight, inextensible, has a uniform cross-section, and does not suffer shear deformation. For more details on modelling we refer to [3,9].

Let $e_1(s), e_2(s), e_3(s)$ be a rod-centred co-ordinate system, where s denotes the axial arclength. e_3 is chosen to be everywhere tangent to the axis of the rod and the three axes align with fixed Cartesian (x, y, z) axes when the rod is unstrained. The strain of the deformed rod is measured by the vector

$$\Omega(s) = \kappa_1(s)e_1(s) + \kappa_2(s)e_2(s) + \tau(s)e_3(s),$$

where τ is the twist about the body axis of the rod and $\kappa_{1,2}$ are the principal curvatures. Let $\boldsymbol{F} = F_1e_1 + F_2e_2 + F_3e_3$ and $\boldsymbol{G} = G_1e_1 + G_2e_2 + G_3e_3$ be the respective resultant force and couple exerted by the material in the direction of increasing s. Here $F_{1,2}$ are shear forces, F_3 is tension, $G_{1,2}$ are bending moments about the axes $e_{1,2}$ and G_3 is the twisting moment about the centre line. We assume linear constitutive relations of the form

$$\kappa_1 = G_1/A, \quad \kappa_2 = G_2/B, \quad \tau = G_3/C,$$

where A and B are the principal bending stiffnesses (about the e_1 and e_2 axes respectively) and C is the torsional stiffness of the rod.

We can now write down, via force and moment balance, equilibrium equations with the components of $\boldsymbol{F}$ and $\boldsymbol{G}$ as the dependent variables. As boundary conditions we assume that a moment M and tension T is applied in the direction e_3

D. H. van Campen (ed.), IUTAM Symposium on Interaction between Dynamics and Control in Advanced Mechanical Systems, 143–150.
© 1997 *Kluwer Academic Publishers.*

at $s = \pm\infty$. In the non-dimensional variables

$$t = (M/B)s, \quad x_1 = F_1/T, \quad x_2 = F_2/T, \quad x_3 = (F_3 - T)/T$$
$$x_4 = G_1/M, \quad x_5 = G_2/M, \quad x_6 = (G_3 - M)/M,$$

the equilibrium equations then read [3]

$$
\begin{aligned}
\dot{x}_1 &= (1+\nu)x_2(1+x_6) - (1+x_3)x_5 \\
\dot{x}_2 &= (1+\rho)(1+x_3)x_4 - (1+\nu)x_1(1+x_6) \\
\dot{x}_3 &= x_1x_5 - (1+\rho)x_2x_4 \\
\dot{x}_4 &= \nu x_5(1+x_6) + x_2/m^2 \\
\dot{x}_5 &= (\rho-\nu)x_4(1+x_6) - x_1/m^2 \\
\dot{x}_6 &= -\rho x_4 x_5
\end{aligned}
\tag{1}
$$

subject to the boundary conditions

$$(x_1, x_2, x_3, x_4, x_5, x_6) \to (0,0,0,0,0,0) \quad \text{as} \quad t \to \pm\infty. \tag{2}$$

Here, '$\cdot$' denotes differentiation with respect to the scaled arclength t and

$$m = M/\sqrt{BT}, \qquad \rho = (B/A) - 1, \qquad \nu = (B/C) - 1$$

are dimensionless parameters. Note that the equations depend on the single composite load parameter m only, showing the equivalence between increasing moment and reducing tension. ρ is an anisotropy parameter which is zero for rods whose principal bending stiffnesses are equal, as would be the case for either a circular or square cross-section. ν measures the ratio of bending to torsional stiffness: in a circular rod ν is equal to Poisson's ratio. We shall keep $\nu = \frac{1}{3}$ throughout the rest of this paper, as this is a typical value for, e.g., rubber (see [10]).

Finally, to interpret solutions of (1) in terms of the fixed co-ordinate system (x, y, z) we have to solve (1) along with the Frenet-Seret equations of differential geometry (cf. [9], art. 253)

$$\dot{e}_i = \tilde{\Omega} \times e_i, \qquad i = 1, 2, 3, \tag{3}$$

as well as the defining equation for the centre line

$$\dot{r} = e_3, \tag{4}$$

where $\tilde{\Omega}$ is the non-dimensionalised strain vector and $r = (r_1, r_2, r_3)$ denotes the dimensionless vector position of the center line in the fixed co-ordinates.

If one solves an initial-value problem, that is if, in place of (2), one specifies the initial conditions for x exactly, then notice that (3) and (4) are slaved to (1). They merely post-process the solutions of (1) to give the actual configuration of the rod. Treated in this way, (1) is a conservative dynamical system (Kirchhoff's celebrated kinetic analogue – see [9], art. 260) with Hamiltonian

$$H = 2x_3 + m^2[(1+\rho)x_4^2 + x_5^2 + (1+\nu)(x_6+1)^2], \tag{5}$$

which is related to the strain energy function of elasticity theory. It is known that (1) has two integrals of the motion corresponding to the conservation along the axial length of the rod of the magnitude of force and the component of torque about the loading axis. In our notation,

$$x_1^2 + x_2^2 + (x_3 + 1)^2 = \text{const.}, \tag{6}$$

$$x_1 x_4 + x_2 x_5 + (1 + x_3)(1 + x_6) = \text{const.} \tag{7}$$

We should remark that (5), (6) and (7) do not represent three independent isolating integrals of (1) and therefore the Hamiltonian system is not necessarily completely integrable. However, in the case of a circular rod, i.e. when $\rho = 0$, we have the obvious additional integral

$$x_6 = \text{const.}$$

which *is* independent of (6) and (7) and hence the system is completely integrable. Physically, this integral corresponds to the conserved torque about the body axis e_3 of the rod.

Another property that will be important later is the invariance of (1) under the following two involutions:

$$R_1 : \quad (x_1, x_2, x_3, x_4, x_5, x_6) \rightarrow (-x_1, x_2, x_3, -x_4, x_5, x_6), \quad t \rightarrow -t,$$

$$R_2 : \quad (x_1, x_2, x_3, x_4, x_5, x_6) \rightarrow (x_1, -x_2, x_3, x_4, -x_5, x_6), \quad t \rightarrow -t.$$

In addition, the system is invariant under the transformation

$$Z : \quad (x_1, x_2, x_3, x_4, x_5, x_6) \rightarrow (-x_1, -x_2, x_3, -x_4, -x_5, x_6).$$

The consequence of this $\mathbb{Z}_2$-symmetry is that all non-trivial solutions come in pairs of Z-images of each other. Physically, this corresponds to the rod being invariant under a rotation of 180° about its centre line.

2. Analytical Considerations

It can be verified that the trivial solution $x = 0$ has always two zero eigenvalues. The corresponding eigenvectors are aligned along the x_3 and x_6 axes, due to the existence of other equilibria with arbitrary non-zero constant values of x_3 and x_6. The remaining four eigenvalues have eigenspaces which are orthogonal to these axes. Their behaviour as parameters are varied is depicted in the ρ-m plane of Fig. 1.

Note that the branch indicated by m_{c_2} corresponds to Hamiltonian-Hopf bifurcations. It is this branch which describes the classical buckling of the rod. Specifically, for the circular symmetric rod ($\rho = 0$) the critical load condition $m = m_{c_2} = 2$ is in agreement with the Timoshenko buckling condition for an infinitely long rod. Buckling of the rod into a helix at this critical load has been known for a long time; the alternative of a localised buckling mode, which corresponds to a homoclinic orbit of the dynamical system, has been the subject of more recent studies [4,10].

146

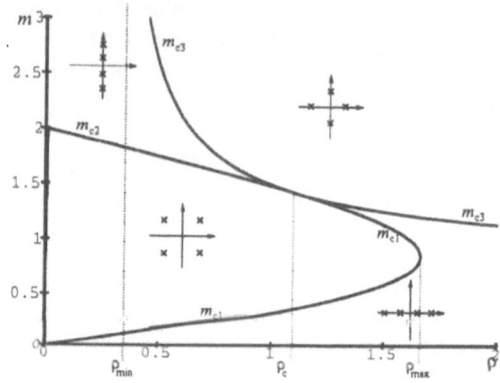

Figure 1. Loci of bifurcation points in the ρ-m parameter plane for $\nu = \frac{1}{3}$.

The trivial equilibrium configuration is physically stable (under dead loading) for $m < m_{c_2}$, and unstable for $m > m_{c_2}$. Note the distinction between stability of physical configurations (an infinite-dimensional problem) and the stability of states of the finite-dimensional dynamical system (1).

Normal form analysis [8] shows that at a Hamiltonian-Hopf bifurcation in a reversible system, given a condition on the sign of a certain coefficient, generically a pair of symmetric homoclinic orbits bifurcate. Given the two involutions R_1 and R_2 above, and the symmetry Z, this could in principle lead to the emergence of eight homoclinic orbits at a Hamiltonian-Hopf point in our system. Numerical results to be presented below strongly suggest that actually four homoclinic orbits appear, two for each of R_1 and R_2, which are mapped into each other under Z.

The present setting of a homoclinic orbit to a fixed point with complex eigenvalues allows us to apply the analysis of Devaney [5] to assert, generically, the existence of a Smale horseshoe, provided the homoclinic orbit forms the transverse intersection of the stable and unstable manifolds of the fixed point. In addition, we can infer the existence of infinitely many secondary homoclinic orbits [1].

3. A Multitude of Homoclinic Orbits

In order to numerically compute approximations to localised buckling responses for $m < m_{c_2}$, we will use a shooting method developed and implemented for the similar problem of a strut on a nonlinear foundation [2]. The method consists in truncating to a finite interval and replacing the boundary value problem by a combined inital/boundary value problem exploiting the reversibility of the system. For the initial conditions we take

$$\boldsymbol{x}(0) = \varepsilon(\boldsymbol{v}_1 \cos \delta + \boldsymbol{v}_2 \sin \delta), \qquad (8)$$

where $\boldsymbol{v}_1 \pm i\boldsymbol{v}_2$ are eigenvectors corresponding to the eigenvalues $\lambda \pm i\omega$, ε is small, and $0 \leq \delta < 2\pi$.

Because our system is reversible an appropriate right-hand boundary condition is to place the solution in the symmetric section of either of the reversing transformations R_1 or R_2, i.e.,

$$x_1(T) = x_4(T) = 0, \qquad \text{or} \qquad x_2(T) = x_5(T) = 0, \tag{9}$$

where $2T \gg 1$ is the length of our truncated rod.

To compute load-deflection bifurcation diagrams for localising solutions it is necessary to also measure the end displacement D and end rotation R from the straight-rod position. Solving (3) and (4) with initial conditions $e_1(0) = (1, 0, 0)$, $e_2(0) = (0, 1, 0)$, $e_3(0) = (0, 0, 1)$ and $r(0) = (0, 0, 0)$, these can be obtained as

$$D = T - r_3(T), \quad \cos R = <e_1(T), (1, 0, 0)>, \quad \sin R = <e_1(T), (0, 1, 0)> .$$

(Note that R is only defined modulo 2π.)

Fixing $\rho = 0.5$, $m = 1.7$ (the eigenvalues corresponding to the unstable manifold of the origin then are $\lambda \pm i\omega$, with $\lambda = 0.167015$, $\omega = 0.675632$), our shooting algorithm can now be applied to compute a sample of homoclinic orbits. Varying δ and T in order to satisfy either of the right-hand boundary conditions in (9), we find four primary homoclinic orbits, denoted $P_1, \ldots, P_4$. See Fig. 2 for their data. Note that for each involution we have a pair of orbits with δ-values differing by π, reflecting the $\mathbb{Z}_2$-symmetry of the problem. Hence, the multiplicity of homoclinic orbits found is consistent with the normal form theory mentioned in Section 2.

For the same parameter values we also find whole families of related so-called multi-modal homoclinic orbits. Table 1 lists the first 10 members of a family of

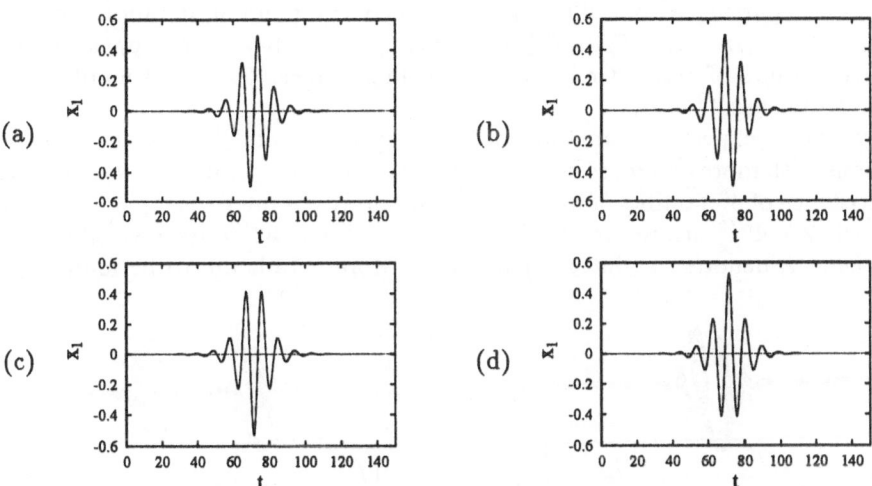

Figure 2. x_1-t diagrams for the four primary buckling modes: (a) P_1 ($\delta = 6.066428$, $T = 71.498592$), (b) P_2 ($\delta = 2.924836$, $T = 71.498592$), (c) P_3 ($\delta = 1.331789$, $T = 71.467436$), (d) P_4 ($\delta = 4.473381$, $T = 71.467436$). P_1 and P_2 are reversible under R_1; P_3 and P_4 are reversible under R_2. In (8) we have used $v_1 = (0.696351, 0.0562830, 0, 0.381032, -0.0412567, 0)$, $v_2 = (0, 0.542187, 0, 0.0857199, 0.252439, 0)$ and $\varepsilon = 10^{-5}$. ($\nu = \frac{1}{3}$, $\rho = 0.5$, $m = 1.7$.)

n	δ_n	T_n	$T_n - T_{n-1}$
1	6.2382	99.1416	
2	5.9794	101.0853	1.9437
3	6.1042	103.5940	2.5087
4	6.0486	105.8371	2.2431
5	6.0745	108.2001	2.3630

n	δ_n	T_n	$T_n - T_{n-1}$
6	6.0627	110.5076	2.3075
7	6.0681	112.8406	2.3329
8	6.0656	115.1618	2.3212
9	6.0668	117.4884	2.3266
10	6.0663	119.8126	2.3242

Table 1. Sequence of R_1-reversible bi-modal homoclinic orbits with labels (P_1, n, P_1) for $\nu = \frac{1}{3}$, $\rho = 0.5$, $m = 1.7$.

R_1-reversible bi-modals with increasing mid-point T. They resemble two copies of the primary homoclinic orbit P_1 separated by a finite number of small-amplitude oscillations, and we shall accordingly label them as (P_1, n, P_1), where n is a measure of the number of small oscillations.

Notice that the δ-values converge to that of the primary orbit P_1 given in Fig. 2, while the difference between successive T-values tends to $\frac{\pi}{2\omega} = 2.324928$. This behaviour (also reported for the strut) can be understood by imagining each successive homoclinic orbit in the sequence making an extra turn around the origin immediately prior to hitting the symmetric section.

A similar sequence of R_1-reversible bi-modals with the initial hump of the form of the primary orbit P_3 is labelled (P_3, n, P_4) (notice that $R_1 : P_3 \to P_4$). In addition, we have the R_2-reversible families with P_1 and P_3 as initial humps: (P_1, n, P_2) and (P_3, n, P_3). Finally, to each of the solutions in the above series there is a Z-partner with δ differing by π. The corresponding families have labels (P_2, n, P_2), (P_4, n, P_3), (P_2, n, P_1) and (P_4, n, P_4). Thus there appear to be 8 distinct families of *reversible* bi-modal buckling modes. The total number of bi-modal families is $4^2 = 16$.

We would expect this method of counting multi-modal orbits to extend to solutions with more and more humps. Thus for any positive integer n we conjecture the existence of 4^n families of n-modals, each family parametrised by $n-1$ integers, of which $2 \times 4^{n/2}$ are reversible if n is even and $4 \times 4^{[n/2]}$ are reversible if n is odd (here $[i]$ denotes the integer part of i). More details are to be found in our

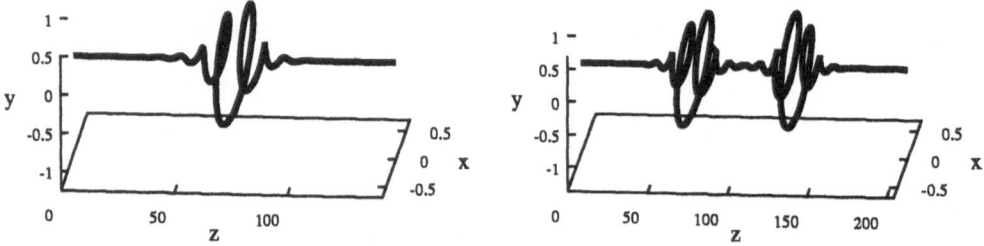

Figure 3. Spatial configuration of two localised buckling modes for $\nu = \frac{1}{3}$, $\rho = 0.5$, $m = 1.7$: the primary $P1$ (left), the bi-modal $(P_1, 3, P_2)$ (right).

forthcoming paper [7].

Fig. 3 shows a couple of rod configurations in physical space obtained by integrating the centre line equation (4).

4. Continuation of Homoclinic Orbits

Bifurcation diagrams, produced using the continuation code AUTO [6], are given in Fig. 4, showing load-deflection characteristics for the primary solutions of Fig. 2 as well as the first eight bi-modals of Table 1. Notice that the curves for the primary orbits terminate in the Hamiltonian-Hopf point, whereas the bi-modal curves show limit points to the right.

In the D-m diagram the curves for bi-modal solutions approximately fall on top of each other: individual bi-modals only differ in the number of small oscillations made around the fixed point before reaching the symmetric section, and these have only little effect on D. By similar reasoning it should not be surprising that D-values for the bi-modals are roughly double those for the primaries.

In the limit points solutions with different labels coalesce. Specifically, solutions with odd humps coalesce with solutions with even humps (when viewed in x_1-t plots), and vice versa. Moreover, the precise pairing of solutions depends on whether n is even or odd. More precisely, we were able to deduce the following coalescence rules for bi-modal homoclinic orbits:

For the R_1-reversible bi-modals the following pairs of solutions share branches and coalesce at limit points upon continuation in m:

$$(P_1, 2k+1, P_1) \longleftrightarrow (P_3, 2k+1, P_4), \qquad (P_1, 2k+2, P_1) \longleftrightarrow (P_4, 2k+2, P_3),$$
$$(P_2, 2k+1, P_2) \longleftrightarrow (P_4, 2k+1, P_3), \qquad (P_2, 2k+2, P_2) \longleftrightarrow (P_3, 2k+2, P_4),$$

Similarly, for R_2-reversible bi-modals the following pairs of solutions coalesce:

$$(P_1, 2k+1, P_2) \longleftrightarrow (P_3, 2k+1, P_3), \qquad (P_1, 2k+2, P_2) \longleftrightarrow (P_4, 2k+2, P_4),$$
$$(P_2, 2k+1, P_1) \longleftrightarrow (P_4, 2k+1, P_4), \qquad (P_2, 2k+2, P_1) \longleftrightarrow (P_3, 2k+2, P_3).$$

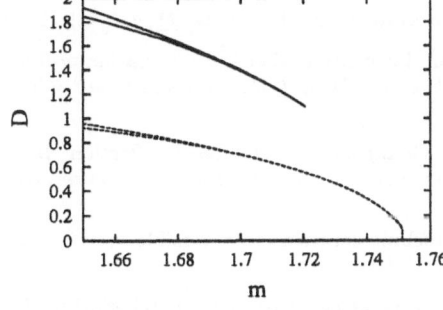 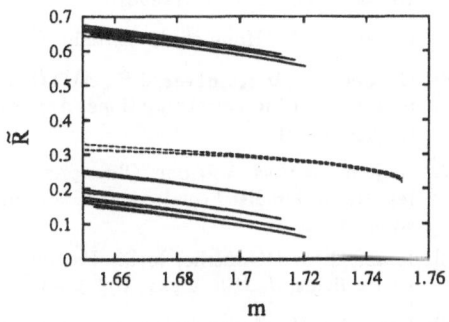

Figure 4. End shortening D and (relative) end rotation $\bar{R} := R/2\pi$ versus load m for the first eight bi-modals (P_1, n, P_1) of Table 1. Curves for the primary orbits are included in dashed lines (P_1 and P_2 yield the same curves, as do P_3 and P_4). Bi-modals with even (odd) n have $\bar{R}$-values which are higher (lower) than those of the primaries. ($\nu = \frac{1}{3}$, $\rho = 0.5$, implying $m_{c_2} = 1.751187$.)

150

Similar rules were found to apply for higher-order multi-modals. For more results (also on using ρ as a bifurcation parameter) the reader is referred to [7].

5. Discussion

We have numerically shown the existence of a multitude of multi-modal buckling modes in rods with non-circular cross-section. Indeed, our results provide a posteriori evidence of the transverse intersection of the stable and unstable manifolds along the primary homoclinic orbits. It follows that Devaney's theory can be applied to give the existence of infinitely many localised buckling modes.

The results described here were for fixed cross-section ($\rho = 0.5$, corresponding to an aspect ratio of 1.22:1 in rods with rectangular cross-section). We should remark that for large ρ ('tapes') buckling no longer occurs along the line m_{c_2} but along m_{c_3} in Fig. 1. It is then no longer described by a Hamiltonian-Hopf but instead by a Hamiltonian pitchfork bifurcation. See [7] for more on this case.

To assess the practical importance of the multitude of multi-modal buckling solutions in our model, it is essential to know their stability properties. Unfortunately, our methods do not give information on this. It seems likely that most, if not all, multi-modals are unstable but, depending on the boundary conditions applied and the loading sequence taken, perhaps some of them can be stabilised.

6. References

[1] Belyakov, L.A., Šil'nikov, L.P., Homoclinic curves and complex solitary waves, *Selecta Mathematica Sovietica* 9, 219-228 (1990).

[2] Champneys, A.R., Spence, A., Hunting for homoclinic orbits: a shooting technique, *Adv. Comp. Math.* 1, 81-108 (1993).

[3] Champneys, A.R. & Thompson, J.M.T., A multiplicity of localised buckling modes for twisted rod equations, to appear (1996).

[4] Coyne, J., Analysis of the formation and elimination of loops in twisted cable, *IEEE J. Ocean. Eng.* 15, 72-83 (1990).

[5] Devaney, R.L., Homoclinic orbits in Hamiltonian systems, *J. Diff. Eqns.* 21, 431-438 (1976).

[6] Doedel, E.J. & Kernévez, J.P., AUTO: Software for continuation and bifurcation problems in ordinary differential equations, *Applied Mathematics Report*, California Institute of Technology (1986).

[7] Heijden, G.H.M. van der, Champneys, A.R., Thompson, J.M.T., Load-deflection characteristics of localised torsional buckling modes in rods with non-circular cross-section, to be submitted.

[8] Iooss, G. & Pérouème, M.C., Perturbed homoclinic solutions in reversible 1:1 resonance vector fields, *J. Diff. Eqns.* 102, 62-88 (1993).

[9] Love, A.E.H., *A treatise on the mathematical theory of elasticity*, 4th ed. (Cambridge University Press, Cambridge, 1927).

[10] Thompson, J.M.T. & Champneys, A.R., From helix to localized writhing in the torsional post-buckling of elastic rods, *Proc. R. Soc. Lond.* A452, 117-138 (1996).

CONTROLLING CHAOTIC MOTION OF A MECHANICAL SYSTEM WITH A SET-UP ELASTIC STOP

H. Y. Hu

Institute of Vibration Engineering Research
Nanjing University of Aeronautics and Astronautics
210016 Nanjing, China

Abstract

It is shown that the Poincare mapping of the mechanical system with a set-up elastic stop is not smooth near some fixed points so that the current control strategies, such as OGY control, fail to stabilize the chaotic motion of the system. Thus, a piecewise linear control strategy is proposed, which is based on the piecewise linearized Poincare mapping reconstructed from sampled data and on the pole assignment in two regions near the fixed point. The efficacy of the strategy is demonstrated through an example of controlling chaos of a harmonically forced oscillator with a set-up elastic stop.

1. Introduction

Numerous studies in recent years have highlighted the dynamics of mechanical systems with discontinuous restoring force due to rigid stops, set-up elastic stops and dry friction. Compared with dynamical systems having smooth vector field, these systems with discontinuous vector field behave more complicatedly. As stated in Nordmark (1991), the motion of the system may become chaotic all of a sudden when it grazes a switching plane of the discontinuous vector field with variation of a control parameter.

For most mechanical systems at present, the chaotic motion is not desired and should be controlled. Since 1990's, great efforts have been made to control the chaos of dynamical systems. Of increasing number of publications, the strategy proposed by Ott, Grebogi and Yorke has drawn great attention. The key idea of the strategy is to stabilize the chaotic motion to an unstable periodic motion embedded in the chaotic attractor, rather than any other steady-state motions considered by previous researchers, so that only very small time-dependent perturbation of a control parameter is required. This strategy has been referred to as OGY control and generalized to several forms. However, the

151

D. H. van Campen (ed.), IUTAM Symposium on Interaction between Dynamics and Control in Advanced Mechanical Systems, 151–158.

152

studies have focused mainly on the dynamical system characterized by a differentiable mapping. The primary aim of this study is to develop a strategy for controlling the chaotic motion of the mechanical system with discontinuous restoring force, which corresponds to a piecewise differentiable mapping.

2. Piecewise Differentiability of the Poincare Mapping

Consider a periodically forced oscillator in the state space

$$\dot{u} = \frac{d}{dt}\begin{bmatrix} x \\ y \end{bmatrix} = \begin{bmatrix} y \\ p(t) - g(x,y) \end{bmatrix} = f(u,t) \tag{1}$$

where $p(t)$ is the excitation of period T_0 and $g(x,y)$ is the restoring force that undergoes a finite jump on the switching plane $s(u) = 0$.

Let Σ denote the Poincare section at excitation phase φ and define the Poincare mapping on Σ as follows

$$u_{k+1} = P(u_k), \qquad u_k = u((k + \frac{\varphi}{2\pi})T_0), \qquad k = 0, 1, 2, \cdots \tag{2}$$

The differentiability of the Poincare mapping is just that of the system state at time T_0 with respect to the initial state. Let $U(t)$ be the Jacobian of state $u(t)$ with respect to initial state u_0. As shown in Filippov (1988), if trajectory $u(t)$ passes through switching plane $s(u) = 0$ when $t = \bar{t}_s$, $U(t)$ jumps from $U(t_s^-)$ to $U(t_s^+)$

$$U(t_s^+) = \{I + \frac{[f(u(t_s^+),t_s^+) - f(u(t_s^-),t_s^-)]D_u s(u)}{D_u s(u) \cdot f(u(t_s^-),t_s^-)}\}U(t_s^-) \tag{3}$$

For example, if a set-up elastic stop under pre-load g_0 is mounted at $x = x_s$, the vector field is discontinuous on the plane $s: x - x_s = 0$ so that

$$U(t_s^+) = \begin{bmatrix} 1 & 0 \\ \frac{g_0}{y_s} & 1 \end{bmatrix} U(t_s^-) \tag{4}$$

This implies that if $u(t)$ passes through the switching plane at velocity y_s, an entry in $U(t)$ will undergo a finite jump at time t_s and that if $u(t)$ grazes the switching plane, the entry in $U(t_s)$ will jump to infinity so that $u(t_s)$ is not differentiable with respect to u_0. Hence, the Poincare mapping is not differentiable at some states on Σ if the trajectory starting from them grazes a switching plane. As the system state is continuous with respect to the initial state, those states on Σ form a continuous curve. As a result, the Poincare mapping is piecewise differentiable on Σ.

The chaotic motion of a discontinuous system is often caused by grazing phenomena of various periodic motions, so it is of great importance to stabilize the chaotic motion to the periodic motion, which grazes or almost grazes a switching plane, by slightly adjusting a control parameter of the system. Because the Poincare mapping is not smooth in the neighborhood of the fixed point corresponding to such a periodic motion, all control strategies based on linearized mappings fail to do so. An intuitive way to solve this problem is to construct the piecewise linearized mapping and piecewise control strategy according to the piecewise differentiability of the Poincare mapping.

3. Piecewise Linearized Poincare Mapping

Consider the series of system states $u_i = [x_i, y_i]^T$, $i = 0, 1, 2, \dots N$ acquired at sampling rate $\Delta t = T_0 / m$. Let Γ be an unstable periodic motion of period nT_0 embedded in the chaotic attractor and $u_F \in \Gamma \cap \Sigma$ be one of corresponding fixed points. We shall study how to reconstruct the piecewise linearized mapping from the sampled data in two steps. The first is to determine the switching condition of the mapping and the second is to fit the mapping in two regions from the sampled data.

We begin with searching all extrema of the displacement from the condition $y_i y_{i+1} \leq 0$

$$\bar{x}_i = (1 - \theta_i)x_i + \theta_i x_{i+1}, \qquad \theta_i = \frac{y_i}{y_i - y_{i+1}} \tag{5}$$

and then find the index j corresponding to $min|\bar{x}_i - x_s|$. By using linear interpolation, we have the time $t_s = (j + \theta_j)\Delta t$, which the periodic motion spends when it evolves from u_F to grazing states. We define a mapping in the neighborhood of u_F

$$v_i = Q(u_i) = (1 - \theta_j)u_{i+j} + \theta_j u_{i+j+1} \qquad u_i \in \delta(u_F) \tag{6}$$

By collecting all v_i, $i = 1, 2, \dots \bar{n}$ in the same side of switching plane as v_F, we get a number of mapping pairs in a linearized form

$$v_i - v_F = J(u_i - u_F), \qquad i = 1, 2 \dots \bar{n} \tag{7}$$

where the mapping matrix J yields the least squared solution

$$J = [\sum_{i=1}^{\bar{n}} (v_i - v_F)(u_i - u_F)^T][\sum_{i=1}^{\bar{n}} (u_i - u_F)(u_i - u_F)^T]^{-1} \tag{8}$$

As shown in Figure 1, the states v_A and v_B on the switching plane can be inversely mapped by J

$$u_A = u_F + J^{-1}(v_A - v_F), \qquad u_B = u_F + J^{-1}(v_B - v_F) \tag{9}$$

The equation of line L passing through u_A and u_B is

$$\alpha(x - x_F) + \beta(y - y_F) + \gamma = 0 \tag{10}$$

where

$$\alpha = y_B - y_A, \qquad \beta = x_A - x_B, \qquad \gamma = (x_A - x_F)(y_A - y_B) + (x_B - x_A)(y_A - y_F) \tag{11}$$

If the set $L \cap \delta(u_F)$ is not empty, the trajectory starting from $L \cap \delta(u_F)$ grazes the switching plane. If this is the case, the Poincare mapping is not differentiable on L, which serves as the switching line of the piecewise differentiable Poincare mapping on Σ. According to the distance from u_F to L, the condition when $L \cap \delta(u_F)$ is not empty simply reads

$$|\gamma| / \sqrt{\alpha^2 + \beta^2} < \delta \tag{12}$$

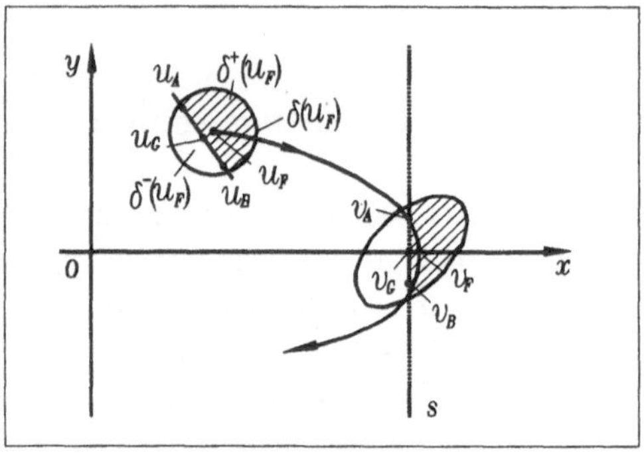

Figure 1. Fixed point and an almost grazing trajectory

If this inequality holds and $\beta \neq 0$, L divides $\delta(u_F)$ into upper and lower parts. Let these two parts be $\delta^+(u_F)$ and $\delta^-(u_F)$, respectively. It is easy to find the following conditions for identifying which part a system state in $\delta(u_F)$ locates

$$\begin{cases} \forall (x,y) \in \delta^+(u_F) & \Leftrightarrow & \beta[\alpha(x - x_F) + \beta(y - y_F) + \gamma] > 0 \\ \forall (x,y) \in \delta^-(u_F) & \Leftrightarrow & \beta[\alpha(x - x_F) + \beta(y - y_F) + \gamma] < 0 \end{cases} \tag{13}$$

If inequality (13) holds but $\beta = 0$, L divides $\delta(u_F)$ into left and right parts. The sign of $x - x_A$ indicates the location of the system state.

Now we reconstruct the piecewise linearized mapping. Taking the case of $u_F \in \delta^+(u_F)$ as an example, we pick $u_i \in \delta^+(u_F)$, $i = 1, 2, \ldots n^+$ from the sampled states and form a series of linearized mapping pairs near u_F

$$u_{i+mn} - u_F = A^+(u_i - u_F), \qquad i = 1, 2, \ldots n^+ \tag{14}$$

Then, by using the least squared fitting we obtain the mapping matrix

$$A^+ = [\sum_{i=1}^{n^+}(u_{i+mn} - v_F)(u_i - u_F)^T][\sum_{i=1}^{n^+}(u_i - u_F)(u_i - u_F)^T]^{-1} \qquad (15)$$

Obviously, the Poincare mapping in $\delta^-(u_F)$ can not be linearized around u_F if $u_F \in \delta^+(u_F)$. We alternatively consider the state u_G that is on L and most close to u_F

$$u_G = [x_F - \frac{\alpha\gamma}{\alpha^2 + \beta^2}, y_F - \frac{\beta\gamma}{\alpha^2 + \beta^2}]^T \qquad (16)$$

and map it to $\hat{u}_G$ by using Eq.(15). We pick $u_i \in \delta^-(u_F)$, $i = 1, 2, ... n^-$ from the sampled states and form linearized mappings with respect to u_G

$$u_{i+mn} - \hat{u}_G = A^-(u_i - u_G), \qquad i = 1, 2, ... n^- \qquad (17)$$

There follows the mapping matrix A^- similar to Eq.(15).

4. Piecewise Linear Control Strategy

Suppose that ρ is an adjustable parameter and the system state is chaotic when $\rho = \bar{\rho}$. The purpose of control is to stabilize the chaotic motion to an unstable periodic motion, which grazes or almost grazes the switching plane, by slightly regulating ρ. From the piecewise linearized mapping and the small perturbation of ρ, we have

$$u_{k+n} = \begin{cases} u_F + A^+(u_k - u_F) + b(\rho_k - \bar{\rho}), & u_k \in \delta^+(u_F) \\ \hat{u}_G + A^-(u_k - u_G) + b(\rho_k - \bar{\rho}), & u_k \in \delta^-(u_F) \end{cases} \qquad (18)$$

where $b \in R^2$ is the derivative vector of u_F with respect to ρ at $\bar{\rho}$.

Define the linear feedback in $\delta^+(u_F)$ and $\delta^-(u_F)$, respectively

$$\rho_k = \bar{\rho} - \begin{cases} s_+^T(u_k - u_F), & u_k \in \delta^+(u_F) \\ s_-^T(u_k - u_G), & u_k \in \delta^-(u_F) \end{cases} \qquad (19)$$

where $s_+ \in R^2$ and $s_- \in R^2$ are the control stiffness to be determined. By substituting Eq.(19) into Eq.(18), we have

$$u_{k+n} = \begin{cases} u_F + (A^+ - bs_+^T)(u_k - u_F), & u_k \in \delta^+(u_F) \\ \hat{u}_G + (A^- - bs_-^T)(u_k - u_G), & u_k \in \delta^-(u_F) \end{cases} \qquad (20)$$

If s_+ and s_- are so selected that all eigenvalues of matrices $A^+ - bs_+^T$ and $A^- - bs_-^T$ fall into the unit circle, u_{k+n} will stay in $\delta(u_F)$ when k increases. Furthermore, if matrix $A^- - bs_-^T$ has a pair of negative real eigenvalues, it will map state u_k from $\delta^-(u_F)$ to $\delta^+(u_F)$ and then u_{k+n} will approach to u_F at last.

As stated in Romeiras (1992), the selection of s_+ and s_- from given eigenvalues of matrices $A^+ - bs_+^T$ and $A^- - bs_-^T$ can be made according to the pole assignment formula in control theory. For instance, the control stiffness s_+ yields

$$s_+^T = \left[b_2^+ - a_2^+, \ b_1^+ - a_1^+\right](CW)^{-1} \tag{21}$$

where

$$C = \left[b, \ A^+ b\right], \qquad W = \begin{bmatrix} a_1^+ & 1 \\ 1 & 0 \end{bmatrix} \tag{22}$$

a_1^+ and a_2^+ are the coefficients of the characteristic polynomial of matrix A^+, while b_1^+ and b_2^+ are those of matrix $A^+ - bs_+^T$.

5. Numerical Simulation

The benchmark in the simulation is a harmonically forced oscillator with a set-up elastic stop. The dimensionless differential equation of motion reads

$$\ddot{x} + 2\zeta\dot{x} + g(x) = \rho \sin \lambda t \tag{23}$$

where

$$g(x) = \begin{cases} x, & x \le 1 \\ x + \mu(x + e - 1), & x > 1 \end{cases} \tag{24}$$

In the simulation, the damping ratio, the stiffness ratio and the ratio of set-up amount to clearance were fixed at $\zeta = 0.02$, $\mu = 0.5$ and $e=1$, respectively. At first, the steady-state motion of the system was calculated for various combinations of excitation amplitude ρ and excitation frequency λ. Shown in Figure 2a is a bifurcation diagram of the displacement on Poincare section $\varphi = 0$ with respect to ρ when $\lambda = 0.692$.

Then, the excitation amplitude ρ was taken as an adjustable parameter around $\bar{\rho} = 0.51$ to stabilize a chaotic motion caused by grazing. The chaotic motion was acquired at sampling rate $\Delta t = T_0 / 128$ and a Poincare section of the motion is shown in Figure 2b. Subsequently, a number of unstable periodic motions embedded in the chaotic attractor were extracted by searching the minimal distance between two states with time delay of desired period (Hu, 1995). Shown in Figure 2c and 2d are the phase portraits of the chaotic motion and an unstable motion of period $2T_0$, respectively. They both touched the set-up elastic stop at low velocity. The periodic motion delayed the excitation approximately $7\pi / 128$ in phase, so the Poincare section of the chaotic attractor in Figure 2b was fixed at $\varphi = 7\pi / 128$. Afterwards, one of fixed points indicated by circles in Figure 2b was taken as the target of control. In its neighborhood of radius $\delta = 0.08$, the linearized mapping matrices A^+ and A^- were reconstructed and the derivative vector b was evaluated by disturbing ρ from $\bar{\rho} = 0.51$ to 0.508 and 0.512, respectively. Finally, the poles of controlled mapping were assigned to 0.1 and 0.2 in $\delta^+(u_F)$, and -0.1 and -0.2 in $\delta^-(u_F)$. Thus, s_+ and s_- were determined.

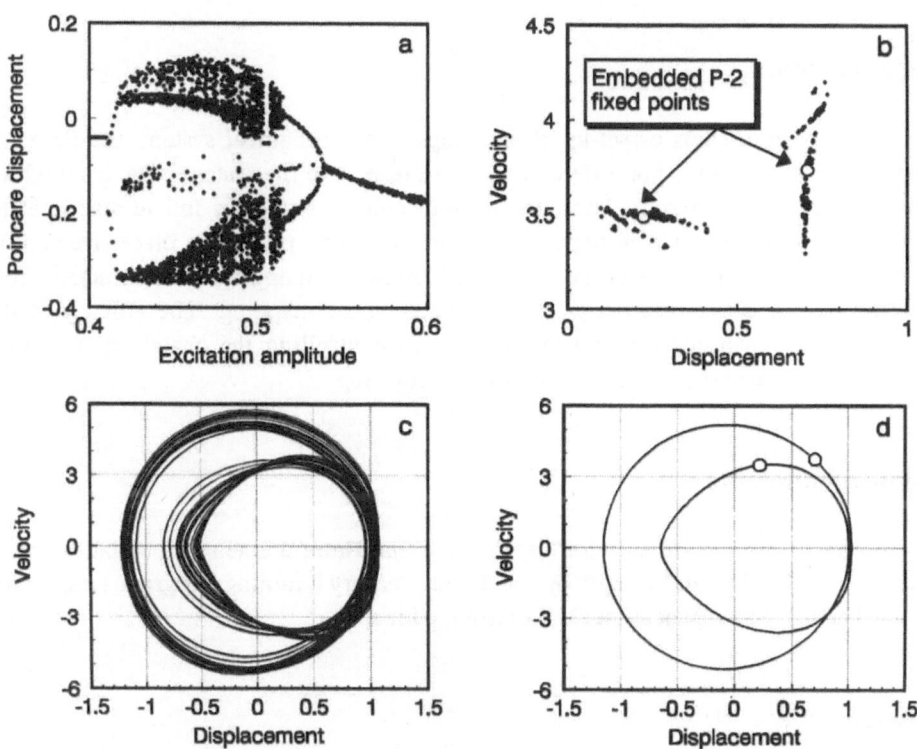

Figure 2. Bifurcation of original system and detailed dynamics when $\bar{\rho} = 0.51$
(a. Bifurcation diagram,   b. Poincare section,   c. Chaotic motion,   d. P-2 unstable motion)

After the preparation for control was made, the system was excited again and the control strategy was initiated as soon as the system state fell into $\delta(u_F)$. In Figure 3 are shown the variations of the displacement on the Poincare section and the excitation amplitude in control process. They well supported the efficacy of the control strategy.

158

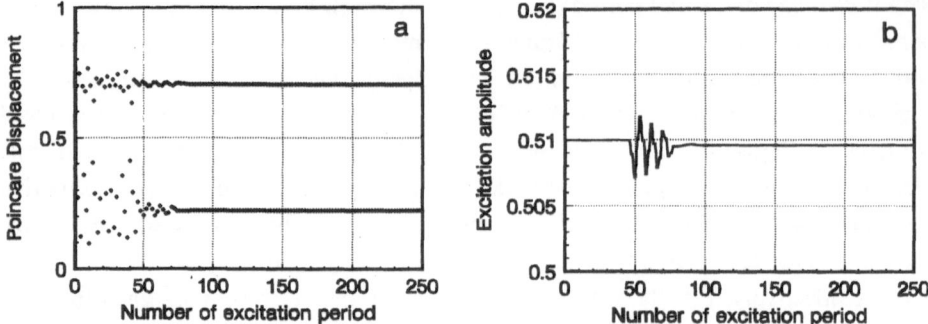

Figure 3. Variation of the Poincare displacement and excitation amplitude in control

6. Conclusions

If there exist rigid stops or set-up elastic stops in a mechanical system, the Poincare mapping of the system is not differentiable with respect to the system state, from which the trajectory grazes a stop. Thus, the current control strategies fail to stabilize the chaotic motion to an unstable periodic motion. The paper presents a piecewise control strategy based on the piecewise linearized Poincare mapping reconstructed from sampled state and the pole assignment for each linear mapping. The efficacy of the strategy is demonstrated through an example of controlling the chaotic motion of a harmonically forced oscillator with a set-up elastic stop.

Acknowledgments

This research was supported in part by the National Natural Science Foundation under Grant No. 19472033 and in part by the Trans-Century Training Program Foundation for the Talents, State Education Commission, China.

References

1. Filippov A. F. (1988) *Differential Equations with Discontinuous Right-hand Sides*, Kluwer Academic Publishers, Dordrecht.
2. Hu H. Y. (1995) Controlling Chaos of a Periodically Forced Non-smooth Mechanical System, *Acta Mechanica Sinica,* **11,** 251-258.
3. Nordmark A. B. (1991) Non-Periodic Motion Caused by Grazing Incidence in an Impact Oscillator, *Journal of Sound and Vibration,* **145,** 279-297.
4. Romeiras F. J. et al. (1992) Controlling Chaotic Dynamic Systems, *Physica D,* **58,** 165-192.

OUTPUT ANNIHILATION AND OPTIMAL H_2–CONTROL OF PLATE VIBRATIONS BY PIEZOELECTRIC ACTUATION

H. IRSCHIK
Institute of Technical Mechanics and Foundations of Machine Design
Johannes Kepler University of Linz, A-4040 Linz, Austria

K. SCHLACHER
Institute of Automatic Control and Electrical Drives
Johannes Kepler University of Linz, A-4040 Linz, Austria

AND

W. HAAS
Institute of Automatic Control and Electrical Drives
Johannes Kepler University of Linz, A-4040 Linz, Austria

1. Introduction

This paper is concerned with the piezoelectrically induced actuation and control of plates. Flexural vibrations of linear elastic laminated plates with layers made of ceramic or polymeric piezoelectric materials are considered, see (Rao *et al*, 1994). In the present formulation, the mechanical modelling of such a type of smart or intelligent plates is related to the fundamental duality between load-stresses, induced by imposed forces, and self-stresses due to eigenstrains (Reißner, 1931). The distributed actuating effect of piezoelectric layers is considered as a distribution of eigenstrains counteracting the load-stresses. Assuming perfect bond between the piezoelectric layers and their substrate, this eigenstrain problem is formulated within the context of the classical lamination theory of thin plates (Jones, 1975). A convolution integral is derived for the piezoelectrically induced vibrations by using dynamic bending moments due to single forces as kernel functions. This Maysel-type integral (Irschik *et al*, 1993), (Irschik *et al*, 1995) is accompanied by an analogous integral statement for the load-induced vibrations in the case of simply supported rectangular plates. Comparing these two formulations, an annihilation problem is solved. It is shown, what kind of distribution of the piezoelectric actuators is able to annihilate precisely a dynamic plate deflection generated by a prescribed distribution of imposed forces. Thus, the problem of distributed actuation by piezoelectric effects is concisely connected to the control by distributed forces. For the analogous problem of bending of beams see (Irschik *et al*, 1994).

The control law uses the information of the so called "natural output" (Nijmeijer *et al*, 1989), which is closely related to the plate deflection and to the input energy caused by the space–wise distribution of the load. In order to guarantee the stability the time derivative of the natural output is measured and used for control, too. The controller design method

D. H. van Campen (ed.), IUTAM Symposium on Interaction between Dynamics and Control in Advanced Mechanical Systems, 159–166.

is based on a frequency domain H_2–approach for multiple–input–multiple–output systems (Haas, 1995), where the 2–norm of a generalized error is minimized. It is well known that the frequency domain description of the mechanical model leads to transfer functions of infinite order. These functions are approximated by functions of low order and so a H_2– design can be applied in a straight forward way. Such approach reduces not only the costs of the theoretic analysis drastically but also the costs of the design of the controller and its implementation. The presented H_2–approach offers a number of new and interesting features in order to achieve a better fitting method to practical control problems. For instance the use of deterministic signals as inputs and the algebraic decoupling of the tracking and disturbance behaviors can be easily implemented in the design.

The stability problem is discussed for the approximating and the original system in the frequency domain. As a result of these investigations the control algorithm and numerical simulations are presented.

2. Piezoelectric–Mechanical Modelling of Laminated Plates

Consider thin, laminated plates with a perfect bond between the layers. For the sake of simplicity, the layers of the plate are assumed to be isotropic and to be symmetrically built up with respect to the mid-plane of the plate, $x_3 = 0$. Let the plate to be actuated by means of a field of piezoelectrically produced eigenstrains $\varepsilon_{\alpha\beta}^*$, $\alpha, \beta = 1, 2$, which are assumed to be odd functions of x_3. (Greek indices refer as the coordinates x_1, x_2 of the mid-plane of the plate.) The kinematic assumptions of Kirchhoff and Love are used in order to approximate the corresponding odd distribution of in-plane strains induced in the laminated plate by these eigenstrains. Within this classical lamination theory (Jones, 1975), (Ugural, 1981) only a state of plane stress, $\sigma_{\alpha\beta}$, $\alpha, \beta = 1, 2$, is taken into account in the modelling. The following convolution integral can be shown to hold for the piezoelectrically induced deflection of the laminated plate

$$w^{(e)}\left(\mathbf{x}; t\right) = \int\limits_0^t \int\limits_V \varepsilon_{\alpha\beta}^*\left(\boldsymbol{\xi}; t\right) \tilde{\sigma}_{\alpha\beta}\left(\boldsymbol{\xi}, \mathbf{x}; t - \tau\right) \mathrm{d}V\left(\boldsymbol{\xi}\right) \mathrm{d}\tau. \tag{1}$$

For the three-dimensional case see (Irschik *et al*, 1993), (Irschik *et al*, 1995). In Eq. 1, $w^{(e)}$ denotes the deflection at the point $\mathbf{x} = [x_1, x_2, x_3]^T$ of the plate at time t. The dummy space and time variables are denoted by $\boldsymbol{\xi}$ and τ, respectively. The plate volume is V. Load-stresses occurring in the point $\boldsymbol{\xi}$ at time t due to a unit single impulsive force applied in $\mathbf{x}$ at time $\tau \leq t$ are used as kernel functions in Eq. 1. They are denoted by $\tilde{\sigma}_{\alpha\beta}\left(\boldsymbol{\xi}, \mathbf{x}; t - \tau\right)$. The piezoelectrically produced eigenstrains in Eq. 1 are contributed to the presence of non-vanishing components E_3 of the electrical field vector in the actuating layers (Tzou *et al*, 1992). As a special piezoelectric behavior, the eigenstrains are modelled in the form

$$\varepsilon_{\alpha\beta}^* = d\, E_3\, \delta_{\alpha\beta}, \tag{2}$$

where d is the proper isotropic piezoelectric constant, and $\delta_{\alpha\beta}$ denotes Kronecker's delta. Eq. 1 now becomes

$$w^{(e)}\left(\mathbf{x}; t\right) = \int\limits_0^t \int\limits_V d\, E_3\left(\boldsymbol{\xi}; t\right) \tilde{s}\left(\boldsymbol{\xi}, \mathbf{x}; t - \tau\right) \mathrm{d}V\left(\boldsymbol{\xi}\right) \mathrm{d}\tau, \tag{3}$$

with the first invariant $\tilde{s} = \tilde{\sigma}_{11} + \tilde{\sigma}_{22}$ of the in-plane stresses. In the classical lamination theory under consideration and for isotropic layers, $\tilde{s}$ is connected to the weightened sum of corresponding bending moments $\tilde{M}$ by

$$\tilde{s}\left(\boldsymbol{\xi}; t\right) = \xi_3 \frac{Y\left(\xi_3\right)}{1 - \nu\left(\xi_3\right)} \frac{\tilde{M}\left(\xi_1, \xi_2; t\right)}{D}, \quad \tilde{M} = \left(\tilde{M}_1 + \tilde{M}_2\right) \frac{D}{D + \bar{D}} \tag{4}$$

where Y and ν denote Young's modulus and Poisson's ratio, respectively. The material parameters Y, ν and d may vary from layer to layer and are thus written as functions of the transverse coordinate ξ_3 in Eq. 4, however they are assumed to be constant within each layer. The moment-sum in Eq. 4 is connected to the corresponding deflection $\tilde{w}$ by

$$\tilde{M} = -D \left[\frac{\partial^2 \tilde{w}}{\partial \xi_1^2} + \frac{\partial^2 \tilde{w}}{\partial \xi_2^2} \right]. \tag{5}$$

The effective stiffnesses in Eqs. 4 and 5 are

$$D = \int\limits_{-h/2}^{h/2} \xi_3^2 \frac{Y\left(\xi_3\right)}{1 - \nu\left(\xi_3\right)^2} \, \mathrm{d}\xi_3, \quad \bar{D} = \int\limits_{-h/2}^{h/2} \xi_3^2 \frac{Y\left(\xi_3\right) \nu\left(\xi_3\right)}{1 - \nu\left(\xi_3\right)^2} \, \mathrm{d}\xi_3. \tag{6}$$

Inserting Eq. 5 into Eq. 3 and integrating over the transverse coordinate, the volume integral of Eq. 3 is converted into an integral over the plate area A:

$$w^{(e)}\left(x_1, x_2; t\right) = \int\limits_0^t \int\limits_A \frac{M^*\left(\xi_1, \xi_2; \tau\right)}{D} \tilde{M}\left(\xi_1, \xi_2, x_1, x_2; t - \tau\right) \mathrm{d}\xi_1 \, \mathrm{d}\xi_2 \, \mathrm{d}\tau, \tag{7}$$

where the piezoelectrically actuating "moment" is

$$M^*\left(\xi_1, \xi_2; \tau\right) = \int\limits_{-h/2}^{h/2} \xi_3 \frac{Y\left(\xi_3\right)}{1 - \nu\left(\xi_3\right)} d \, E_3\left(\xi_1, \xi_2, \xi_3; \tau\right) \mathrm{d}\xi_3. \tag{8}$$

Note that Eq. 7 holds for arbitrary boundary conditions of the plate. In a consistent manner, it connects the dynamic deflection due to piezoelectric actuation of Eq. 2 with the load-stresses due to single unit impulsive forces.

3. Output-Annihilation of Load-Induced Deflections in Simply Supported Polygonal Plates by Piezoelectric Actuation

Consider a plate under the disturbing action of imposed lateral forces $p(x_1, x_2; t)$. Given the space- and time-wise distribution of $p(x_1, x_2; t)$, the question arises, what piezoelectrically induced moment $M^*(x_1, x_2; t)$ of Eq. 8 must be imposed in order to annihilate the load-induced flexural vibrations $w^{(p)}(x_1, x_2; t)$. In the following, this output-annihilation problem is solved for simply supported plates of polygonal shape. Following lines similar to a frequency-domain study given by (Nowacki, 1975) again considering homogeneous initial conditions, and transforming into the time-domain, the following convolution integral formulation is obtained:

$$w^{(p)}\left(x_1, x_2; t\right) = \int\limits_0^t \int\limits_A \frac{M_s^{(p)}\left(\xi_1, \xi_2; \tau\right)}{D} \tilde{M}\left(\xi_1, \xi_2, x_1, x_2; t - \tau\right) \mathrm{d}\xi_1 \, \mathrm{d}\xi_2 \, \mathrm{d}\tau, \tag{9}$$

where the quasi-static weightened moment-sum $M_s^{(p)}$ is the solution of the Poisson equation

$$\left[\frac{\partial^2}{\partial \xi_1^2} + \frac{\partial^2}{\partial \xi_2^2}\right] M_s^{(p)}(\xi_1, \xi_2; \tau) = -p(\xi_1, \xi_2; \tau) \tag{10}$$

with homogeneous boundary conditions. Note that in the cited original statement of (Nowacki, 1975) a quasi-static kernel and a dynamic load-induced moment is used. Comparing Eqs. 7 and 9, it is immediately seen that the load–induced vibrations are annihilated by the piezoelectric actuators

$$w^{(p)}(x_1, x_2; t) + w^{(e)}(x_1, x_2; t) = 0, \tag{11}$$

if

$$M^*(\xi_1, \xi_2; \tau) = -M_s^{(p)}(\xi_1, \xi_2; \tau). \tag{12}$$

Therefore, what has to be done for an exact annihilation of the load-induced vibrations, is to solve the quasi-static problem, Eq. 10, with homogeneous boundary conditions and to provide the corresponding piezoelectrically actuating moment according to Eq. 8. In practically important situations, the disturbing force may often be modelled as a function separated in space and time

$$p(\xi_1, \xi_2; \tau) = P(\xi_1, \xi_2) f(\tau), \tag{13}$$

where the space-wise distribution of P is a known one, while f represents a disturbance to be damped by a control system. Taking

$$M^*(\xi_1, \xi_2; \tau) = -M_s^{(P)}(\xi_1, \xi_2) u(\tau), \tag{14}$$

where $M_s^{(P)}$ denotes the static moment due to P, the controller output u will be able to reach the flat plate position exactly, even if f does not vanish in the stationary limit.

4. Control of Simply Supported Square Plates

The total deflection of the controlled plate due to disturbing forces and piezoelectric actuation can be written as

$$w(x_1, x_2; t) = \int_0^t (f(\tau) - u(\tau)) \int_A \frac{M_s^{(P)}(\xi_1, \xi_2)}{D} \tilde{M}(\xi_1, \xi_2, x_1, x_2; t - \tau) \, d\xi_1 \, d\xi_2 \, d\tau, \tag{15}$$

see Eqs. 7, 9 and 12. Applying the Laplace-transformation to Eq. 15, the convolution integral leads to

$$\hat{w}(x_1, x_2; s) = G(x_1, x_2; s)\left(\hat{f}(s) - \hat{u}(s)\right), \tag{16}$$

where the symbol $\hat{\ }$ denotes the Laplace-transform, e.g. $\mathcal{L}\{f(t)\} = \hat{f}(s)$. The transition function G is given by

$$G(x_1, x_2; s) = \int_A \frac{M_s^{(P)}(\xi_1, \xi_2)}{D} \mathcal{L}\left\{\tilde{M}(\xi_1, \xi_2, x_1, x_2; t)\right\} d\xi_1 \, d\xi_2. \tag{17}$$

In the control law the natural output signal is used, which is related to the deflection via the following equation

$$\hat{y}(s) = \int_A \hat{w}(\xi_1, \xi_2; s) P(\xi_1, \xi_2) \, d\xi_1 \, d\xi_2, \tag{18}$$

or by

$$\hat{y}(s) = H(s) \left(\hat{f}(s) - \hat{u}(s) \right), \tag{19}$$

where the transfer function $H(s)$ is as follows

$$H(s) = \int_A G(\xi_1, \xi_2; s) P(\xi_1, \xi_2) \, d\xi_1 \, d\xi_2. \tag{20}$$

As a matter of fact, the product of the time derivative of the natural output y and the control input u is the input power caused by the controller.

In the case of a space-wise constant disturbing load, $P(\xi_1, \xi_2) = P_0 = $ constant, the transition function of a square plate, $0 \le x_1 \le l$, $0 \le x_2 \le l$ takes the form

$$G(x_1, x_2; s) = \frac{16 P_0}{\pi^2} \sum_{n=1}^{\infty} \sum_{m=1}^{\infty} \frac{\sin(m\pi x_1/l) \sin(n\pi x_2/l)}{\left(D\left((m\pi/l)^2 + (n\pi/l)^2 \right)^2 + \mu s^2 \right) mn}, \tag{21}$$

where μ is the mass per unit. Further on for the sake of simplicity a non dimensional

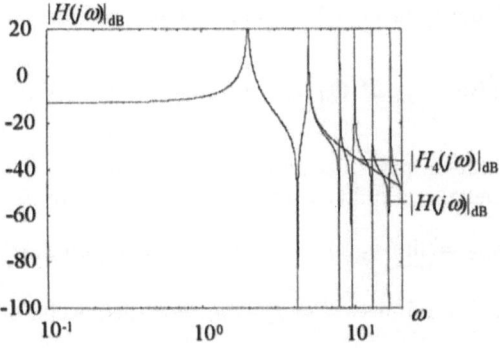

Figure 1. Magnitude Bode Plots of $H(s)$ and $H_4(s)$.

model H is used. Due to the nature of this problem the transfer function H has infinite order, but a simple H_2–design requires a plant transfer function of low order. It turns out that the choice

$$H_4(s) = 0.27 \frac{\left(1 + (s/4.2)^2 \right)}{\left(1 + (s/2)^2 \right) \left(1 + (s/5)^2 \right)} \tag{22}$$

leads to good results. Fig. 1 shows the bode plot of the magnitude of H and H_4.

To guarantee the stability the time derivative $v(t) = \dfrac{\partial y(t)}{\partial t}$ of y is measured and used for control, too. Then the input–output relation in the frequency domain

$$\hat{\mathbf{z}}(s) = \begin{bmatrix} \hat{y}(s) \\ \hat{v}(s) \end{bmatrix} = \begin{bmatrix} H_4(s) \\ sH_4(s) \end{bmatrix} \left(\hat{f}(s) - \hat{u}(s) \right) \tag{23}$$

follows with the vector $\mathbf{z}$ of measurement variables and the matrix $\mathbf{H}_4 = \begin{bmatrix} H_4 & sH_4 \end{bmatrix}^T$. Figure 2 shows the structure of the control system, where $\mathbf{C}$ denotes the matrix of the

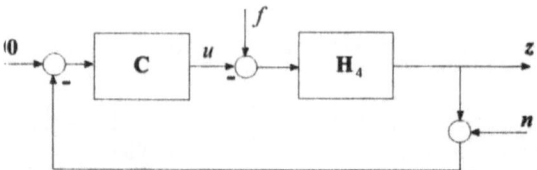

Figure 2. One Parameter Controller Scheme.

controller and $\mathbf{n} = \begin{bmatrix} n_1 & n_2 \end{bmatrix}^T$ is the vector of measurement noise variables. Now a nonstandard H_2-approach is used to derive an optimal controller. The key to the proposed control synthesis problem is the following one: The input signals of the control system are combined in the vector $\mathbf{x} = \begin{bmatrix} f & n_1 & n_2 \end{bmatrix}^T$. All input signals are chosen to be zero except one which is set to $q_i\sigma$ with the unit step σ and the real number q_i. For this special choice the control system produces the signals $\mathbf{z}_i$ and u_i which are used in the cost functions

$$J_i = \int\limits_0^\infty \left((\mathbf{z}_i - \mathbf{z}_{i,\infty})^T \mathbf{Q}_d (\mathbf{z}_i - \mathbf{z}_{i,\infty}) + (u_i - u_{i,\infty})^2 \, r_d^2 \right) e^{2at} dt, \tag{24}$$

where the positive definite Matrix $\mathbf{Q}_d$ and the real number r_d denote weights while the term e^{2at}, $a > 0$, specifies the stability margin. The existence of J_i is ensured by

$$u_{i,\infty} = \lim_{t\to\infty} u_i(t) \qquad \text{and} \qquad \mathbf{z}_{i,\infty} = \lim_{t\to\infty} \mathbf{z}_i(t). \tag{25}$$

The optimal controller $\mathbf{C}$ minimizes $J = \sum_{i=1}^{3} J_i$. For the special choice

$$\mathbf{Q}_d = \begin{bmatrix} 6 & 0 \\ 0 & 4 \end{bmatrix}, \quad r_d = 1, \quad q_1 = 1, \quad q_2 = q_3 = 0.5, \quad a = 0.75$$

the optimal controller can be found by

$$\mathbf{C}(s) = \left[\frac{2747.664 + 1998.248s}{d(s)} \quad \frac{1719.523s + 286.732s^2 + 41.3s^3 + s^4}{d(s)} \right]$$

with the denominator polynomial

$$d(s) = 417.647s + 103.647s^2 + 17.191s^3 + s^4.$$

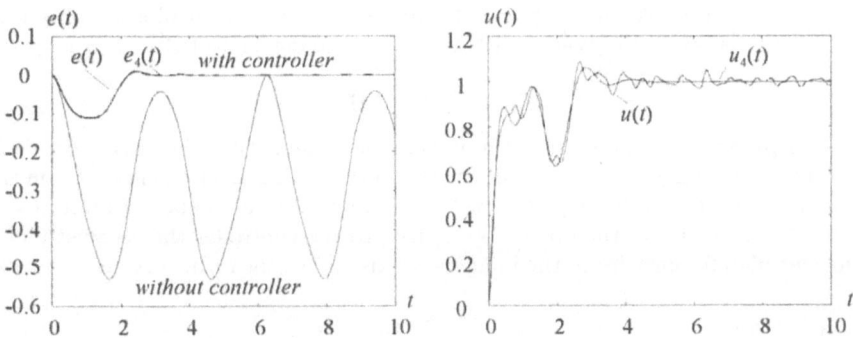

Figure 3. Output and Actuator Signals.

Fig. 3 (left hand side) presents the unit step response ($\hat{f} = 1/s$) of the control system. It shows the simulation of the control error $e_4 = -y_4$ for the approximating design model $\mathbf{H_4}$, the simulation of the control error $e = -y$ for the plant $\mathbf{H}$ of infinite order and the simulation of the plant without controller. It should be noted that for the simulation of $\mathbf{H}$ an approximation of sufficient high order is used. Any further increase of the order will not produce a difference in the results. The difference between e and e_4 is less than the plotting accuracy. Therefore the controller guarantees an excellent damping of the higher eigenfrequencies, too. The corresponding plant inputs u and u_4 are given on the right hand side of Fig. 3.

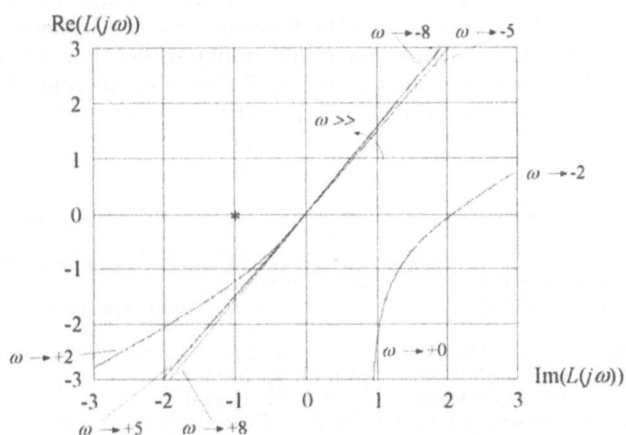

Figure 4. Nyquist Plot of the Open Loop.

166

To prove the stability for the plant of infinite order a Nyquist plot stability criterion for such plants is used (Vidyasagar, 1993): If the Nyquist plot of the open loop transfer function

$$L(s) = -\mathbf{C}(s)\,\mathbf{H}(s)$$

does not encircle the critical point -1 in the complex plane, then the closed loop is stable. As a matter of fact, the Nyquist plot of the transfer function H lies completely on the real axis. Furthermore the Nyquist plot of H splits in an infinite number of branches. Fig. 4 shows the Nyquist plot of the open loop L. Due to the controller the Nyquist plot tends for sufficient high frequencies to the imaginary axis and so the critical point -1 is avoided.

5. Conclusion

In this contribution the distributed actuating effect of piezoelectric layers was considered as a distribution of eigenstrains counteracting the load-stresses. A convolution integral, Eq. 7, has been derived for the piezoelectrically induced vibrations by means of dynamic bending moments due to single forces as kernel functions. This integral was compared with an analogous integral statement for the load-induced vibrations in the case of simply supported rectangular plates. So it was shown what kind of distribution of the piezoelectric actuators is able to annihilate precisely a dynamic plate deflection generated by a prescribed distribution of imposed forces, Eq. 14. The external force p has been separated in a given space–wise distribution P and an unknown time depending function f , Eq. 13. Because of Eq. 14 the space–wise distribution of the controller has been computed and a further task was to find a control law for the time depending function u.

The goal of the controller design was not only to ensure zero steady state error for the step response but also a sufficient decay of the vibrations. Here the natural output and its time derivative has been used for the controller design. The design method is based on a nonstandard H_2–approach in the frequency domain where a generalized error is minimized. Since the transfer function H has infinite order it was approximated by a transfer function H_4 of low order for the design. Stability of the control system with the plant H has been discussed and proved by a Nyquist plot criterion.

References

Haas, W. (1995) H_2–Entwurf für Mehrgrößensysteme im Frequenzbereich, Ph.D. thesis, Johannes Kepler University of Linz, Austria.

Irschik, H., Fotiu, P. and Ziegler, F. (1993) Extension of Maysel's formula to the dynamic eigenstrain problem, *J. Mechanical Behavior of Materials* **5**, 59–66.

Irschik, H. and Ziegler, F. (1995) Dynamic processes in structural thermo-viscoplasticity, *Applied Mechanics Reviews* **48**, 301-315.

Irschik, H., Schlacher, K. and Belyaev A.K. (1994) Distributed Control of Structures Using Eigenstrain Analysis, *Int. Association for Structural Control* **1**, WP3, 73- 82.

Jones, R.M. (1975) *Mechanics of Composite Materials*, Mc Graw-Hill.

Nijmeijer, H. and Van der Schaft, A.J. (1989) *Nonlinear Dynamical Control Systems*, Springer Verlag.

Nowacki, W. (1975) *Baudynamik*, Springer–Verlag.

Rao, S.S. and Sunar, M. (1994) Piezoelectricity and its use in disturbance sensing and control of flexible structures: A survey, *Applied Mechanics Reviews* **47**, 113-123.

Reißner, H. (1931) Eigenspannungen und Eigenspannungsquellen, *ZAMM* **11**, 1-30.

Tzou, H.S. and Anderson, G.L. Intelligent Structural Systems, Kluwer, Dordrecht.

Ugural, A.C. (1981) *Stresses in Plates and Shells*, Mc Graw-Hill.

Vidyasagar, M. (1983) *Nonlinear System Analysis*, Prentice Hall.

Ziegler, F. (1989) *Mechanics of Solids and Fluids*, Springer Verlag.

OBJECTORIENTED MODELLING AND SIMULATION OF MECHATRONIC SYSTEMS

R. KASPER

Faculty of Mechanical Engineering, Institute of Machine Systems and Drive Technology. Magdeburg University, P.O. Box 4120, 39016 Magdeburg, Germany. E-mail: roland.kasper@masch-bau.uni-magdeburg.de

Abstract

In many technical applications mechatronic products offer superior functionality and improved flexibility by integrating features of mechanical, electronic and control systems into one device. But traditional design and modelling methods well known for mechanical, electronic and control systems do not support a complete interdisciplinary design process.

In this paper a new objectoriented design methodology based on a structured top down design strategy is presented. The key idea is the definition of communicating simulation objects acting as reusable building blocks for a computer aided design of mechatronic systems. This approach supports both, the structural top down design as well as the simulation of the behaviour of complete mechatronic devices involving mechanical, electronic and control systems. A very economic implementation based on software processes, allows to distribute large simulation models across multiple processors.

1. Mechatronic Systems

In the last years mechatronic systems have attracted much interest from both sides industry and science. The great promise of mechatronic products is to deliver an optimised function and construction by combining mechanical, electronic and information processing (control) elements into <u>one device</u>. Today we have a large number of successful mechatronic products from the producer to the consumer industry as well as from automotive or aircraft industry. Despite this variety of applications, mechatronic products share a common base structure and a similar scheme of interaction as given in figure 1. This mechatronic approach overcomes the traditional borders of classical engineering disciplines. Today there is a strong trend to replace pure mechanical functions by more flexible solutions built up by the interaction of sensor, control and actuator systems.

167

D. H. van Campen (ed.), IUTAM Symposium on Interaction between Dynamics and Control in Advanced Mechanical Systems, 167–174.
© 1997 *Kluwer Academic Publishers.*

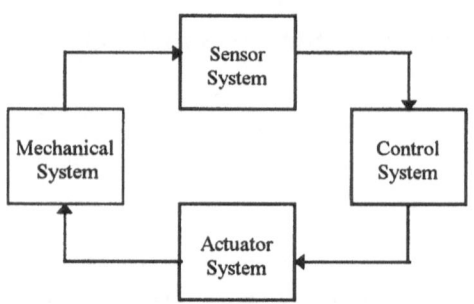

Figure 1: Base structure of mechatronic systems

Following this approach, new functions become possible. Moreover, due to their inherent programmability, mechatronic systems are able to adapt automatically to their environment and to changing tasks. But this trend will reflect only the most obvious advantages. There are much more degrees of freedom for example, how to integrate sensors and actuators optimally into the mechanical system etc. Further information processing tasks can be realised electronically in the sensor or actuator system or by software in the control system. Designing a mechatronic system, consequently needs engineering knowledge from very different disciplines, if globally optimised function and construction of the device is the goal. In order to use all degrees of freedom during the design process, to reduce development time and to improve product quality, modelling and simulation of complete mechatronic systems are key strategies. This need for interdisciplinary thinking, working and designing however leads to a number of unsolved problems.

2. Traditional approach

Traditionally the mechanical design engineer is faced with a design task that is concerned with questions about outline and shape of mechanical components. Consequently the modelling of mechanical systems in most cases is based on rigid or elastic bodies connected together via joints and linkages. There exist a number of modelling and simulation tools that assist this mechanical view during the specification of models and the calculation of simulation results.

The electrical engineer does his job in an equivalent way. His point of interest is to select the optimal electronic circuits and components and to connect them to complete electronic systems. Consequently the modelling of electric and electronic systems traditionally is based on standardised component models and net lists describing the connections between them. There exists a large number of powerful simulators that are very comfortably integrated into the CAE framework used by the design engineer for his work.

The control system of mechatronic products in most cases is realised by software. The software engineer uses a programming language for specification and compilers, linkers and so on to produce final object code /1/. Simulation is not supported by typical software development tools. Self made test programs are the common approach to validate the correct working of a design idea. As a result of this separation of disciplines and tools, it is not possible to develop and validate the functions of a complete mechatronic system using one of the traditional modelling and simulation tools.

One possibility to avoid this restriction is to couple different tools together. A standardised method to accomplish the necessary tasks of exchanging and synchronising design and simulation data between different tools, offer so called simulation backplanes /2,3,4/. Another possibility is to import and export models between different tools using a standardised modelling language. In this area VHDL-A, a powerful modelling language covering problems described by continuous or discrete differential or algebraic equations, has a good chance to establish as a standard /5,6,7/. Actually both approaches suffer from bad performance and it will take some more time until satisfying tools are available.

3. Top down functional design

Even if a complete and satisfactory tool support would be available, at this point should be asked, if a design approach can really manage the task that is focused so strong on the component level only. Today electronic design methodology moves rapidly toward a straight forward top down approach: in a first design step only the demanded functions of the component under design are specified, without restrictions concerning the method or technology of implementation of one of these functions. If this step leads to one or more functions that are too complex to be implemented, it is repeated iteratively and the complex functions are broken down into less complex ones. The result of this process will be a function hierarchy with the complete product functionality at the top and a collection of less complex, easier to implement base functions at the bottom.

During a second synthesis or implementation step each base function is built up by existing components and circuits. For digital systems, today this step is completely automated. Only this second level of implementation needs simulation support on the component level. The first design level only deals with functional models that are represented by their input output behaviour, comparable to well known control devices but with more complex interfaces to represent the more complex interactions between the functions.

This idea can be generalised and is well suited to establish a modelling and simulation method that consequently supports functional top down design of mechatronic systems. From that point of view a mechanical component for example will no longer be represented by bodies and joints but by a dynamic transfer function and interfaces representing the behaviour of the joints. The physical meaning of the transfer functions and the interfaces in the case of mechanical systems can be made clear using the example of a multibody system consisting of a chain of several bodies.

Investigating one of the well known recursive algorithms available today to simulate the dynamic behaviour of tree structured multibody systems, gives a clear sight on the data flow at the interfaces of the i-th body /8, 9/. In a first step kinematic information k_i delivered by the predecessor is taken by the i-th body and locally transformed to get the kinematic information k_{i+1} at the output. This process runs from the root to the top

170

body. In a second step all forces and masses are collected beginning at the top body and then transmitted down to the root. The i-th body receives forces and masses J_{i+1} from its successor and transforms it to its output J_i acting on its predecessor. In a final step all accelerations are transmitted from the root to the top of the chain. The i-th body takes the accelerations acting at the joints to its predecessor, integrates its local states and delivers the accelerations needed by its successor. Based on this method a very universal interface for mechanical subsystems can be formulated. Of course each subsystem has its own internal structure and an interface dependent on the number and type of joints. Further the orientation of the chain has to be taken into account, when the subsystem is used. From a design point of view

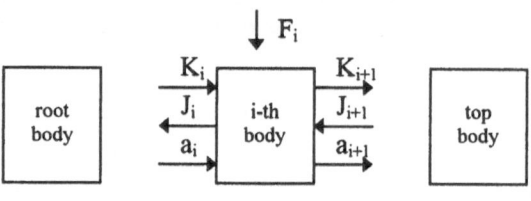

Figure 2: Data flow along a chain of rigid bodies

this demands are no relevant restrictions, because they reflect typical design considerations. For real systems only compatible joints (interfaces) can be connected together. Even the functional orientation at a component (e.g. gear, clutch, arm of a robot) is fixed and cannot be changed in practice. It should be mentioned that it is not possible to close kinematic loops directly using this method. For design purposes this will not be necessary, because the conditions of a kinematic loop usually should be handled inside a functional unit!

A similar approach allows to model functional units of electric and electronic systems. An example for a typical application consisting of a power supply, a switching controller and a consumer, is given in figure 3. For electrical systems the laws of Kirchhoff leads to interfaces that carry information about voltages U_i, currents I_i and impedances Z_i. Generally the structure of interfaces will be problem specific. But for typical applica-

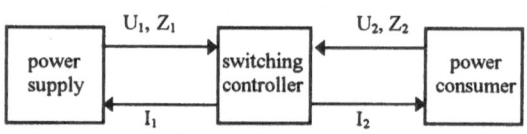

Figure 3: Example of electrical functions and interfaces

tions, it is possible to standardise the interfaces for functions like power supplies or switching controllers etc. Based on this method it is possible to create a library of building blocks for electric and electronic functions necessary to construct the electronic part of mechatronic systems.

Modelling and simulation of the control system can be done very straight forward, because only unidirectional signals have to be handled. The typical method to define control systems is a block diagram. The interface of a model for a control system consists simply of the signals representing the measured physical quantities at the controller input and the logic signals to drive the actuators at the controller output. Although the structure of these interfaces is very clear, the timing dependencies may be very complicated. Sensor signals for examples in most cases are converted sequentially at different points in time. On the other side the signals for the actuator systems also can be calculated only one after the other /10/.

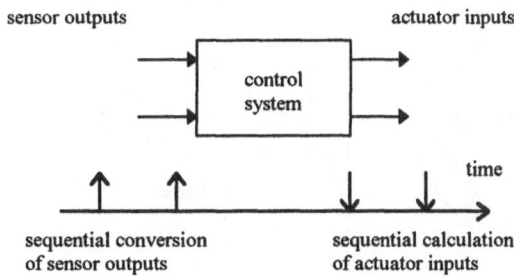

Figure 4: Timing dependencies of control system interface

Additional complications arise, because many control systems of mechatronic products are implemented using one of the standard low cost microcontrollers. These devices only support integer arithmetic, so the effect of quantisation and rounding of controller signals cannot be neglected but have to be simulated.

4. Simulation objects

Investigating the models resulting from this approach shows that they have a very similar structure. Although being specific for different types of applications, all models have well defined interfaces that carry domain specific interface data. They have internal data like model parameters and states. And there is a specific structure of the equations that always have to be computed in a problem dependent given order. This common structure of the mathematical models allows for a common method of description using object-oriented techniques /11, 12/.

The mapping of model entities into the world of objects is explained in figure 5. Each model of a basic function will be represented as a class. The instance variables of this class keep the internal states and parameters of the basic function. The equations are packed together and made available by methods. Instances of these classes are called simulation objects.

The mechanical interface described earlier for example, can be implemented using 3 methods , one for each step. They have to be activated in correct order up and down the chain. Interface objects to represent complex interface data like vectors of forces, mass matrices ands so on.

To improve the reusability of the simulation objects they are implemented as software processes. Their interfaces are built by communication channels that transport the interface objects and activate the associated methods. These processes can be connected

172

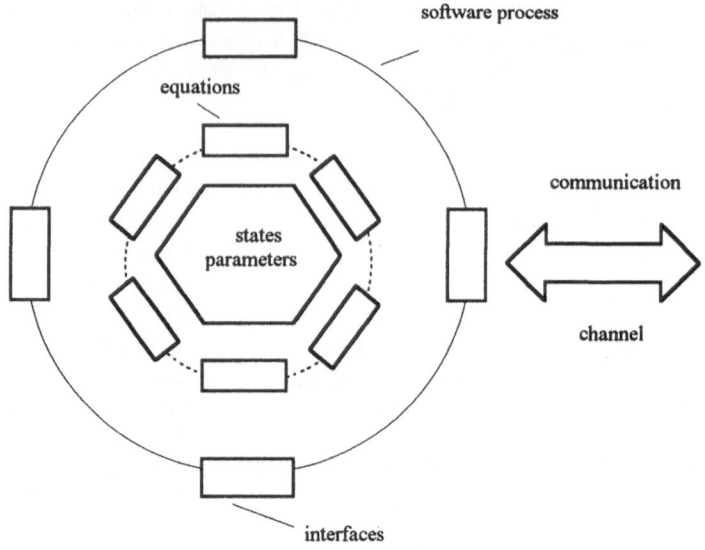

Figure 5: Communicating simulation objects

together very flexible like blocks of a block-diagram. Due to their more complex internal structure and much more powerful interfaces, complete mechatronic products, including mechanical, electrical and control parts, can be constructed very easy. Following the idea of functional top down design consequently, <u>all</u> functions will be implemented as classes. As already stated, classes corresponding to base functions implement computational models. The classes on the higher levels are necessary to represent the complete functional hierarchy of a mechatronic product and carry only structural information. They manage the function's interfaces and keep how functions are built up by sub-functions. These objects efficiently support the installation of the complete process and communication model on the base level and help to observe variables or to change design parameters across the hierarchy. The total computational work however is done on the base level without any overhead of hierarchical function calls or communication overhead. In this manner communicating simulation objects help to combine a very structured and consequent top down specification, even for very large systems containing many levels of hierarchy, with a very economic computation and communication doing the work on the base level.

The transformation of today used component oriented descriptions into functional models with well defined input-output behaviour can be carried out for example by means of symbolic computation /13/. One of the tools that can be used to manage this step is Dymola /14/. Dymola supports a powerful object-oriented formulation of basic components for different disciplines like rigid bodies and joints for mechanical systems and capacitors, resistors an inductors etc. for electrical systems. These basic components then can be used to build up complete base functions in the usual way by connecting electrical components by wires or by connecting mechanical bodies by joins. The symbolic computation capabilities of Dymola helps to generate the equations of a mathematical model for each function. The generation of symbolic equations has to be done in such way that interfaces for mechanical and electrical functions can be established. Following the ideas of top down design the base functions used as building blocks ideally are not too large. This is an ideal situation for symbolic computation because the complexity of symbolic expressions remains restricted.

To specify control systems, powerful tools like Matlab/Simulink are available /15, 16/. They allow to generate equations out of block diagrams that can be packed into simulation objects with little effort.

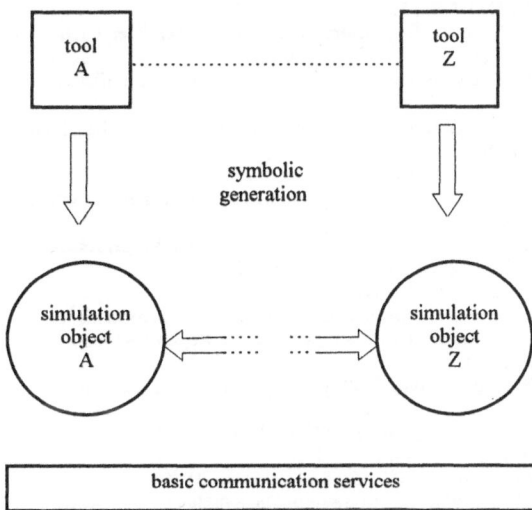

Figure 6: Modelling and simulation environment

To set the simulation objects into action, a platform of basic communication services is needed. Implementing these services can be done with standard operating systems like Windows NT or with real-time operating systems like PSOS. Implementing simulation objects as software processes allows the distribution of large simulation models on multiple processors. Splitting up the functions this way reduces communication overhead, because only a moderate amount of data has to be transferred across the interfaces. Thus a high speed-up is possible, when the number of processors is increased. Simulation speed offered by communicating simulation objects is very high. They are well suited to manage design and optimisation tasks for nonlinear mechatronic devices based on off-line simulation. For many applications even real-time simulation is possible /17, 18, 19/.

5. Conclusions

The paper presents a powerful modelling and simulation approach based on a structured top down functional design idea. Objectoriented methodology is used to represent the complete design hierarchy as classes that can be reused in different designs very

174

easily. The internal structure is encapsulated strictly and behaviour is available only via problem specific interfaces. The base of the hierarchy is built up by communicating simulation objects, representing the behaviour of implemented functionality. Implementing these objects as software processes guarantees a very economic and fast simulation procedure. Objects can be distributed across multiple processors with little overhead. Computation speed of the simulation objects is very high due to symbolic generation procedure so that large models can be simulated in real-time.

6. References

/1/   Microsoft Corporation (1994) Introducing Visual C++.
/2/   Long, D. I. and Mehat, S. S. (1993) Mixed signal simulators: tools for mixed-signal 'design for test'? Proceedings Mixed Signal VLSI Test, London, GB, 13. Dec. 1993, Vol. 1993, 240, 11.1-11.5.
/3/   Die Korrespondenz der A- und D-Simulatoren. Simulation Backplane fürs Mixed-Signal-Systemdesign. Markt und Technik, (1993) Heft 25, S. 14.
/4/   Cosimulation auf dem Supersimulator. Die Integration mehrer Simulatoren und anderer Tools. Markt und Technik, (1993) Heft 6, S. 61-63.
/5/   IEEE Standard VHDL Language (1988). Reference Manual. Institute of Electrical and Electronics Engineers, Inc., New York.
/6/   Patterson, A. and Weigert, S.(1993) Analoge Beschreibungselemente in VHDL integrierbar. Elektronik, München, Band 42 Heft 4, S. 34-38.
/7/   Bergé, J.-M. (1994) Analogue VHDL. The VHDL Newsletter # 12, Januar 1994, 7-9.
/8/   Kasper, R. and Hagel, S. (1994) Object-oriented modelling and simulation of mechatronic systems and their application to automotive industry. Mathmod 94, Vienna.
/9/   Kekcskeméthy, A. (1993) Objektorientierte Modellierung der Dynamik von Mehrkörpersystemen mit Hilfe von Übertragungselementen, VDI-Verlag, Düsseldorf, Reihe 20, Nr. 88
/10/   ETAS GmbH (1994): ASCET User Manual., Schwieberdingen.
/11/   Ahmed, S. et. Al. (1991) A Comparison of Object-Oriented Database Management Systems for Engineering Applications, Massachusetts Institute of Technology, Research Report R91-12.
/12/   Digitalk Inc. (1993) Smalltalk/V User's Guide.
/13/   Wolfram, S. (1991) Mathematica: A System for Doing Mathematics by Computer, Addison-Wesley.
/14/   Dynasim AB (1994) Dymola - Dynamic Modelling Language User's Manual.
/15/   The MathWorks Inc. (1992) MATLAB Reference Guide.
/16/   Simulink User's Guide (1992) The MathWorks, Inc. Natick, Massachusetts 01760.
/17/   Eppinger, A., Kasper, R. and Heinkel H.M. (1990) Hardware-in-the-loop design techniques with ASCET. Esprit CIM, CIM-Europe workshop on computer integrated design of controlled industrial systems, Paris.
/18/   Eppinger, A. and Kasper, R. (1992) Schnelle Echtzeitsimulation mit dem Softwarewerkzeug AS-CET. Bosch-Zünder, Ausgabe 5, Robert Bosch GmbH, Stuttgart.
/19/   Kasper, R. and Koch ,W. (1995) Object-oriented behavioural modelling. Third Conference on Mechatronics and Robotics. Paderborn, Teubner, Stuttgart.

ROBUST DECENTRALIZED CONTROL
OF MULTIBODY SYSTEMS

P.K.KIRIAZOV

Institute of Mechanics, Bulgarian Academy of Sciences
Acad.G.Bonchev Str., bl.4, BG-1113 Sofia, Bulgaria

Abstract. The paper presents a unified approach for decentralized feedback control of multibody systems (MBS) in the face of bounded parameter inaccuracies and random disturbances. The approach is based on optimal trade-off relations between the given bounds of perturbations, the system output accuracy and the control force limits. Several examples will be given to show applicability of the proposed control design concepts to various-type MBS. Decentralized sliding-mode controllers are designed for a half-car model with active suspensions to illustrate how robust such controllers can be.

1. Introduction

Most MBS like vehicles, robots, and mechanical structures with active vibration damping are complex dynamic systems, and all the dynamics model parameters are difficult if not impossible to identify. Besides model uncertainties, measurement and environment noises exist that further deteriorate the performance of system responses, especially in the case of higher velocities or heavier manipulation loads.

There is growing interest in designing efficient control systems for MBS in the presence of dynamic uncertainties. It is widely recognized that conventional approaches such as the well-known computed torque method are fairly sensitive to modelling errors. Another approach, adaptive control, is attractive in the sense that it is possible to improve the tracking accuracy with time even in the case of large parameter variations. Unfortunately, the guarantee of estimate convergence and the avoidance of long-term drift in parameter estimates are two major problems, especially in the case of measurement and environment noises. In general, controllers, using

175

D. H. van Campen (ed.), IUTAM Symposium on Interaction between Dynamics and Control in Advanced Mechanical Systems, 175–182.

full dynamic models or involving extensive computations, can hardly produce the required fast and accurate response.

Decentralized control techniques, based on simplified models, have been proposed in Bellini *et al.* (1988), Gavel and Hsia (1988), Oh and Jamshidi (1987), Singh (1990). The decentralization is in the sense that, during the motion, each controller refers only to the corresponding measured output. Such controllers have the advantages of low demand on computer resources and high speed operation, but, in most of the papers, the inertia coupling effects are viewed as disturbances and their bounds can not be *a priori* determined. A new design approach for robust decentralized control is therefore needed which takes into account these interactions and which should give optimal trade-off relations between the known bounds of model errors, the required system output accuracy and the existing control force limits.

Recently, a diagonal dominance condition on the control input matrix has been found (Kiriazov, 1992, 1994) to be necessary and sufficient for a decentralized control system to be robust to arbitrary, but otherwise bounded, disturbances. In our control system design method, we use the theory of generally diagonally dominant (GDD) matrices as presented in (Lunze, 1992). As numerous verifications show, most properly designed MBS have control input matrices which are GDD and can therefore be robustly controlled in a decentralized manner. For that purpose, we apply sliding-mode control (Utkin, 1992), the stability of which is easily verified by Lyapunov theory. With known bounds on modelling errors and external disturbances, the minimum control force magnitudes can be found which are sufficiently large to achieve the required system output accuracy. In this way, the proposed control design method can give an optimal compromise between the contradicting requirements of high accuracy, fast response, and low energy consumption.

The layout of the paper is as follows. In the next section, a general MBS dynamics model is introduced in order to define robust decentralized controllability and to present the necessary and sufficient condition for robustness. To show the domain of applicability of the proposed design approach, several examples of MBS which do or do not satisfy this condition are given in Section 3. Besides two-degree of freedom MBS, there are various-type MBS of higher dimensionality which, if properly designed, can be controlled in a decentralized manner. For all such MBS, a sliding-mode control design is next proposed which is based on the optimal trade-off relations. In Section 5, a system of decentralized sliding mode controllers is designed for a half-car model with active suspensions to show how robust they can be. Finally, conclusion are drawn.

2. Decentralized Control of Multibody Systems

2.1. DYNAMICS MODELLING OF CONTROLLED MBS

There is always a trade-off between full dynamic modelling, accurate identification and robust control design. MBS, though actually flexible, can be approximated by a composition of rigid bodies connected by joints, springs, dampers, and actuator forces (Schiehlen, 1990). For simplicity, we consider controlled MBS with collocated sensors/ actuators that can be modelled by the following system of differential equations

$$\ddot{q}_i(t) = \sum_{j=1}^{n} A_{ij}(q,t)u_j(t) + a_i(q,\dot{q},t) \ , \quad i=1,...,n \tag{1}$$

where A is the control input matrix (i.e., the product of the inverse inertia matrix and the control gain matrix), u and q are the input force and controlled output, respectively. We assume that a can be feedforward compensated to some extent, and, for the purpose of feedback stabilization, we shall use the following error model

$$\ddot{e}_i(t) = -\sum_{j=1}^{n} A_{ij}(q,t)u_j(t) + d_i(t) \ , \quad e_i = q_i^{ref} - q_i, \quad i=1,...,n \tag{2}$$

where the disturbance vector d stands for all uncompensated and neglected terms in (1) as well as the the existing measurement and environment noises. As a measure of the tracking precision, we take the absolute value of $s_i = \dot{e}_i + \lambda e_i$, where $\lambda \geq 0$.

2.2. DECENTRALIZED FEEDBACK CONTROL

We prefer decentralized control for its main advantages over centralized control: fast response, higher reliability, and minimum computation effort.

Definition: In the presence of arbitrary disturbances d_i with known upper bounds d_i^+, a system of decentralized feedback controllers $u_i = u_i(s_i)$ is robust when it gets every local subsystem state $(q_i, \dot{q}_i)$ to track the desired state $(q_i^{ref}, \dot{q}_i^{ref})$ with prespecified, maximum allowable values δ_i of errors $|s_i|$, $i=1,...,n$.

We have to deal with the following control system design problem:

178

With known bounds of disturbances and desired system output accuracies, design a system of decentralized controllers that need minimum actuator forces to robustly stabilize the motion.

Main Result (Kiriazov, 1992, 1994): A necessary and sufficient condition for MBS (1) to be robustly controlled by a system of decentralized controllers is that the control input matrix A be GDD.

When the control input matrix A is quadratic and symmetric, then, according to (Lunze, 1992, p.320) and (Kiriazov, 1992, 1995), A is GDD if and only if its comparison matrix is positive definite.

3. Examples of MBS with GDD Control Input Matrices

In case of actuator/sensor colocation, the necessary and sufficient condition stated above will be satisfied if only the inverse inertia matrix is GDD. The quantities of this matrix can be expressed in terms of parameters presenting geometrical dimensions, mass distribution, and control force locations. We give examples of two-, three-, and six-degree of freedom MBS which, if their parameters are properly designed, can have GDD inverse inertia matrices.

3.1. TWO-DEGREE-OF-FREEDOM MBS

It is evident that in this case MBS have GDD inverse inertia matrices as this property for 2x2 matrices is equivalent to the fact that the inertia matrix is positive definite. For example, such MBS can be all two-joint manipulators, or any rigid beam-like mechanical systems with two active suspensions.

3.2. THREE-DEGREE-OF-FREEDOM MBS

3.2.1. *Platform with Three Active Suspensions*

Depending on the actuator location, such a platform can have GDD inverse inertia matrix or not (Fig.1).

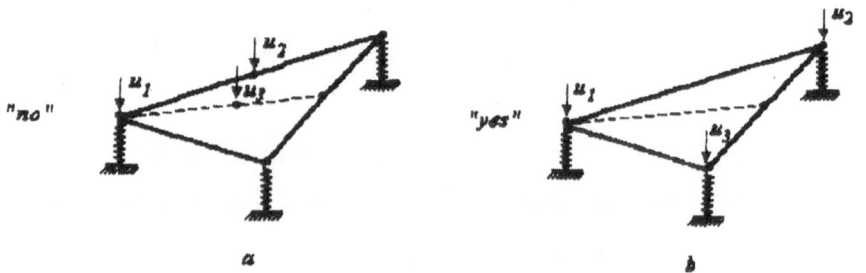

Fig.1: Platform with active suspensions

3.2.2. *Planar Manipulator with Three Rotational Joints* (Fig.2)

For such a redundant manipulation system in the plane, the inverse inertia matrix (in the worst case of stretched configuration) may not be GDD if the mass distribution is not appropriate. Two geometrical designs are considered: (*a*) with equal link lengths and (*b*) with link lengths decreasing in progression from the base to the tip. Accordingly, the conditions on how the link masses should decrease (in order to have GDD property) are derived.

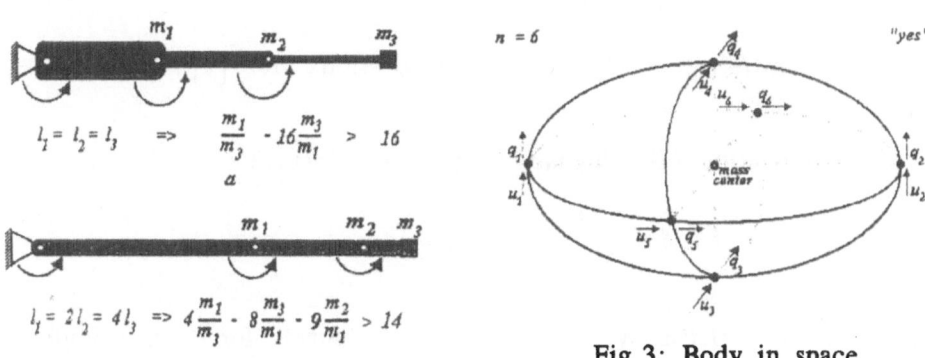

$$l_1 = l_2 = l_3 \quad \Rightarrow \quad \frac{m_1}{m_3} - 16\frac{m_3}{m_1} > 16$$

a

$$l_1 = 2l_2 = 4l_3 \quad \Rightarrow \quad 4\frac{m_1}{m_3} - 8\frac{m_3}{m_1} - 9\frac{m_2}{m_1} > 14$$

b

Fig.2: Planar manipulator

Fig.3: Body in space

3.3. SIX-DEGREE-OF-FREEDOM MBS

A rigid body ellipsoid with six generalized coordinates controlled by six actuators is considered. According to (Kiriazov, 1995) where det$A$ is defined as a measure of the robust decentralized controllability, the best actuator/sensor location is obtained, as shown in Fig.3.

4. Sliding-Mode Control Design

Introduce matrix B: $B_{ij} = -|A_{ij}|$ if $i \neq j$, and $B_{ii} = A_{ii}$. When A is GDD, it follows from the theory in (Lunze, 1992) that the corresponding system of equations

$$\sum_{j=1}^{n} B_{ij} u_i^+ = d_i^+ \quad , \quad i = 1, \ldots, n \tag{3}$$

will have positive solutions with respect to the bounds u_i^+ of $|u_i(s_i)|$. Existence of such solutions enables robust decentralized control system design with prespecified magnitudes of the control functions. For that purpose, we apply sliding-mode control which stability can be easily verified applying Lyapunov theory.

As was reported in (Slotine & Shastry, 1983), sliding modes in motion control exhibit the so-called chattering imposed by the discontinuity of the control action. To avoid this, the authors propose a "boundary layer" method in which continuous control functions of saturation type can be used:

$$u_i(s_i) = u_i^+ \operatorname{sat}(s_i/\delta_i) \ , \quad i = 1, \ldots, n \tag{4}$$

$$\operatorname{sat}(y) = y \text{ for } |y| < 1 \text{ and } \operatorname{sat}(y) = \operatorname{sgn}(y) \text{ for } |y| > 1 \tag{5}$$

The conditions for boundary layer attractiveness

$$|s_i| \geq \delta_i \ \Rightarrow \ \dot{s}_i s_i \leq -\mu_i |s_i| \tag{6}$$

are satisfied (Kiriazov, 1994) if the control force magnitudes are determined from the following system of equations

$$\sum_{j=1}^{n} B_{ij} u_i^+ = d_i^+ + \mu_i, \quad \mu_i > 0, \quad i = 1, \ldots, n \tag{7}$$

Relations (4) and (7) present optimal trade-offs between the tracking precision, the range of disturbances, and the amount of feedback control effort. With given δ and d^+, Eqs. (7) give the minimum values of the upper control bounds that are necessary for the robust stability of MBS with decentralized control structure.

Besides linear spline forms (4), we can employ other bounded but smoother control functions. For example, it is rather convenient to use, as in (Kreuzer & Pinto, 1995), the following control laws:

$$u_i(s_i) = (2/\pi) u_i^+ \arctan(p_i s_i) \ , \quad i = 1, \ldots, n \tag{8}$$

5. Half-Car Model with Active Suspensions

According to (Gordon et al., 1994), (Gordon, 1995), (Hac, 1994), controllers for the body (slow) motion and the wheels (fast) motion can be designed separately and then combined into one composite controller. In these papers, optimal control methods are proposed based on LQR formalism with mixed state/control cost functional.

In this study, the sliding-mode control design (Section 3) is applied to the body two-dimensional motion (pitch & heave or roll & heave) subject to random disturbances with known bounds. The control force magnitudes needed to robustly stabilize the car motion are determined from the optimal trade-off relations (7).

In the numerical considerations, the half-car model given in (Krtolica & Hrovat, 1992) is used. Random disturbances with non-zero mean value were introduced in this model. Their amplitudes were increased up to 90% of the nominal accelerations that can be produced by the maximum control forces. The robust transient behaviour from a non-zero initial state, in the presence of such persistent disturbances, is shown in Fig.4.

$$
A = \frac{1}{MJ} \begin{bmatrix} J+Mf^2 & J-Mfr \\ J-Mfr & J+Mr^2 \end{bmatrix}
$$

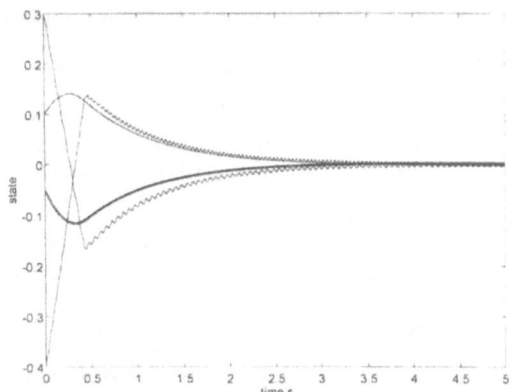

M - sprung mass
J - moment of inertia
f/r - distances between
front/rear wheels and
center of mass

Fig.4: System response

6. Conclusion

A simple approach for robust decentralized control of various-type MBS has been proposed. The only requirement to MBS is to have GDD control input matrices. Any properly designed MBS can satisfy this condition. The control system design approach is based on optimal trade-off relations between the given bounds of perturbations, the system output accuracy and the control force limits. Decentralized sliding-mode controllers are designed for a half-car model with active suspensions to illustrate how robust such controllers can be.

Thus the GDD condition should play the central role in the integrated design of MBS having decentralized control systems (Kiriazov, 1995). Further, research on the robust control design in the discrete case has to be done to take into account the other important robustness parameter: the sampling time.

182

Acknowledgement. Financial support from the IUTAM and the National Science Foundation of Bulgaria (Project MM-426/94) is gratefully acknowledged.

7. References

Bellini, A., G. Fidalli and G.Ulivi (1988) Robust decoupling control of an industrial manipulator using sliding mode, *Int. J. Robotics and Automation*, No.1, pp.42-50.

Gavel, D.T. and T.C. Hsia (1988) Decentralised adaptive control experiments with the PUMA robot arm, *Proc. IEEE Conf. Rob. and Auto.*, pp. 1022-1025

Gordon, T.J. (1995) An integrated strategy for the control of complex mechanical systems based on sub-system optimality criteria, *Proc. IUTAM Symposium on Optimization of Mechanical Systems*, Eds. D. Bestle and W. Schiehlen, Kluwer Academic Publishers, pp.97-104.

Gordon, T.J., Palkovics, L., Pilbeam,C. and Sharp, R.S. (1994) Second generation approaches to active and semi-active suspension control system design, *The Dynamics of Vehicles on Roads and Tracks, Proc. of the 13th IAVSD Symp.*, China, 1993, *Supplement to Vehicle System Dynamics*, 23, pp. 158-171.

Hac, A. (1994) Decentralized control of active suspensions with preview, *Proc. American Control Conf.*, pp. 1952-1956

Kiriazov, P. (1992) A decentralized controllability measure for robotic manipulators, Proc. IEEE Conf. Rob. and Auto., pp.2141-2145.

Kiriazov, P. (1994) Necessary and sufficient condition for robust decentralized controllability of robot manipulators, *Proc. American Control Conf.*, pp.2285-22S7.

Kiriazov, P. (1995) Robust integrated design of controlled multibody systems, *Proc. IUTAM Symposium on Optimization of Mechanical Systems*, Eds. D. Bestle and W. Schiehlen, Kluwer Academic Publishers, Dordrecht, pp.155-162.

Krtolica, R. and Hrovat, D. (1992) Optimal active suspension control based on a half-car model: an analytical solution, *IEEE Trans. Auto. Control*, No.4, pp. 528-532.

Kreuzer, E. and Pinto, F. (1995) Remotely operated vehicle - mechatronic system, *Proc. 3rd Conf. on Mechatronics and Robotics*, Paderborn, Germany.

Lunze, J. (1992) *Feedback Control of Large-Scale Systems*, Prentice Hall, UK.

Oh, B.J. and M. Jamshidi (1987), Decentralized adaptive feedforward/feedback robot manipulator control, *Proc. IEEE Conf. Rob. and Auto.*, pp. 355-359.

Schiehlen, W. (1990) (ed.): *Multibody Systems Handbook*. Berlin: Springer.

Singh, S.K. (1990) Decentralized variable structure control for tracking nonlinear systems, *Int. J. Control*, No. 10, pp. 811-831.

Slotine, J.J. and S.S. Shastry (1983) Tracking control of nonlinear systems using sliding surfaces with application to robotic manipulators, *Int. J. Control*, No.2, pp.465-492.

Utkin, V.I. (1992) *Sliding Modes in Control and Optimization*. Springer-Verlag.

DYNAMIC MODELING OF IMPEDANCE CONTROLLED DRIVES FOR POSITIONING ROBOTS

K. GR. KOSTADINOV, G. V. BOIADJIEV
Institute of Mechanics - Laboratory "Mechatronic Systems"
Acad. G. Bonchev St., Block 4, 1113-Sofia, Bulgaria

1. Introduction

Secondary positioning robots appear to be essential for the automation of precise operation in various branches of industry. The positioning robots are used for precise product orientation in the work space and successive feeding during given technological operations, such as drilling, grinding, assembling, deburring and dimensional quality control. To provide for the accuracy of the positioning robots is the main goal of the control engineer. The most appropriate method of controlling the dynamic properties of the positioning robots is the method of impedance control(Hogan, 1985). Impedance control consists in modifying specified dynamic properties of the servo drive in addition to controlling the position and the velocity of its output link. In this work a dynamic modelling of impedance controlled drives (Kostadinov, 1994) is considered. A third approach for impedance control is used, which is realised by redundancy of drives. The dynamic modelling of impedance controlled drives for these robots is based on graph theory and the Orthogonality Principle.

2. Problem description.

The drives discussed in this work can be treated as components of a more complex system, which consists of a mechanical subsystem, an electrical subsystem and a servocontroller. The substance of the method(Boiadjiev, 1991) lies in the decomposition of the complex system into relatively simple components, and the derivation of characteristic terminal equations for each component. The system is assigned a general graph, which reflects the structure of the interrelations between individual components. The system is discussed jointly with its general energy space , which consists of energy subspaces, depending on the types of energy in the particular problem. The energy is considered as a general measure of the motion of the material objects and is also used as a quantitative estimate of the material interactions. Two main characteristics of energy are energy flow and energy potential. When a physical system is analysed, the flow and potential are expressed by a number of variables, called *through* and *across* variables. These are parameters of the main energy subspaces and form their basis. The general

D. H. van Campen (ed.), IUTAM Symposium on Interaction between Dynamics and Control in Advanced Mechanical Systems, 183–190.
© 1997 *Kluwer Academic Publishers.*

184

graph is used to obtain the equations of the flow and potential (also called the *cutset* and *circuit* equations) in the main energy subspaces. The three groups of equations - the *terminal* equations and the cutset and circuit equations - are supplemented by the fourth group, the equations that describe the transformation of energy from one kind to another (called *connection* equations). The four groups of equations defined in the analysis of a particular system are used to obtain its dynamic equations.

3. Impedance controlled drives for positioning robots.

The idea to introduce redundancy of drives for each degree of freedom of the positioning robots is based on the method (Kostadinov and Parushev, 1987) for transition of motion. It consists of introducing a two-zone dynamic controllable gearing at the output link by an opposing drive unit, identical to the primary one for each joint. In this way the kinematic chain is symmetrical and closed with the additional drive. In these drives, the differences between control voltages of both motors set up the necessary knot mechanical impedance, while the control voltage of the drive motor sets up the necessary output link velocity. The structure of these impedance controlled drives is shown in Fig. 1.

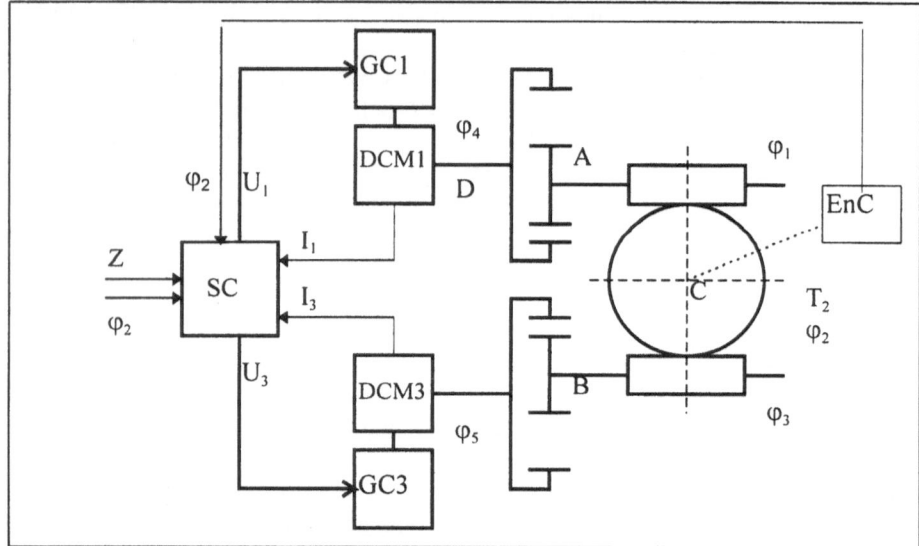

Figure.1. Structure of impedance controlled drives.

4. Dynamic model of impedance controlled drives.

The drives discussed in this work can be treated as components of a complex system consisting of mechanical and electrical subsystems and a servocontroller.

4.1. EQUATIONS OF THE MECHANICAL SUBSYSTEM.

The mechanical subsystem consists of a worm wheel with two worm gears attached.

Each worm gear is connected by a conventional gear internal gearing to the shaft of a DC motor (Fig. 1.).

The mechanical subsystem is divided into two components, with one formed by the worm gear, and the other formed by conventional gears with internal gearing.

4.1.1. Equations of the worm gear unit.

The worm gear unit is connected to the remaining part of the mechanical subsystem at the contacting points A, B and C (Fig. 1.). It is assigned a graph G_R (Fig. 2.a.), whose arcs are related to the following variables (O is the origin of the inertial frame).

arcs 1, 2, 3: through variables T_1, T_2, T_3 (D'Alembert torques) (Boiadjiev, 1991); across variables $\dot{\varphi}_1, \dot{\varphi}_2, \dot{\varphi}_3$ (angular velocities).

arcs 4, 5 : the relative angular velocities $\dot{\varphi}_4, \dot{\varphi}_5$ and the torques T_4, T_5.

arcs 6, 7, 8: external dimensions of the across and through variables that describe the connection of the unit with the remaining part of the system. They are noted by $\dot{\varphi}_6, \dot{\varphi}_7, \dot{\varphi}_8$ and T_6, T_7, T_8, respectively.

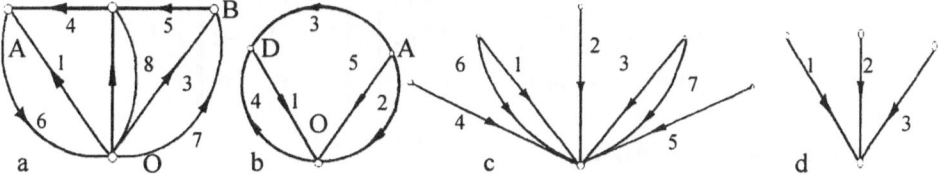

Fig.2. Graphs of the mechanical subsystem.

As the vectors T_k and $\dot{\varphi}_k$ $(k=1,2,...,8)$ have fixed direction, the variables can be described by scalar quantities. The following terminal equations can be specified for the components of the unit - objects 1, 2 and 3 (worms and worm-wheel) (Boiadjiev and Lilov, 1993):

$$\left[J_i \frac{d^2}{dt^2} + B_i \frac{d}{dt} + C_i \right] \varphi_i = -T_i, \quad i=1,2,3 \tag{1}$$

Here J_i ($i = 1, 2, 3$) designates the inertia moments of the worms 1 and 3 and worm wheel 2. B_i and C_i are the coefficients of friction and the stiffness, respectively.

The following relation is obtained for the geometric parameters that describe the unit where m is the gear ratio between the worm gear and the wheel:

$$\varphi_1 = m \, \varphi_2, \qquad \varphi_3 = m \, \varphi_2, \tag{2}$$

In the graph (Fig. 2.a.) a tree with branches formed by the arcs 6, 7, 8 is chosen. The cutset equations for it have the following form:

$$\left[U_3 \quad Q \right] \begin{bmatrix} Y_b \\ Y_c \end{bmatrix} = 0, \tag{3}$$

where $\mathbf{U}_3$ is a unit (3x3) matrix, $\mathbf{Y}_b = [T_1, T_2, T_3]^T$ contains the through variables associated with the branches and $\mathbf{Y}_c = [T_4, ..., T_8]^T$ the through variables associated with the chords. $\mathbf{Q}$ is a matrix with scalar elements $0, -1, +1$, shown below

$$\mathbf{Q} = \begin{bmatrix} -1 & 0 & 0 & -1 & 0 \\ 0 & 0 & 1 & 0 & 1 \\ 0 & 1 & 0 & -1 & -1 \end{bmatrix} \tag{4}$$

For the sake of simplicity, let L_k $(k = 1,2,3)$ denote the operator $J_k \dfrac{d^2}{dt^2} + B_k \dfrac{d}{dt} + C_k$.

Then, after substituting (1) into (3), excluding the torques T_4 and T_5 and changing all the variables with subscript 6, 7, 8 into new 1, 3, 2, we obtain the terminal equations of the unit in the form:

$$\begin{bmatrix} 1 & -m & 0 \\ 0 & m(L_1 + L_3) + L_2 & 0 \\ 0 & -m & 1 \end{bmatrix} \begin{bmatrix} \varphi_1 \\ \varphi_2 \\ \varphi_3 \end{bmatrix} = \begin{bmatrix} 0 & 0 & 0 \\ 1 & -1 & -1 \\ 0 & 0 & 0 \end{bmatrix} \begin{bmatrix} T_1 \\ T_2 \\ T_3 \end{bmatrix} \tag{5}$$

4.1.2. Equation of the gear-drive.
Similarly, the gear-drive (Fig. 1.) treated as a unit, is assigned a graph (Fig. 2.b). For it the terminal equations are:

$$\begin{bmatrix} 1 & -n \\ 0 & nP_1 + P_2 \end{bmatrix} \begin{bmatrix} \varphi_1 \\ \varphi_2 \end{bmatrix} = \begin{bmatrix} 0 & 0 \\ 1 & -1 \end{bmatrix} \begin{bmatrix} T_1 \\ T_2 \end{bmatrix}, \tag{6}$$

where (k =1,2) denotes the operators $P_k = J_k^* \dfrac{d^2}{dt^2} + B_k^* \dfrac{d}{dt} + C_k^*$. Here J_k^*, B_k^*, C_k^* have a meaning similar to (1) and n is the gear-drive ratio.

4.1.3. Equation of the entire subsystem with the two gear-drives.
According to the conditions considered in this particular case, the two gears A and B (Fig.1.) operate in different modes. From (6) for the gear A after renumbering the indices 1,2 of the variables to 4,6 and for gear B after renumbering, the indices to 5,7 we obtain:

$$T_6 = T_4 - (nP_1 + P_2)\varphi_6, \qquad T_7 = T_5 + (nP_1 + P_2)\varphi_7 \tag{7}$$

Next, as the parameters of the two reducers are equal in this particular case, the stiffnesses C_1^* and C_2^*, will be denoted by $\tilde{C}$, whereas the stiffnesses C_1, C_2 and C_3 are denoted by C. The graph G_R^* of entire mechanical subsystem is shown in Fig. 2.c.

This leads to the of the cutset and the circuit equations

$$-T_1 = T_6, \quad -T_3 = T_7, \quad \varphi_1 = \varphi_6, \quad \varphi_3 = \varphi_7 \ .$$

For a unit which participates in an electromechanical drive system the following terminal presentation is obtained (the arcs 4 and 5 and their variables are renumbered to 1 and 3):

terminal graph - the same as it is shown in Fig. 2.d., but with the only arcs 1, 2, 3 terminal equations:

$$\begin{bmatrix} 1 & -mn & 0 \\ 0 & L_2+2m(L_1+P_2)+2mnP_1 & 0 \\ 0 & -mn & 1 \end{bmatrix} \begin{bmatrix} \varphi_1 \\ \varphi_2 \\ \varphi_3 \end{bmatrix} = \begin{bmatrix} 0 & 0 & 0 \\ 1 & -1 & -1 \\ 0 & 0 & 0 \end{bmatrix} \begin{bmatrix} T_1 \\ T_2 \\ T_3 \end{bmatrix},$$

where the parameters for the worm gears and gears are given by

$$L_1 = J_r \frac{d^2}{dt^2} + B_r \frac{d}{dt} + C; \quad P_1 = J_{1R}\frac{d^2}{dt^2} + B_{1R}\frac{d}{dt} + \tilde{C};$$

$$L_2 = J_2 \frac{d^2}{dt^2} + B_2 \frac{d}{dt} + C; \quad P_2 = J_{2R}\frac{d^2}{dt^2} + B_{2R}\frac{d}{dt} + \tilde{C};$$

The following notation is used: J_r - worm gear inertia moment, B_r - worm gear friction coefficient (when engaged to the wheel), C - worm gear stiffness which is assumed to be equal to that of the wheel, J_2, B_2, C - similar parameters related to the wheel, J_{1R}, B_{1R}, $\tilde{C}$ - parameters on the gear's input, J_{2R}, B_{2R}, $\tilde{C}$ - parameters on the gear's output (via the worm gear).
The differential equation that describes the motion of the mechanical subsystem is:

$$\begin{aligned} \left[J_2+2m(J_r+J_{2R})+2mnJ_{1R}\right]\ddot{\varphi}_2 + \left[B_2+2m(B_r+B_{2R})+2mnB_{1R}\right]\dot{\varphi}_2 + \\ +\left[C+2m(C+\tilde{C})+2mn\tilde{C}\right]\varphi_2 = T_1 - T_2 - T_3 \end{aligned} \quad (8)$$

4.2. EQUATIONS OF THE ELECTROMECHANICAL UNIT CONSISTING OF THE MECHANICAL SUBSYSTEM AND THE DC MOTORS.

Let two motors be connected to the mechanical subsystem discussed above. They are described by (Kostadinov, 1994).

$$U_1 = k_e\omega + i_1R + L\frac{di_1}{dt}, \quad T_1 = J\frac{d\omega}{dt} + T_{mm} + T_T, \quad (9)$$

$$T_1 = k_Ti_1, \quad T_2 = k_Ti_2, \quad U_1 = U_1 - \frac{R}{k}(T_1-T_3) = k_e\omega + Ri_3 + L\frac{di_3}{dt}.$$

Here the usual notation of U, i, T, ω for voltage, current, torque and angular velocity is adopted. The other symbols denote the parameters characterising the two connected DC motors in their working mode.

In the following, the parameters of the first motor will be subscripted by 1 and those of the second motor by 3. The equations (9) after some development lead to the terminal equations (10) of the motors, where J means the motor inertia moment:

$$T_s - J\ddot{\varphi}_s + \frac{k^2}{R}\dot{\varphi}_s = \left(\frac{k}{R}U_s - L\frac{d}{dt}i_s\right), \quad (s = 1,3), \tag{10}$$

4.3. EQUATIONS OF THE GATE CONVERTORS.

We suppose that the gate converters are added to the motors, and consider them as 4-terminal components with terminal equations

$$U_1^* k_{cvT1} = U_1 + T_{cvT}dU_1 / dt, \quad U_3^* k_{cvT2} = U_3 + T_{cvT2}dU_3 / dt \tag{11}$$

After similar transformations, (11) leads to the terminal equation of the unit which is taken into account in the final equation. The coefficients of the gate converters are assumed equal in this particular case, i.e.: $k_{cvT1} = k_{cvT2} = k_{vT}; \quad T_{cvT1} = T_{cvT2} = T_{vT};$):

4.4. EQUATIONS OF THE SERVOCONTROLLER.

A servocontroller is connected to the electromechanical unit. It is considered as a 5-terminal component. The following variables are connected to the terminals:

- the correction voltages ΔU_1^* and ΔU_3^*, added to the pre-set voltages $U_1^{(r)}$ and $U_3^{(r)}$. The voltages U_1 and U_3 are fed to the gate converter inputs.
- the signal depending on the angular velocity of the output shaft of the electromechanical unit. Its value is given by the equation (14) which gives the relation between $\dot{\varphi}_2$ and ΔU_1^*.
- the currents i_1 and i_3 in the motor anchor circuits 1 and 3.

By means of these variables, the servo controller forms the output corrections ΔU_1^* and ΔU_3^*, on the basis of equations (12-14):

$$U_1^* = U_1^{(r)} + \Delta U_1^*, \qquad U_3^* = U_3^{(r)} + \Delta U_3^* \tag{12}$$

$$\Delta U_3^* = k_{p2}\left[I^{(r)}(t) - i_3^*(t)\right] + k_{g2}\left[\frac{d}{dt}I^{(r)}(t) - \frac{d}{dt}i_3^*(t)\right] \tag{13}$$

$$\Delta U_1^* = k_{p1}\left[\dot{\varphi}_2^{(r)}(t) - \dot{\varphi}_2(t)\right] + k_1 \int_0^t\left[\dot{\varphi}_2^{(r)}(\tau) - \dot{\varphi}_2(\tau)\right]d\tau \tag{14}$$

The current i_1^* is implicitly involved in $\dot{\varphi}_2(t)$, k_{p1}, k_{p2}, k_{g2}, k_1 are constant coefficients, $I^{(r)}(t)$ and $\dot{\varphi}_2^{(r)}(t)$ are pre-set functions of the current and the angular velocity and $U_1^{(r)}$ and $U_3^{(r)}$ are pre-set functions of the input voltages.

The equation (12) involves pre-set functions $U_1^{(r)}$ and $U_3^{(r)}$, which are described by:

$$U_1^{(r)}(t) = mnk_e\dot{\varphi}_2^{(r)}(t) \tag{15}$$

$$U_3^{(r)}(t) = \frac{mn}{k_z}\left[k_e\dot{\varphi}_2^{(r)}(t) + R_a I^{(r)}(t) + k_r\frac{d}{dt}I^{(r)}(t)\right], \tag{16}$$

Here, the derivative of $I^{(r)}(t)$ can be neglected for simplicity.

4.5. EQUATION OF THE DRIVE SYSTEM.

If the right-hand side of equation (8) is designated by F, then the equation of motion of the electromechanical drive will be taken as:

$$\hat{L}\left(\varphi_2,\dot{\varphi}_2,\ddot{\varphi}_2\right) = F\left(\varphi_2^{(r)}, I^{(r)}, i_3^*, \dot{\varphi}_2^{(r)}, \frac{d}{dt}i_3^*\right), \tag{17}$$

The right-hand side contains the known functions of time $\varphi_2^{(r)}$, $I^{(r)}$ and the function i_3^* that describes the current of the second opposing motor. After some analysis (eliminating the currents taking into account their relation with the voltages applied to the motors), equation (17) can be written as

$$\left[J_2 + 2m(J_r + J_{2R}) + 2mn(J_{1R} + J) - \frac{kk_{vT}\cdot L}{R^2}\left(k_p + \frac{k_e mn}{L + k_{vT}\cdot k_g}\right)\right]\ddot{\varphi}_2 +$$

$$+ \left[B_2 + 2m(B_r + B_{2R}) + 2mnB_{1R} - \frac{kk_{vT}}{R}\left(\frac{Lk_g}{R} - k_p\left(1 + \frac{k_e mn}{R + k_{vT}\cdot k_p}\right)\right)\right]\dot{\varphi}_2 +$$

$$+ \left[C + 2m(C + \tilde{C}) + 2mn\tilde{C} + \frac{kk_{vT}\cdot k_g}{R}\right]\varphi_2 = \tag{18}$$

$$= \frac{kk_{vT}}{R}\left(1 + \frac{k_p}{k_e} + \frac{k_p}{R + k_{vT}\cdot k_p} - \frac{Lk_g}{Rk_e}\right)U_1^{(r)} + \frac{kk_{vT}\cdot k_g}{R}\int_0^t U_1^{(r)}d\tau - \frac{kk_{vT}\cdot L}{R^2}\left(1 + \frac{k_p}{k_e} + \frac{Lk_p}{R(L + k_{vT}\cdot k_g)}\right)\dot{U}_1^{(r)}$$

$$- \frac{kk_{vT}}{R + k_{vT}k_p}\left(1 + \frac{k_p k_z}{R}\right)U_3^{(r)} + \frac{kk_{vT}\cdot L^2}{R^2(L + k_{vT}\cdot k_g)}\left(1 + \frac{k_p k_z}{R}\right)\dot{U}_3^{(r)} - T_2$$

Let us consider the equation of motion of the mechatronic feeding (rotating) device with impedance controlled drive, which is characterised by the following parameters

$L = 1,37.10^{-3}\,H;\quad R = 1,14\Omega;\quad m = 52;\quad n = 62/30;$

$T_{vT} = 0,1ms;\quad J_2 = 2,2136.10^{-3}\,kgm^2;\quad J_r = 2,429.10^{-3}\,kgm^2;$

$J_{2R} = 1,321.10^{-3} kgm^2;$ $J_{1R} = 2,415.10^{-5} kgm^2;$ $J = 4,11.10^{-3} kgm^2;$

$B_2 = B_r = B_{2R} = B_{1R} = 0,1;$ $C = 9,6962.10^6 Nm / rad;$ $\tilde{C} = 2,424.10^5 Nm / rad;$

$k_{p1} = k_{p2} = 0,1;$ $k_{g1} = k_{g2} = 0,05;$ $k_{vT} = 1,5;$ $k = k_e = 0,172 Vs/rad;$

A control synthesis is made by a program which works under the MATLAB package. The control voltages of the drive motor for two values of the drive mechanical impedance are shown on Fig.3. The corresponding output velocities have variation less than 0,5% in comparison with the reference velocity (Fig.4).

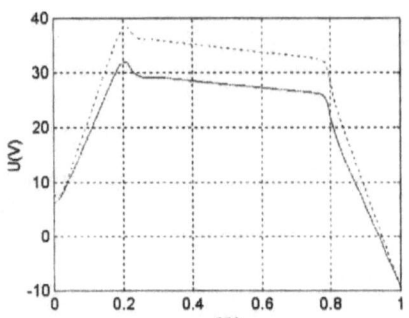
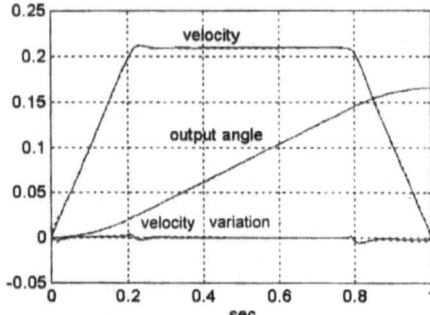

Figure 3. Control voltages of the drive motor.   Figure 4. Output link parameters for 2 impedances.
(Dotted line is for the case with higher impedance)

5. Conclusion.

Dynamic modelling of impedance controlled drives is done by using graph theory. This allows investigation of the separated parts (mechanical part, electrical part and servocontroller) as well as the whole. The equation of motion is obtained as a function of the control voltages U_1^* and U_3^* of the pair of opposing motors for the pre-set output link velocity $\dot{\varphi}_2$ and torque T_2. The difference $U_1^* - U_3^*$ - sets up the knot mechanical impedance so that the desired smooth motion is achieved.

Acknowledgements: The authors gratefully acknowledge the support from the EC Copernicus Program under the ROQUAL Project CIPA CT94 0109.
   The contribution of N. Alexeeva, in modelling and programming is also acknowledged.

6. References

Boiadjiev, G.V., and Lilov, L. (1993), Dynamics of multi component systems based on the orthogonality principle, Sofia, *Theoretical and Applied Mechanics*, 1, 11-26.

Boiadjiev, G.V., (1991), "Modelling of electromechanical systems", (in Bulgarian), *Ph.D. Thesis*, Sofia University, Sofia.

Hogan, N., (1985), Impedance control an approach to manipulation, Trans. ASME *Dynamics of System , Measurement and Control*, 107, 1, 1-23.

Kostadinov, K. Gr., (1994), Synthesis and investigation of electromechanical drives for positioning robots, *PhD. Thesis*, Institute of Mechanics and Biomechancis, Sofia, (in Bulgarian).

Kostadinov, K. Gr. and Parushev, P.R., (1987), "Method of Motion Transition", *Bulg. Patent* N 44365,

PARAMETER-IDENTIFICATION IN NONLINEAR MODELS USING PERIODIC EQUILIBRIUM STATES

A. DE KRAKER, G. VERBEEK AND N. VAN DE WOUW
Department of Mechanical Engineering,
Eindhoven University of Technology,
P.O. Box 513, 5600 MB Eindhoven, The Netherlands

1. Introduction

For the identification of parameters in numerical models for nonlinear dynamical systems often transient trajectories (for example resulting from drop-tests for aircraft landing gears) or chaotic trajectories are used as input information for the identification procedure. Instead of these types of trajectories the use of stable or even unstable periodic solutions has some clear advantages such as the possibility of use for a future design optimization for a landing gear or the possibility of a huge data-reduction in the case of a chaotic signal. In the following, first a method for calculating **stable** as well as **unstable** periodic solutions and a Bayesian estimation procedure will be presented. Next two applications of parameter identification will be given, referring to a landing gear with stable, non-smooth periodic solutions and to a pendulum exhibiting a chaotic trajectory, in which case multiple, smooth unstable periodic solutions will be applied.

2. Periodic Equilibrium States

We consider the ordinary differential equations g of a structural system in the degrees of freedom q, and corresponding output equations $\hat{y}$:

$$g(q, q', q'', F, \tau, \theta) = 0 , \qquad \hat{y} = f(q, q', q'', F, \tau, \theta) . \tag{1}$$

The l-dimensional vector θ stands for the uncertain model parameters. The external forcing frequency is f_e (period $T_e = 1/f_e$), F stands for the periodic excitation force. A nondimensional time variable τ, $\tau \in [0, 1\rangle$ is introduced as $\tau = f_e\, t$, where the prime ($'$) stands for differentiation with respect to τ. For details see [1] and [4]. An approximate periodic response can be ob-

D. H. van Campen (ed.), IUTAM Symposium on Interaction between Dynamics and Control in Advanced Mechanical Systems, 191–198.
© *1997 Kluwer Academic Publishers.*

tained by application of an equidistant discretization procedure. The time variable τ is discretized using n time intervals: $\tau_\mu = (\mu - 1)\Delta\tau$, $\mu \in \{1, 2, \ldots, n\}$, $\Delta\tau = 1/n$. All time dependent variables are also discretized and will be subscripted by μ on time step τ_μ. The discretized degrees of freedom q_μ are stored in the column $z = [q_1^T, q_2^T, .., q_n^T]^T$.

Subsequently, the derivatives of the discretized degrees of freedom q_μ can be approximated by any difference scheme. For the landing gear identification a second order backward scheme for both the first and second derivatives appeared to perform best (i.e. showed no occurrence of numerical instabilities due to the non-smooth character of the response, for sufficiently fine time grids with $n \approx 800$). For the pendulum problem a 2nd order central difference scheme was succesfully applied. For periodic solutions we have the additional boundary condition:

$$q_i = q_{i \pm n} \quad , \quad i \notin \{1, 2, \ldots, n\} \quad \wedge \quad i \pm n \in \{1, 2, \ldots, n\} . \tag{2}$$

Now for each time step τ_μ a set of approximate difference and discrete output equations can be derived $\tilde{g}_\mu(z, x_\mu, \tau_\mu, \theta) = 0$, $\hat{y}_\mu = \tilde{f}_\mu(z, x_\mu, \tau_\mu, \theta)$ leading for all time steps to a set of nonlinear algebraic equations

$$\tilde{g}(z, x, \tau, \theta) \equiv [\tilde{g}_1^T, \tilde{g}_2^T, \ldots, \tilde{g}_n^T]^T = 0 , \tag{3}$$

which is solved for the discrete periodic solution z from an initial static equilibrium guess by a modified Newton procedure, making use of the decomposition technique and a robust step halving technique as proposed in [1]. Substitution of z in the discretized output equations will give the requested output values $\hat{y}_{\mu\kappa}$.

For the Bayesian estimator used in this paper, the first derivatives of the outputs $\hat{y}_{\mu\kappa}$ to the model parameters θ are required and can be derived from

$$d\hat{y}_{\mu\kappa}/d\theta = \tilde{f}_{\mu\kappa,\theta} + [d\tilde{f}_{\mu\kappa}/dz]z_{,\theta} , \quad d\tilde{g}/d\theta = \tilde{g}_{,\theta} + [d\tilde{g}/dz]z_{,\theta} = 0. \tag{4}$$

The derivatives $\tilde{f}_{\mu\kappa,\theta}$ and $d\tilde{f}_{\mu\kappa}/dz$, can be derived by hand or by symbolic computation. The remaining unknown partial derivatives $z_{,\theta}$ follow from the so-called sensitivity equation $d\tilde{g}/d\theta$. The partial derivatives $\tilde{g}_{,\theta}$ should also be derived by hand or by symbolic computations from the differential equations. The derivatives $d\tilde{g}/dz$ follow directly from the last Newton iteration of the periodic solver.

3. Bayesian Estimation

We use a Bayesian estimator, see [4], for estimating the model parameters as well as unknown distribution parameters of the residuals, i.e. the difference between measured outputs y and predicted outputs $\hat{y}$

$$e_{\mu\kappa}(\theta) \equiv y_{\mu\kappa} - f_{\mu\kappa}(q, \dot{q}, \ddot{q}, F(f_e), t, \theta) = y_{\mu\kappa} - \hat{y}_{\mu\kappa} . \tag{5}$$

We assume normally distributed $\mathcal{N}_s(0, \mathrm{diag}(v))$ residuals and also normally distributed $\mathcal{N}_l(\bar{\theta}, \mathrm{diag}(\sigma^2))$ a priori knowledge of the model parameters. Measurement errors and modelling errors are not distinguished and the total errors are assumed to be statistically independent over the n samples μ, the N experiments κ, and the s measurement channels β. Now the Bayesian estimation problem leads to a maximization problem

$$\Phi_B(\theta, v_1, .., v_s) = -\frac{1}{2}\left[nN \ln \prod_{\beta=1}^{s} v_\beta + \sum_{\beta=1}^{s} \frac{M_{\beta\beta}(\theta)}{v_\beta} + \sum_{\alpha=1}^{l} \frac{(\theta_\alpha - \bar{\theta}_\alpha)^2}{\sigma_\alpha^2} \right] \quad (6)$$

$$\text{with} \qquad M_{\beta\beta}(\theta) = \sum_{\kappa=1}^{N} \sum_{\mu=1}^{n} e_{\mu\kappa\beta}^2(\theta) \ ,$$

in which $M_{\beta\beta}(\theta)$ stands for the diagonal terms of the moment matrix of the residuals. The estimation problem is solved iteratively by Newton-Gauss approximation of the Taylor series of the object function (7) by staged optimization. Firstly, from (7) the unknown measurement channel variances can be solved with bias correction for each iteration

$$v_\beta = \frac{s}{nNs - l} \sum_{\kappa=1}^{N} \sum_{\mu=1}^{n} e_{\mu\kappa\beta}^2 \ . \quad (7)$$

Secondly the model parameter updating formula reads

$$\Delta\theta_i = \left(-H_{\theta\theta}^{-1} q_\theta\right)_i = -\left[\sum_{\kappa=1}^{N} \sum_{\mu=1}^{n} e_{\mu\kappa,\theta}^T \mathrm{diag}(v)^{-1} e_{\mu\kappa,\theta} + \mathrm{diag}(w)^{-1}\right]_i^{-1}$$

$$\times \left(\sum_{\kappa=1}^{N} \sum_{\mu=1}^{n} e_{\mu\kappa,\theta}^T \mathrm{diag}(v)^{-1} e_{\mu\kappa}\right)_i \quad \text{with } w = (\sigma_1^2, \ldots, \sigma_l^2)^T. \quad (8)$$

in which $H_{\theta\theta}$ stands for the θ-partition of the Hessian of the object function (7) and q_θ is the corresponding partition of the gradient. The first order derivatives $e_{\mu\kappa,\theta} = -\hat{y}_{\mu\kappa,\theta}$ can be computed efficiently from a large set of periodic algebraic sensitivity equations (4). For more information on the estimator, the solution procedure or the computation of derivative information the reader is referred to [4].

4. Landing Gear Identification

4.1. MODEL DESCRIPTION AND EXPERIMENTAL SET-UP

The identification procedure will first be illustrated for identifying parameters of a simplified F-16 nose landing gear damper under periodic excitation. With hydraulic equipment, a periodic axial displacement has been exerted on the landing gear, measuring the force acting on the wheel axle and the internal gas pressure of the damper. For more information see [4],[5]. The basic model for this damper reads:

$$\theta_1(\ddot{q} + \theta_2) + \theta_3\dot{q}|\dot{q}| + \theta_4 p + (\theta_8 + \theta_9 p)\arctan(\theta_{10}\dot{q}) + \theta_{11} - F_e = 0,$$
$$\theta_{15}\dot{T} + \theta_{16}(1 + \theta_6 q)(T + \theta_{17}) - (\theta_4 p + \theta_{11})\dot{q} = 0,$$
$$\text{with} \quad p = \theta_5 T/(1 + \theta_6 q + \theta_7 T), \quad \hat{y} = q + \theta_{12}. \tag{9}$$

The mechanical model (1st. equation) includes inertia, velocity squared oil damping, a nitrogen gas spring approximated by polytropic ideal gas behaviour and a continuous approximation of Coulomb friction, whereas the thermo-dynamical model (2nd. equation) includes a gas heat capacity, cylinder-wall heat conduction and a source term due to gas compression and extension. Some of the 15 parameters θ_k can be determined rather accurately in advance, some of them could not be identified due to limited hydraulic power (especially the damping parameter θ_3) and some parameters haven been extracted directly from the measurement data. The identification-procedure has been applied to estimate the 3 most important and uncertain parameters, namely θ_4, θ_6 and θ_{16}. Due to limited hydraulic

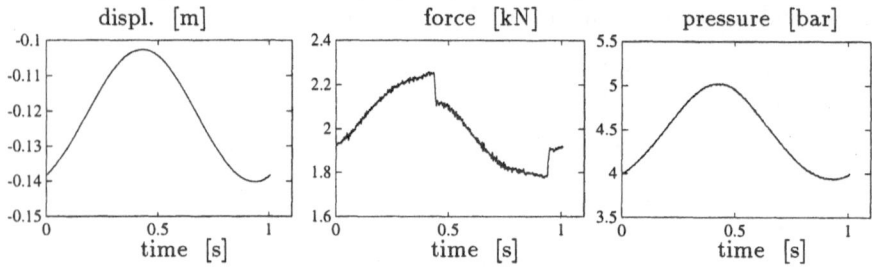

Figure 1. Frequency 1 Hz, small amplitude ($\pm$ 18 mm), experiment 920713m1.

power but also for reasons of efficiency a set of experiments was defined that covers the total available system state and parameter space. This resulted in a set of 14 large-stroke-low-frequency up to small-stroke-high-frequency experiments.

For the excitation a controlled sinusoidal axial displacement was selected. Typical time histories of the measurements are shown in Fig. 1. The figure shows the nearly perfect sinusoidal displacement signal, the force signal with significant force discontinuities at 0.5 and 1 s (friction), and smooth relative gas pressure signals.

4.2. IDENTIFICATION RESULTS

In general the large amplitude residuals are less by half than the small amplitude residuals, also the spreads in parameters θ_4 and θ_6 are much smaller and acceptable. The heat conduction parameter θ_{16} is not really constant, and for some small amplitude inputs the parameter is even negative which means that the heat transfer is still not modelled accurately

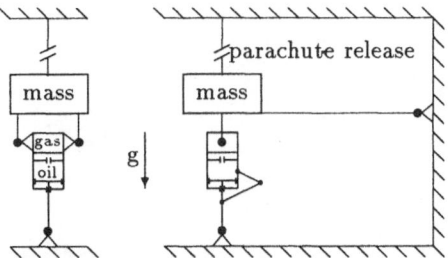

Figure 2. Schematic view of the laboratory drop-test set-up. Dimensions between the joints: landing gear height 1.16 m, width 0.58 m, pivoting frame length 2.02 m.

enough. For the verification of the feasibility of the shaker-test identification method a small number of (simplified) drop-test experiments have been performed, see Fig. 2 for a schematic view. The measurements start when a mass is suddenly set free, loading the damper. The experimental results and predicted outputs for a representative landing experiment are shown in Fig. 3. The measured quantities are plotted with solid lines and the predictions with dashed lines. The figures show that the amplitudes

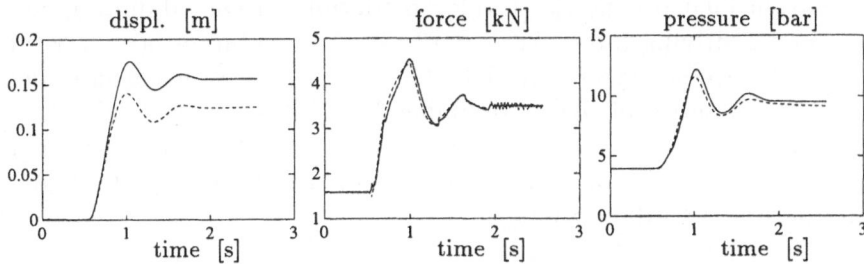

Figure 3. Predictions for drop-test experiment 930426c0

and the frequency of the oscillations in all three signals are predicted very well, but the displacement signal shows a large static offset component. However, at 2.5 s the gas temperature is not yet equal to the environmental temperature. Simulation for a longer time interval shows that the static equilibrium value is predicted correctly after approximately 30 s. Despite the good overall predictions for the force and pressure signals, it must be concluded that the thermo-mechanical model needs further research.

5. A Parametrically Excited Pendulum

5.1. MODEL DESCRIPTION

To illustrate the use of chaotic data in the estimation of structural parameters from a nonlinear mechanical system we investigated a parametrically excited pendulum as given in Fig. 4. In the top position the instantaneous angular position ϕ and the angular velocity $\dot{\phi}$ of the pendulum are measured, giving a point $(\phi, \dot{\phi})$ in the Poincaré section; phase space points are

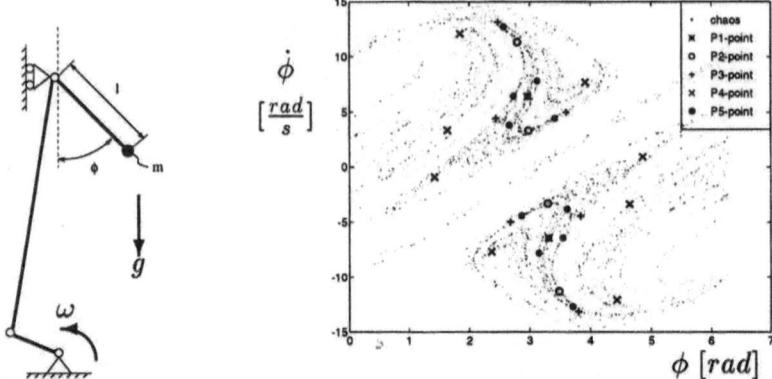

Figure 4. The parametrically driven pendulum

Figure 5. Poincaré section of the chaotic attractor and some low order, unstable periodic points

thus obtained with time intervals of $\frac{2\pi}{\omega}$ s. The mathematical model reads:

$$\ddot{\phi} + \frac{k_1}{ml^2} sgn(\dot{\phi}) + \frac{k_2}{ml^2}\dot{\phi} + \frac{k_3}{ml^2}(\dot{\phi})^2 sgn(\dot{\phi}) + f_{exc}(\omega)\sin(\phi) = 0 \quad (10)$$

This equation includes inertia, Coulomb friction, viscous damping, air resistance and a driving force. At $\omega = 13 \frac{rad}{s}$ the pendulum is in a rotating motion whose frequency is locked to the excitation. When decreasing ω, we encounter a series of period-doubling bifurcations that finally lead to a large chaotic attractor at $\omega = 9.09 \frac{rad}{s}$, extending over the full angular range and shown in Fig. 5. For more details of the experimental set-up and nominal parameter values see [3].

5.2. PERIODIC ORBIT ANALYSIS

The structure of the chaotic attractor can be characterized by a set of **unstable** periodic orbits (solutions of the differential equation), see [2]. The identification procedure will be based on a selection of these unstable periodic orbits. In order to extract periodic orbits from a Poincaré section of the (experimental) chaotic time series the phase space is partitioned in small boxes of linear size ε (1 % of the attractor size). A point that returns to either its own box or to a neighbouring box is called a recurrent point. The time it takes to return, gives the length of a candidate cycle. The list of candidate cycles is further reduced by requiring that a box contains at least a few points that ε-return as a p cycle. These points are used to determine a local linear approximation A^p of the dynamical system. A minimum of 32 neighbouring points y_t, taken from the neighbourhood of $x_t{}^p$, is included in the fit. The estimated stable and unstable eigenvalues of the periodic orbits are the eigenvalues of A^p. Fig. 5 shows the periodic points belonging to the cycles with period times up to length 5 that were found from an experimental time series of 42,754 points, see [2]. Note that a huge reduction of data has been obtained by extracting a few periodic points and their eigenvalues from a long chaotic time series. These unstable periodic points are now used to estimate parameters in the differential equation (10). It should be remarked that starting solutions for the periodic solver have to be chosen carefully otherwise convergence to other periodic solutions may result. Until now only the position in phase space (ϕ and $\dot{\phi}$ values) is used in the object function. In the future also the difference between estimated- and computed eigenvalues will be added to the object function.

5.3. RESULTS

The basis for the identification is equation (10) with ω supposed to be known. From this equation it is clear that the parameters m, l, k_1, k_2 and k_3 can not be estimated simultaneously, because these parameters do not represent independent terms in the differential equation. The identification will therefore be split up in two parts namely (a) the estimation of m and l with fixed values for k_i, $i = 1, 3$ and (b) estimation of k_3 with known m, l and k_i, $i = 1, 2$. We only estimated friction parameter k_3 because this is the largest term in the differential equation and variation of k_1 and k_2 showed to have only a small influence on the calculated unstable periodic solutions. As a reference we use a priori data for the friction parameters (k_{1exp}, k_{2exp}, and k_{3exp}) as determined by [2] and values for m_{exp} and l_{exp} simply calculated from geometrical data. For the estimation, unstable periodic orbits up to period 5 were used. The results are presented in Table 1. From this

TABLE 1. Estimates of m, l and k_3 with fixed k_1 and k_2

Initial state	$\frac{m_0 - m_{exp}}{m_{exp}} = 51\%$	$\frac{l_0 - l_{exp}}{l_{exp}} = 18.9\%$	$\frac{k_{30} - k_{3exp}}{k_{3exp}} = 100\%$
Final state	$\frac{m - m_{exp}}{m_{exp}} = 9.6\%$	$\frac{l - l_{exp}}{l_{exp}} = 0.3\%$	$\frac{k_3 - k_{3exp}}{k_{3exp}} = 11.6\%$

table, it can be concluded that l can be estimated very accurately, using the experimental data. However, the estimation of m and k_3 are less accurate. This can be caused by the fact that the experimental time series was relatively short and contaminated by noise and the (experimental) periodic points unfortunately had to be picked by hand from Fig. 5.
The accuracy of the estimate l is very good and the estimates m and k_3 are very acceptable given the level of measurement noise on the data.

6. Conclusions

In general it can be concluded that a nonlinear parameter identification method based on a set of stable and/or unstable periodic solutions com-

198

bined with the Bayesian estimation procedure has proven to be a powerful tool for the identification of parameters in a broad range of nonlinear dynamic systems. For the landing gear holds:

- A procedure based on periodic orbits is computationally far more efficient than a procedure based on transient signals.
- The thermo-dynamical part of the model should be improved.
- The identification procedure can be used as a quality-control instrument for landing gear production.

For the pendulum problem the following remarks can be made:

- By using periodic points instead of a chaotic trajectory, huge data reduction has been accomplished without loosing essential information.
- A more comprehensive method in finding suitable starting solutions for the periodic solver is desirable.
- The introduction of the eigenvalues of the unstable periodic orbits in the object function should be investigated.
- The difficult numerical computation (due to positive Lyapunov exponents) of chaotic trajectories for very long times is not needed.

Acknowledgements

The landing gear research was partially supported by DAF Special Products, Eindhoven, The Netherlands, and the authors thank Willem van de Water for helpful discussions and putting the pendulum experimental data at their disposal.

References

1. Fey, R.H.B.,Van de Vorst, E.L.B., Van Campen, D.H., De Kraker, A., Meijer, G.J., Assinck, F.H., 1994. *Chaos and Bifurcations in a multi-dof beam system with nonlinear support*, In: Nonlinearity and Chaos in Engineering Dynamics, J.M.T. Thompson and S.R. Bishop, editors, John Wiley & Sons, 125–139
2. Van de Water, W., Hoppenbrouwers, M., & Christiansen, F., 1991, *Unstable periodic orbits in the parametrically excited pendulum*, Physical Review A, 44, 6388–6398.
3. Van de Wouw, N., Verbeek,G.,Van Campen, D.H., 1995, *Nonlinear Parametric Identification Using Chaotic Data*, Jnl. of Vibration and Control, 1, 291–305
4. Verbeek, G. 1993, *Nonlinear parametric identification using periodic equilibrium states, with application to a landing gear damper*, Ph.D. thesis, Eindhoven University of Technology.
5. Verbeek, G., de Kraker, A., & van Campen, D.H., 1995, *Nonlinear parametric identification using periodic equilibrium states, with application to a landing gear damper*, Nonlinear Dynamics, 7, 499–515.

OPTIMIZATION OF AN ACTIVELY STEERED PEOPLE MOVER

RALF KRAUSE
Daimler-Benz AG, Goldsteinstr. 235, Frankfurt, Germany,
DIETER BESTLE
Institute of Machine Dynamics, Brandenburg Technical University,
P.O.Box 101344, 03013 Cottbus, Germany,
JEFF SCHWALM
ABB Daimler-Benz Transportation (North America) Inc., 1501 Lebanon
Church Road, Pittsburgh, Pennsylvania 15236-1491,USA

Abstract

The goal of the investigations is to reduce the lateral steering forces caused by the passive steering mechanism of the people mover. To accomplish this goal, an additional degree of freedom was added to each suspension. The motion of this degree of freedom was controlled by an active element. That active element includes sensors, an actuator, and a controller. This paper demonstrates simulation and optimization, and how they can support prototyping by predicting the optimized dynamic behavior of the system with respect to various parameters of a system, especially the controller coefficients.

1. Introduction

Designing a dynamic system like an actively controlled vehicle by computer aided engineering (CAE) involves several steps: *(i)* modeling the complete dynamic system, *(ii)* defining performance criteria and constraints, *(iii)* choosing design variables, *(iv)* solving the optimization problem resulting in optimal values for the design variables, and *(v)* benchmarking the optimized vehicle model with experimental results.

Although well known and obvious from an academic point of view, such a straight forward design process is not used in practice. The main reason is that acceptable tools for real world problems are not combined in one software package. In this paper, the software program AIMS [1] based on symbolic computer-aided modeling of multibody systems, design sensitivity analysis, and numerical optimization is tested for designing a controller for the people mover.

D. H. van Campen (ed.), IUTAM Symposium on Interaction between Dynamics and Control in Advanced Mechanical Systems, 199–206.

2. Modeling the dynamic system

A people mover is a fully automated, electrically powered, rubber tired vehicle, frequently used on airports to take people from one terminal to another. An outline plot of the vehicle is shown in Fig.1.

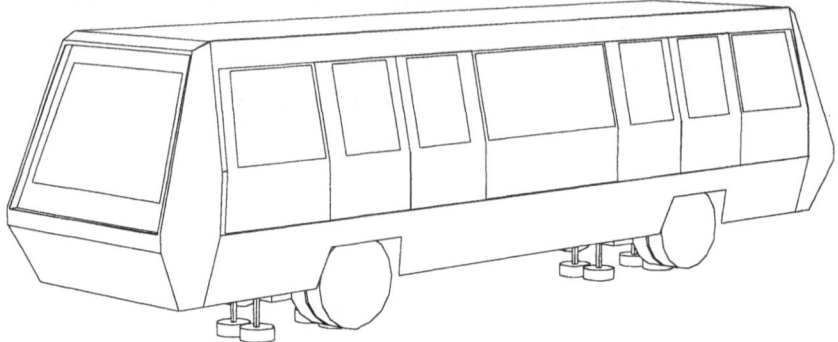

Figure 1. Outline plot of the people mover.

The vehicle is supported and guided by two suspension assembles (bogies) located at both ends of the vehicle. The bogies are separately mounted to the vehicle chassis by means of large diameter flanged ring bearings. The ring bearing allows a bogie to rotate about its geometric center and provides positive steering similar to steel wheel trailer suspensions, but is guided from the wayside by a centrally located I-beam called the guidebeam, as seen in Fig. 2. The unsprung portion of the bogie consists of a lower guidance structure which supports four guide axles with rubber tires and the main drive axle with dual pneumatic rubber drive tires and run-flat safety discs.

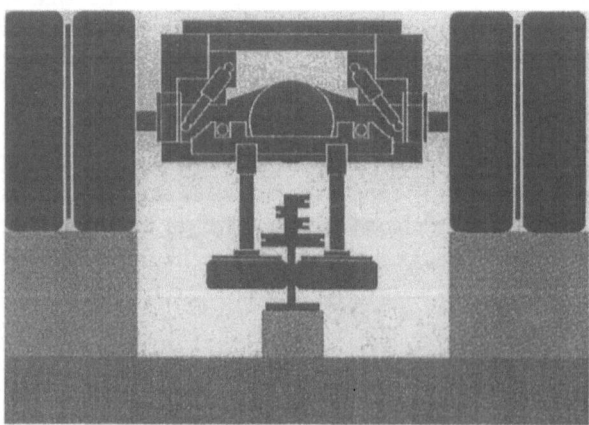

Figure 2. Bogie of a people mover

The unsprung components are connected to the upper suspension frame with leaf springs and upper radius arms. Attached to the lower guidance structure are four guide tires arranged in a rectangular pattern where two guide tires, one on each side of the guidebeam, are located in front of the axle and a second pair is located behind the axle. The guidebeam is fixed to the guideway. It is centrally located between the drive tires and is positioned just below the running surface

In a curve, and of course also on a straight track, the steering mechanism keeps the drive tires always in a tangential position with respect to the track. The drive tires therefore, do not keep any lateral forces, and all the forces, e.g. resulting from centrifugal forces, must be taken by the guide tires, see Fig. 3a. In general, the lateral forces of the drive tires depend on their cornering stiffness, and the slip angle, which is the difference between the longitudional drive tire axle, and the direction of the speed [2]. In order to reduce the guide tire forces and move the forces to the drive tires, the angle between the drive axle centerline and the guide tire alignment is changed from $\varphi_0 = 0^0$, see Fig. 3b.

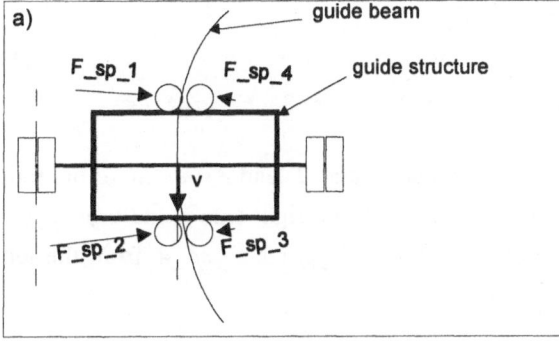

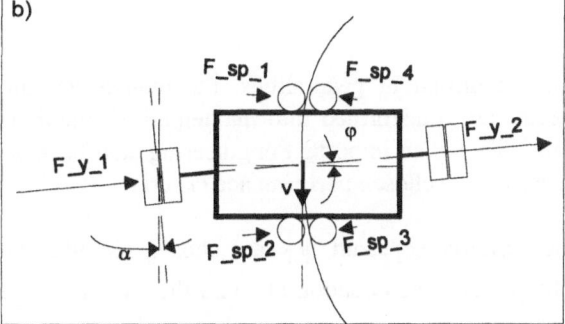

Figure 3. Concept of passive and active steering mechanism.

The guide structure is split in two parts and attached by another ring bearing and each can rotate about the vertical centerline. One part is connected to the guide tires, and the other part is connected to the axle housing of the drive tires. An actuator is installed between the two parts controlling the relative angle φ and thus the slip angle α. The actuator control is based on the remaining guide tire forces F_sp_1, F_sp_2, F_sp_3, and F_sp_4.

The controller design has to be based on a dynamic model of the technical system. Therefore, the more complex model [3] was reduced to a five degree-of-freedom model. The carbody has three degrees of freedom, which are the longitudinal, lateral, and yaw motion. Each bogie has one yaw degree of freedom. The actuators are considered as ideal positioning elements, described by the angle φ_1 for the front bogie and φ_2 for the rear bogie, respectively. The sensor dynamic is neglectable.

The controller approach depends highly on the restrictions of the already installed hardware equipment for the automatic travel of the vehicle. Therefore, a fast control algorithm with the structure of a P-, or PI-controller has to be realized.

$$\varphi = K \, \Delta l \ ,$$
$$\varphi = K \, \Delta l + T \int \Delta l \ dl \ , \tag{1}$$
with
$$\Delta l = (F\_sp\_1 + F\_sp\_2 - F\_sp\_3 - F\_sp\_4)/4 \ .$$

The controller output φ results from the constants K and T and a sum Δl of the four guide tire forces. The Forces F_sp_1, F_sp_2 from the right side guide tires of the guide beam are considered as positive and the forces F_sp_3, F_sp_4 from the left sided guide tires are considered as negative.

3. Defining performance criteria

Goals or demands are often verbally identified in generalities, i.e. reduce steering forces. In this design process, the goals are transformed into mathematical functions and referred to as performance criteria. This is an important engineering step because all the following investigations are based on the chosen performance criteria.

In our case, two integral performance criteria ψ_1, and ψ_2 are chosen /3/, like the square of the sum of the guide tire forces Δl, see equation (1), and the square of the yaw acceleration $\ddot{\beta}$ of the bogie for reducing the oscillation of the bogie:

$$\psi_1 = \int \Delta l^2 \ dt \ ,$$
$$\psi_2 = \int \ddot{\beta}^2 \ dt \ . \tag{2}$$

Minimizing each of the criteria (2) individually represents a nonlinear programming problem which may be solved by any general purpose optimization algorithm. In general, the resulting optimal parameters will be different from each other. Multicriteria optimization theory offers a concept to handle several criteria simultaneously and an optimal design in conflicting situations is identified. Several strategies exist for computing these optimal designs. The most commonly used strategie is the weighted objective method which belongs to the class of scalar-ization methods. Instead of vector criterion, scalar weighted - sum criterion has to be minimized. The multicriteria problem then reduces to a nonlinear programming problem with the coefficients α and γ :

$$\psi\_all = \alpha\,\psi_1 + \gamma\,\psi_2,$$
$$\alpha + \gamma = 1 . \tag{3}$$

4. Choosing design variables

The design variables are system parameters which are calculated to optimal values with respect to the performance criteria. Choosing the right design variables which will have some influence on the dynamic behavior, is also an important engineering task.

In our case, the people mover itself is an existing vehicle. Only the damper coefficient d_gt and the stiffness coefficient c_gt of the guide tires can be chosen as design variables, as well as the unknown controller parameters K and T for the P- and I-part of the control law, respectively. They may be summarized in a vector

$$p = [d\_gt, c\_gt, K, T]^T . \tag{4}$$

5. Solving the optimization problem resulting

Solving technical problems by optimization does not provide an optimal solution in the first attempt. An adequate problem formulation has to be determined by trial and error in several runs. In order to facilitate such a process, the setup of the problem equations and the solution process has to be highly automated.

Over the last two decades multibody formulas have been developed to automatically generate the equations of motion in numerical or symbolical form [6]. If sensitivity analysis has to be supported, the use of a symbolic code like NEWEUL [7] is recommended.

Advanced optimization algorithms like SQP [4] are based not only on function evaluations but also on gradient information of the objective and constraint functions.

Some algorithms offer the possibility of computing gradients by numerical differentiation. Applying difference formulas to the optimization of multibody systems will yield poor results due to the rather large error in evaluating ψ_all by numerical integration. Therefore, it is advisable to use sensitivity analysis methods for generating analytical information on the gradients. The adjoint variable method [1] yields efficient and reliable gradient information. Numerical studies have shown the adjoint variable method to be superior to numerical differentiation [5].

The use of symbolic equations of motion provides the opportunity to choose any of the model parameters as a design variable. The partial derivatives with respect to the design variables and the state variables required for sensitivity analysis are computed automatically by the programmable formula manipulation package MAPLE [8]. The resulting equations of motion and the adjoint differential equations are then solved by numerical integration [5].

Table 1 shows some results of the optimization step. In order to not converge to local minima, the optimization algorithm was started several times with different initial values of the design variables and initial conditions of the multibody system model. The results, however, were very close together indicating that the optimization problem is well behaved. The controller parameter have a big influence on the guide tire forces, but a neglectiable influence on the yaw accalaration of the bogy. The yaw accalaration is strongly influenced by the damping coefficient d_gt of the guide tires. When the value goes to the upper limit, which is 5 times higher than the original one, the optimization criterion Ψ_2 is 3 times lower.

TABLE 1. Comparison of optimized parameters in respect to different optimization criteria.

Optimization criteria	Improvement of actively steered vehicle ($\Psi_{1,2}$) versus passive People Mover ($\Psi_{1\_o,2\_o}$):	Calculated design parameters: dimensionless control parameters T and K, stiffness c_gt und damper d_gt versus coefficients of the passive People Mover
Ψ_1	$\Psi_1 / \Psi_{1\_o} = 76\%$, $\Psi_2 / \Psi_{2\_o} = 105\%$	K=0.51, T=1.28
	$\Psi_1 / \Psi_{1\_o} = 27\%$, $\Psi_2 / \Psi_{2\_o} = 300\%$	K=0.57, T=0.76, C_GT=0.7, D_GT=1/6
Ψ_2	$\Psi_1 / \Psi_{1\_o} = 97\%$, $\Psi_2 / \Psi_{2\_o} = 95\%$	K=0.04, T=0.10
	$\Psi_1 / \Psi_{1\_o} = 87\%$, $\Psi_2 / \Psi_{2\_o} = 35\%$	K=0.33, T=0.31, C_GT=1,0, D_GT=5

6. Benchmarking the optimized vehicle model with experimental results.

Since the optimization process was based on a five DOF design model, the resulting control laws were first tested on a more complex 18 DOF model [3]. Numerical simulations, however, showed very similar behavior compared to the design model.

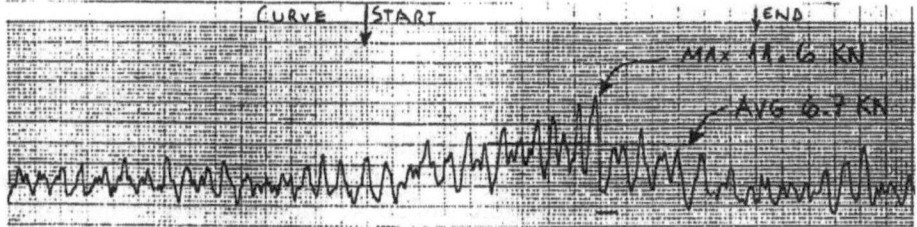

Figure 4a. Experimental guide tire forces for the passive system.

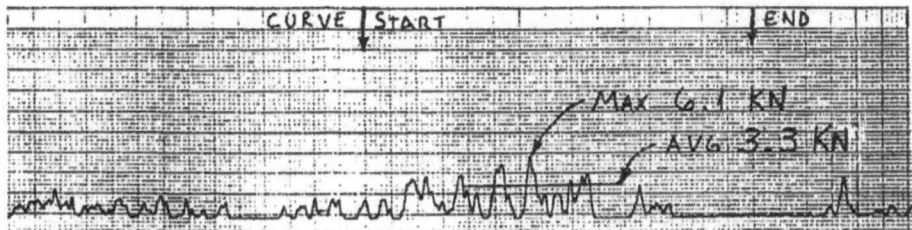

Figure 4b. Experimental guide tire forces for the active system with K= 0.3 and T= 0.5.

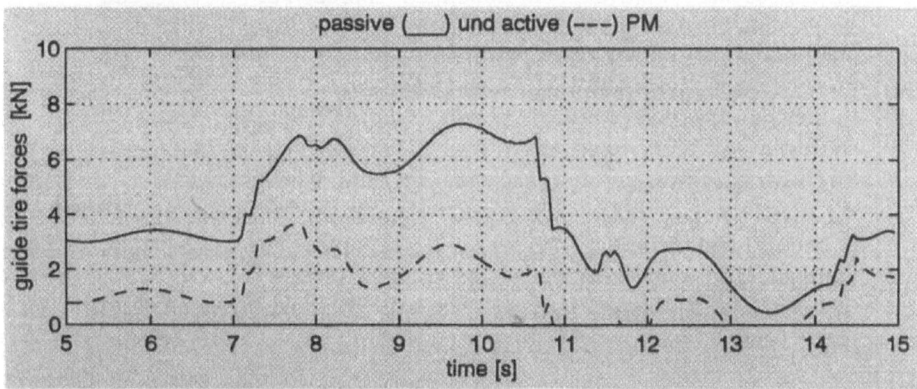

Figure 4c. Simulation results for the passive and the actively steered People Mover with K= 0.3 and T= 0.5.

A benchmark with a real actively steered people mover was done on the test track in Pittsburgh, USA. A clothoide curve with a minimum radius of 60 m was traversed with a speed of 20 km/h. The experiments show a guide tire force reduction of 50% for the PI-controller, compared to the original pure passive system, as seen in Fig. 4a and Fig. 4b. The simulation of the complex 18 DOF model shows the same guide tire reduction in Fig. 4c.

7. Conclusion

A computer-aided design process involving modeling, analysis by simulation, numerical optimization of symbolical equations and experimental benchmarking has been applied to an existing people mover for developing active steering components. The results indicate highly accuracy and correlation. The predicted optimal values are close to the final values which saved considerable time in experimental fine tuning and results in a reduction of the guide tires forces of 50 %. Moreover, this computer-aided process predicted success prior to any mechanical or testing investments. This offers a possibility to make design decisions in an early design phase before setting up expensive experiments.

8. References

[1]   D. Bestle: Analyse und Optimierung von Mehrkoerpersystemen. Berlin: Springer ..., 1994.
[2]   J. Reimpell und K. Hoseus. Fahrzeugmechanik. Vogel Buchverlag, Würzburg, 1989.
[3]   R. Krause and J. Schwalm: An Integrated Approach of Modeling and Improving the dynamical Behavior of the People Mover C-100. 5th international conference on automated people mover, June, 12-14 1996. Proceedings will be published. Paris, 1996.
[4]   R. Fletcher: Practical Methods of Optimization. Chichester: Wiley, 1987.
[5]   D. Bestle and P. Eberhard: Analyzing and Optimizing Multibody Systems. Mechanics of Structures and Machines 20 (1992) 67-92.
[6]   W. Schiehlen (ed.): Multibody Systems Handbook. Berlin: Springer, 1990.
[7]   E. Kreuzer and G. Leister: Programmsystem Neweul'90, User's Guide AN-24. Stuttgart: University, Institute B of Mechanics, 1991.
[8]   B.W. Char et. al.: Maple-Reference Manual. Waterloo: Waterloo Maple Publ., 1990.

CONTROLLING TORSIONAL VIBRATIONS THROUGH PROPER ORTHOGONAL DECOMPOSITION

E. KEUZER AND O. KUST
Ocean Engineering Section II
Technical University Hamburg-Harburg
Eißendorfer Str. 42, 21071 Hamburg, Germany

Abstract. Nonlinear excitations cause angular vibrations in torsional strings. As an example for long torsional strings, we consider a drill string for drilling deep holes for the exploration and production of oil and gas. Due to the nonlinear torque at the bit, self-excited oscillations can occur by means of a supercritical Hopf bifurcation. Such vibrations are highly unwelcome, and in severe cases they could damage the string. In order to minimize these vibrations, we present an active damping system by employing the drive via feedback of the system state. The feedback law is formulated by a projection of the infinite-dimensional system state onto a real input variable. The basis of the projection will be determined by proper orthogonal decomposition (POD). The design of a POD-based controller and the determination of the controller parameters by optimization will be presented. The improvement of the dynamics will be verified by simulations.

1. Introduction

In many technical systems, torsional strings serve as power transmission elements between the drive and the output. Examples are the drive-shafts in cars and ships. In this paper we consider a drill string for deep drilling, Fig. 1. With diameters of only a few decimeters and length up to 5000 m, drill strings are very slender structures. The upper end of the drill string is held by a hoist in the derrick. The most common way of producing the drill torque at the bit is by means of a rotary table, that is a large horizontal gear wheel which is driven by an electric or hydraulic motor and for which the drill pipes serve as transmission elements. Drill collars are heavy, thick-walled pipes designed to transfer load to the bit and to produce tension in the drill pipes, which otherwise would bend.

D. H. van Campen (ed.), IUTAM Symposium on Interaction between Dynamics and Control in Advanced Mechanical Systems, 207–214.

Self-excited vibrations can be induced by the nonlinear torque at the bit, and have been known for a long time in practice, but less investigated analytically (s. Zamudio *et al.* (1987), Clayer *et al.* (1990)). The oscillations depend on time and space, and result in different trajectories over the length, Fig. 2. The boundary conditions are given by the speed of the drive at the surface and by the nonlinear torque at the bit.

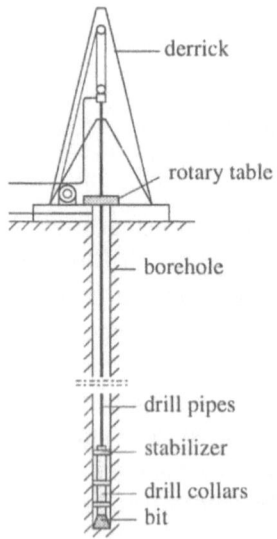

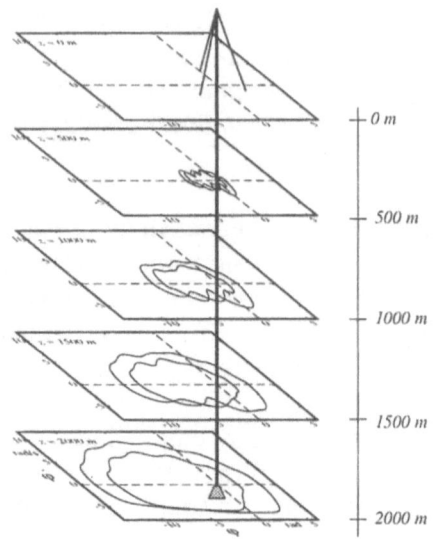

Fig. 1. Drilling platform

Fig. 2. Phase portraits of self-excited torsional vibrations

We first introduce a system for a long torsional string and then describe the mathematical model. Next we give an introduction to the proper orthogonal decomposition for discrete signals. In a further section we discuss the application of POD to vibration control.

2. Modeling

A simplified model of a long torsional string has the following properties: The drill string is considered as a one-dimensional elastic body. Due to the large torsional stiffness of the drill collars in relation to the drill pipes, the drill collars are appropriately modeled as a single rigid body at the lower end of the drill pipes.

2.1. EQUATIONS OF MOTION

By visualizing the equilibrium of an infinitesimal thin disk of the drill string, we can describe the dynamics by a partial differential equation (PDE) with density ρ, polar second moment of area I_p, shear module G, inner and outer damping d_i and d_a, respectively, length l of the string, and an applied torque per unit length

$m(z,t)$:

$$\rho I_p \ddot{\varphi} + d_a \dot{\varphi} - G I_p \varphi'' - d_i G I_p \dot{\varphi}'' = m(z,t) \quad , \quad z \in \Omega = [0, l] \quad . \tag{1}$$

The boundary conditions are given by the torque of the electric motor T_m, the moment of inertia of the motor and the rotary table (Θ_m, Θ_{rt}) at the upper end, the nonlinear bit torque T_{bit}, and the angular momentum of the disc of the drill collars (moment of inertia Θ_{dc}) at the lower end:

$$\partial\Omega : \begin{cases} z = 0 & : \quad (\Theta_{rt} + n_{gear}^2 \Theta_m)\ddot{\varphi} = -G I_p \varphi' + T_m \quad , \\ z = l & : \quad G I_p [\varphi' + d_i \dot{\varphi}'] = -T_{bit} - \Theta_{dc}\ddot{\varphi} \quad , \end{cases} \tag{2}$$

with n_{gear} the gear ratio is described. The dynamics of the electric motor is given by an ordinary differential equation (with impedance L, resistance R, motor voltage U_m, motor constants k_V and k_T):

$$L\dot{I} + RI = U_m - k_V n_{gear} \dot{\varphi}(z=0) , \tag{3}$$
$$T_m = k_T I . \tag{4}$$

The nonlinear function T_{bit} is known from field experiments (Brett, 1992) and depends on the speed of the bit, Fig. 3.

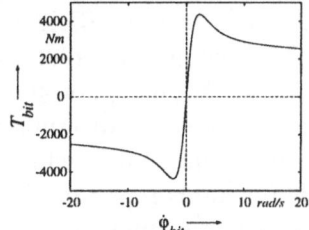

Fig. 3. Torque on bit

2.2. SIMULATION

The equations of motion are given by (1) and (3). In order to solve the PDE, the spatial domain Ω will be discretized with central differences for the spatial derivatives. Expanding the spatial derivatives on the boundary with Taylor series and transforming all second order ODEs into a set of first order ODEs, we obtain the complete set of equations of motion

$$\dot{x} = A(\mu)x + f(x, \mu) \quad , \quad x = \begin{bmatrix} \varphi^T & \dot{\varphi}^T \end{bmatrix}^T \quad , \tag{5}$$

with A and f denoting the linear and non-linear part of the system. The parameter μ is employed to enable variation of characteristics such as the motor voltage U_m or the length l of the string.

Finally, the initial-value problem is solved with a standard variable-order variable-step routine. Time simulations of self-excited vibrations are depicted in Fig. 4 for a typical working speed of the rotary table and a string-length of 2000 m.

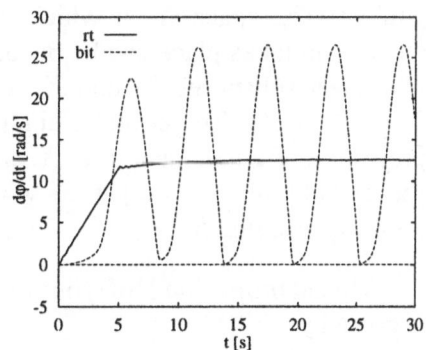

Fig. 4. Time simulation of self-excited vibrations

3. Proper Orthogonal Decomposition

For analyzing spatio-temporal dynamics the proper orthogonal decomposition (POD) which is often also called Karhunen-Loève (KL) transformation has been recently used in several disciplines as an efficient tool. POD is used in the context of turbulence, picture recognition, data compression, and chemistry. The POD goes back to Loève (1945), (1978) and Karhunen (1946).

For technical problems the question arises: how can a complex set of data be understood in a simple manner? The POD is a method for extracting phase space information out of spatio-temporal data. The procedure leads to a set of orthonormal basis functions and corresponding amplitudes. For applying POD, no a-priori knowledge of the system and no ansatzfunctions are necessary, and often the dynamics can be approximated very well by only a few basis functions. Due to this fact an efficient reduction of the system dimension can be reached. Furthermore, the coherent spatial structures will be optimal for a given data-set.

With POD, the spatio-temporal dynamics of linear as well as nonlinear systems will be projected onto a subset of the state space in which the most dominant dynamics take place, and in which the most kinetic energy is stored. The time-invariant basis functions represent the most persistent structures in the system, while the corresponding amplitudes are uncorrelated. Non-typical patterns like noise in the dynamics are fade out. The variance of time-functions serve as a measure for their dominance. Therefore, even nonlinear dynamics can be investigated. The basis functions are comparable with the well known eigenmodes of linear systems.

The spatio-temporal dynamics may be given as a vector-valued function $x(t) \in \mathbb{R}^n$ with $x_i(t) = x(z_i, t), i = 1, ..., n$, and can be split into a mean part $\bar{x}$ and a second part η with zero time average. Then the dynamics has the following properties:

$$x(t) = \bar{x} + \eta(t) \ , \ x, \bar{x}, \eta \in \mathbb{R}^n \ , \ \mathrm{E}\{x\} = \bar{x} \ , \ \mathrm{E}\{\eta\} = 0 \ , \qquad (6)$$

thereby $\mathrm{E}\{\bullet\}$ denotes the time average of its argument. The space X in which the dynamics of the system takes place can be decomposed into the two linear subspaces X_1 and X_2 as the direct sum $X = X_1 \oplus X_2$ (Troger and Steindl, 1991). For the projection of an element $x \in X$ onto the two subspaces hold, that $x = x_1 + x_2$ with $x_1 \in X_1$ and $x_2 \in X_2$, see Fig. 5.

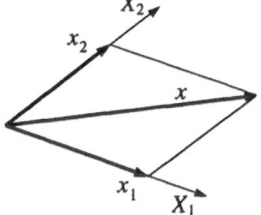

Fig. 5. Decomposition of X

A linear projection P of η onto a subspace with orthonormal basis $\{\psi_1, ..., \psi_n\}$ is given by

$$P\eta = \sum_{i=1}^{n} <\psi_i, \eta> \psi_i \ , \qquad (7)$$

where $< \bullet, \bullet >$ denotes the inner or scalar product of two vectors. The data-set x can be expanded by the series

$$x = \bar{x} + \sum_{i=1}^{n} a_i(t)\psi_i \quad . \tag{8}$$

For the time functions $a_i(t)$ equations (7) give

$$a_i = < \psi_i, \eta > = < \psi_i, x - \bar{x} > \quad . \tag{9}$$

The data-set can be written in matrix-notation with $a(t) = [a_1(t), ..., a_n(t)]^T$, $\Psi = [\psi_1, ..., \psi_n]$ as

$$x(t) = \bar{x} + \Psi a(t) \quad , \qquad \Psi \in \mathbb{R}^{n \times n} \quad . \tag{10}$$

It is the aim of the KL transformation to find the basis vectors in such a way that the corresponding time-functions are uncorrelated. Assuming $E\{a\} = 0$ means all covariances of a vanish, and the covariance matrix becomes diagonal:

$$C_{aa} = E\{aa^T\} = \text{diag}\{\lambda_1, ..., \lambda_n\} \quad . \tag{11}$$

In order to determine a basis Ψ, the second moment $E\{xx^T\}$ of the data x is calculated. Equations (10), (11) and $E\{a\} = 0$ leads to

$$E\{xx^T\} = \bar{x}\bar{x}^T + \Psi C_{aa} \Psi^T \quad . \tag{12}$$

Together with the rules for calculating covariance matrices,

$$C'_{xx} = E\{xx^T\} - E\{x\}E\{x^T\} = E\{xx^T\} - \bar{x}\bar{x}^T = \Psi C_{aa} \Psi^T \tag{13}$$

and the orthonormality of the basis vectors, $< \psi_i, \psi_j > = \delta_{ij}$, we obtain an eigenvalue problem for the basis vectors ψ_i:

$$(C_{xx} - \lambda_i E)\psi_i = 0 \quad , \quad i = 1, ..., n \quad , \tag{14}$$

where E denotes the $n \times n$ identity matrix. Since the covariance matrix C_{xx} of a data-set is real, symmetric and of order n, there exist exactly n real eigenvalues λ_i with n corresponding real orthogonal eigenvectors ψ_i. If the eigenvalues are labelled in decreasing order, $\lambda_1 \geq \lambda_2 \geq ... \geq \lambda_n$, the eigenfunction ψ_1 represents the most persistent spatial structure corresponding to a time-function $a_1(t)$ with greatest variance. The first four eigenfunctions $\psi_1, ..., \psi_4$ of the position of the drill string are shown in Fig. 6.

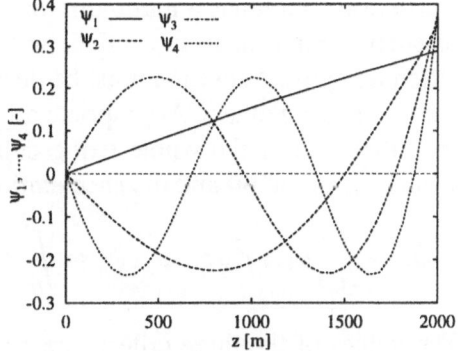

Fig. 6. First four eigenfunctions

4. Controlling Vibrations

In linear control theory the eigenmodes are used to project the system state to get real-valued amplitudes. As shown above, the first basis vectors $\psi_i, i = 1, ..., n$, represent the most persistent structures in linear as well as nonlinear systems.

The projection of the system state x onto the time-independent basis vector ψ_i results in a real single valued time-dependent amplitude $a_i(t)$ which is given in (9). Here the state x represents the measured positions and/or velocities of the system. All amplitudes vanish if the difference vector $(x - \bar{x})$ becomes zero. For employing the amplitude a_i as a control variable, it is necessary that the amplitude vanishes if the desired state is reached. Therefore, the projection of the shifting vector from the average state $\bar{x}$ to the desired state w_0 onto the basis vectors had to be subtracted from the amplitude. Then, the linear control law for the input variable $u(t)$ is given by (with proportional gain term k_{POD})

$$
\begin{aligned}
u &= k_{POD} \left[<\psi_i, x - \bar{x}> - <\psi_i, w_0 - \bar{x}> \right] \\
&= k_{POD} <\psi_i, x - w_0> \quad .
\end{aligned}
\tag{15}
$$

The damping of the vibrations in the string will be achieved by controlling the speed of the drive by varying the motor voltage U_m. A servo controller and a POD controller are set in parallel.

The motor voltage is then given by (with servo-controller gain term k_I and gain term k_{POD} of the POD controller):

$$
U_m = U_0 + k_I \int_{t_0}^{t} \left[\omega_{rt}(\tau) - \omega_0 \right] d\tau + k_{POD} <\psi_1, x - w_0> \quad .
\tag{16}
$$

Investigations have shown that the velocities can lead to an unstable system behavior due to in-phase oscillations with the bit vibrations. Best results have been obtained by applying the POD only to the position state.

The two controller parameters k_I and k_{POD} have been determined by optimization. One way to describe mathematically the desired optimal state is to define a criterion which has a minimum for the optimal solution. Obviously, the choice of such a criterion is subjective, and depends on the goal of the optimization. Often an optimal solution must be described by various criteria (which can even contradict each other). Appropriate criteria for the dynamics of the drill-string are the vibrations of the whole string defined by the amplitude function $a_1(t)$, the vibrations of the bit and the accelerations of the rotary table:

$$
\psi_1 = \int_{t_0}^{t_1} a_1(\tau)^2 d\tau \quad , \quad \psi_2 = \int_{t_0}^{t_1} (\dot{\varphi}_{bit} - \omega_0)^2 d\tau \quad , \quad \psi_3 = \int_{t_0}^{t_1} \ddot{\varphi}_{rt}^2 d\tau \quad .
\tag{17}
$$

The values of the three criteria are depicted in Fig. 7 for the varying parameter k_{POD}. They are scaled being equal to one in the minimum to be comparable. The value with $k_{POD} \equiv 0$ indicates the system without POD-controller.

Results of a numerical simulation are shown in Fig. 8. In comparison with Fig. 4 showing the uncontrolled system, the improvement of the dynamics is obvious. The results were compared with other controllers like servo-controller feeding back the velocity of the bit, but no better performance could be obtained.

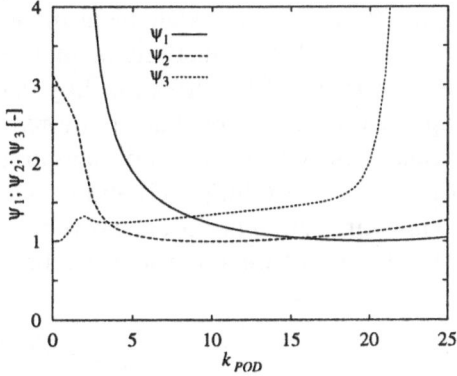

Fig. 7. Values of the criterions for the controller parameter k_{POD}

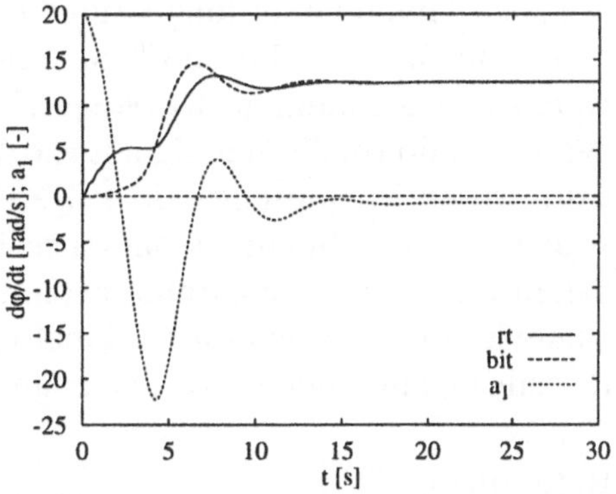

Fig. 8. Speed of the bit and the drive of the controlled system with servo- and POD-controller

For the design of controllers, a rather important question is how robust is a realization? Due to the nonlinearity of the system, design techniques developed from linear cases do not work. In nonlinear system theory (s. Bergé *et al.* (1984), Kreuzer (1987)) bifurcation analysis is used to obtain insight into the changes of behavior of a system due to parameter changes.

In the present case the speed of the drive ω may be such a varying parameter. Without going into much detail, we present just a few results. Figure 9 shows the phase displacement of the bit and the rotary table against the drive speed. Additionally, the phase condition is that the speed of the bit is equal to the desired speed. A single line stands for no vibrations (constant phase displacement), two lines denote torsional vibrations. The solid line corresponds to the POD-controller, the dashed line to the servo-controller.

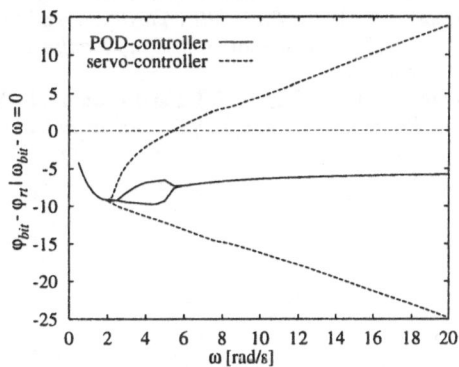

Fig. 9. Phase displacements of the stable solutions

Figure 9 shows that the servo controlled system oscillates for almost all drive speeds. On the other hand, the POD controlled system shows vibrations of very small amplitudes and these only for a small range of speeds (which are not typical working speeds). Therefore, the controlled system is robust against disturbances, and stable in a wide range of working speeds.

5. Conclusions

In this paper the application of proper orthogonal decomposition for controlling torsional vibrations in long strings has been presented. The mechanical model and its mathematical description were introduced and the equations of motion were simulated numerically. We gave a short overview of the proper orthogonal decomposition of discrete signals and their application to the torsional vibrations of a drill string. We presented the POD as a means of projecting the spatio-temporal structure of the vibrations onto a real variable for controlling oscillations. The determination of the controller parameters via optimization has been shown with examples of different criteria. Finally, the robustness of the controller has been investigated through bifurcation analysis.

References

Bergé, P., Pomeau, Y., and Vidal, Ch. (1984) *Order within chaos*, John Wiley & Sons, New York/... .

Brett, J. F. (1992) The genesis of torsional drillstring vibrations. *SPE Drilling Engineering*, September, 168 – 174.

Clayer, F., Aquitaine, E., Vandiver, J. K., and Lee, H. Y. (1990) The effect of surface and downhole boundary conditions on the vibration of drillstrings. In: *Proceedings of the 65th Annual Technical Conference and Exhibition of the SPE*, 431 – 442.

Karhunen, K. (1946) Zur Spektraltheorie stochastischer Prozesse. *Annales Academiæ Scintiarum Fennicæ. Series A - I. Mathematica - Physica* **34**.

Kreuzer, E. (1987) *Numerische Untersuchung nichtlinearer dynamischer Systeme*, Springer-Verlag, Berlin/... .

Loève, M. (1945) Functions aléatoires de second ordre. *C. R. Acad. Sci Paris*, **220**.

Loève, M. (1978) *Probability Theory*. 4th edition, Springer-Verlag, New York/... .

Troger, H. and Steindl, A. (1991) *Nonlinear Stability and Bifurcation Theory*, Springer-Verlag, Wien/... .

Zamudio, C. A., Tlusty, J. L., and Dareing, D. W. (1987) Self-excited vibrations in drillstrings. In: *Proceedings of the 62th Annual Technical Conference and Exhibition of the SPE*, 117 – 124.

DYNAMICS AND CONTROL OF DISCRETE ELECTROMECHANICAL SYSTEMS

P. MAISSER, G. KIELAU, O. ENGE AND H. FREUDENBERG
Institute of Mechatronics
at the Technical University of Chemnitz,
Reichenhainer Straße 88, D-09126 Chemnitz

1. Introduction

Dynamics and control of complex mechatronic systems can be investigated efficiently by using a suitable mathematical standard model. Based on such a mathematical standard model, well-known approaches to dynamics and control can be used in different application fields. Some of the most important mechatronic systems are electromechanical systems which can be regarded as physical structures characterized by interaction of electromagnetic fields with inertial bodies. The equations governing discrete electromechanical systems are obtained by combining Kirchhoff's theory with the appropriate constitutive equations. The motion of an electromechanical system will be understood as the motion of its representing point in its configuration space. Based on the principle of virtual work, the equations of motion are Lagrange's equations of the second kind (Maißer and Steigenberger, 1979). The analytical mechanics and its application to multibody dynamics can be regarded as a suitable starting point (Maißer, 1991). Lagrange's equations of mixed type for constrained mechanical systems constitute a mathematical standard model which is often used in dynamics as well as in force and position control of robot manipulators (Arimoto *et al.*, 1993). Often the so called disturbed equations of motion obtained by linearization of Lagrange's equations near a nominal trajectory $q(t)$ are used to design optimal controllers in mechatronics.

2. Dynamics of Multibody Systems

For a holonomic Multibody System (MBS) with kinematical tree structure and n degrees of freedom generalized coordinates q^a will be intro-

D. H. van Campen (ed.), IUTAM Symposium on Interaction between Dynamics and Control in Advanced Mechanical Systems, 215–222.
© 1997 *Kluwer Academic Publishers.*

duced according to the Denavit-Hartenberg-notation: $q^a = s_a \tau + (1 - s_a)\varphi$, $a = 1, \ldots, n$. Here, τ, φ denote complementary joint variables (displacement, angle) and $s_a = 1, 0$ describes the distribution of translational and rotational joints. $q = (q^a)$ is said to be the representing point of the MBS. The motion of an MBS can now be understood as the motion of its representing point in the configuration space $\mathbb{R}^n$. Based on the principle of virtual work, the motion equations of an MBS in $\mathbb{R}^n$ are Lagrange's equations (LE) of the second kind. Starting from the kinetic energy $T = \frac{1}{2}g_{ab}(q)\dot{q}^a\dot{q}^b$ of a holonomic scleronomic MBS, we can write LE $(\partial_a T)\dot{} - \partial_a T = Q_a$ explicitly:

$$g_{ab}(q)\ddot{q}^b + \Gamma_{abc}(q)\dot{q}^b\dot{q}^c = Q_a. \tag{1}$$

The generalized mass-matrix g_{ab} (symmetric, positive definite) defines a Riemannian metric in $\mathbb{R}^n$; $\mathbb{R}^n$ becomes a Riemannian space V^n. The Christoffel symbols of the first kind $\Gamma_{abc} = \frac{1}{2}(\partial_b g_{ac} + \partial_c g_{ab} - \partial_a g_{bc})$ define (in the language of Newtonian mechanics) Coriolis and centrifugal forces, Q_a denote the generalized applied forces.

Introducing the nonholonomic translational and angular velocities of the body B_k: $\underset{k}{v}_i = \underset{k}{v}_{ia}(q)\dot{q}^a$, $\underset{k}{\omega}_i = \underset{k}{\Omega}_{ia}(q)\dot{q}^a$, we can represent the generalized forces Q_a, the metric components g_{ab} and the Christoffel symbols of the first kind as linear, quadratic and cubic forms, respectively, related to the so-called kinematic basic functions $\underset{k}{v}_{ia}$, $\underset{k}{\Omega}_{ia}$ of the MBS:

$$Q_a = \sum_{k=a}^{K}[\underset{k}{K}^i\underset{k}{v}_{ia} + (1 - s_a)\underset{k}{M}^i\underset{k}{\Omega}_{ia}],$$

$$g_{ab} = \sum_{k=b}^{K}[\underset{k}{m}\underset{k}{v}_{ia}\underset{k}{v}_{ib} + (1 - s_a)(1 - s_b)\underset{k}{\Theta}^{ij}\underset{k}{\Omega}_{ia}\underset{k}{\Omega}_{jb}], \quad a \leq b, \tag{2}$$

$$\Gamma_{abc} = (1 - s_b)\sum_{k=c}^{K}\varepsilon_i{}^{qr}[\underset{k}{m}\,\delta^{ij}\underset{k}{v}_{ja}\underset{k}{\Omega}_{qb}\underset{k}{v}_{rc} + (1 - s_a)(1 - s_c)\underset{k}{\vartheta}^{ij}\underset{k}{\Omega}_{ra}\underset{k}{\Omega}_{jb}\underset{k}{\Omega}_{qc}],$$

$a < b \leq c$ or $b \leq a < c$, where $\underset{k}{m}$ denotes the mass of the body B_k, $\underset{k}{\vartheta}^{ij}$ Binet's inertia tensor ($\underset{k}{\vartheta}^{ij} = \underset{k}{\vartheta}^r_r\delta^{ij} - \underset{k}{\Theta}^{ij}$), $\underset{k}{K}^i$, $\underset{k}{M}^i$ are the applied forces/torques acting on the body B_k, and $g_{ab} = g_{ba}$, $\Gamma_{abc} = \Gamma_{acb}$, $\Gamma_{abc} = -\Gamma_{cba}$ for $b \leq a, c$.

The kinematic basic functions can be generated recursively and in a derivative free manner (the Lagrange-formalism for MBS is algebraized) (Maißer, 1988).

3. Electrical Multipole Networks

Using some basic concepts and methods of multibody dynamics as before, we can describe the electrical system dynamics in a similar way.

3.1. MULTIPOLES

The application of Lagrange's approach to electrical systems (ES) with lumped parameters is based on the concept of a multipole and its representation by abstract 2-poles, as well as on Kirchhoff's theory and the principle of virtual work (Maißer and Steigenberger, 1979; Enge *et al.*, 1995).

3.2. KIRCHHOFF'S THEORY

Definition 1 *An electrical system (ES) is a finite set of galvanically connected electrical multipoles (finite multipole network).*

After defining all pole-graphs, we represent an ES by a finite network of abstract 2-poles with the network graph Γ (containing B branches, N nodes, p components). Let G denote an arbitrary (but fixed) frame of Γ and $H(G)$ the coframe of G in Γ. The corresponding fundamental cut and the fundamental loop matrices are denoted by Q and A, respectively.

Then, *Kirchhoff's laws* become:

$$A^i{}_j V_i = 0, \quad j \in H, i \in \Gamma \quad (\text{"voltage law"}), \tag{3}$$

$$Q_i{}^j I^i = 0, \quad j \in G, i \in \Gamma \quad (\text{"current law"}). \tag{4}$$

A and Q are related to each other by

$$A^i{}_j = \left\{ \begin{array}{ll} \delta^i{}_j , & i, j \in H \\ -Q_j{}^i , & i \in G, j \in H. \end{array} \right. \tag{5}$$

Hence, $A^T Q = 0$.

3.3. LAGRANGE'S EQUATIONS FOR ELECTRICAL SYSTEMS IN CHARGE FORMULATION

(4) yields

$$I^j = -Q_i{}^j I^i, \quad i \in H, j \in G,$$

and due to (5)

$$I^j = A^j{}_\mu i^\mu, \quad \mu \in H, j \in \Gamma. \tag{6}$$

This means, that each current I^j, $j \in \Gamma$, can be represented by a linear combination of coframe currents i^μ, $\mu \in H$. (6) is called the *mesh-transformation* (MT), it defines the kinematics of the ES.

Generalized coordinates of ES are introduced as follows:
Defining functions

$$\bar{q}^j(t) := \int_0^t I^j(\tau)\, d\tau, \quad \bar{\lambda}_j(t) := \int_0^t V_j(\tau)\, d\tau \tag{7}$$

as *charge* and *flux* of the abstract 2-pole j, $j \in \Gamma$, respectively; its branch relation has the form

$$\dot{\bar{\lambda}}_j \equiv V_j = f_j(\ddot{\bar{q}}, \dot{\bar{q}}, \bar{q}, t) \quad \text{or} \quad \bar{\lambda}_j = f_j(\dot{\bar{q}}, \bar{q}, t) \quad \text{or} \quad \bar{q}^j = g_0^j(t), \qquad (8)$$

where f_j and g_0^j are given, differentiable functions (notation: $\bar{q} := (\bar{q}^1, \ldots, \bar{q}^B)$, $\bar{\lambda} := (\bar{\lambda}_1, \ldots, \bar{\lambda}_B)$). $\bar{q}^j$ denotes the charge which has been moved in (0,t) caused by the current I^j; $\bar{\lambda}_j$ has not necessarily the meaning of *magnetic flux*. An abstract 2-pole with the relation $\bar{q}^j = g_0^j(t)$ is called a *current generator*, $g_0^j(t)$ denotes its charge source. Hence, the equations (8) are the constitutive equations of an ES (in the sense of mechanics).

Due to the mesh-transformation (6), a frame G must not contain current generators. Consequently, all current generators belong to the coframe $H(G)$ and decomposite the index set

$$H(G) \quad = \quad H^* \cup H_0, \qquad (9)$$
$$H^* : \text{coframe branches not containing current generators,}$$
$$H_0 : \text{coframe branches containing current generators,}$$

so that (6) with (7) yields

$$\bar{q}^j = A^j{}_\mu q^\mu + q_0^j(t), q_0^j := A^j{}_\lambda \bar{q}^\lambda \equiv A^j{}_\lambda g_0^\lambda(t), \, j \in \Gamma, \, \mu \in H^*, \, \lambda \in H_0. \quad (10)$$

The set of branch charges $\{\bar{q}^j, j \in \Gamma\}$ is called a *configuration* of the ES. All charges $\bar{q}$ fulfilling Kirchhoff's current law in integrated form $a^i{}_j \bar{q}^j = 0$ ($a^i{}_j$ - node-branch incidence matrix) at time t determine the set

$$\mathfrak{L}_t := \{\bar{q}^j \mid \bar{q}^j = A^j{}_\mu q^\mu + q_0^j(t); \, q^\mu \in \mathbb{R}\} \qquad (11)$$

of all *admissible configurations* of the ES at time t.

Definition 2 *The ES is said to be* holonomic, *having the* quasi degree of freedom m. $q = (q^\mu, \, \mu \in H^*)$ *is called its* representing point, $q \in \mathbb{R}^m$; $\mathbb{R}^m$ *is called its* configuration space, *the* q^μ, $\mu \in H^*$, *are called (topologically generated)* generalized coordinates *of the ES.*

The motion of the ES is described by C^2-functions $q = q(t)$. The state of the ES is given by $(\dot{q}, q)$. The ES is called *scleronomic* if $\mathfrak{L}_t$ does not depend on t explicitly, otherwise it is called *rheonomic* ($H_0 \neq \emptyset$).

A *virtual displacement* of an ES is defined by a set of differential increments of charges $\bar{q}^j$ belonging to a variation δq^μ, $\mu \in H^*$, at fixed time t:

$$\{\delta \bar{q}^j \mid \delta \bar{q}^j = A^j{}_\mu \delta q^\mu, \quad j \in \Gamma, \, \delta q^\mu \text{ arbitrary}\}. \qquad (12)$$

(Because $A^j{}_\mu = \delta_\mu^j \, \forall j, \mu \in H$, it is $\delta \bar{q}^j = 0 \quad \forall j \in H_0$.) Motion and state of an ES are defined analogously to mechanical systems (Maißer, 1988).

Axiom 1 *(Principle of virtual work in the charge approach)*
Let the constitutive equations be given as $\dot{\lambda}_i \equiv V_i = f_i(\ddot{\bar{q}}, \dot{\bar{q}}, \bar{q}, t)$, $i \in \Gamma \setminus H_0$.
Then $\delta' A := -V_i \delta \bar{q}^i = 0$ *($\forall \delta \bar{q}^i$ virtual) defines the actual motion of an ES.*

For constitutive equations satisfying Helmholtz-conditions, the axiom 1 yields Lagrange's equations of motion of the ES

$$(\dot{\partial}_\mu \Lambda)^{\cdot} - \partial_\mu \Lambda + \dot{\partial}_\mu D = Q_\mu^{(S)}, \tag{13}$$

where $\Lambda(\dot{q}, q, t) := W'_m - W_e - V^h$ is the Lagrangian (W'_m - magnetic coenergy, W_e - electric energy, V^h - generalized potential) and $D(\dot{q}, q, t) := D^{(0)} + D^{(1)}$ denotes the dissipation function of the ES. $Q_\mu^{(S)} := -A^i{}_\mu V_i^{(S)}|_{MT}$ are voltages not described by Λ or D (Maißer and Steigenberger, 1979).

4. Discrete Electromechanical Systems

EMS can be regarded as physical structures characterized by interactions of electromagnetic fields with inertial bodies. The interaction is expressed by constitutive equations describing the coupling between Maxwell's theory and mechanics. Constitutive equations describing the coupling of multibody dynamics with Kirchhoff's theory (as quasi stationary approximation of Maxwell's theory) define discrete EMS. In the following, only such systems will be considered.

Definition 3 *An EMS is a finite set of physical objects with mechanical and/or electrical multipole-properties interacting among themselves by electrodynamical and/or electro-magneto-mechanical coupling.*

4.1. KINEMATICS OF ELECTROMECHANICAL SYSTEMS

The kinematics of an EMS is defined by the geometric constraints between the rigid bodies and the topology of the electrical network represented by abstract 2-poles. The set $\{\bar{q}^i, \bar{x}^s_k | i \in \Gamma, s = 1,\ldots,6; k = 1,\ldots,K\}$ is called a *configuration* of the EMS. All branch charges $\bar{q}^i$ and mechanical coordinates $\bar{x}^s_k$, which fulfil all constraints of the EMS at given time t (i.e. geometric constraints and Kirchhoff's current law), determine the set $\mathfrak{L}_t := \{\bar{q}^i, \bar{x}^s_k | \bar{q}^i = A^i{}_\mu q^\mu + q^i_0(t), \bar{x}^s_k = \bar{x}^s_k(q^\kappa, t); i \in \Gamma, s = 1,...,6; k = 1,...,K; (q^\mu, q^\kappa) \in \mathbb{R}^n, \mu \in H^*, \kappa \in J\}$ of all *admissible configurations* of an EMS at time t. J is an index set of mechanical coordinates q^κ; $J \cap H^* = \emptyset$. Convention: $\mu, \nu, \omega \in H^*$; $\kappa, \lambda, \rho \in J$; $a, b, c \in H^* \cup J$; $\alpha, \beta \in H^* \cup J \cup \{0\}$, and $q^0 = t$, $\dot{q}^0 = 1$. $\mathfrak{L}_t$ is a 1-1 map to a cylindrical domain $D \subset \mathbb{R}^n$ of the configuration space, $n := |H^*| + |J|$. n is called quasi degree of freedom of the EMS, and $q = (q^a) = (q^\mu, q^\kappa)$ denotes its representing point.

The mesh charges q^μ and the mechanical coordinates q^κ are the generalized coordinates of the EMS. The motion of the EMS is given by C^2-functions $q = q(t)$, and the state of the EMS is given by $(\dot{q}, q)$. The EMS is called *holonomic* if there are no nonintegrable kinematic constraints, otherwise it is called *nonholonomic*. The EMS is called *scleronomic* if $\mathfrak{L}_t$ does not depend on t explicitly, otherwise it is called *rheonomic*.

4.2. CONSTITUTIVE EQUATIONS

Let $d\mathfrak{k}(\xi)$ be the applied forces acting on the bodies of the EMS, V_i denote the voltages of abstract 2-poles, and Q_κ and v_μ are the generalized forces and the mesh voltages, respectively. Hence,

$$Q_\kappa := S\partial_\kappa \mathfrak{x} d\mathfrak{k}, \quad v_\mu := A^i{}_\mu V_i|_{MT}. \tag{14}$$

Assumption: The constitutive equations of an EMS

$$Q_\kappa = Q_\kappa(\dot{q}^a, q^a, t), \quad v_\mu = v_\mu(\ddot{q}^a, \dot{q}^a, q^a, t) \tag{15}$$

are given by sufficient smooth functions Q_κ and v_μ. The simultaneous presence of q^λ and q^ν and their derivatives in these equations indicates the electromechanical interaction. The class representation

$$
\begin{aligned}
Q_\kappa &= & Q_\kappa^{(0)} &+& Q_\kappa^{(1)} &+& Q_\kappa^{(2)}, \\
-v_\mu &= & -\dot{\psi}_\mu &-& u_\mu^{(0)} &-& u_\mu^{(1)} &-& u_\mu^{(2)}
\end{aligned} \tag{16}
$$

with

$$
\begin{aligned}
\psi_\mu &= \dot{\partial}_\mu \Psi, & u_\mu^{(0)} &= -\partial_\mu \Psi, & Q_\kappa^{(0)} &= \partial_\kappa \Psi, & (Q_a^{(0)} &= \partial_a \Psi), \\
& & u_\mu^{(1)} &= -\delta_\mu V, & Q_\kappa^{(1)} &= \delta_\kappa V, & (Q_a^{(1)} &= \delta_a V), \\
& & u_\mu^{(2)} &= \dot{\partial}_\mu D, & Q_\kappa^{(2)} &= -\dot{\partial}_\kappa D, & (Q_a^{(2)} &= -\dot{\partial}_a D)
\end{aligned} \tag{17}
$$

leads to the following state functions of an EMS:

the magnetomechanical copotential: $\quad \Psi := \int \psi_\mu d\dot{q}^\mu - u_\mu^{(0)} dq^\mu + Q_\kappa^{(0)} dq^\kappa$
the generalized
electromechanical potential: $\quad V = \omega_\alpha(q,t)\dot{q}^\alpha \equiv \omega_a \dot{q}^a + \omega_0$
the dissipation function: $\quad D := \int u_\mu^{(2)} d\dot{q}^\mu - Q_\kappa^{(2)} d\dot{q}^\kappa$

Here, ω_a and ω_0 are defined by a PDE-system (Maißer and Steigenberger, 1979).

4.3. KINETICS OF ELECTROMECHANICAL SYSTEMS

The kinetics of EMS is based on the principle of virtual work in Lagrange's notation.

Axiom 2 *The actual motion of an EMS is characterized by the vanishing of the virtual work*

$$\delta' A := -V_i \delta \bar{q}^i + S\delta \mathfrak{x}(d\mathfrak{k} - \ddot{\mathfrak{x}}dm) = 0 \quad \forall \delta \bar{q}^i, \delta \mathfrak{x} \text{ virtual} \tag{18}$$

at any time t.

(18) yields Lagrange's equations of an EMS

$$(\dot{\partial}_a \Lambda)^{\cdot} - \partial_a \Lambda + \dot{\partial}_a D = 0, \quad a \in J \cup H^*, \tag{19}$$

$$\Lambda := T - \Omega = T(\dot{q}^\lambda, q^\lambda, t) + \Psi(\dot{q}^\nu, q^a, t) - V(\dot{q}, q, t). \tag{20}$$

Starting from a Lagrange model $\{\Lambda, D\}$ $\Lambda = \frac{1}{2}g_{\alpha\beta}(q,t)\dot{q}^\alpha\dot{q}^\beta$, $D = \frac{1}{2}s_{\alpha\beta}(q,t)\dot{q}^\alpha\dot{q}^\beta$ $(\alpha, \beta = 0, 1, \ldots, n; q^0 = t, \dot{q}^0 = 1)$, we obtain LE (19) in explicit form:

$$g_{ab}(q,t)\ddot{q}^b + \Gamma_{a\alpha\beta}(q,t)\dot{q}^\alpha\dot{q}^\beta = Q_a, \tag{21}$$

$(g_{ab}) := (\dot{\partial}_a\dot{\partial}_b\Lambda) = (g_{\mu\nu}) \bigoplus (g_{\kappa\lambda})$, $g_{\mu\nu} = \dot{\partial}_\mu\dot{\partial}_\nu\Psi = l_{\mu\nu} = A^i{}_\mu A^j{}_\nu L_{ij}$ - generalized inductivities, $g_{\kappa\lambda}$ - defined by (2).

A special EMS (with inductivities, permanent magnets, resistors, current and voltage generators) with the constitutive equations

$$\Psi_i = L_{ij}\dot{\bar{q}}^j + \Psi_{i0}, \quad L_{ij} = L_{ij}(q^\kappa), \quad \Psi_{i0} = \Psi_{i0}(q^\kappa), \quad i,j \in H,$$

$$V^{(R)} = R_{ij}\dot{\bar{q}}^j, \qquad V_i = V_{i0}(t), \qquad I^i = I_0^i(t), \qquad i \in H$$

has Lagrange's equations of the following explicit form:

$$A^i{}_\mu A^j{}_\nu[L_{ij}\ddot{q}^\nu + \partial_\lambda L_{ij}\dot{q}^\lambda\dot{q}^\nu] + A^i{}_\mu\partial_\kappa[L_{ij}\dot{q}_0^j(t) + \Psi_{i0}]\dot{q}^\kappa$$
$$+A^i{}_\mu V_{i0}(t) + A^i{}_\mu L_{ij}\ddot{q}_0^j(t) = -A^i{}_\mu A^j{}_\nu R_{ij}\dot{q}^\nu - A^i{}_\mu R_{ij}\dot{q}_0^j(t) \tag{22a}$$

$$g_{\kappa\lambda}\ddot{q}^\lambda + \Gamma_{\kappa\lambda\varrho}\dot{q}^\lambda\dot{q}^\varrho - \frac{1}{2}A^i{}_\mu A^j{}_\nu\partial_\kappa L_{ij}\dot{q}^\mu\dot{q}^\nu - A^i{}_\mu\partial_\kappa[L_{ij}\dot{q}_0^j(t) + \Psi_{i0}]\dot{q}^\mu$$
$$= \sum_{k=\kappa}^{K}[\underset{k}{K^i}\underset{k}{\psi_{i\kappa}} + \underset{k}{M^i}\underset{k}{\Omega_{i\kappa}}] + \frac{1}{2}\partial_\kappa L_{ij}\dot{q}_0^i(t)\dot{q}_0^j(t) + \partial_\kappa\Psi_{i0}\dot{q}_0^i(t). \tag{22b}$$

(22a) are Kirchhoff's mesh equations, (22b) are mechanical equations, both containing interaction terms.

5. Constraint by kinematical control

Generalized coordinates prescribed as functions of time define special constraints of an EMS, so-called constraints by kinematical control:

$$q^{a_1} = q_0^{a_1}(t) \quad \sim \quad f^{a_1}(q,t) \equiv q^{a_1} - q_0^{a_1}(t) = 0, \quad \{a_1\} \subseteq \{a\}. \tag{23}$$

From (21) the constraint motion equations

$$
\begin{aligned}
g_{a_2 b_2}(q^{c_2}, t)\ddot{q}^{b_2} &= Q_{a_2}(\dot{q}^{b_2}, q^{b_2}, t) - 2\Gamma_{a_2 b_1 c_2}(q^{b_2}, t)\dot{q}_0^{b_1}(t)\dot{q}^{c_2} \\
&\quad -\Gamma_{a_2 b_1 c_1}(q^{b_2}, t)\dot{q}_0^{b_1}(t)\dot{q}_0^{c_1}(t) - g_{a_2 b_1}(q^{b_2}, t)\ddot{q}_0^{b_1}(t)
\end{aligned}
\tag{24}
$$

and the reaction forces

$$
R_{a_1} := g_{a_1 b}(q^{b_2}, t)\ddot{q}^b + \Gamma_{a_1 bc}(q^{b_2}, t)\dot{q}^b \dot{q}^c - Q_{a_1}(\dot{q}^{b_2}, q^{b_2}, t)
\tag{25}
$$

result separately. R_{a_1} denote mechanical forces/torques if $a_1 \in J$ or voltages across the current generators if $a_1 \in H_0 = H \setminus H^*$.

For a current controlled MAGLEV ($\dot{q}^\mu = \dot{q}_0^\mu(t) \; \forall \mu \in H$), $H_0 = H$, $H^* = \emptyset$ holds. In this case, there exist only "mechanical" motion equations (22b) with $\kappa \in J$; $i, j \in H$:

$$
g_{\kappa\lambda}\ddot{q}^\lambda + \Gamma_{\kappa\lambda\varrho}\dot{q}^\lambda\dot{q}^\varrho - \partial_\kappa[\frac{1}{2}L_{ij}\dot{q}_0^i(t)\dot{q}_0^j(t) + \Psi_{i0}\dot{q}_0^i(t)] = Q_\kappa.
\tag{26}
$$

(26) defines a nonlinear control problem. A PID-controller

$$
\dot{q}_0(t) := K_R \left[(\bar{x} - x) + T_V(\bar{x} - x)^\cdot + \frac{1}{T_N} \int_{t_0}^{t} (\bar{x} - x)\, d\tau \right] + \dot{q}_0(t_0)
$$

($\bar{x} - x$: distance between actual and reference position of the magnets) yields good results. In general, a force controller can be designed based on the so-called disturbed equations of motion:

$$
g_{ab}\ddot{x}^b + (2\Gamma_{abc}\dot{q}^c - \partial_b Q_a)\dot{x}^b + (\partial_b g_{ac}\ddot{q}^c + \partial_b \Gamma_{acd}\dot{q}^c \dot{q}^d - \partial_b Q_a)x^b = \mu\Phi_a,
\tag{27}
$$

x - disturbance, q - nominal trajectory: $q \to \bar{q} = q + x$.

The motion equations (22a, 22b) as well as the disturbed equations (27) can be generated automatically by using the software tool **alaska** developed at the Institute of Mechatronics.

References

Arimoto, S., Naniva, T. and Tsubuchi, T. (1993) Principle of Orthogonalization for Hybrid Control of Robot Manipulators, in Takamori, T. and Tsuchiya, K. (eds.), *Robotics, Mechatronics and Manufacturing Systems*, Elsevier Science Publishers B.V., North-Holland, pp. 295–302.

Enge, O., Kielau, G. and Maißer, P. (1995) Modelling and Simulation of Discrete Electromechanical Systems, in Lückel, J. (editor), *Proceedings of the Third Conference on Mechatronics and Robotics, October 4-6 1995*. B.G.Teubner Stuttgart, pp. 302–318.

Maißer, P. (1988) Analytische Dynamik von Mehrkörpersystemen, *ZAMM* **68** (10), 463–481.

Maißer, P. (1991) A Differential-Geometric Approach to the Multi Body System Dynamics, *ZAMM* **71** (4), T116–T119.

Maißer, P. and Steigenberger, J. (1979) Lagrange-Formalismus für diskrete elektromechanische Systeme, *ZAMM* **59** 717–730.

Paper presented at: IUTAM Symposium on Interaction between Dynamics and Control in Advanced Mechanical Systems
Eindhoven University of Technology, Eindhoven, The Netherlands
April 21–26, 1996

A TWO-LEVEL CONTROL-DESIGN METHODOLOGY
AND A SOFTWARE TOOLSET FOR MECHATRONICS

H. MANN
Computing Centre, Czech Technical University
Zikova Street 4, CZ-166 35 Prague 6, Czech Republic

AND

Z. ÚŘEDNÍČEK
EVPÚ Research Institute
SK-018 51 Nová Dubnica, Slovak Republic

1. Introduction

The available control synthesis methods, as well as the synthesis software tools, however sophisticated, are capable of dealing only with rather simple controlled system models, mostly linear. In practice, however, the systems tend to be complex, nonlinear, and subject to many constraints. In addition, the characteristics of the system components may not be precisely known. Also the controller and sensor characteristics are usually far from ideal. Thus, the design of any practical control system always involves many trial-and-error procedures.

After the control synthesis has been completed, a real prototype of the controlled system is usually constructed, and its behavior in response to a sufficient set of various signals and disturbances in the whole range of the assumed system operation is tested to see whether or not it is satisfactory. If not, the system model is corrected, the control is redesigned, the prototype is modified and tested again.

Practical experience shows that costly and time consuming production and experimental verification of system prototypes can be lessened considerably by computer-aided analysis of sufficiently realistic system models. It appears that such models can be formed very easily using the multipole approach adopted and elaborated by the first author. He and his students also developed a software tool for the analysis of such models called DYNAST.

D. H. van Campen (ed.), IUTAM Symposium on Interaction between Dynamics and Control in Advanced Mechanical Systems, 223–230.
© 1997 *Kluwer Academic Publishers.*

When the second author, a control-design researcher in the industrial environment, started to use DYNAST, he soon realized that the multipole models are too complex for the available control synthesis methods. Hence, he arrived at the idea of a control-design methodology using two levels of system modeling, differing from each other by the degree of the system idealization and abstraction.

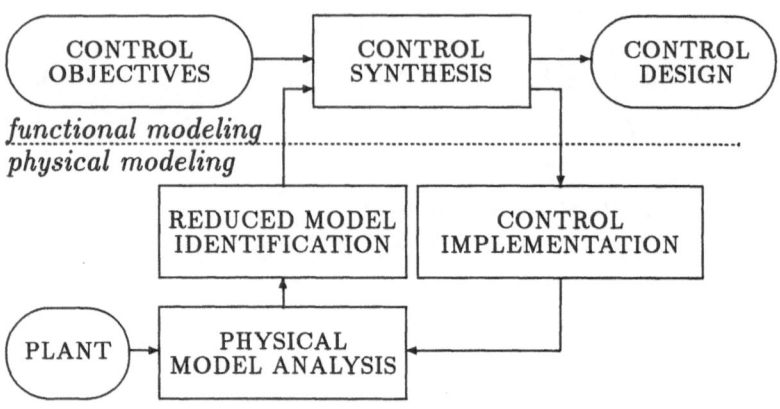

Figure 1. Two-level control-design procedure.

2. Physical vs. functional models

The proposed control-design methodology is illustrated by Fig. 1. The design procedure starts by setting up a sufficiently detailed physical model of the plant to be controlled. *Physical models* allow for a very faithful mimicking of real systems as they reflect dynamic interactions between system components in terms of flows of energy governed by the physical laws for such interactions.

In the next step, the physical model is reduced to a plant functional model corresponding to the constraints imposed on its form and complexity by the applied control-synthesis approach. *Functional models* are concerned only about transformations between system inputs, outputs, states and disturbances approximating them by means of mathematical relations (represented by block diagrams, transfer functions or state-space equations) ignoring any physical laws.

After the control synthesis has been completed, the plant physical model is augmented by physical models of the proposed controllers and sensors implemented in terms of available real devices. The resulting physical model of the complete controlled system is then subjected to thorough verification, taking into consideration the control-design objectives. If the verification results are satisfactory, a real prototype is built and verified experimentally.

If this is not the case, the system functional model is corrected and the control synthesis is repeated. Such an iterative procedure goes on until the complete system physical model meets the design objectives.

Obviously, the ability to construct, analyze and modify the realistic physical models in a rapid and flexible way plays a decisive role in the proposed methodology. The block diagrams, used so commonly in control design, represent the functional models well, but they are not suited to the physical models. They give a graphical representation of sets of equations, not the physical structure of real systems. The block interconnections respect algebraic rules, but not any physical laws. Besides this, the common block-diagram oriented simulation tools form the underlying differential equations in an explicit form and solve them in a consecutive way. This imposes additional restrictions on the block-diagram structure and computational robustness.

3. Multipole physical models

The need for realistic physical models is served very well by the multipole modeling approach which applies to all the energy domains involved in mechatronic systems (mechanical, electrical, magnetic, acoustic, fluidic, thermal, thermodynamic, electrochemical, etc.). The most important feature of the multipole models is in the isomorphism between their structure and the structure of the real systems.

The *multipole* model of a system component models the total energetic interaction between the component and the rest of the system assuming that the interactions take place just in a limited number of component *energy entrances* like electrical terminals, pipe inlets, shafts or other mechanical contacts, etc. Each such entrance is represented by a multipole *pole*, and the energy flow through the entrance into the component is approximated by the product of two variables – a *pole-through variable*, and a *pole-across variable*. Such variable pairs are, for example, force – velocity, torque – angular velocity, volume flow – pressure, electrical current – voltage, entropy flow – temperature, etc.

The two complementary variables differ from each other in the way they can be measured. To measure a pole-through variable at an entrance, the entrance must be disconnected first from the system, and the measuring instrument must be inserted then between the entrance and the system. The pole-across variables are measured between an entrance and a reference (like the absolute mechanical frame for velocity, the electrical ground for voltage, the open atmosphere for pressure, etc.) without any disconnection.

A system multipole model can be displayed graphically by a *multipole diagram* in which the mutual interaction of several energy entrances is indicated by the coalescence of the corresponding multipole poles. Each set of mutually coalesced poles forms a *node* of the diagram, which is associated with a *node-across variable* identical to the pole-across variables of the coalesced poles.

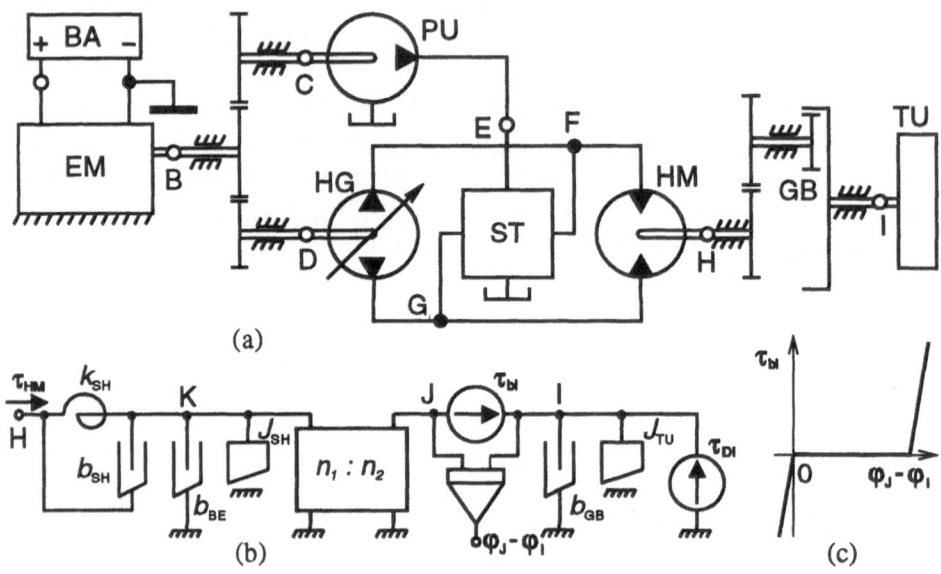

Figure 2. (a) Example of a dynamic system, (b) detailed multipole diagram of the gearbox and turret, (c) characteristic of the gearbox backlash and teeth flexibility.

Fig. 2a shows an example of a dynamic system [2] consisting of the turret TU driven by the hydraulic motor HM via the gearbox GB, while the oil intake into the motor is governed by the controlled hydrogenerator HG. The hydraulic subsystem is supplied by the pumpe PU, the oil pressure is limited by the hydraulic circuitry ST. Both the pumpe PU and the hydrogenerator HG are driven by the battery BA supplied DC motor EM.

The system scheme given Fig. 2a can be taken directly as a multipole diagram of the system. The diagram nodes are denoted there by single letters. For example, the node A is associated with the electrical-terminal voltage V_A, the node B with the shaft angular velocity ω_B, the node E with the oil pressure p_E, etc.

4. Multipole postulates

The multipole model of a component is thus based on the assumption, that the rate of the total *energy flow* into the component by means of all its energy entrances can be approximated as

$$P(t) = \sum_k a_k(t)\, u_k(t) \tag{1}$$

where $a_k(t)$ is the k-th pole-across variable and $u_k(t)$ is the k-th pole-through variable complementing $a_k(t)$.

Multipole models respect the continuity and compatibility postulates. According to the *continuity postulate*, the pole-through variables of a multipole satisfy the relation

$$\sum_k u_k(t) = 0 \tag{2}$$

within each energy domain separately.

The *compatibility postulate* states that within each energy domain the across variable between any two poles of a multipole can be expressed as a difference between the related pole across variables. Hence,

$$a_{kj}(t) = a_k(t) - a_j(t) \tag{3}$$

assuming that $a_k(t)$ and $a_j(t)$ are the pole across variables of the k-th and j-th poles while $a_{kj}(t)$ is the across variable of the k-th with respect to the j-th pole.

The former postulate corresponds to the physical laws of conservation of energy, matter, electrical charge, mechanical momentum, etc. The latter postulate is related to the geometrical connectedness of real components.

A multipole itself can be usually characterized by a *constitutive relation* in the form of a set of algebro-differential equations

$$\mathbf{f}(\mathbf{a}(t), \mathbf{u}(t), \mathbf{w}(t), \dot{\mathbf{w}}(t), t) = \mathbf{0} \tag{4}$$

where $\mathbf{f}(.)$ is a vector function, $\mathbf{a}(t)$ and $\mathbf{u}(t)$ are vectors of the multipole across and through variables respectively, $\mathbf{w}(t)$ is a vector of auxiliary variables (which may be identical to the pole variables), and $\dot{\mathbf{w}}(t)$ is its derivative with respect to t.

The constitutive relation of a multipole is independent of the constrains imposed on its variables by interconnecting it with other multipoles in a particular system. On the other hand, the postulates are concerned only with the structure of the interactions between multipoles regardless of their character. The relations (2) and (3) are valid also for nonlinear, time-variable and parameter-distributed multipole models.

This separation facilitates modeling and analysis of real dynamic systems considerably. It allows for setting up the multipole model of a system in a kit-like manner from the multipole models of the individual system components in the same way in which the real system is assembled from these components. Thus the modeling procedure can be based on a mere inspection of the real system.

To analyze a system-multipole model means to simultaneously solve the constitutive relations (4) of all the component multipoles in the system together with all the relations (2) and (3) characterizing the multipole interconnections. However, as the constitutive relations (4) are independent of the interconnections, they can be formed for the individual component models just once and then reused later. On the other hand, all the equations (2) and (3) related to the multipole interconnections can be formed by the analysis program automatically without asking its user to derive some additional equations or to construct a bond graph, etc.

Once the multipole model of a system has been set up, various modifications of the system concept can be easily investigated. For example, replacement of the hydraulic drive by an electrical one in Fig. 2a requires only replacing the corresponding multipoles in the system model. Also, the system model can be gradually refined by 'plugging in' more and more more sophisticated component-multipole models.

5. Control-design toolset

The multipole modeling approach has been implemented in the DYNAST software package [4]. It admits also block diagrams and sets of implicit nonlinear algebro-differential or algebraic equations in a natural textual form (without converting them into a block diagram). Multipoles, blocks and equations can be freely combined regardless of any 'algebraic loops'. Diagrams can be submitted in a graphical form using a screen editor.

Multipole and block models of various typical system components can be stored in independent files in a component-model library. Forming a model of a particular system is then very easy. It is sufficient to specify which multipole should be retrieved from the library to model each component, what should be the component-model parameters, and with what system-model nodes the component-model poles are incident.

The individual library multipoles and blocks can be characterized very freely by sets of algebro-differential equations (e.g., of the Lagrange's form with multipliers), or they can be set up from multipoles and/or blocks, combined with equations. Fig. 2b gives an example of a multipole submodel for the gearbox GB and the turret TU from the system shown in Fig. 2a. The submodel is mostly set up from elementary pure twopoles. The gearbox shaft driven by the hydraulic-motor torque τ_{HM} is modeled by the pure torsional spring of torsional stiffness k_{SH}, by the damper of damping factor b_{SH}, and by the rotary inertor of moment of inertia J_{SH}. The damper b_{BE} models the friction of the shaft bearing.

Both the static and kinematic transfer features of the idealized gearbox are represented by the pure transformer of the angular-velocity ratio $n_1 : n_2$. The gearbox unsymmetrical backlash is modeled by the source of torque τ_{bl} controlled by the backlash angle $\varphi_J - \varphi_I$. Its characteristic can be seen in Fig. 2c, the slope of which outside the backlash corresponds to the gear-teeth compliance. The inertor J_{TU} represents the turret moment of inertia, the source of torque τ_{DI} models the external disturbances acting on the turret.

To analyze nonlinear problems, DYNAST uses a very efficient and robust stiff-stable integration routine, varying both its step length and order to proceed as fast as possible without exceeding the admissible truncation error. (As shown in [3], DYNAST was about 2000-times faster than MATLAB when solving a comparison set of stiff nonlinear equations.) For linear or automatically linearized problems, DYNAST provides transfer functions as well as frequency and time characteristics in a semisymbolic form. A fast Fourier transform is available also.

In a control-design toolset supporting the proposed two-level design methodology, DYNAST can be easily combined, for example, with the well-tried and widespread MATLAB control-design toolboxes within the MS Windows environment. DYNAST allows for a rapid and flexible physical model construction and modification, whereas the MATLAB tools can be exploited to reduce physical models into functional and to the control synthesis.

6. Example

Experience from designing numerous real-life mechatronic systems has proven that the proposed methodology and toolset can design the control faster, more economically and also more thoroughly than the traditional way. Due to lack of space, we shall demonstrate it only for the example system shown in Fig. 2a. The objective of the design was the stabilization and tracking control for the turret.

First, a detailed multipole model of the complete system was constructed. Multipole parameters for the individual system components were determined partly from data given by the component manufacture, partly from the component geometry and material constants, and, in some cases, also from measurements of the component dynamic characteristics. The physical model of the complete system was verified by comparing the responses predicted by simulation with laboratory experiments.

The MATLAB *System Identification Toolbox* was used to derive a linear functional model of a reduced complexity from the simulated responses of the system physical model. Then, using the MATLAB *Control System*

230

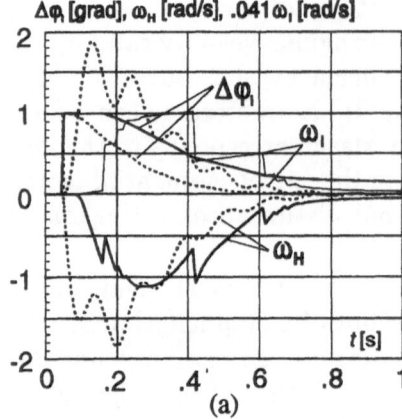
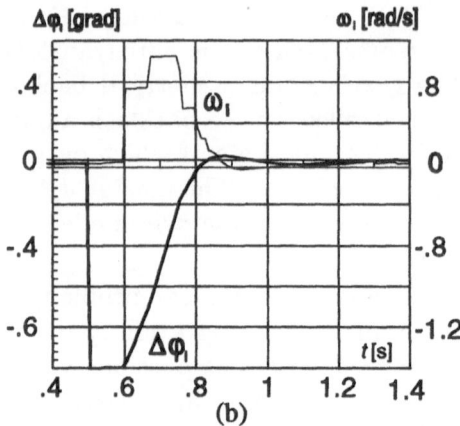

Figure 3. (a) Physical (dotted) vs. functional (full) model step-responses, (b) physical model step-responses for the enhanced system.

Toolbox the full state-space controller was designed with the state vector $x^t = [\varphi_I, \omega_I, \omega_H, \tau_{bl}]$, where φ_I and ω_I is the turret angular displacement and velocity, respectively, ω_H is the hydraulic-motor angular velocity, and τ_{bl} is the torque transferred across the gearbox backlash. As an input for the control, the hydraulic-motor pressure difference $\Delta p = p_F - p_G$ was considered.

As only the variables φ_I and ω_I are measurable directly, the controller has been modified to take this fact into consideration. The results of the design are shown in Fig. 3a. The step responses of the linear functional model (dotted) are compared there with those of the physical model (full). We can see that for small $\Delta\varphi_I = \varphi_{des} - \varphi_I$, where φ_{des} is the desired position of the turret, the gearbox backlash and the dry friction in several system components have important influences on the responses. (The physical-model simulated and measured responses were very close to each other.)

As shown in Fig. 3b, this problem was solved by tuning up the controller parameters and by enhancing the control using a discontinuously operating integrator.

References

1. Mann H. (1985) *Computer-aided design of dynamic systems in mechanical and electrical engineering* (in Czech). ČSVTS-FEL-ČVUT, Prague.
2. Úředníček Z. and Ondrejíčka J. (1995) Utilization of simulation experiments in the control design of mechatronic systems. An unpublished report. EVPÚ, Nová Dubnica.
3. *EUROSIM Simulation News Europe*, November 1991.
4. *DYNAST documentation.* DYN Co., Nad lesíkem Street 27, CZ-160 00 Prague 6, Czech Republic, dyn@vc.cvut.cz, http://icosym.cvut.cz/dyn/.

ENTRAINMENT CONTROL OF CHAOS NEAR UNSTABLE PERIODIC ORBITS

R. METTIN
Institut für Angewandte Physik
Technische Hochschule Darmstadt
Schloßgartenstr. 7
64289 Darmstadt, Germany[†]

Abstract. It is demonstrated that improved entrainment control of chaotic systems can maintain periodic goal dynamics near unstable periodic orbits without feedback. The method is based on the optimization of goal trajectories and leads to small open-loop control forces.

1. Introduction

In recent years, a large number of investigations has been made into the control of chaotic dynamics, and many techniques have been applied in simulations and experiments (see, for instance, [1] for an overview).

Two techniques have been proposed for an *open-loop* control of chaos. The first approach is related to vibrational methods, as a scalar periodic perturbation is applied to the chaotic system. Usually, the control signals are sinusoidal [2] or two-mode forces [3]. Recently it has been shown that the method can be improved by use of optimized multimode signals which are more complex [4].

The second method uses equations of motion and a specific goal dynamics to derive vector control forces. If the goal trajectory is suitably chosen, the chaotic system under control converges to the goal. Therefore, this method is often referred to as entrainment control [5]. In this paper it is shown how entrainment control can be improved with respect to small control forces. For this purpose, a specific property of chaotic systems is exploited: dense unstable periodic orbits (UPOs).

[†]Current address: 3. Phys. Inst., Univ. Göttingen, Bürgerstr. 42-44, D-37073 Göttingen

D. H. van Campen (ed.), IUTAM Symposium on Interaction between Dynamics and Control in Advanced Mechanical Systems, 231–238.
© *1997 Kluwer Academic Publishers.*

In fact, UPOs are common goal orbits of most feedback techniques employed for control of chaos (see, e.g., [6]). Because a UPO is a natural, but unstable motion of the system, control forces have to be applied only for transfer to and stabilization of the orbit. In the ideal (noiseless) case, the stabilization forces tend to zero once the UPO is actually reached. Therefore, feedback control of UPOs can be maintained with very small forces in low noise systems. Despite the power of such methods, however, there are situations where one wants to or has to dispense with a feedback from the system. Then, open-loop techniques are needed.

While there exists a theory for open-loop stabilization of unstable fixed points (vibrational control, see [7]), a counterpart for stabilization of unstable periodic orbits is still lacking to the author's knowledge. Although it is supposed that the scalar periodic perturbation methods mentioned above usually stabilize periodic dynamics in the vicinity of UPOs, this has not been shown yet explicitly. The underlying mechanism, possibly some resonance phenomenon, as well as the exact final dynamics are still unknown. This is different for entrainment control, where the appropriate dynamics is given *a priori*, and which is described in the following.

2. Entrainment control

We start with a nonlinear dynamical system in the continuous time domain, which is influenced by a control signal vector. The equations of motion are supposed to be known, according to the ordinary differential equation

$$\frac{d\mathbf{x}(t)}{dt} = \dot{\mathbf{x}}(t) = \mathbf{f}(\mathbf{x}(t), \mathbf{u}(t)) \tag{1}$$

with the system state $\mathbf{x}$ in the state space $\mathbb{R}^n$, the control vector $\mathbf{u}$ in the control signal space $\mathbb{R}^s$ and the vector field $\mathbf{f} : \mathbb{R}^n \times \mathbb{R}^s \to \mathbb{R}^n$. Let $\mathbf{z}$ be a goal dynamics generated by a vector field $\mathbf{g}$, $\dot{\mathbf{z}}(t) = \mathbf{g}(\mathbf{z}(t))$, $\mathbf{z}(t_0) = \mathbf{z}_0$. To introduce the goal dynamics as solutions of Eq. (1), we have to apply forces $\mathbf{u}(t)$ which solve the equation

$$\dot{\mathbf{x}}(t) = \mathbf{f}(\mathbf{x}(t), \mathbf{u}(t)) = \mathbf{g}(\mathbf{x}(t)). \tag{2}$$

Thus, the vector field $\mathbf{f}$ is simply changed to $\mathbf{g}$ by the control forces. To do this, however, feedback from the system is necessary, as the actual system state $\mathbf{x}(t)$ appears in Eq. (2). The main point of the entrainment control method is to eliminate $\mathbf{x}(t)$ by the assumption that the system is already located in the initial goal state $\mathbf{z}_0$ when the control is started at t_0. If so, the correct control signal $\mathbf{u}(t)$ is given by the solution of the equation

$$\mathbf{f}(\mathbf{z}(t), \mathbf{u}(t)) = \mathbf{g}(\mathbf{z}(t)) = \dot{\mathbf{z}}(t). \tag{3}$$

Equation (3) can be solved without any system state measurement; in fact, not even a generating vector field **g** has to be given: it is sufficient to know the goal trajectory itself and its time derivative (velocity) for the control time interval, $\{\mathbf{z}(t), \dot{\mathbf{z}}(t)\}_{t \in [t_0, t_1]}$.

There are several conditions that have to be fulfilled to make the entrainment control scheme work. First of all, Eq. (3) has to be solvable for $\mathbf{u}(t)$. This is trivially true for simple vector additive forces:

$$\mathbf{f}(\mathbf{z}(t), \mathbf{u}(t)) = \tilde{\mathbf{f}}(\mathbf{z}(t)) + \mathbf{u}(t) \quad \Rightarrow \quad \mathbf{u}(t) = \dot{\mathbf{z}}(t) - \tilde{\mathbf{f}}(\mathbf{z}(t)) \qquad (4)$$

We restrict our discussion to such forces, which are the most common in the literature on entrainment control. However, problems immediately arise if, e.g., the control space dimension s is less than the state space dimension n (a treatment of general control influence with $s=d$ can be found in [8]).

The next condition to be satisfied is asymptotic stability of the goal trajectory. While control forces according to Eq. (3) ensure that the goal trajectory is a solution of the controlled system Eq. (1), there is no statement about whether nearby located system states are attracted by it – in other words, whether entrainment occurs in a vicinity of $\mathbf{z}(t)$ according to $\lim_{t \to \infty} |\mathbf{x}(t) - \mathbf{z}(t)| = 0$. Even if this is the case, one needs a large basin of attraction (ideally the whole phase space) to make the method work for a large set of possible initial states distant from $\mathbf{z}(t)$. Statements about basins are very hard to find (compare [9]), but a discussion of stability can more easily be made. For simple vector additive control, Eq. (4), we call all points in phase space where all eigenvalues of the Jacobian of $\tilde{\mathbf{f}}$ have negative real part, *convergent regions* (see also [8, 9]). Goal trajectories entirely located in convergent regions turn out to be asymptotically stable, if their time derivative $\dot{\mathbf{z}}(t)$ is sufficiently bounded.

3. Optimization of the goal trajectory

Because of the stated stability aspects, goal trajectories are usually chosen to be located in convergent regions. This results in a typical drawback of the method: Due to very little overlap of convergent regions and unperturbed chaotic attractor, such a goal dynamics is quite different to the natural system dynamics. Consequently, the system is strongly altered by control, and control forces are large. In fact, they have to be of about the magnitude of the velocities appearing in the uncontrolled system in order to pull the movement into convergent regions.

To attack this problem, one has to realize that location in convergent regions is not a necessary condition for stability of a goal trajectory. The resulting dynamics of chaotic systems controlled by periodic perturbation methods indeed suggest that stable dynamics can also be achieved in the chaotic attractor region, especially near a UPO. Control forces are smaller then, as the natural dynamics is only slightly altered. An extreme case would be to consider a UPO itself as a goal for entrainment control: we get zero control forces according to Eq. (3). However, such a goal is of course unstable. It has to be at least slightly changed to result in a stable one. To this end, a family of deformations of a UPO is considered in the following.

Let $z_{UPO}(t)$ denote a known UPO of the chaotic system with period $T = 2\pi/\omega$. We chose a finite Fourier series as a deformation $\Delta z(t)$, and also include a linear time transformation by a factor η. The family of goal trajectories now reads

$$\begin{aligned} \mathbf{z}(t) &= \mathbf{z}_{UPO}(\eta t) + \Delta \mathbf{z}(\eta t) \\ &= \mathbf{z}_{UPO}(\eta t) + \sum_{m=0}^{M} \left(\mathbf{a}^m \cos(m\omega\eta t) + \mathbf{b}^m \sin(m\omega\eta t) \right). \end{aligned} \quad (5)$$

It is parametrized by $\eta \in \mathbb{R}$ and the Fourier coefficient vectors $\mathbf{a}^m$, $\mathbf{b}^m \in \mathbb{R}^n$. Since $\mathbf{b}^0$ has no effect on $\mathbf{z}(t)$, a deformation (or a goal) is characterized by a total of $d = (2M+1)n + 1$ real numbers.

Now, we formulate the determination of advantageous deformation parameters that lead to a stable goal trajectory with small control forces as an optimization problem. A real number according to a cost function is assigned to each probed set of parameters. The cost assesses the stability of the chosen goal trajectory (which is determined numerically) as well as the magnitude of the resulting control forces:

$$cost = \begin{cases} \|\mathbf{u}(t)\|_{max} + \gamma[\exp(\mu) - 1] & : \quad \mu < 1, \\ \|\mathbf{u}(t)\|_{max} + \gamma[\exp(\mu) - 1] + C & : \quad \mu \geq 1 \end{cases} \quad (6)$$

Here, $\mu = \max_i\{|\mu_i|\}$ is the maximum absolute value of the characteristic (Floquet) multipliers of the goal orbit. These are well defined, as the goal is periodic, and they are calculated by integration of the variational equations [10]. Instability is indicated by $\mu > 1$ and causes high cost via a large positive penalty term C. The cost function further includes an $\exp(\mu)$ and the maximum norm of the resulting forces. The weight of stability with respect to magnitude of forces can be adjusted by γ. The global minimum of the cost function in the deformation parameter space corresponds to the best goal trajectory in sense of the chosen balance between stability and small forces.

For various reasons, a direct analytic treatment of the given optimization problem is usually not possible: the UPO is not known in analytic form, a direct expression for stability of a deformed UPO is missing (the variational equations have to be integrated), and the cost function is not continuous. Consequently, numerical methods are employed. The UPO is represented by a periodic cubic spline interpolation, and μ is calculated via numerical integration of the variational equations. The optimization is done by a numerical technique that can handle high-dimensional problems with rough and rapidly varying cost functions. For this purpose, the stochastically guided algorithm **amebsa** from [11] was chosen; this is a combination of simulated annealing and the downhill simplex method.

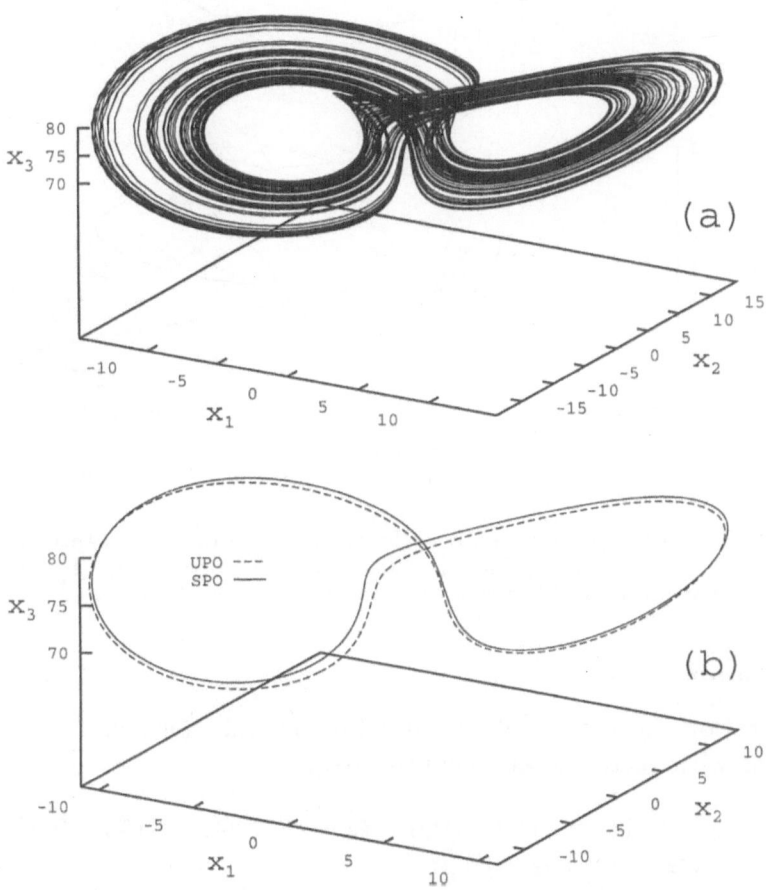

Figure 1. (a): Chaotic attractor of the Lorenz system ($\sigma = 10.0$, $r = 75.0$, $b = 0.4$). (b): Embedded unstable periodic orbit (UPO, interrupted line) and the optimized deformation of it that turns out to be a globally stable goal trajectory (SPO, solid line).

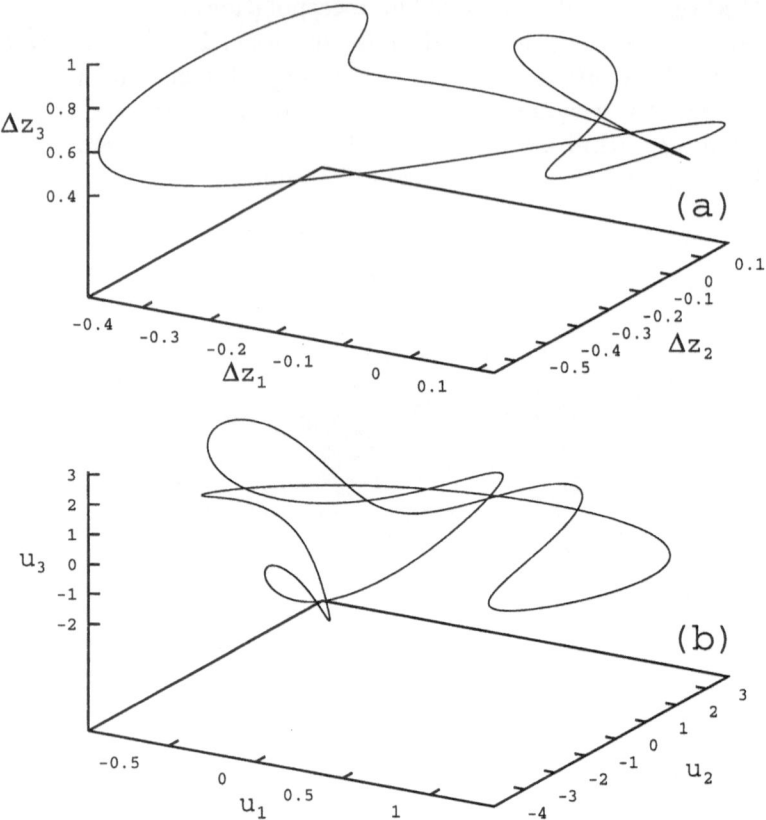

Figure 2.   (a): Orbit in phase space of the difference vector Δz between the UPO of the Lorenz system and the optimized stable goal trajectory (SPO). (b): Path of the corresponding vector control forces in the control signal space.

4. Example

In this section, the control of a chaotic Lorenz system is demonstrated. The equations with vector additive control read

$$\dot{x}_1 = \sigma(x_2 - x_1) + u_1(t), \quad \dot{x}_2 = rx_1 - x_2 - x_1 x_3 + u_2(t),$$
$$\dot{x}_3 = x_1 x_2 - bx_3 + u_3(t), \tag{7}$$

where time dependence of the forces is explicitly written. A given periodic goal trajectory $\{z(t), \dot{z}(t)\}_{t \in [0,T]}$ yields control forces according to Eq. (4). Parameters of the Lorenz system are set to $\sigma = 10$, $r = 75$, and $b = 0.4$. The resulting chaotic attractor is shown in Fig. 1(a). An embedded UPO which corresponds to z_{UPO} in Eq. (5) is given in Fig. 1(b) by the interrupted line.

The deformation is defined by five Fourier modes ($M{=}5$) which leads to a 34-dimensional search space for optimization; γ is set to 5, C to 100 in

Eq. (6). The best result of several optimization runs is shown in Fig. 1(b) by the solid line. It is a stable goal trajectory and therefore a stable periodic orbit (SPO) of the controlled system. The actual values of deformation parameters can be found in Tab. 1 together with additional data of the UPO and the SPO. The deformation lies in the range of some percent, and it is plotted in Fig. 2(a). The resulting forces, shown in Fig. 2(b), change the vector field of the chaotic system less than about 10%. This is an improvement of more than a magnitude if compared to goals in convergent regions [9].

Numerical tests indicated that the SPO is globally asymptotically stable; the basin is the whole phase space. However, transient times until control is established depend strongly on the initial state, and range from just a few up to a few hundred control periods. A typical behavior is presented in Fig. 3. After control is turned on, an intermittent transient appears. Finally, the system settles down on the desired goal orbit, which is maintained.

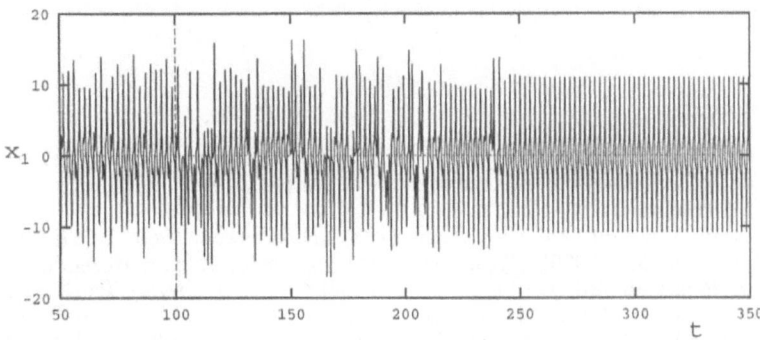

Figure 3. Controlled Lorenz system, first coordinate $x_1(t)$: Control is turned on at $t_0 = 100$, and the goal trajectory is reached after transients at about $t = 250$ (after approximately 75 control periods).

5. Conclusion

It has been shown how open-loop entrainment control in the vicinity of unstable periodic orbits can be realized. The search for suitable goal trajectories has been formulated in terms of an optimization problem with respect to UPO deformations. Feasibility has been demonstrated in an example, where a chaotic Lorenz system has been successfully controlled to an optimized distortion of a UPO. The locations of such goal orbits are independent of convergent regions, and thus the required forces are small compared to hitherto used goal dynamics far from the chaotic attractor.

This work was supported by the Deutsche Forschungsgemeinschaft (Sonderforschungsbereich 185).

238

TABLE 1. Coordinates $z_{1,2,3}$ give a point of the original UPO of the Lorenz system, T its period. Maximum absolute values of the Floquet multipliers are given by μ for the UPO and for the optimized deformation (SPO). Parameters of the SPO are the time transformation coefficient η and Fourier coefficients a_n^m, b_n^m. Subscripts of numbers indicate a decimal shift, i.e., $1.234_{(-2)}$ stands for 1.234×10^{-2}.

$z_1 = 8.0$	$z_2 = 11.59618$	$z_3 = 68.87129$	$T = 2.153957$
$\mu_{UPO} = 2.555$	$a_1^0 = -9.794703_{-2}$	$a_1^1 = 2.303546_{-2}$	$b_1^1 = -1.946829_{-1}$
$\mu_{SPO} = 0.719$	$a_2^0 = -1.126072_{-1}$	$a_2^1 = -4.441229_{-2}$	$b_2^1 = -1.543789_{-1}$
$\eta = 1.001267$	$a_3^0 = 6.741204_{-1}$	$a_3^1 = -9.082832_{-2}$	$b_3^1 = 5.702340_{-2}$
$a_1^2 = 1.063004_{-1}$	$b_1^2 = 2.904178_{-5}$	$a_1^3 = 5.343983_{-2}$	$b_1^3 = -9.527600_{-3}$
$a_2^2 = 1.553288_{-1}$	$b_2^2 = -4.403841_{-2}$	$a_2^3 = 6.312158_{-2}$	$b_2^3 = -2.940390_{-2}$
$a_3^2 = -2.747584_{-1}$	$b_3^2 = -7.345061_{-2}$	$a_3^3 = 9.633147_{-2}$	$b_3^3 = 8.743705_{-2}$
$a_1^4 = 2.634346_{-2}$	$b_1^4 = 8.791274_{-2}$	$a_1^5 = 2.069052_{-3}$	$b_1^5 = 1.429860_{-2}$
$a_2^4 = 9.267735_{-2}$	$b_2^4 = 6.849559_{-2}$	$a_2^5 = 2.874568_{-2}$	$b_2^5 = 2.781748_{-2}$
$a_3^4 = 7.009472_{-2}$	$b_3^4 = 2.443292_{-2}$	$a_3^5 = -2.478029_{-2}$	$b_3^5 = 3.759263_{-2}$

References

1. Chen, G. and Dong, X. (1993) From chaos to order – perspectives and methodologies in controlling chaotic nonlinear dynamical systems, *Int. J. of Bifurcation and Chaos* **3**(6), 1363-1409.
2. Alekseev, V.V and Loskutov, A.Yu. (1987) *Dokl. Akad. Nauk. SSSR* **293**; Lima, R. and Pettini, M. (1989) *Phys. Rev. A* **41**; Azevedo, A. and Rezende, S.M. (1991) *Phys. Rev. Lett.* **66**; Braiman, Y. and Goldhirsch, I. (1989) *Phys. Rev. Lett.* **66**; Fronzoni, L. et al. (1991) *Phys. Rev. A* **43**.
3. Salerno, M. (1991) *Phys. Rev. B* **44**; Farrelly, D. and Milligan, J.A. (1993) *Phys. Rev. E* **47**; Cicogna, G. and Fronzoni, L. (1993) *Phys. Rev. E* **47**.
4. Mettin, R. and Kurz, T. (1995) Optimized periodic control of chaotic systems, *Phys. Lett. A* **206**, 331-339.
5. Hübler, A. and Lüscher, E. (1989) *Naturwissenschaften* **76**; Plapp, B.B. and Hübler, A. (1990) *Phys. Rev. Lett.* **65**; Jackson, E.A. (1990) *Physica D* **50**; Shermer, R. et al. (1991) *Phys. Rev. A* **43**; Breeden, J.L. (1994) *Phys. Lett. A* **190**.
6. Ott, E., Grebogi, C., and Yorke, Y.A. (1990) Controlling chaos, *Phys. Rev. Lett.* **64**, 1196-1199.
7. Bellman, R.E., Bentsman, J., and Meerkov, S.M. (1986) Vibrational control of nonlinear systems: Vibrational stabilizability, *IEEE Trans. Automat. Contr.* **AC-31**, 710-716.
8. Mettin, R., Hübler, A., Scheeline, A., and Lauterborn, W. (1995) Parametric entrainment control of chaotic systems, *Phys. Rev. E* **51**, 4065-4075.
9. Jackson, E.A. (1991) Controls of dynamic flows with attractors, *Phys. Rev. A.* **44**, 4839–4853.
10. Parker T.S. and Chua, L.O. (1989) *Practical Numerical Algorithms for Chaotic Systems*, Springer-Verlag, New York.
11. Press, W.H., Teukolsky, S.A., Vetterling, W.T. and Flannery, B.R. (1992) *Numerical Recipes in C*, 2nd ed., Cambridge University Press, Cambridge.

NEAR-TIME-OPTIMAL FEEDBACK CONTROL OF

MECHANICAL SYSTEMS WITH

FAST AND SLOW MOTIONS

S.A. MIKHAILOV AND P.C. MÜLLER
Safety Control Engineering
University of Wuppertal, Gauss str. 20, D-42097 Wuppertal,
Germany

Abstract. In the paper a new method for precise feedback control of singularly perturbed systems is developed; it is based on the separation of slow and fast motions. The switching surface (locus) is constructed in the space of slow variables. The proposed composite control algorithm steers the slow and fast modes to the origin. The cost (minimum time) slightly increases with respect to reduced system minimum-time.

1. Introduction

Problems of optimal control of singularly perturbed systems have been intensively studied(see the surveys of the literature on singular perturbation in control theory [3],[4]). These systems frequently occur in applications. Examples are: drives, actuators, robots and electronic circuits. The small parameter ε in these systems may represent small masses, small time constants, large stiffness or large gains. Consider a system described in state-space form by the set of equations

$$\dot{x_1} = A_{11}x_1 + A_{12}x_2 + B_1 u \tag{1.1}$$
$$\varepsilon \dot{x_2} = A_{21}x_1 + A_{22}x_2 + B_2 u \tag{1.2}$$

where ε is a small positive scalar, $x_1 \in R^n$ is the slow state vector, $x_2 \in R^m$ is the fast state vector, $u \in R^k$ is the vector of control variables, and $\dot{(\ )} = d/dt$. The control function is subject to the constraints

$$u(t) \in U \tag{1.3}$$

where U is a compact set in R^k.

The problem of time-optimal synthesis is considered, where the state variables in the initial moment of time can take arbitrary values in the state space

$$x_1(0) = x_1^0, \ x_2(0) = x_2^0. \tag{1.4}$$

D. H. van Campen (ed.), IUTAM Symposium on Interaction between Dynamics and Control in Advanced
Mechanical Systems, 239–246.
© 1997 Kluwer Academic Publishers.

The task is to steer the initial state x_1^0, x_2^0 to the origin

$$x_1(T) = x_2(T) = 0 \tag{1.5}$$

in minimum time, taking into account the restrictions (1.3).

It should be mentioned that the answer of the considered synthesis problem is the switching surface in the state space. The control u must be defined as a function of state co-ordinates $u = u(x_1, x_2)$. Exact analytical solutions for optimal feedback control exist only for specific problems, such as linear systems with integral quadratic cost criteria. Numerical solutions are possible for optimal programs or open-loop controls, but are very difficult for determining the numerical solutions for synthesis (feedback) controls if the dimension of the system is high. Therefore, we implement methods based on the small parameter technique.

The problem of open-loop time-optimal control is considered in a number of studies [1]-[3]. We generalize these results to time-optimal synthesis (feedback) control.

The time-optimal synthesis in singularly perturbed systems has at least two special properties.

1. The switching surface is singularly perturbed (Fig.1)

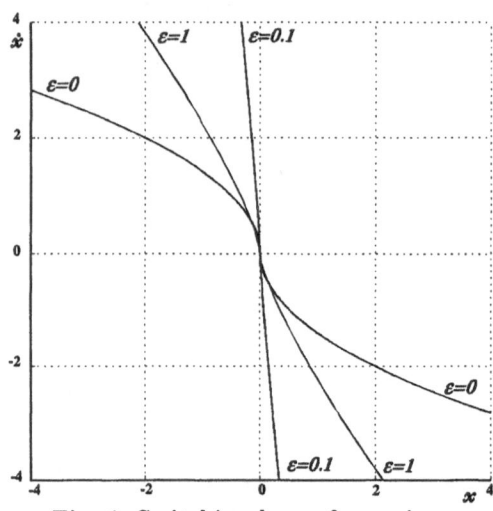

Fig. 1. Switching locus for various ε .

To illustrate this property, time-optimal synthesis for the following system is considered

$$\varepsilon\ddot{x} + \ddot{x} = u(x, \dot{x}, \ddot{x}, \varepsilon), \quad |u| \leq 1, \quad T \Rightarrow min_u \tag{1.6}$$

where $u(x, \dot{x}, \ddot{x}, \varepsilon)$ is the synthesizing function for the feedback optimal control. Fig. 1 shows the projection of the switching curve (two optimal trajectories which lead to zero) on the plane x, $\dot{x}$ for various ε. It is clear that this switching curve cannot be constructed by regular expansions in nonnegative powers of ε in a neighborhood of the point $\varepsilon = 0$. It is said that such a function is singular in ε .

2. The application of the synthesis obtained in the so-called reduced system ($\varepsilon = 0$) will lead to a limit cycle in the original system. To demonstrate this

property consider the following example based on the system (1.6). For the reduced system $\ddot{x} = u(x, \dot{x})$ the switching curve (locus) has the explicit form $\dot{x} = -sgn(x)\sqrt{2|x|}$. If this synthesis is applied in the original system (1.6), the results given in Fig. 2 are obtained.

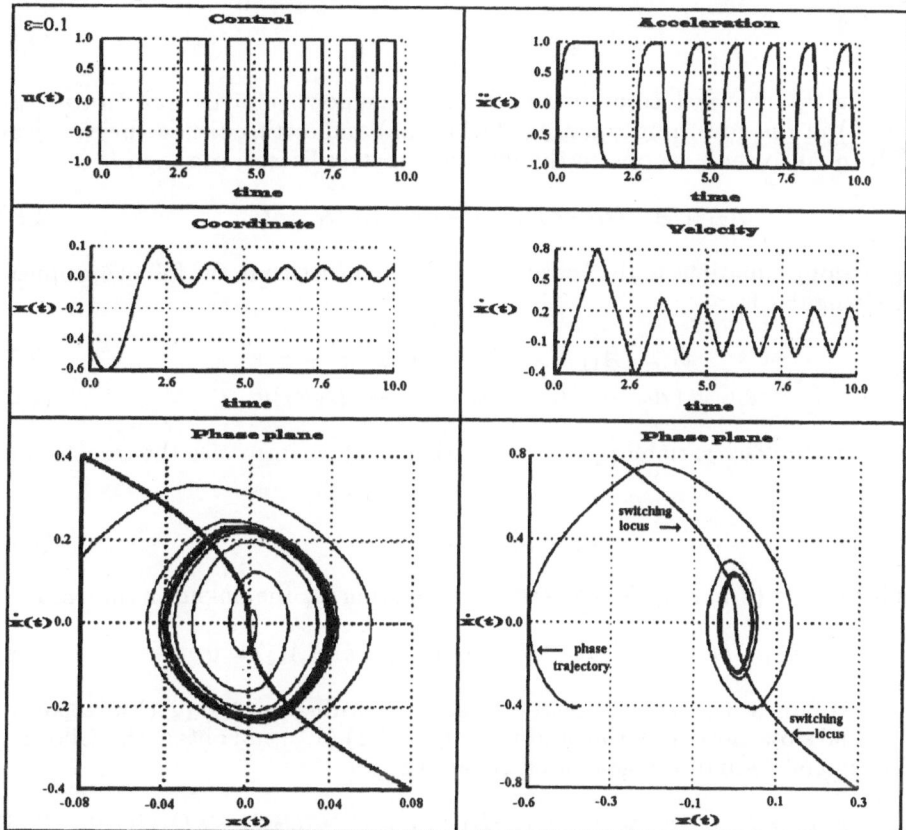

Fig. 2 The results of numerical simulation .

Fig. 2 shows the switching curve for the reduced system $\ddot{x} = u$ and the projection of the phase trajectory ($\varepsilon = 0.1$) on the phase plane $x, \dot{x}$. For high precision systems, the limit cycle is inadmissible. Therefore, we need to correct the feedback given by the reduced system in order to obtained a more precise control algorithm for steering the state trajectory to the origin.

2. The decoupling of the slow and fast variables

To alleviate the difficulties described above, we decouple motions in the system

$$\dot{x_1} = A_{11}x_1 + A_{12}x_2 + B_1 u \qquad (2.1)$$
$$\varepsilon \dot{x_2} = A_{21}x_1 + A_{22}x_2 + B_2 u \qquad (2.2)$$

The Riccati transformation is used to obtain the uncoupled equations for the slow and fast variables. The separation of motions in the system (2.1),(2.2) can be made in two steps. First, the slow variables x_1 are eliminated from equation (2.2) by means of a linear transformation

$$x_2 = y_2 + D_1 x_1. \tag{2.3}$$

Here y_2 is the new fast co-ordinate, D_1 is an unknown matrix. Substituting x_2 from (2.3) in the governing equations (2.1), (2.2) and setting the coefficient of x_1 in (2.2) equal to zero, we obtain the following matrix Riccati equation

$$A_{21} + A_{22}D_1 - \varepsilon D_1 A_{11} - \varepsilon D_1 A_{12} D_1 = 0. \tag{2.4}$$

The governing equations in the variables x_1, y_2 are rewritten in the following upper-block triangular form

$$\dot{x}_1 = (A_{11} + A_{12}D_1)x_1 + A_{12}y_2 + B_1 u \tag{2.5}$$
$$\varepsilon \dot{y}_2 = (A_{22} - \varepsilon D_1 A_{12})y_2 + (B_2 - \varepsilon D_1 B_1)u. \tag{2.6}$$

To separate the fast variables y_2 from equation (2.5) we introduce the transformation

$$x_1 = y_1 + \varepsilon D_2 y_2 \tag{2.7}$$

By substituting (2.7) in (2.5) we derive the equation for the unknown matrix D_2 :

$$A_{12} + \varepsilon(A_{11} + A_{12}D_1)D_2 - D_2(A_{22} - \varepsilon D_1 A_{12}) = 0. \tag{2.8}$$

By means of linear transformations of the state co-ordinates (2.3),(2.7) we separate the fast and slow motions in the original system (1.1),(1.2) and obtain the following block-diagonal form of the governing equations

$$\dot{y}_1 = E_1 y_1 + F_1 u, \quad E_1 = A_{11} + A_{12}D_1, \quad F_1 = B_1 - D_2 B_2 + \varepsilon D_2 D_1 B_1 \tag{2.9}$$
$$\varepsilon \dot{y}_2 = E_2 y_2 + F_2 u, \quad E_2 = A_{22} - \varepsilon D_1 A_{12}, \quad F_1 = B_2 - \varepsilon D_1 B_1. \tag{2.10}$$

The system (2.9),(2.10) is equivalent to the original equations (1.1),(1.2) but is simpler. It consists of two subsystems of order n - slow mode and order m - fast mode interacting by means of control. Note that the system for the slow variable is regularly perturbed and this is certainly the most important simplification. Looking for the solution of (2.4) and (2.8) in the form of expansions in power series of ε

$$D_1 = D_1^0 + \varepsilon D_1^1 + \cdots, \quad D_2 = D_2^0 + \varepsilon D_2^1 + \cdots,$$

and assuming that matrix A_{22} is regular, we obtain the solution up to the first order of ε:

$$D_1 = -A_{22}^{-1}A_{21} + \varepsilon A_{22}^{-2}A_{21}(-A_{11} + A_{12}A_{22}^{-1}A_{21}) + O(\varepsilon^2) \tag{2.11}$$
$$D_2 = A_{12}A_{22}^{-1} + \tag{2.12}$$
$$\varepsilon(A_{11}A_{12}A_{22}^{-1} - A_{12}A_{22}^{-2}A_{21}A_{12} - A_{12}A_{22}^{-1}A_{21}A_{12}A_{22}^{-1})A_{22}^{-1} + O(\varepsilon^2).$$

The decomposition into separate slow and fast subsystems suggests that separate slow and fast control laws are designed for each subsystem, and then combined into a composite control of the original system. These ideas have produced numerous two-time-scale designs in linear state feedback, output feedback, observers and optimal control. It should be mentioned that this procedure for the system decomposition has been extended to linear time-varying systems.

3. The control algorithm

In this section we propose a composite control algorithm. The time interval $[0, T]$ is divided into two parts. In the first interval $0 \le t \le T - \varepsilon \tau_0$ the slow variables are steered from y_1^0 to $\hat{y}_1$ (the values $\hat{y}_1$ are calculated in advance), where $\varepsilon \tau_0$ is the duration of the terminal boundary layer. In the second time interval (the terminal boundary layer) $T - \varepsilon \tau_0 < t \le T$ we implement a control steering the fast and slow modes to the origin.

Let y_1^*, y_2^*, u^* be the variables and control in the terminal boundary layer. Since the system (2.9), (2.10) is autonomous, we can construct the solution that terminates at the origin by a process of backing out of the origin in the inverse time $\vartheta = T/\varepsilon - \tau$. Equation (2.10) for the fast mode in the inverse time ϑ is rewritten in the following form

$$y_2^{*\prime} = -E_2 y_2^* - F_2 u^* \tag{3.1}$$

$$y_2^*(0) = 0, \quad y_2^*(\tau_0) = -E_2^{-1} F_2 u. \tag{3.2}$$

The first boundary condition (3.2) in the inverse time corresponds to the final state of the system in the original time scale (we intend to steer state variables to zero). The second boundary condition conforms to the asymptotically stable equilibrium for equation (3.1).

The solution of the boundary value problem (3.1) (3.2) can be written in the form of a convolution

$$\int_0^{\tau_0} \exp(E_2 \xi) F_2 u^*(\xi) \, d\xi = \exp(E_2 \tau_0) E_2 F_2 u. \tag{3.3}$$

This is an integral equation for the unknown function $u^*(\xi)$. To construct a simple solution of (3.3) we assume that the control function $u^*(\xi)$ in the final boundary layer is a polynomial

$$u^*(\xi) = a_0 + a_1 \xi + \cdots + a_l \xi^l; \tag{3.4}$$

here $a_0, \ldots, a_l$ are the unknown vectors $a_i \in R^k$, $i = 0, \ldots, l$. The polynomial order depends on the dimension m of fast variables and dimension k of control vector.

$$l = \begin{cases} 0, & if \quad m < k \\ p - 1, & if \quad m \ge k, \; m = pk \\ p, & if \quad m > k, \; m = pk + s, \; 0 < s/k \le 1. \end{cases}$$

By substituting (3.4) in the integral equation (3.3) we obtain a linear system for the unknown coefficients $a_0, \ldots, a_l$. The solution of this linear system depends on the parameter τ_0. The optimal value of τ_0 can be found by minimizing this solution with respect to τ_0, taking into account the constraints on the control function $u(t) \in U$.

Let us consider two examples.

Example 1.

$$y_2' = -y_2 + u \quad y_2 \in R^1, \quad u \in R^1, \quad |u| \leq 1 \tag{3.5}$$

The polynomial order is $l = 0$ and $u^*(\xi) = a_0$. Using equation (3.3) we obtain

$$a_0 \int_0^{\tau_0} \exp(-\xi)\, d\xi = -\exp(-\tau_0)u, \quad u^* = a_0 = -u/(\exp(\tau_0) - 1) \tag{3.6}$$

The optimal value of τ_0 is: $\tau_0 = \ln 2$, $u^*_{1,2} = -u = \mp 1$. Note, that the slow mode optimal control at the beginning of terminal layer can take two values. Therefore, we obtain two solutions $u^*_{1,2}$ for the terminal layer. The upper minus corresponds to $u = 1$, the lower plus to $u = -1$.

Example 2.

$$y_2 \in R^2, \quad u \in R^1, \quad |u| \leq 1$$

$$E_2 = \begin{bmatrix} 0 & 1 \\ -1 & -1, \end{bmatrix} \quad F_2 = \begin{bmatrix} 0 \\ 1 \end{bmatrix}$$

Here is $l = 1$, $u^*(\xi) = a_0 + a_1\xi$. Substituting $u^*(\xi)$ in the (3.3) we obtain a linear system for the coefficients a_0, a_1

$$(1 + \exp(\tau_0/2)\sin(\sqrt{3}\tau_0/2)/\sqrt{3} - \exp(\tau_0/2)\cos(\sqrt{3}\tau_0/2))a_0 + \tag{3.7}$$
$$(1 + \tau_0 - \exp(\tau_0/2)\sin(\sqrt{3}\tau_0/2)/\sqrt{3} - \exp(\tau_0/2)cos(\sqrt{3}\tau_0/2))a_1 = u$$

$$-2\exp(\tau_0/2)\sin(\sqrt{3}\tau_0/2)a_0/\sqrt{3} + \tag{3.8}$$
$$(-1 - \exp(\tau_0/2)\sin(\sqrt{3}\tau_0/2)/\sqrt{3} + \exp(\tau_0/2)cos(\sqrt{3}\tau_0/2))a_1 = 0.$$

The solution of (3.7),(3.8) is $u^*_{1,2}(\xi) = \pm 0.95 \mp 0.94\xi$, where $\tau_0 = 2.03$ is the minimum time.

Let us consider the variation of the slow variable in the final boundary layer. Then the control is given by (3.4)

$$\dot{y}_1^* = -E_1 y_1^* - F_1 u^* \tag{3.9}$$
$$y_1^*(0) = 0. \tag{3.10}$$

The initial condition (3.10) in the inverse time corresponds to the final state of the system in the original time. Solving the equation (3.9) with initial condition (3.10) we obtain the values of the slow variables at the beginning of the final boundary layer $t = T - \varepsilon\tau$

$$y_1^* = -\int_0^{\varepsilon\tau_0} \exp(E_1\xi - \varepsilon\tau_0)F_1 u^*(\xi)\, d\xi = \hat{y}_1. \tag{3.11}$$

For the slow mode the time-optimal synthesis is considered:

$$\dot{y}_1 = E_1 y_1 + F_1 u. \tag{3.12}$$

The state variables at the initial moment of time are given by $y_1(0) = y^0$. The task is to steer the slow variables to the value $\hat{y}_1$ in a minimum time, taking into account the restrictions on the control function.

Thus the composite control algorithm consists of two phases: first, we consider the terminal boundary layer and determine the control $u^*(\xi)$ in accordance with the formulas (3.3),(3.4), then we calculate $\hat{y}_1$ (3.11); second, we solve the time-optimal problem with the final condition (3.11) for the slow mode. This problem is easier than the original one because we construct the synthesizing function in the space of slow variables, and the synthesis for the slow variables is regularly perturbed.

4. Example

Now we consider an example which illustrates the proposed procedure.

$$\varepsilon \dddot{x} + \ddot{x} = u, \quad |u| \le 1 \tag{4.1}$$

The third order differential equation (4.1) may be rewritten in the state space form

$$\dot{x} = y, \quad \dot{y} = z, \quad \varepsilon \dot{z} = -z + u. \tag{4.2}$$

The task is to steer x, y and z to the origin in minimum time. In this example x, y are the slow variables and z is the fast variable. The separation of slow and fast modes can be made by means of the transformation

$$x = \xi + \varepsilon^2 z, \quad y = \eta - \varepsilon z.$$

After transformation, we obtain the uncoupled equation for the slow variables ξ, η and the fast variable z

$$\dot{\xi} = \eta - \varepsilon u, \quad \dot{\eta} = u, \quad \varepsilon \dot{z} = -z + u. \tag{4.3}$$

The system (4.3) has one remarkable property. By the transformation

$$\eta = \beta, \quad \xi = \alpha - \varepsilon \beta$$

the slow part can be reduced to a nonperturbed form

$$\dot{\alpha} = \beta, \quad \dot{\beta} = u, \quad \varepsilon \dot{z} = -z + u.$$

The control in the terminal boundary layer is defined by the expression (3.6), where $u_{1,2}^* = \mp 1$, $\tau_0 = \ln 2$. The values $\hat{\alpha}$, $\hat{\beta}$ calculated by means (3.11) are equal to

$$\hat{\alpha}_{1,2} = \mp(\varepsilon \ln 2)^2/2, \quad \hat{\beta}_{1,2} = \pm \varepsilon \ln 2 \tag{4.4}$$

In this example the switching locus has a complex analytical form and consists of two curves S_+, S_- (see Fig.3).

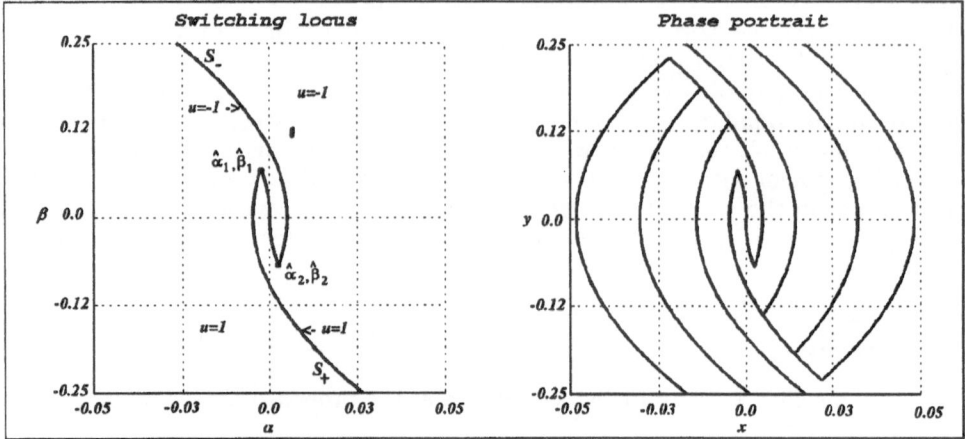

Fig. 3. The switching locus and phase portrait.

The switching locus divides the phase plane into two parts. For the phase points above the switching locus, the feedback control is $u(\alpha, \beta) = -1$; for the phase points below the switching locus, $u(\alpha, \beta) = 1$. In order to test the developed synthesis, the system (4.2) has been numerically simulated. Fig. 3 shows the projections of the state trajectories on the plane (x, y) for various initial conditions $x(0)$, $y(0)$ $z(0)$. The results of numerical simulations show that the composite control algorithm can steer the phase point with arbitrary initial conditions to the origin in a period of time which is slightly longer than the slow mode minimum-time.

5. Conclusions

A composite algorithm has been developed for near-time-optimal feedback control of singularly perturbed linear systems. The main problems concerned with these systems are the difficulties with high-order systems, singularly perturbed switching surface (locus) and limit cycles. These difficulties have been successfully overcome by means of decoupling the slow and fast variables, and the construction of a composite algorithm. The control goal is the reduction of dimensionality to the dimension of the slow part. It should be pointed out that the synthesis for the slow part is regularly perturbed. The proposed control algorithm steers both the slow and fast variables to the origin. Simulation results have shown the effectiveness of the composite algorithm, compared with the reduced order control.

REFERENCES

1. Collins, W.D. (1973) Singular perturbations of linear time-optimal control. *In Recent Mathematical Developments in Control, D. J. Bell, Ed. New York: Academic.*

2. Halanay, A., Mirica, St. (1979) The time-optimal feedback control for singular perturbed linear systems . *Rev. Roum. Math. Pures et Appl.*, 24, pp. 585-596.

3. Kokotovic, P.V., Khalil, H.K., and O'Reilly, J. (1986) *Singular perturbation methods in control: analysis and design.* Academic Press. 371 pp.

4. O'Malley, R.E. (1991) *Singular Perturbation Methods for Ordinary Differential Equations.* Springer-Verlag, Berlin. p. 225.

OPTIMAL CONTROL OF MECHANICAL DESCRIPTOR SYSTEMS

P. C. MÜLLER

Safety Control Engineering
University of Wuppertal
D-42097 Wuppertal, Germany

1. Introduction

In recent years the analysis and synthesis of linear control systems in descriptor form has been established. One class of descriptor systems are mechanical descriptor systems, i.e. mechanical systems with explicit holonomic and/or nonholonomic constraints. This special form of description of mechanical systems according to Lagrange's equations of the first kind is important for many applications such as vehicle dynamics, machine dynamics, dynamics of robots etc. Although well-known methods of analytical mechanics exist to eliminate the constraints and form a regular set of differential equations, aspects of physical transparency and subsystem techniques are reasons to deal with the original descriptor system.

The paper deals with the design of optimal control for mechanical descriptor systems. For this, the calculus of variations and Pontryagin's maximum principle have to be generalized for descriptor systems. The simultaneous handling of differential equations and algebraic constraints of the optimization problem will be overcome by transferring the differential-algebraic equations to the set of underlying ordinary differential equations on a manifold. In general, this procedure includes the occurrence of time derivatives of the control depending on the size of the index and on the causality of the system. Causality will be defined and a related criterion will be developed. Causality plays an important role, whether Pontryagin's maximum principle can be applied as usual or not. Causality and non-causality distinguish between whether the descriptor system is exclusively governed by the control input, or by its higher-order time-derivatives additionally. Non-causal descriptor systems are really influenced by time derivatives of the control input. In this

D. H. van Campen (ed.), IUTAM Symposium on Interaction between Dynamics and Control in Advanced Mechanical Systems, 247–254.
© 1997 *Kluwer Academic Publishers.*

unusual case a quite different problem of optimal control design has to be considered.

For causal descriptor systems, the usual optimization methods can be applied, introducing adjoint variables and Hamilton's function. The optimal control is characterized by an adjoint descriptor system. For non-causal descriptor systems, it is recommended to use the representation of the system by the set of underlying ordinary differential equations on a manifold. Introducing additional variables to deal with the higher-order time-derivatives of the control input, the optimization methods will be applied to an extended system. But in this case, the original control constraints are transferred into state constraints which complicates the optimization procedure very much.

The general theory of optimal descriptor systems will be applied to mechanical descriptor systems having regard to their special structure. Controlled mechanical descriptor systems with control-free holonomic and/or nonholonomic constraints are causal systems. But if the mechanical system is controlled kinematically, i.e. the constraints are controlled, then a non-causal descriptor system appears, including the above mentioned difficulties.

2. Problem Statement

Controlled time-invariant finite-dimensional holonomic mechanical systems can be described by a set of f ordinary differential equations

$$\mathbf{M}(\mathbf{z})\,\ddot{\mathbf{z}} + \mathbf{k}(\mathbf{z}, \dot{\mathbf{z}}) = \mathbf{F}^T(\mathbf{z})\,\boldsymbol{\lambda} + \mathbf{T}(\mathbf{z})\,\mathbf{u} \tag{1}$$

and of p holonomic constraints

$$\mathbf{f}(\mathbf{z}) + \mathbf{R}\,\mathbf{u} = \mathbf{0}. \tag{2}$$

$\mathbf{z}$ denotes an f-dimensional vector of displacements, and $\mathbf{u}$ is the r-dimensional vector of control inputs. The p-dimensional vector $\boldsymbol{\lambda}$ represents constraint forces due to the holonomic constraints (in the calculus of variations they are Lagrange multipiers). The matrix of inertia $\mathbf{M}(\mathbf{z})$ is assumed to be symmetric, bounded and (uniformly) positive definite. The vector function $\mathbf{k}(\mathbf{z}, \dot{\mathbf{z}})$ includes Coriolis and centrifugal forces and uncontrolled applied forces as well. $\mathbf{T}$ is the input matrix of suitable dimension. The $p \times f$-matrix $\mathbf{F}(\mathbf{z})$ is the Jacobian matrix of the constraint (2):

$$\mathbf{F}(\mathbf{z}) = \frac{\partial \mathbf{f}}{\partial \mathbf{z}^T}. \tag{3}$$

It is assumed that the constraints are independent. For linear constraints, this assumption is guaranteed by the condition

$$\operatorname{rank} \mathbf{F} = p. \tag{4}$$

The matrix **R** represents a constant control input matrix which characterizes a possible control of the constraints. For simplicity, only holonomic constraints are considered, but nonholonomic constraints can be handled in a modified manner, too.

Mechanical descriptor systems (1,2) are a special case of general descriptor systems

$$\dot{x}_1 = f_1(x_1, x_2, u), \quad 0 = f_2(x_1, x_2, u), \tag{5}$$

where $x_i, i = 1, 2$, are n_i-dimensional vectors and $n_1 + n_2 = n$. System (1,2) is represented in the form of (5) by defining

$$x_1 = \begin{bmatrix} z \\ \dot{z} \end{bmatrix}, \quad x_2 = \lambda, \quad n_1 = 2f, \quad n_2 = p + q, \tag{6}$$

$$f_1 = \begin{bmatrix} \dot{z} \\ -M^{-1}(k - F^T\lambda - Tu) \end{bmatrix}, \quad f_2 = f + Ru. \tag{7}$$

For the following discussion of optimal control design, we assume that the vector functions are sufficiently smooth.

3. Representation of Descriptor System (5)

To study the control problem of descriptor systems carefully it is well to have in mind different forms of representation of the dynamical system (5) which are related either to the generation of differential equations for the x_2-vector by differentiating the algebraic equations as many times as necessary, or to the elimination of redundant coordinates generating the state space differential equations. In both cases the key for that is the index k of the system (5) which is roughly speaking the number of differentiations of the algebraic equations to get the underlying set of ordinary differential equations (ODE) on a manifold. For simplicity we assume that all algebraic equations have a uniform index. Then we have in case of index k:

$$\frac{d^j f_2}{dt^j} \equiv L^j(f_2) = 0, \quad j = 0, \ldots, k-1, \tag{8}$$

$$\dot{x}_2 = -\left(\frac{\partial}{\partial x_2^T} L_{f_1}^{k-1}(f_2)\right)^{-1} L^k(f_2) = \bar{f}_2(x_1, x_2, u, \ldots, u^{(k)}) \tag{9}$$

where

$$L(\cdot) = L_{f_1}(\cdot) + L_{\dot{u}}(\cdot) + \tfrac{\Delta}{\Delta t}(\cdot), \quad L_{f_1}(\cdot) = \tfrac{\partial(\cdot)}{\partial x_1^T} f_1,$$
$$L_{u^{(j)}}(\cdot) = \tfrac{\partial(\cdot)}{\partial u^T} u^{(j)}, \quad \tfrac{\Delta}{\Delta t} L_{u^{(j)}}(\cdot) = L_{u^{(j+1)}}(\cdot). \tag{10}$$

The operator L is defined by the other operators where two of them are certain Lie derivatives and the last one is only applied on $L_{u^{(j)}}$, differentiating the control input function with respect to time. The vector functions $L^j(\mathbf{f_2})$, $j = 0, \ldots, k-2$, depend on $\mathbf{x_1}$ but not on $\mathbf{x_2}$. The function $L^{k-1}(\mathbf{f_2})$ depends on $\mathbf{x_2}$ so that the related Jacobian matrix is regular, and the differential equation (9) can be derived. $L^{k-1}(\mathbf{f_2}) = \mathbf{0}$ is a first integral of (9). Additionally the functions $L^j(\mathbf{f_2})$ depend generally on the time derivatives of the control input $\mathbf{u} : \mathbf{u}, \dot{\mathbf{u}}, \ldots, \mathbf{u}^{(j)}$, $j = 0, \ldots, k$. Therefore, the descriptor system (5) may be also represented by the differential equations (5) and (9) on a manifold described by the invariants (8).

Additionally, in principle it is possible to use the invariants (8) to eliminate $k\,n_2$ redundant variables and to end up with a set of $n_1 - (k-1)\,n_2$ ordinary differential equations in the state space. But for nonlinear problems this elimination procedure is very cumbersome in general so that we do not follow this way. In the following we try to analyse the system by manipulating the differential-algebraic equations (5) directly, or by having in mind the representation (5,9) with invariants (8). The initial conditions of (5) are consistent if they satisfy the invariants (8).

4. Optimal Control

The purpose of this section is to derive conditions for the design of optimal control of descriptor systems. Looking for the system description (5,8,9) instead of (5) we encounter a problem. Generally the system behaviour may depend not only on the control input $\mathbf{u}$ but also on its higher order time-derivatives $\dot{\mathbf{u}}, \ddot{\mathbf{u}}, \ldots, \mathbf{u}^{(k)}$. This is an unconvenient problem statement. Therefore, first it is necessary to clarify this unusual situation. For this, the notion of "causality" is introduced.

4.1. CAUSALITY

According to a definition by Dai(1989) for time-discrete systems, the time-continuous system (5) or (5,8,9) is called "causal", if the solution $[\mathbf{x_1}(t), \mathbf{x_2}(t)]$ does not depend on $\dot{\mathbf{u}}(t), \ldots, \mathbf{u}^{(k-1)}(t)$ but on $\mathbf{u}(t)$ only.

This definition does not include $\mathbf{u}^{(k)}$ as it may be expected by the notation of (9), because $L^{(k-1)}(\mathbf{f_2}) = \mathbf{0}$ is a first integral of (9) only depending on $\dot{\mathbf{u}}(t), \ldots, \mathbf{u}^{(k-1)}(t)$. It should be mentioned that descriptor systems of index $k = 1$ are always causal. Non-causal systems may appear only for higher index problems $k \geq 2$.

Now we must consider how causality can be checked. For linear descriptor systems characterized by regular matrix pencils, a necessary and sufficient condition is available if the system is represented in its Weierstrass-Kronecker canonical form, cf. Dai(1989). But this solution does not help us for nonlinear descriptor systems (5). Therefore, a new criterion has to be developed. Following the description (5,8,9) of the system we can derive

Theorem 1. The descriptor system (5) or (5,8,9) is causal iff the conditions

$$\frac{\partial}{\partial \mathbf{u}^T}\left(L^j_{f_1}(\mathbf{f}_2)\right) = \mathbf{0}, \quad j = 0, \ldots, k-2, \tag{11}$$

hold. In this case the invariants (8) are

$$L^j(\mathbf{f}_2) \equiv L^j_{f_1}(\mathbf{f}_2) \equiv \mathbf{f}_{2j}(\mathbf{x}_1, \mathbf{u}) = \mathbf{0}, \quad j = 0, \ldots, k-2, \tag{12}$$

$$L^{(k-1)}(\mathbf{f}_2) \equiv L^{(k-1)}_{f_1}(\mathbf{f}_2) \equiv \mathbf{f}_{2,k-1}(\mathbf{x}_1, \mathbf{x}_2, \mathbf{u}) = \mathbf{0}. \tag{13}$$

They do not depend on higher-order time-derivatives of the control input.

The proof follows by explicit calculations.

4.2. OPTIMIZATION

The notion of causality allows us to distinguish two cases, whether Pontryagin's maximum principle can be applied as usual or not. Assuming a performance criterion

$$J = \int_0^T f_0(\mathbf{x}_1, \mathbf{x}_2, \mathbf{u}) \, dt \quad \longrightarrow \quad \underset{\mathbf{u} \in U}{\text{minimum}} \tag{14}$$

we have the following result.

Theorem 2. For causal descriptor systems (5) the maximum principle of Pontryagin can be applied as usual. Necessarily there are nontrivial adjoint vectors λ_1 and λ_2 such that with the Hamiltonian

$$H = \lambda_1^T \mathbf{f}_1 + \lambda_2^T \mathbf{f}_2 - f_0 \tag{15}$$

the adjoint differential-algebraic equations

$$\dot{\lambda}_1 = -\frac{\partial H}{\partial \mathbf{x}_1}, \quad \mathbf{0} = -\frac{\partial H}{\partial \mathbf{x}_2} \tag{16}$$

hold and the optimal control satisfies

$$\underset{\mathbf{u} \in U}{\max} \, \mathbf{H} = H_{opt}. \tag{17}$$

The proof only considers variations of motion which are consistent with the invariants (12) so that under this restriction usual optimization procedures can be applied. For simplicity, the boundary conditions of the adjoint variables are not considered explicitly, but they have to be chosen corresponding to the usual conditions.

The optimal control problem looks quite different for non-causal descriptor systems because of the influence of the higher-order time-derivatives of the control input. We recommend the following procedure. Having regard to $\dot{\mathbf{u}}, \ldots, \mathbf{u}^{(k-1)}$, we prefer the representation (5,8,9) of the descriptor system. Introducing the vectors

$$\boldsymbol{\xi}_1 = \mathbf{u}, \quad \boldsymbol{\xi}_2 = \dot{\mathbf{u}}, \quad \ldots, \boldsymbol{\xi}_k = \mathbf{u}^{(k-1)}, \quad \mathbf{v} = \mathbf{u}^{(k)} \tag{18}$$

we define an extended state vector

$$\mathbf{x}_e = \begin{bmatrix} \mathbf{x}_1^T & \mathbf{x}_2^T & \boldsymbol{\xi}_1^T & \boldsymbol{\xi}_2^T & \cdots & \boldsymbol{\xi}_k^T \end{bmatrix}^T \tag{19}$$

satisfying the ordinary differential equation

$$\dot{\mathbf{x}}_e = \begin{bmatrix} \mathbf{f}_1^T(\mathbf{x}_1, \mathbf{x}_2, \boldsymbol{\xi}_1) & \bar{\mathbf{f}}_2^T(\mathbf{x}_1, \mathbf{x}_2, \boldsymbol{\xi}_1, \boldsymbol{\xi}_2, \ldots, \boldsymbol{\xi}_k, \mathbf{v}) & \boldsymbol{\xi}_2^T & \cdots & \boldsymbol{\xi}_k^T & \mathbf{v}^T \end{bmatrix}^T \tag{20}$$

and the algebraic equations

$$L^j(\mathbf{f}_2) \equiv \mathbf{f}_{2j}(\mathbf{x}_1, \mathbf{x}_2, \boldsymbol{\xi}_1, \boldsymbol{\xi}_2, \ldots, \boldsymbol{\xi}_{j+1}) = \mathbf{0}, \quad j = 0, \ldots, k-1. \tag{21}$$

The original control constraints $\mathbf{u} \in U$ now appear as state constraints $\boldsymbol{\xi}_1 \in U$. We question whether the control problem is stated properly. A reasonable problem statement demands additional constraints on $\boldsymbol{\xi}_2 = \dot{\mathbf{u}}, \ldots, \boldsymbol{\xi}_k = \mathbf{u}^{(k-1)}$ and particularly on $\mathbf{v} = \mathbf{u}^{(k)}$. Now, the corresponding Hamilton function will be introduced:

$$H = \boldsymbol{\lambda}_1^T \mathbf{f}_1 + \boldsymbol{\lambda}_2^T \bar{\mathbf{f}}_2 - f_0 + \boldsymbol{\psi}_1^T \boldsymbol{\xi}_2 + \boldsymbol{\psi}_2^T \boldsymbol{\xi}_3 + \cdots + \boldsymbol{\psi}_{k-1}^T \boldsymbol{\xi}_k + \boldsymbol{\psi}_k^T \mathbf{v}$$

$$+ \sum_{j=0}^{k-1} \boldsymbol{\mu}_j^T L^j(\mathbf{f}_2). \tag{22}$$

For the extended system (20) and the Hamilton function (22), the procedures of the calculus of variations or of Pontryagin's maximum principle can be applied, leading to an optimal control design of non-causal descriptor systems.

5. Optimal Control of Mechanical Descriptor Systems

Let us apply the results of the preceding section to mechanical descriptor systems (1,2). First of all, we must check the index and causality. For this, the algebraic constraints are differentiated, leading to the following result.

Theorem 3. If the algebraic constraints (2) are independent then the holonomic mechanical descriptor system (1,2) has index $k = 3$. Additionally the system is causal iff $\mathbf{R} = \mathbf{0}$, i.e. the constraints are not controlled. If at least one constraint is controlled, i.e. it depends explicitly on **u**, then the system is non-causal.

In the following we illustrate the design of optimal control by two simple examples. The first example consists of two spring-mass-oscillators where the masses are connected by a rigid bar such that the constraint is not controlled. The second example differs from the first by placing the control to the constraint such that we have the typical problem of a controlled mechanism. The two systems are governed by Lagrange's equation of the first kind:

$$m_1 \ddot{z}_1 + c_1 z_1 = \lambda + u_1, \quad m_2 \ddot{z}_2 + c_2 z_2 = -\lambda, \quad z_1 - z_2 + u_2 = 0. \quad (23)$$

If $u_1 \not\equiv 0, u_2 \equiv 0$ we have the first example; the second is obtained by $u_1 \equiv 0, u_2 \not\equiv 0$. The description (5) runs as

$$\mathbf{x}_1^T = [\, z_1 \quad z_2 \quad \dot{z}_1 \quad \dot{z}_2 \,], \quad x_2 = \lambda, \quad (24)$$

$$\mathbf{f}_1 = \begin{bmatrix} x_3 \\ x_4 \\ -\frac{c_1}{m_1} x_1 + \frac{\lambda}{m_1} + \frac{u_1}{m_1} \\ -\frac{c_2}{m_2} x_2 - \frac{\lambda}{m_2} \end{bmatrix}, \quad f_2 = x_1 - x_2 + u_2. \quad (25)$$

In the first case, $u_2 \equiv 0$, the control should be designed with respect to a quadratic performance criterion:

$$J = \frac{1}{2} \int_0^\infty \left[z_1^2 + z_2^2 + \varepsilon(\dot{z}_1^2 + \dot{z}_2^2) + r_1 u_1^2 \right] dt \rightarrow \text{minimum}. \quad (26)$$

We assume that there are not restrictions on the control such that we have the LQ-regulator design problem. According to theorem 3, the system is causal and, therefore, theorem 2 can be applied to give the Hamiltonian (15) and the adjoint differential-algebraic equations (16). Manipulating these equations, we find the optimal control

$$u = \left(c - \sqrt{c^2 + \frac{2}{r_1}} \right) x_1 - \sqrt{2\frac{\varepsilon}{r_1} - 2mc + 2m\sqrt{c^2 + \frac{2}{r_1}}} \, x_3. \quad (27)$$

In the second case, $u_1 \equiv 0$, the system is non-causal according to theorem 3. Then the extension (18) has to be introduced:

$$\xi_1 = u, \quad \xi_2 = \dot{u}, \quad v = \ddot{u}: \quad \dot{\xi}_1 = \xi_2, \quad \dot{\xi}_2 = v. \quad (28)$$

To get a proper problem statement it is necessary to modify the performance criterion (26) to

$$J^* = \frac{1}{2} \int_0^\infty \left[z_1^2 + z_2^2 + \varepsilon(\dot{z}_1^2 + \dot{z}_2^2) + r_2 u_2^2 + \rho v^2 \right] dt \rightarrow \text{minimum} \quad (29)$$

254

where $v = \ddot{u}$ is also weighted. Then the calculus of variations can be applied to the extended system (23-25,28,29). The solution of the two-point-boundary-value-problem is cumbersome, but can be performed even by hand. The formulas become very lengthy, and are not shown explicitly.

It should be mentioned that the solution of the problem by a generalized Riccati matrix equation is not always recommended. For causal systems, a generalized Riccati matrix equation can be defined and solved numerically, e.g. Schüpphaus(1995) and Hou(1995). But for non-causal systems there are still many problems in understanding and solving the Riccati matrix equation, cf. Mehrmann(1991).

6. Conclusions

The description of nonlinear dynamical systems in mechanics by so-called descriptor systems, i.e. by Lagrange's equations of the first kind, is becoming more and more popular. Therefore, tools for the analysis and the control design of such systems are needed. In this contribution, some results on the optimal control design have been reported. Here it was necessary to introduce the notion of causality and to distinguish between causal and non-causal descriptor systems. In the first case the calculus of variations and Pontryagin's maximum principle can be applied as usual, but in the second case one has to take care and it is recommended that the problem be solved on the basis of the so-called underlying ordinary differential equation on a manifold. Two examples of constrained spring-mass oscillators illustrated the different approaches. There are still many open problems of the optimal control design of mechanical descriptor systems. Particularly efficient numerical methods are missing, especially for non-causal systems.

7. References

Dai, L. (1989) *Singular Control Systems*, Lecture Notes in Control and Information Sciences Vol. 118, Springer-Verlag, Berlin-Heidelberg.

Hou, M. (1995) *Descriptor Systems: Observers and Fault Diagnosis*, Fortschr.-Ber. VDI, Reihe 8, Nr. 482, VDI-Verlag, Düsseldorf.

Mehrmann, V. L.. (1991) *The Autonomous Linear Quadratic Control Problem*, Lecture Notes in Control and Information Sciences Vol. 163, Springer-Verlag, Berlin-Heidelberg.

Schüpphaus, R. (1995) *Regelungstechnische Analyse und Synthese von Mehrkörpersystemen in Deskriptorform*, Fortschr.-Ber. VDI, Reihe 8, Nr. 478, VDI-Verlag, Düsseldorf.

ADAPTIVE/ROBUST CONTROL OF CHAOTIC SYSTEMS

HENK NIJMEIJER
Department of Applied Mathematics
University of Twente
P.O. Box 217
7500 AE Enschede
The Netherlands
fax: +31-53-4340733
e-mail: h.nijmeijer@math.utwente.nl

Abstract. This note deals with feedback controllers for chaotic systems such as the forced Duffing equation and forced van der Pol equation. Both state feedback controllers as well as output feedback controllers are developed. Also robust and adaptive controllers are obtained in case model uncertainties are present.

1. Introduction

Recently an increasing interest has arisen in controlling complex or chaotic nonlinear systems that appear in physics, engineering and other sciences. The number of relevant references is growing rapidly, see Chen (1996). An essential element in the control of chaos is that in many cases the ultimate goal of control is to decrease random effects and to stabilize the system at an equilibrium or reference trajectory. Among the various methods that can be used in controlling (chaotic) systems, a prominent role is played by the so-called Lyapunov-type methods. Typically Lyapunov controllers were developed for mechanical systems such as rigid robots, see e.g. Berghuis and Nijmeijer (1993) and references therein. In a previous paper (Nijmeijer and Berghuis (1995)) we have shown that similar techniques can be used in the trajectory control of the chaotic forced Duffing equation. Two basic observations in light of this work are worth mentioning. First, whether or not the system under consideration exhibits chaotic motion is not relevant.

255

D. H. van Campen (ed.), IUTAM Symposium on Interaction between Dynamics and Control in Advanced Mechanical Systems, 255–262.

Moreover, tracking control towards *any* reference trajectory is possible, the price being that possibly large control actions are needed.

The purpose of the present note is to extend our previous work in two directions. Namely, we introduce, as an alternative way of dealing with parameter uncertainty in the Duffing equation, an adaptive feedback controller that asymptotically achieves trajectory tracking. In addition, we show that our methodology can be extended without much difficulty to other – similar – (chaotic) systems like the forced van der Pol equation.

Acknowledgements The cooperation with dr. Harry Berghuis has been very useful in the research which is reported here. Part of this work has been done when Michael Stuckings was partially completing his MSc. thesis during his traineeship at the University of Twente in 1995. Finally, the cooperation with Erjen Lefeber has been useful in this paper.

2. Feedback control of the forced Duffing equation

Consider the controlled forced Duffing

$$\ddot{x} + p\dot{x} + p_1 x + x^3 = q\cos(\omega t) + u \tag{1}$$

The uncontrolled version of (1), i.e. $u \equiv 0$, may exhibit chaotic motion for certain parameter values for p, p_1, q and ω, see Chen and Dong (1993). The problem we are interested in is to find a feedback controller

$$u = \varphi(x, \dot{x}, x_d, t) \tag{2}$$

which is such that, independent of what the initial state $(x(0), \dot{x}(0))$ of (1) is, the closed loop system (1), (2) has the property that $(x, \dot{x})$ converges asymptotically to $(x_d, \dot{x}_d)$, where x_d is an arbitrary smooth desired trajectory. Contrary to much work in the context of control of chaos, we do not demand that x_d is a trajectory of the uncontrolled system (1). The latter may simplify things in some cases, but is of true importance only when we consider the tracking problem under input constraints, i.e. $\mid u \mid \leq M$. We will not address the input saturation problem here, but postpone it to a forthcoming paper.

A state feedback controller that solves the above tracking problem is given by

$$\begin{aligned} u &= \ddot{x}_d + p\dot{x}_d + p_1 x_d + x_d^3 - q\cos(\omega t) \\ &\quad -K_p e - K_d \dot{e} + \nu \end{aligned} \tag{3}$$

with $e = x - x_d$, $K_p, K_d > 0$ and ν properly chosen so as to make the error dynamics asymptotically stable. Besides the choice of $\nu = 3xx_d e + e^3$ which makes the error dynamics linear we can take as in Nijmeijer and Berghuis (1995)

$$\nu = 3xx_d e \tag{4}$$

Typically the controller (3), (4) contains three parts, namely a model based part $\ddot{x}_d + p\dot{x}_d + p_1 x_d + x_d^3 - q\cos(\omega t)$, a P(roportional) D(ifferential) part $-K_p e - K_d \dot{e}$, and a nonlinear part ν given by (4). The closed loop dynamics (1), (3), (4) is described by

$$\ddot{e} + (p + K_d)\dot{e} + (p_1 + K_p)e + e^3 = 0 \tag{5}$$

Standard Lyapunov techniques show that the dynamics (5) are asymptotically stable, that is (see Nijmeijer and Berghuis (1995))

$$\lim_{t \to \infty} (e(t), \dot{e}(t)) = (0,0) \tag{6}$$

provided that

$$K_d > -p, \quad K_p > -p_1 \tag{7}$$

The controller (3), (4) (or with any other suitably chosen ν) requires knowledge of $\dot{x}$, unless $p > 0$. Although this condition might be fulfilled in a special case, we do not want to impose this condition here. An obvious drawback is that we now need to feed back the velocity $\dot{x}$. However we may overcome this problem by including a velocity observer. The following controller-observer combination has been proposed in Nijmeijer and Berghuis (1995)

$$u = \ddot{x}_d + p\dot{x}_d + p_1 x_d + x_d^3 - q\cos(\omega t) - K_p \hat{e} - K_d \dot{\hat{e}} + 3xx_d e \tag{8}$$

$$\begin{cases} \dot{\hat{e}} = w + 2K_d(e - \hat{e}) - pe \\ \dot{w} = 2K_p(e - \hat{e}) - p_1 e - e^3 \end{cases} \tag{9}$$

Indeed, if $K_p = \lambda K_d (\lambda > 0)$, the closed loop system (1), (8), (9) is asymptotically stable provided (7) holds and in addition

$$0 < \lambda < \min(K_d, p + K_d) \tag{10}$$

Since the latter requirement, (10), can always be met, the asymptotic convergence of (8), (9) follows under the same condition as for the state feedback (3), (4). The proof of this result, again using Lyapunov-techniques, is given in the aforementioned reference.

3. Robust and adaptive control of the forced Duffing equation

The (observer-) controller developed in the previous section for the tracking control of the Duffing equation is constructed under the proviso that the complete model (1) is known, and more specifically the characteristic parameters p, p_1, q and ω are given. If this is not true, and only estimates for these parameters are available, there are basically two possible approaches. One way is what might be called a robust control approach, and replaces the controller (3), (4), or the controller-observer (8), (9) by the following

$$u = -K_p e - K_d \dot{e} \tag{11}$$

or, including a linear velocity observer

$$u = -K_p \hat{e} - K_d \dot{\hat{e}} \tag{12}$$

$$\begin{cases} \dot{\hat{e}} & = & w + 2K_d(e - \hat{e}) \\ \\ \dot{\omega} & = & 2K_p(e - \hat{e}) \end{cases} \tag{13}$$

In both cases, either (11), or (12) and (13), both the model based part (depending upon the parameters!) and the nonlinear ν is eliminated. The key result is that, if $K_p = \lambda K_d (\lambda > 0)$, the closed loop system (1) and (11), or (1), (12) and (13), is *practically stable* (also called *ultimately uniformly bounded*), that is, the error states converge in finite time towards a bounded region of the origin. Moreover by increasing the gain K_d (and thus K_p) the neighborhood of 0 can be made arbitrarily small, see Berghuis and Nijmeijer (1994). It should be noted that such high-gain assumption may lead to practical limitations, such as noise amplification. On the other hand, the result is appealingly simple in that the possible chaotic dynamics of (1) is almost completely annihilated under the high-gain controllers (11), or (12) and (13); the chaos is in fact suppressed into the (small) bounded region of the (error)-origin.

Another way to cope with uncertainty in the model (1) is to use an *adaptive* controller for the unknown parameters p, p_1 and q in (1) (assuming for simplicity that ω is known). In this case the controller (3), (4) takes the form

$$u = \ddot{x}_d + Y^T(x, x_d, \dot{x}_d, t)\hat{\theta} - K_p e - K_d \dot{e} + \nu \tag{14}$$

where $\hat{\theta} = (\hat{p}, \hat{p}_1, \hat{q})^T$ is the vector of parameter estimates and Y is the so-called regressor vector given as

$$Y^T(x, x_d, \dot{x}_d, t) = (\dot{x}_d, x_d, \cos(\omega t)) \tag{15}$$

(for the observer-controller (8), (9) the results are similar). By introducing a gradient adaptation law for $\tilde{\theta} = \hat{\theta} - \theta$ of the form

$$\dot{\tilde{\theta}} = -\Gamma Y(x, x_d, \dot{x}_d, t)(\dot{e} + \lambda e) \tag{16}$$

with $\Gamma = \Gamma^T > 0$ a positive definite $(3,3)$-matrix, and $\lambda > 0$ an arbitrary constant, we find that the overall closed-loop system (1), (14), (3), (16) is globally asymptotically stable if the conditions (7) hold and λ is chosen such that $K_d > -p + \lambda$. This result is standard in adaptive control, cf. Slotine and Li (1991) and follows again using Lyapunov-type arguments. However, it is worth mentioning that no parameter error convergence to zero necessarily follows, and that for this reason the analysis of (1), (14), (3), (16) becomes more involved because of the time-varying nature of the closed loop system. The adaptive controller as presented here could be modified in many ways, see e.g. Slotine and Li (1991), or in this context Stuckings (1995) where various adaptation laws have been examined. At this point it is useful to make a comparison between the robust controller (11) and the adaptive controller (14), (3), (16). In Figures 1 and 2 simulations are given with the robust controller (11) (Figure 1) and the adaptive controller (14), (3), (16) (Figure 2). The simulations with SIMNON^{TM} were done for nominal parameter values $p = 0.4$, $p_1 = -1.1$, $q = -2.1$ and $\omega = 1.8$ for which the forced Duffing equation exhibits chaotic behavior (cf. Chen and Dong (1993)). In both settings the PD-gains were chosen as $K_d = 12.5$ and $K_p = 50$; the adaptation parameters were set at $\lambda = 6$, $\Gamma_1 = 0.5$, $\Gamma_2 = 1$, $\Gamma_3 = 2$, with initial parameter estimates as $\hat{p} = \hat{p}_1 = \hat{q} = 0$. The initial state was chosen as $(x_0, \dot{x}_0) = (0,0)$, and in order to clearly show the effect of the controllers, only at $t = 10$ were the controllers switched on. A simple inspection of the simulations shows that, in the present setting, the adaptive controller is superior to the robust controller (11). There are obvious reasons for this. First the robust controller is totally independent of the forcing $q \cos(\omega t)$, whereas in the adaptation law (16) knowledge of the frequency ω is required. A second, related aspect, is that (11) is completely model independent – as we will see in the next section! – in contrast to (14), (15), (16).

4. Controlling the forced van der Pol equation

The previous sections deal with the control of the (chaotic) forced Duffing equation. The purpose of this section is to show that our methodology is by no means limited to the Duffing equation, but can also be developed for other second order systems (like the 'passivity-based' controllers for rigid robots, cf. Berghuis and Nijmeijer (1993)).
Consider the forced van der Pol equation

$$\ddot{x} - \mu(1 - x^2)\dot{x} + x = q\cos(\omega t) + u \qquad (17)$$

The uncontrolled version of (17) again exhibits chaotic motion for certain parameter values μ, q and ω. In order to achieve tracking control of (17)

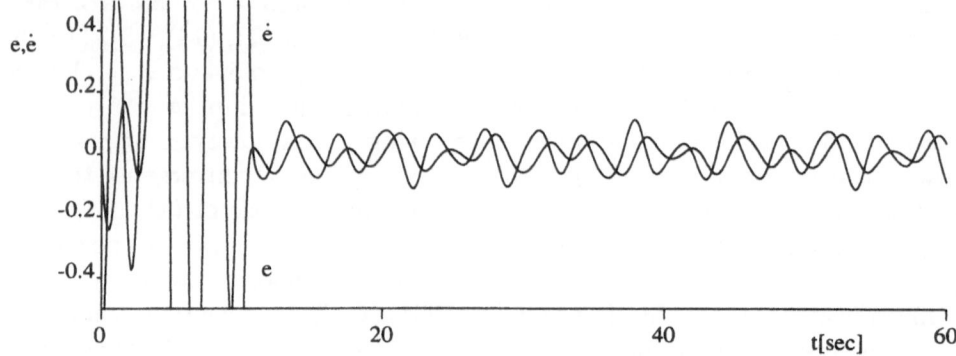

Figure 1. robust controller (10)

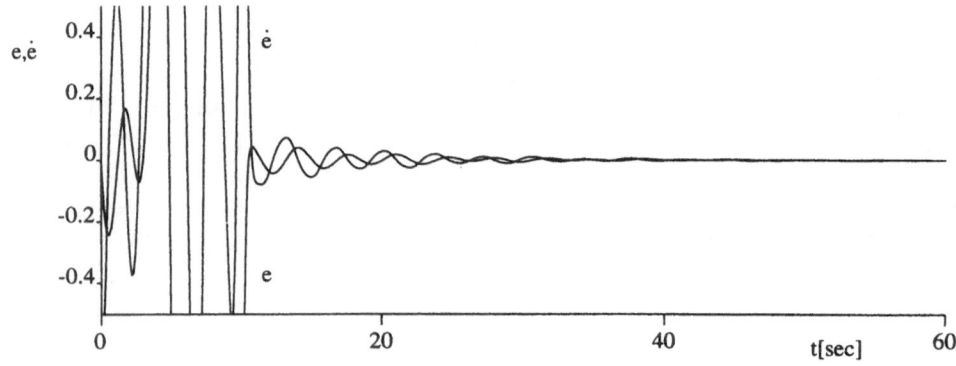

Figure 2. adaptive controller (12,3,14)

towards a desired trajectory x_d we may take the controller

$$u = \ddot{x}_d - \mu \dot{x}_d + x_d - q\cos(\omega t) - K_d \dot{e} - K_p e + \nu \qquad (18)$$

where ν can be chosen in various ways, e.g.

$$\nu = \mu x^2 \dot{x} \qquad (19)$$

or

$$\nu = \mu(2ex_d\dot{e} + x_d^2\dot{e} + x^2\dot{x}_d) \qquad (20)$$

As before, the controller (18), (19) or (18), (20), yields global asymptotic stability, using a suitable Lyapunov-analysis. Also both controllers allow for an observer-controller combination in case velocity measurements are lacking. For instance (18), (20) then becomes

$$
\begin{cases}
u &= \ddot{x}_d - \mu\dot{x}_d + x_d - q\cos(\omega t) - K_d\dot{\hat{e}} - K_p\hat{e} + \hat{\nu} \\
\hat{\nu} &= \mu(2ex_d\dot{\hat{e}} + x_d^2\dot{\hat{e}} + x^2\dot{x}_d)
\end{cases}
\tag{21}
$$

$$
\begin{cases}
\dot{\hat{e}} &= \omega + 2K_d(e - \hat{e}) + \mu e - \frac{1}{3}\mu e^3 \\
\dot{\omega} &= 2K_p(e - \hat{e}) - e
\end{cases}
\tag{22}
$$

which yields semi-global asymptotic stability of the closed loop system. Other controllers and observer-controller combinations are described in Stuckings (1995). It is straightforward to develop an adaptive version of (18), (19) or (18), (20), or (21), (22). Completely analogously, the robust controller (11) or observer-controller (12), (13) can be used to obtain practical stability for the closed loop system (17), (11) or (17), (12), (13), cf. Berghuis and Nijmeijer (1994). Simulations in this situation exhibit a performance like those described for the forced Duffing equation.

5. Concluding remarks

As with a large class of second order mechanical systems (e.g. rigid robot-manipulators) we have derived controllers and observer-controller combinations that achieve asymptotic trajectory tracking for the forced Duffing and forced van der Pol equation. In case exact model information is lacking, we propose two alternative methods, namely robust (output) PD control or adaptive (output) feedback control. The latter method is in case there are only parameter uncertainties superior, but cannot be used when there is uncertainty in the model structure. In that case PD control is the best alternative, automatically leading to a high gain feedback scenario. For that reason we have to search for other robust controllers, as high gain feedback automatically causes noise amplification (and, of course, large control efforts). In a forthcoming paper we will develop bounded feedback tracking controllers that achieve asymptotic trajectory tracking. Especially in that situation, a careful use of the (chaotic) attractors of the (un)controlled system becomes important and in this way a close connection may exist in this case with the method of Ott, Grebogi and Yorke (1990) where small controllers in chaotic systems are developed.

References

Berghuis, H., and Nijmeijer, H. (1993), "A passivity approach to controller observer design for robots", *IEEE Trans. Robotics and Autom.*, **Vol. 9**, pp. 940–954.
Berghuis, H., and Nijmeijer, H. (1994), "Robust control of robots via linear estimated state feedback", *IEEE Trans. Autom. Control*, **Vol. 39**, pp. 2159–2162.

Chen, G. (1996), Control and synchronization of chaos (bibliography), *Dept. of EE, Univ. of Houston, TX, USA*. Available from ftp: `uhoop.egr.uh.edu/pub/TeX/chaos.tex` (login name and password both "anonymous").

Chen, G., and Dong, X. (1993), "On feedback control of chaotic continuous-time systems", *IEEE Trans. Circuits and Systems*, **Vol. 40**, pp. 591–601.

Nijmeijer, H., and Berghuis, H. (1995), "On Lyapunov control of the Duffing equation", *IEEE Trans. Circuits and Systems - I: Fund. Theory and Applic.*, **Vol. 42**, pp. 473–477.

Ott, E., Grebogi, C., and Yorke, J.A. (1990), "Controlling chaos", *Physics Review Letters*, *Vol. 64*, pp. 1196-1199.

Parlitz, U., and Lauterborn, W. (1987), "Period-doubling cascades and devil's staircases of the driven van der Pol oscillator", *Physical Review A*, **Vol. 36**, pp. 1428-1434.

Slotine, J.J.E., and Li, W. (1991), *Applied Nonlinear Control, Prentice-Hall Int. Editions*, Englewood Cliffs.

Stuckings, M.J. (1995), "A study in nonlinear control: the Duffing's and van der Pol equations", *Final year Project*, Dept. Electr. and Comp. Eng., Newcastle University, NSW, Australia.

EXPERIMENTAL IMPLEMENTATION OF SATURATION CONTROL

SHAFIC S. OUEINI AND ALI H. NAYFEH
Department of Engineering Science and Mechanics
Virginia Polytechnic Institute and State University
Blacksburg, Virginia 24061-0219, U.S.A.

Abstract. An approach for implementing an active nonlinear vibration absorber is presented. The technique is based on exploiting the saturation phenomenon exhibited by multidegree-of-freedom systems having quadratic nonlinearities and possessing two-to-one internal or autoparametric resonances. The proposed technique consists of introducing a second-order controller and coupling it to the plant through a sensor and an actuator, where both the feedback and control signals are quadratic. Once the plant is forced near resonance, its response saturates to a small value, and the remaining oscillatory energy is channeled to the controller. We present experimental results of the application of the proposed control technique. The controller is built in electronic components, and the strategy is tested by regulating the responses of a rigid arm and a cantilever beam. In the first case, the arm is actuated by a DC motor, whereas in the second case, the actuator consists of piezoceramic patches attached to the beam.

1. Introduction

The saturation phenomenon has been the subject of extensive theoretical and experimental research. Nayfeh, Mook, and Marshall (1973) first discovered the phenomenon in the context of analyzing coupling between the roll and pitch motion of ships. Haddow, Barr, and Mook (1984), Nayfeh and Zavodney (1988), and Nayfeh and Balachandran (1989) conducted experiments on an L-shaped beam structure and verified the results reported by Nayfeh, Mook, and Marshall (1973). Haxton and Barr (1972) proposed a passive vibration absorber based on autoparametric resonances. Their tech-

D. H. van Campen (ed.), IUTAM Symposium on Interaction between Dynamics and Control in Advanced Mechanical Systems, 263–270.
© 1997 *Kluwer Academic Publishers.*

264

nique hinges on attaching a cantilever beam with a tip mass to the single-degree-of-freedom system under consideration. Once the plant is forced at resonance, they found that by properly tuning the frequency of the beam, the absorber can effectively limit the amplitude of oscillations of the main mass. Hatwall, Mallik, and Ghosh (1982) and Ertas and Cuvalci (1993) attached a pendulum to a single-degree-of-freedom system consisting of a mass and a restoring spring to act as a passive vibration absorber. Mustafa and Ertas (1995) proposed a pendulum based vibration absorber. They conducted experimental studies and demonstrated the existence of quasiperiodicity in the response of the system. Cartmell and Lawson (1994) improved the performance of the pendulum by installing a motor, which actuates the absorber system. Their strategy renders the control scheme active by incorporating a computer into the actuation mechanism. The computer senses the amplitude of the mass and tunes the frequency of the pendulum to optimize the response characteristics. Tuer, Golnaraghi, and Wang (1994) introduced a control scheme based on internal resonance for regulating the free oscillations of lightly damped structures. Their strategy is based on introducing a second-order controller and coupling it to the plant with a quadratic nonlinear feedback control law. Oueini and Golnaraghi (1996) built an electronic circuit to emulate the equation of the controller and experimentally tested the IR strategy by controlling the free oscillations of a rigid arm attached to a DC motor.

2. Development of the Control Law

To describe the strategy, we consider the problem of controlling the response of a second-order damped oscillator to a harmonic excitation. Including the control signal T, we express the motion of the oscillator (plant) as

$$\ddot{p} + f(\dot{p}) + \omega_p^2 p = F \cos(\Omega t) + T, \tag{1}$$

where p is the generalized plant coordinate, $f(\dot{p})$ represents dissipative terms, and ω_p is the plant's natural frequency. We consider the case when $\Omega \approx \omega_p$ and introduce a second-order nonlinear controller and a control law in the form

$$\ddot{u} + 2\zeta_c \omega_c \dot{u} + \omega_c^2 u = \alpha p u \quad \text{and} \quad T = \gamma u^2, \tag{2}$$

where ζ_c and ω_c are, respectively, the controller's damping ratio and natural frequency, and α and γ are positive constants. An autoparametric-resonance condition is created by choosing ω_c so that $2\omega_c \approx \omega_p$. The form of the quadratic coupling in equations (2) is not unique. Nayfeh (1988) and Nayfeh and Zavodney (1988) have demonstrated that coupling involving other combinations of nonlinear terms can induce the saturation phenomenon, and hence they can also be used to formulate adequately the

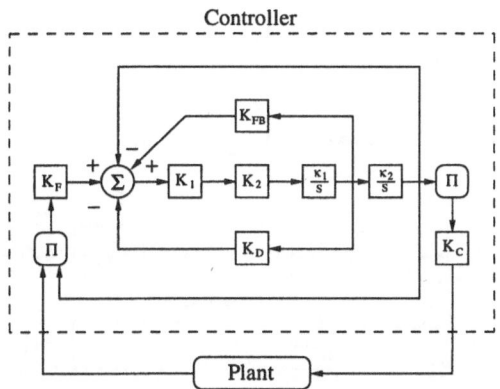

Figure 1. A block diagram of the controller.

control law. The choice of nonlinearities will depend on the available feedback signal from the plant.

To implement experimentally the strategy, an electronic circuit is developed to emulate the controller equation. The circuit is assembled with electronic hardware. The details of its design are presented in Oueini and Golnaraghi (1996). The block diagram depicting the design of the circuit is shown in Figure 1. The inverses of the time constants of the integrators are labeled κ_1 and κ_2. The natural frequency of the controller is adjusted through the two gains K_1 and K_2, where K_1 represents a coarse gain, while K_2 defines a fine tuning gain. The feedback term K_D represents the inherent damping associated with the electronic components. The theory of Nayfeh, Mook, and Marshall (1973) indicates that the amplitude of the beam is minimized by reducing the damping in the controller. This is achieved by incorporating a positive velocity feedback loop in the circuit and adjusting its amplitude with the variable gain K_{FB}. Hence, the inherent damping of the circuit assembly is reduced by properly tuning K_{FB} such that $K_{FB} \approx K_D$. The nonlinear feedback and control signals are generated by the two multipliers Π. Their respective magnitudes are adjusted through the variable gains K_F and K_C. In the next two sections, we present experimental results that demonstrate the performance of the proposed control strategy for suppressing the vibrations of a rigid beam and a flexible cantilever beam.

3. Robotic Arm

The plant under consideration consists of a rigid arm attached to a permanent magnet DC motor where the motor is connected in position feedback mode through a potentiometer. Neglecting the electrical time constant, we

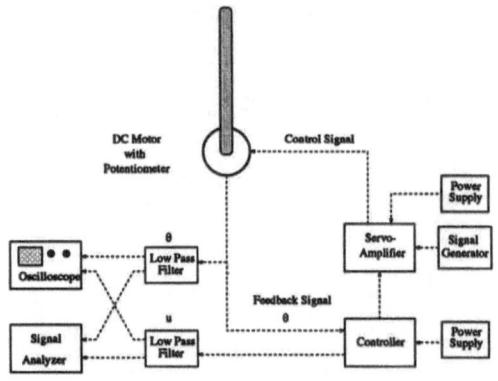

Figure 2. A schematic of the motor experiment.

describe the dynamics of the motor as

$$\ddot{\theta} + 2\zeta\omega\dot{\theta} + \mu\,\mathrm{sgn}(\dot{\theta}) + \omega^2\theta = T + \omega^2\,\frac{v}{K_P}\,\cos(\Omega t), \qquad (3)$$

where θ is the rotation of the motor shaft, ω is the natural frequency, ζ and μ represent viscous and Coulomb damping, respectively, T is the control input, v is the amplitude of the forcing voltage, Ω is the frequency of excitation, and K_P is a potentiometer constant.

The control law is established by introducing a controller and a control signal governed by

$$\ddot{u} + 2\zeta_c\omega_c\dot{u} + \omega_c^2 u = \alpha\,u\,(K_P\theta) \quad \text{and} \quad T = \frac{\gamma}{K_P}\,u^2. \qquad (4)$$

The experimental setup consists of the DC motor, a servo-amplifier, the electronic circuit, a signal generator, and two DC power supplies. A schematic is shown in Figure 2.

The motor position feedback gain was adjusted to obtain a free-oscillation frequency $\omega \approx 88.0\,\mathrm{rad/s}$, which is equivalent to approximately 14 Hz. The experiment involved forcing the motor at its resonance frequency, $\Omega \approx 14\,\mathrm{Hz}$. Initially the forcing amplitude is $v = 60\,\mathrm{mV}$. The controller is activated at $t \approx 35\,\mathrm{sec}$. As illustrated in Figure 3, the controller is very successful in reducing the motor's amplitude of oscillations. To further examine the performance of the controller, we increased the forcing amplitude in steps of 20 mV at $t \approx 75\,\mathrm{sec}$, $t \approx 97\,\mathrm{sec}$, and $t \approx 125\,\mathrm{sec}$ to reach a value of $v = 120\,\mathrm{mV}$. In all of these cases, the amplitude of the motor response remained almost constant while the controller response increased. At $t \approx 135\,\mathrm{sec}$, the controller was turned off for a period of 10 seconds allowing the resonant response to develop at $v = 120\,\mathrm{mV}$. Upon the reactivation of the controller, the motion of the motor was again quenched.

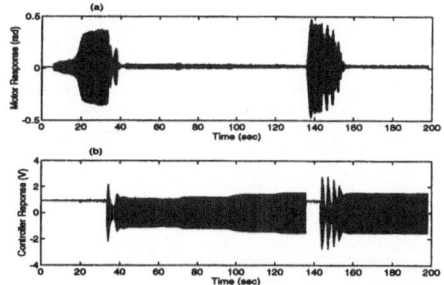

Figure 3. Saturation control of the motor when $K_F = K_C = 0.1$: (a) motor and (b) controller.

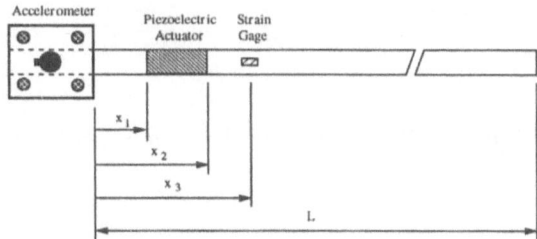

Figure 4. The location of the actuators and the strain gage.

4. Cantilever Beams

Here, the plant consists of a cantilever beam attached to a shaker. The actuation is provided by piezoceramic patches attached near the root of the beam, and the feedback signal is generated by a strain gage. The layout is illustrated in Figure 4.

The equation of motion of the beam is

$$\rho A \frac{\partial^2 w}{\partial t^2} + C \frac{\partial w}{\partial t} + EI \frac{\partial^4 w}{\partial x^4} = -\rho A \frac{d^2 w_0}{dt^2} + \frac{\partial^2 M}{\partial x^2}, \tag{5}$$

where $w(x,t)$ is the deflection of the beam, $w_0(t)$ is the shaker base motion, A is the crossectional area of the beam, ρ is the mass density, EI is the flexural rigidity, C is a damping constant, and $M(x,t)$ is the moment produced by the actuators. It is defined by (Fanson and Caughey, 1987)

$$M(x,t) = K_a V_a(t) [H(x - x_1) - H(x - x_2)], \tag{6}$$

where K_a is a constant, $V_a(t)$ is the control voltage, and $H(x)$ is the Heaviside step function. We introduce a control law of the form

$$\ddot{u} + 2\zeta_c \omega_c \dot{u}_c + \omega_c^2 u_c = \alpha u K_e e(x_3, t) \quad \text{and} \quad V_a(t) = \gamma u^2, \tag{7}$$

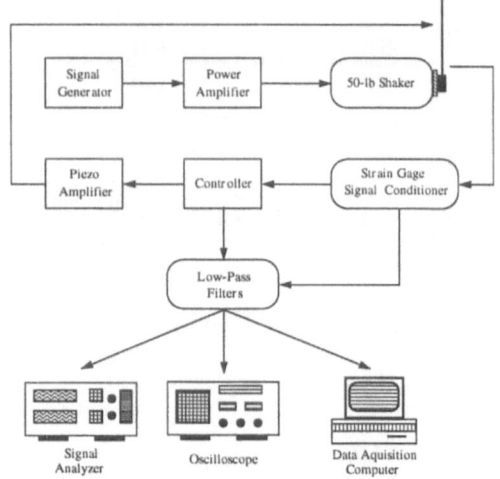

Figure 5. A schematic of the beam experiment.

TABLE 1. Beam and actuator properties

	Beam		Actuator	
Beam No.	1	2	1	2
Length, m	0.21	0.49	2.7×10^{-2}	3.56×10^{-2}
Width, m	1.27×10^{-2}	1.59×10^{-2}	1.07×10^{-2}	1.52×10^{-2}
Thickness, m	5.59×10^{-4}	8.38×10^{-4}	1.91×10^{-4}	
$\omega_1/2\pi$, Hz	11.45	2.98	–	–
$\omega_2/2\pi$, Hz	–	18.15	–	–

where K_e is a constant, and $e(x_3, t)$ is the strain measured at location x_3. The autoparametric resonance condition is achieved by letting $2\omega_c \approx \omega_i$, where ω_i is the frequency of the ith resonant mode.

Two steel beams were used to carry out the experiments. Their dimensions and the dimensions of the patches are shown in Table 1. The experimental setup is depicted in Figure 5. The setup consists of a 50-lb permanent magnet shaker and its driving power amplifier, a piezoamplifier, the controller circuit, a strain gage signal conditioner, an accelerometer, and a signal generator. The first experiment consisted of controlling the response of the first mode of beam No. 1. We subjected the beam to an initial base excitation of 5.8 mg and increased it in five steps to 8.9 mg. Figure 6 illustrates the response of the system when the controller is acti-

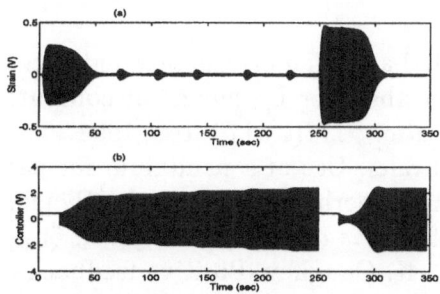

Figure 6. Saturation control of the first mode of beam No. 1 when $\Omega \approx 11.5\,\mathrm{Hz}$, $K_C = 10$, and $K_F = 1$: (a) strain voltage and (b) controller response.

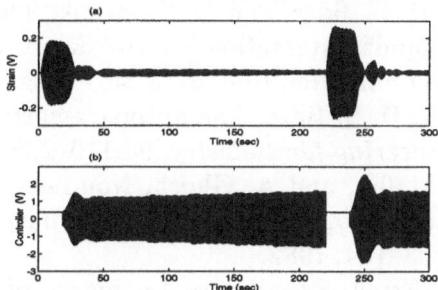

Figure 7. Saturation control of the second mode of beam No. 2 when $\Omega \approx 18.2\,\mathrm{Hz}$, $K_C = 20$, and $K_F = 1$: (a) strain voltage and (b) controller response.

vated at $t \approx 20\,\mathrm{sec}$. At $t \approx 250\,\mathrm{sec}$, the controller was deactivated allowing the response to fully develop at $8.9\,mg$. The controller was activated again at $t \approx 260\,\mathrm{sec}$. In the second experiment, the second mode of beam No. 2 was controlled. The procedure here was similar to the one in the first experiment. The initial base excitation was $15.3\,mg$. It was incremented to a final value of $25.2\,mg$. Figure 7 depicts the closed-loop response when the controller was turned on at $t \approx 20\,\mathrm{sec}$. The controller was turned off at $t \approx 225\,\mathrm{sec}$ and on again at $t \approx 240\,\mathrm{sec}$. It is clear that the strategy is very effective in suppressing the resonant responses of both beams.

Acknowledgement

This work was supported by the Army Research Office under Grant No. DAAH04-96-1-0045, the Air Force Office of Scientific Research under Grant No. F49620–95–1–0254 and the National Science Foundation under Grant No. CMS–9423774.

5. References

M. Cartmell and J. Lawson. Performance enhancement of an autoparametric vibration absorber by means of computer control. *Journal of Sound and Vibration*, 177(2):173–195, 1994.

A. Ertas and O. Cuvalci. Use of a pendulum for passive structural vibrations control: An experimental study. In *Dynamics and Vibrations of Time-Varying Systems and Structures*, Vol. 56, pp. 215–221, 1993.

J. L. Fanson and T. K. Caughey. Positive feedback control for large space structures. In *Proceedings of the 28th AIAA/ASME/ASC/AHS Structures, Structurals Dynamics, and Materials Conference*, pp. 588–598, 1987.

A. G. Haddow, A. D. S. Barr, and D. T. Mook. Theoretical and experimental study of modal interaction in a two degree-of-freedom structure. *Journal of Sound and Vibration*, 97(3):451–473, 1984.

R. S. Haxton and A. D. S. Barr. The autoparametric vibration absorber. *Journal of Engineering for Industry*, 94:119–225, 1972.

H. Hatwall, A. K. Mallik, and A. Ghosh. Non-linear vibrations of a harmonically excited autoparametric system. *Journal of Sound and Vibration*, 81(2):153–164, 1982.

G. Mustafa and A. Ertas. Experimental evidence of quasiperiodicity and its breakdown in the column-pendulum oscillator. *Journal of Dynamic Systems, Measurement and Control*, 117(2):218–225, 1995.

A. H. Nayfeh. On the undesirable roll characteristics of ships in regular seas. *Journal of Ship Research*, 32(2):92–100, 1988.

A. H. Nayfeh and B. Balachandran. Modal interactions in dynamical and structural systems. *Applied Mechanics Review*, 42(11):175–201, 1989.

A. H. Nayfeh, D. T. Mook, and L. R. Marshall. Nonlinear coupling of pitch and roll modes in ship motion. *Journal of Hydronautics*, 7:145–152, 1973.

A. H. Nayfeh and L. D. Zavodney. Experimental observation of amplitude- and phase-modulated responses of two internally coupled oscillators to a harmonic excitation. *Journal of Applied Mechanics*, 55:706–710, 1988.

S. S. Oueini and M. F. Golnaraghi. Experimental implementation of the internal resonance control strategy. *Journal of Sound and Vibration*, 191(3):377–396, 1996.

K. L. Tuer, M. F. Golnaraghi, and D. Wang. Development of a generalized active vibration suppression strategy for a cantilever beam using internal resonance. *Nonlinear Dynamics*, 5:131-151, 1994.

DYNAMICS OF THE OPPOSED PILE DRIVER

F. PETERKA, O. SZÖLLÖS
Institute of Thermomechanics, Academy of Sciences of the Czech Republic
Dolejškova 5, 182 00 Prague 8, Czech Republic

1. Introduction

Many mechanical systems are strongly nonlinear due to the existence of impacts between moving bodies or the action of dry friction forces. These phenomena cause sudden changes of velocities or acting forces during the system motion. Practical application of such systems is in the field of forming machines, mechanical hammers, power picks, impact dampers, gearings, clock mechanisms, relays, vibromotors, etc.

The behaviour of strongly nonlinear systems is characterized by a manifold response of the system on the periodic excitation, bifurcations, nonstabilities, chaotic motions, motion with dead zones (Peterka and Vacik 1992) etc. A theoretical solution of the problems mentioned can be found only for the simplest mechanical systems and for simple regimes of motion. The numerical simulation of the motion of nonlinear systems is a very effective method to solve the laws of motion of complex systems (Peterka 1995a).

Fig.1 shows a two-mass mechanical system which is a model of a forming machine. It has several advantages in comparison with the simple one-mass model, e.g.

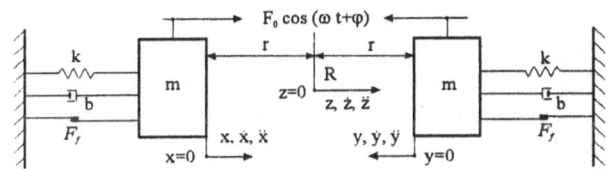

Figure 1. Scheme of the mechanical system.

the isolation of the machine neighbourhood from impact impulses, the lower weight of the machine and higher efficiency of the energy transfer from the excitation into the formed material. Each mass is excited by a force of period $T = 2\pi/\omega$ and is damped by viscous force and by Coulomb dry friction force. The impact of the masses with each other is governed by Newton's law of impact, with coefficient of restitution R; $R=0$ means that the impact is plastic, and that the bodies can move together for some time. Than there is a zone of zero relative motion.

The system motion is simulated by use of the Runge-Kutta integration method. The numerical simulation method is described in more detail in Peterka (1995b).

2. Mathematical model of the system

The impactless motion of the masses is described by the following differential equations (cf. Fig. 1)

271

D. H. van Campen (ed.), IUTAM Symposium on Interaction between Dynamics and Control in Advanced Mechanical Systems, 271–278.
© 1997 *Kluwer Academic Publishers*.

$$m\ddot{x} + b\dot{x} + F_f \text{sign}\dot{x} + kx = F_0\cos(\omega t + \varphi) , \quad m\ddot{y} + b\dot{y} + F_f\text{sign}\dot{y} + ky = F_0\cos(\omega t + \varphi) . \quad (1)$$

An impact of masses appears when the condition

$$x + y \geq 2r \qquad (2)$$

is met ($2r$ is the static clearance between the masses, see Fig.1). Then the velocities of the masses are suddenly changed according to the relations

$$\dot{x}_+ = [(1 - R)\dot{x}_- - (1 + R)\dot{y}_-]/2, \quad \dot{y}_+ = [(1 - R)\dot{y}_- - (1 + R)\dot{x}_-]/2, \qquad (3)$$

where $\dot{x}_+, \dot{y}_+$ are after-impact velocities,
           $\dot{x}_-, \dot{y}_-$ are before-impact velocities,
$0 \leq R \leq 1$ is the restitution coefficient. For plastic impact ($R=0$) the masses have equal after-impact velocities

$$\dot{x}_+ = -\dot{y}_+ = \dot{x}_p = (\dot{x}_- - \dot{y}_-)/2 \qquad (4)$$

and there are two possibilities of the after-impact motion, depending on the polarity of the relative acceleration $\ddot{x} + \ddot{y}$:
a) when the condition

$$\ddot{x} + \ddot{y} \leq 0 \qquad (5)$$

is met, then forces acting on the masses do not press the masses together and the motion is similar to that with elastic impact.
b) when the condition (5) is not met, the masses are pressed together and have to move together. Then the system motion is described by the equation

$$m\ddot{z} + b\dot{z} + F_f \text{ sign } \dot{z} + kz = 0 \qquad (6)$$

with initial conditions

$$z(0) = x(t_i) - r ; \quad \dot{z}(0) = \dot{x}_p, \qquad (7)$$

where t_i is the instant of the plastic impact. This means that there is an after-impact dead zone of relative motion of masses which continues until the instant t_e when the forces vanish, i.e. the condition (5) is met. Then the system motion is simulated again according to eqs. (1) with initial conditions

$$x(0) = z(t_e) + r; \quad y(0) = -z(t_e) + r; \dot{x}(0) = \dot{z}(t_e); \quad \dot{y}(0) = -\dot{z}(t_e). \qquad (8)$$

The mathematical model of the more complex system with impacts and dry friction force is explained in (Peterka 1995c).

In the following sections simulation results will be presented.

3. Regions of different types of system motion

Existence regions of different types of system motion were evaluated depending on the dimensionless static clearance $\bar{r} = rk/F_0$ and the angular excitation frequency $\eta = \omega/\sqrt{k/m}$. The hatched regions in Fig.2 correspond to periodic and chaotic impact motions. The simplest impact motion exists in the oblique hatched region labelled by $z=1$, where z is the quantity expressing the number of impacts per excitation period T. Time histories $x(t)$, $y(t)$ and phase trajectories (x,x'), (y,y') of the one-impact motion are shown in Fig. 3(a). The parameters $\eta, \bar{r}$ of the system correspond to point a in Fig.2. In

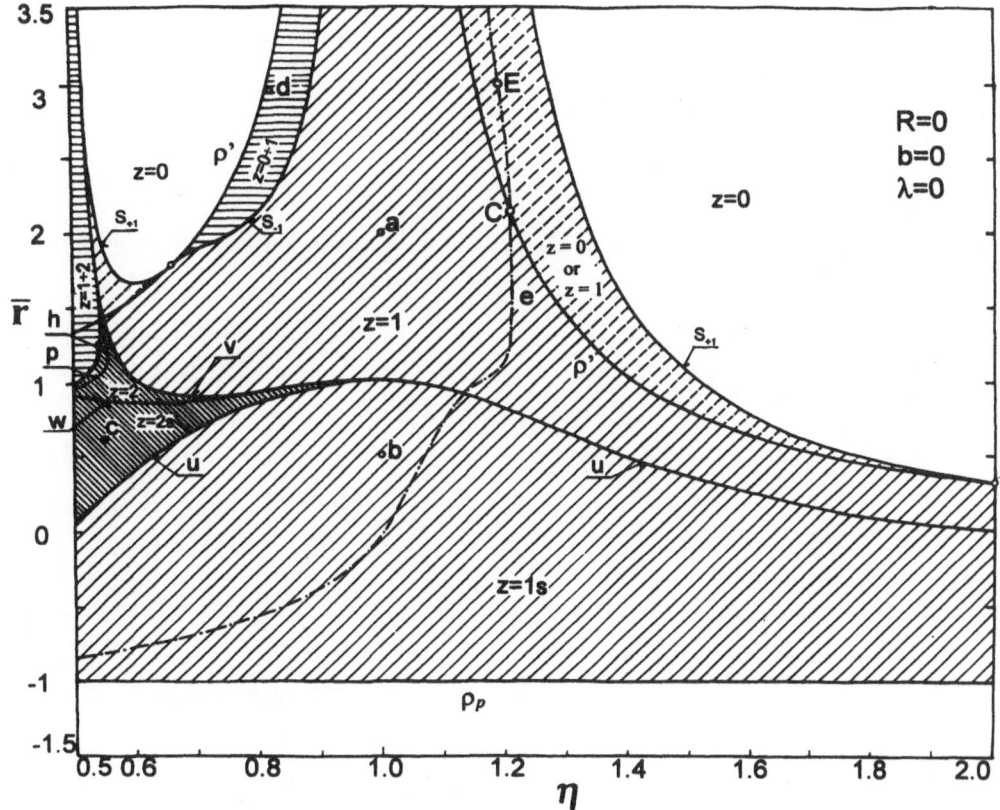

Figure 2. Regions of existence of the motion with plastic impacts

the region under the curve u, labelled $z=1s$, there is one-impact motion characterized by some time interval after the impact, during which the impacting masses are connected (see Fig.3(b) and point b in Fig.2). The length of the time interval approaches the period T, when the clearance $\bar{r}$ decreases to the value $\bar{r} = -1$ (see the boundary ρ_p in Fig.2). Under this boundary the system is motionless and the masses are continuously connected.

Impactless motion exists above the boundary ρ' (see the unhatched regions $z=0$ in Fig.2), where the amplitudes of the excited vibration are less than the static clearance between them. There are also regions of hysteresis between the boundaries ρ' and s_{+1} hatched by dashed lines in Fig.2 (one of them is labelled ($z=0$ or $z=1$)), where both impactless and one-impact motion can exist, depending on the initial conditions.

Impactless motion cannot exist and one-impact motion is unstable in the horizontally hatched region in Fig.2 (see the region between boundaries ρ', s_{-1} in Fig.2 which is labelled $z=0-1$), where different periodic subharmonic impact motions and chaotic motions (see Fig.3 (d)) exist; they are described in more detail in (Peterka 1995d).

274

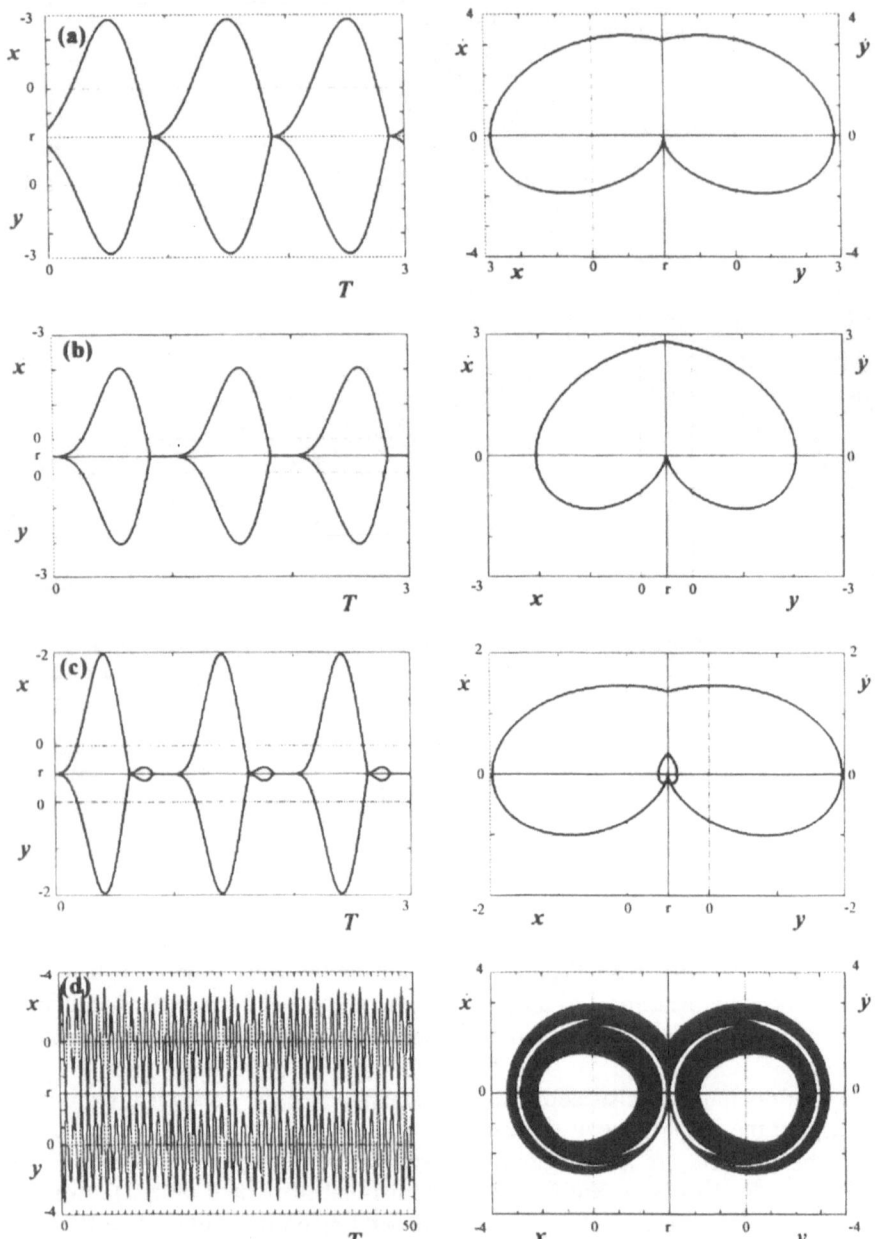

Figure 3. Time histories and phase trajectories of the system motion with plastic impacts.
(a) one-impact motion (z=1; η=1; r=2) without after-impact connection of masses,
(b) one-impact motion (z=1s; η=1; r=0.5) with after-impact connection of masses,
(c) two-impact motion (z=2s; η=0.55; r=0.6) with connection of masses after the second impact in period T.
(d) chaotic impact motion (z=0-1; η=0.817; r=3).

There are also more complex two-impact motions (see Fig. 3(c), the regions of which are labelled $z=2$ and $z=2s$ in Fig.2 and demarcated by the curves h, v, u. They are separated by the curve w, where the two-impact motion ($z=2$) changes to the two-impact motion ($z=2s$) with the connection of masses after the second impact in the period T (see Fig.3(c)).

4. Optimization of system parameters to obtain maximum before-impact velocities

When the system is proposed as a forming machine, it is advisable to choose the type of impact motion and optimal parameters of the system to obtain maximum impact impulses. One-impact motion is the most suitable regime; this exists and is stable for a large region of system parameters (see hatched regions $z=1$, $z=1s$ in Fig.2). Local maxima of the before-impact relative velocities $\dot{x}_- + \dot{y}_-$ of masses lie on the curve e in Fig. 2. The dependence of impact velocities on the dimensionless clearance $\bar{r}$ is shown in Fig.4. When the system motion is not damped by viscous or dry friction, the impact velocities grow with increasing clearance $\bar{r}$ (see curve (c) in Fig.4). Therefore it is desirable to construct the forming machine with minimum damping of the impactless motion. When the viscous damping b increases, the impact velocities decrease (curves (c_1) and (c_2)) and there are local maxima E_1, E_2 of the impact velocities for corresponding optimal clearances $\bar{r}_{1opt}$, $\bar{r}_{2opt}$.

It follows from Fig.2 that the extreme curve e goes through the hysteresis region ($z=0$ or $z=1$) when the clearance $\bar{r}$ grows from a certain value (see cross-section C in Fig.2). So the operating mode can lie in the region where also impactless motion exists.

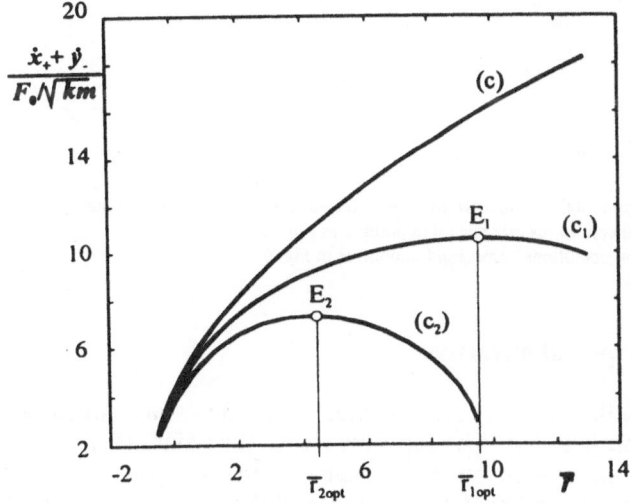

Figure 4. Before-impact velocities as the function of clearance.
(c) - without damping - $b' = 0$; (c_1) - $b' = 0.05$; (c_2) - $b'= 0.1$,
where $b' = b/\sqrt{km}$ is dimensionless damping

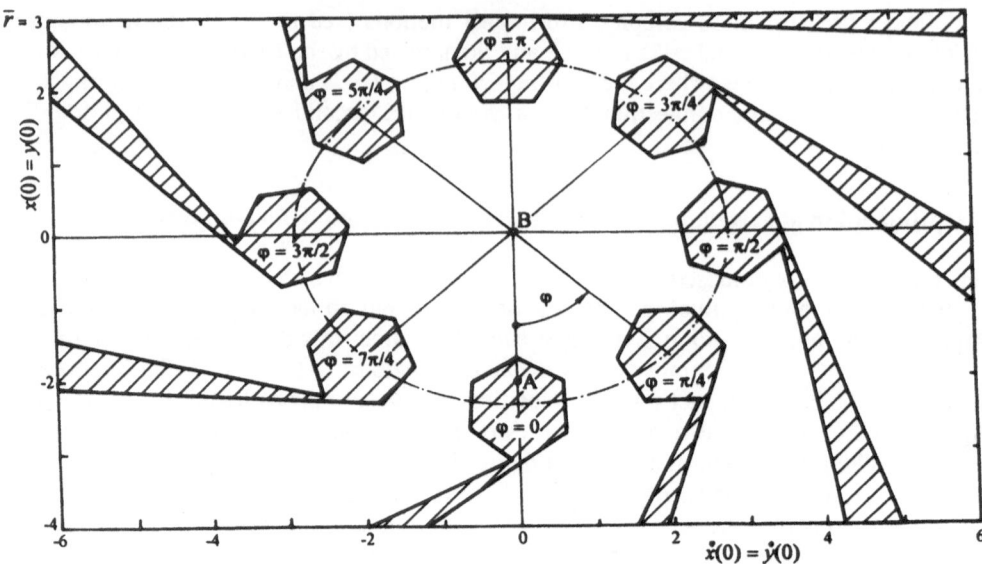

Figure 5. Zones of attraction of one-impact and impactless motion in the plane of symmetrical initial conditions

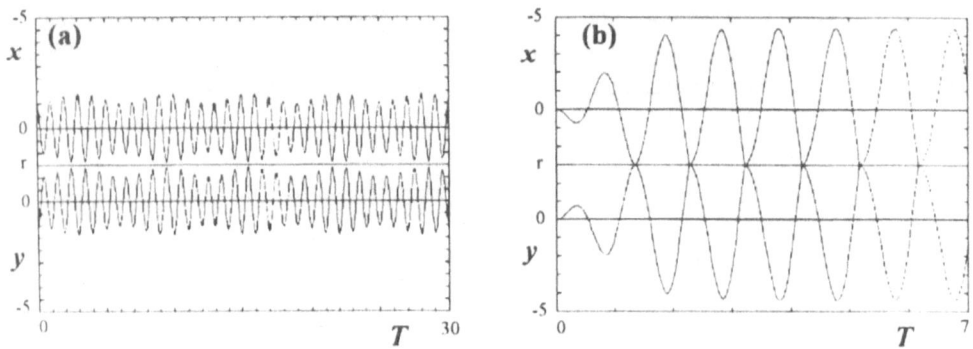

Figure 6. Transition effects of the system motion from certain inital conditions - the stabilization of the:
impactless motion (a) - initial conditions correspond to point A in Fig. 5,
one-impact motion (b) - initial conditions correspond to point B in Fig. 5.

5. Global stability analysis - zones of attraction

Analysis of the global stability of the system motion is necessary when the proper impact regime of the machine has to be set. This means that zones of attraction of both possible motions of the system must be evaluated in the space of initial conditions of motion. There are five initial (t=0) conditions of the system motion: $x(0)$. $y(0)$, $\dot{x}(0)$, $\dot{y}(0)$, φ, but the zones of attraction will be evaluated only in the plane $x(0)$. $\dot{x}(0)$, under the assumption of motion symmetry (i.e, $x(0)=y(0)$ and $\dot{x}(0) = \dot{y}(0)$), for different values of the excitation force phase shift φ. There is one hatched zone of attraction of impactless motion (z=0) for every value of the phase shift φ. Zones of attraction are shown in Fig. 5 for the values φ=0, $\pi/4$, $\pi/2$, ... 2π. When φ is changed, the hatched zones shift and turn around the centre of the hexagonal zone. The hexagon centres move on the dot-and-dashed ellipse. The hatched zones of attraction of impactless motion have both a hexagonal and a sector form, and they are of fractal character (Isomäki *et al.*, 1987).

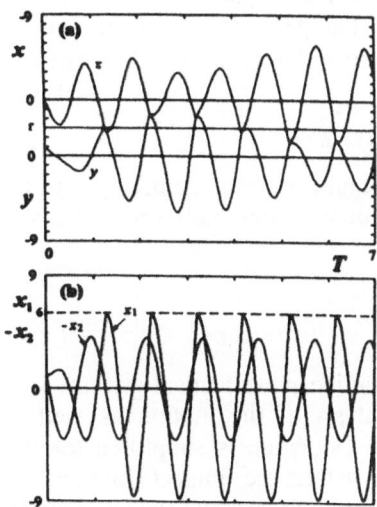

Figure 7. Transition effect from asymmetric initial conditions $x(0) = 0$; $x'(0) = 2$; $y(0) = 1$; $y'(0) = -2$ and $\varphi = 0$ for system parameters $R = 0$; $r = 3$; $\eta = 1.19$; $\lambda = 0$.

The stabilization of the impactless motion from initial conditions corresponding to the hexagon, is impactless (see Fig. 6(a) corresponding to point A in Fig.5); for the sector, it is accompanied by several impacts.

The periodic one-impact motion stabilizes from any point of the remaining unhatched zone (see Fig. 6(b) and point B in Fig.5). It follows from Fig.5, that the zone of attraction of one-impact motion prevails over the zone of attraction of impactless motion. So the one-impact motion can be set very easily. e.g. from the static positions (point B) of masses and arbitrary phase shift φ.

6. Behaviour of the system with asymmetric initial conditions

It was shown that for the system with zero value of viscous damping b and for asymmetric initial conditions, asymmetric oscillations of the system remain. This is a surprising fact, since losses of energy at impacts (especially for plastic impacts) represent a strong damping factor. Impacts have no influence on the symmetrization of motion. This phenomenon was explained by the following elementary analysis (Landa 1996) based on the transformation of the system equations (1) of motion into

anti-phase oscillation $x_1 = x+y$: $\quad \ddot{x}_1 + b\dot{x}_1 + x_1 = 2F_0 \cos(\omega t + \varphi),$ $\quad\quad$ (9)

and in-phase oscillation $x_2 = x-y$: $\quad \ddot{x}_2 + b\dot{x}_2 + x_2 = 0$, $\quad\quad$ (10)

where the dry friction forces are omitted. Similarly relations (3) can be transformed into the following formulas:

$$\dot{x}_+ + \dot{y}_+ = -R(\dot{x}_- + \dot{y}_-) \text{ , or } \dot{x}_{1+} = -R\dot{x}_{1-}. \quad\quad (11)$$

$$\dot{x}_+ - \dot{y}_+ = \dot{x}_- - \dot{y}_- \text{ , or } \dot{x}_{2+} = \dot{x}_{2-}. \qu\quad (12)$$

It is evident that the in-phase oscillation x_2 behaves as free oscillation of the linear impactless system (see Eq. (10)), which is damped only by viscous friction, i.e. this part of the motion causes the preservation of asymmetry of motion when the viscous damping does not act. It is documented on the time history of the variable $-x_2$ in Fig. 7(b).

The anti-phase oscillation x_1 corresponds to the motion of the usual one-degree-of-freedom impact oscillator, the mass of which impacts on the rigid stop with coefficient of restitution R (see Eqs. (9), (11)). The periodic one-impact motion stabilizes during several periods T (see time history of variable $\dot{x}_1$ in Fig.7(b)). The real variables x, y in Fig.7(a) are given by

$$x = (x_1 + x_2) / 2 ; \quad y = (x_1 - x_2) / 2. \tag{13}$$

Every real vibrating system contains some small viscous damping. So the symmetric one-impact motion always stabilizes even though some initial asymmetry is brought into the system motion.

7. Conclusion

This paper shows that optimal control of strongly nonlinear mechanical systems with impacts is possible, but very detailed stability analysis of the motion as well as optimization of system parameters are necessary. The situation is simplified for this special two-degree-of-freedom system, because its symmetric impact motion has dynamics identical to a one-degree-of-freedom system, the mass of which impacts against a rigid stop. The dynamics of this system was solved in more detail in Peterka (1992).

Acknowledgement

The authors gratefully acknowledge the financial support of the Grant Agency of the Czech Republic (No. 101/94/0126).

References

Isomäki. H.M., J. von Boehm, and Räty, R. (1987) Chaotic oscillations and the fractal basin boundaries of an impacting body, In *Proceedings of the ICNO IX.* Budapest.,656-659.
Landa. P. (1996) *Nonlinear Oscillations and Waves in Dynamical Systems,* Kluwer Academic Publishers, Dordrecht, The Netherlands.
Peterka, F. and Vacík J. (1992) Transition to chaotic motion in mechanical systems with impacts. *Journal of Sound and Vibration* **154**(1), 95-115.
Peterka. F. and Formánek P. (1994) Simulation of motion with strong nonlinearities, In *Proceedings CISS - First Joint Conference of International Simulation Societies,* 1994, ETH Zurich, Switzerland, August 22-25, 137-141.
Peterka, F. (1995a) Impact interaction of two heat exchanger tubes, In *Proceedings of the 6th International Conference on Flow-Induced Vibration,* London, Imperial College, April 10-12, Ed. P.W.Bearman. A.A. BALKEMA/ROTTERDAM/BROOKFIELD/1995, 393-400.
Peterka, F. (1995b) Simulation of two-mass-system motion with plastic impacts and dry friction, In *Proceedings ESM 1995 - Modelling and Simulation,* Technical University of Prague, June 5-7, 257-258.
Peterka, F. (1995c) Mathematical model of mechanical sytems with strongly nonlinear couplings, In *Proceedings of the Colloquium Dynamics of Machines '95.* Institute of Thermomechanics of the AS CR. Prague. February 1-2, 149-152, (in Czech).
Peterka, F. (1995d) Different ways from periodic to chaotic motion in mechanical vibro-impact systems. In *Proceedings Ninth World Congress on the Theory of Machines and Mechanisms,* Politecnico of Milano, Italy, August 30-September 2, 1995, Vol.2, 1030-1035.

THE ANALYSIS AND STABILITY OF PIECEWISE LINEAR DYNAMICAL SYSTEMS

N.B.O.L. PETTIT, P.E. WELLSTEAD
The Control Systems Centre
UMIST, P.O.Box 88
Manchester, M60 1QD, UK
E-mail: pettit@csc.umist.ac.uk

R. WILSON-JONES
Advanced Engineering
Centre
Lucas Automotive Ltd
Dog Kennel Lane, Shirley
B90 4JJ, UK

1. Introduction

Piecewise linear (PL) dynamical systems form an important class of engineering systems. This is because a wide range of system nonlinearities, particularly those found in electrical and mechanical systems, are well approximated using piecewise linear functions. Examples of such nonlinearities are saturation, dead-zones, hysteresis and backlash. More importantly a wide range of logic functions can be described by piecewise linear functions. Here rule-based controllers, gain scheduled controllers and programmable logic controllers are examples of PL type logic. Systems that contain the above nonlinearities and use logic in their control are widespread, particularly in advanced electrical and mechanical systems such as found in modern vehicles. The effects of the logic on the system dynamics are almost impossible to analyse and tend to be designed using engineering heuristics coupled with intensive and expensive simulation and prototyping.

This article will review an analysis method for piecewise linear systems developed by the authors. It will then use the example of an Anti-Lock Braking system to illustrate the application of the technique and in particular the ability of the technique to offer insight into the interaction of logic and dynamics. The final part discusses the difference between a prototype idea and the demands imposed to develop tool that can analyse systems of a more realistic complexity.

2. An Analysis Method for Piecewise Linear Systems

The ideas behind the analysis techniques are outlined here. They are described in detail in Pettit (1995) as well as in Pettit and Wellstead (1995b). As discussed in the

D. H. van Campen (ed.), IUTAM Symposium on Interaction between Dynamics and Control in Advanced Mechanical Systems, 279–286.
© 1997 *Kluwer Academic Publishers.*

280

introduction, many system nonlinearities are piecewise linear in nature. Figure 1 shows some typical nonlinearities.

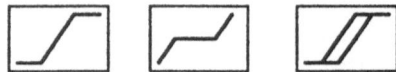

Figure 1. Typical Nonlinearities: Saturation, Dead-Zone and Hysteresis

Similarly many kinds of logic can be represented as piecewise linear functions within the system. Note the term function is used loosely here, as it is possible for the piecewise linear function to be a condition that must be satisfied by the input for some output to be valid. The effect of the output does not have to be related via the function to the input. In its simplest form, this type of PL system is described as a set of convex polytopes $P_i \in \mathfrak{R}^n$ each containing some linear system of the form:

$$\underline{\dot{x}} = \underline{A}_m \underline{x} + \underline{b}_m, \quad \underline{x} \in P_m \tag{1}$$

and P_i form a partition of $\mathfrak{R}^n$ s.t.:

$$\cup P_i = \mathfrak{R}^n, P_i \cap P_j = \varnothing, i \neq j \tag{2}$$

This has a strong geometric interpretation that can be thought of as 'boxes' stacked together in state space with each box containing a different linear dynamic system. Any global analysis must somehow identify the behaviours in each box and then link them together to form a global picture of the dynamics. Figure 2 illustrates this geometric interpretation.

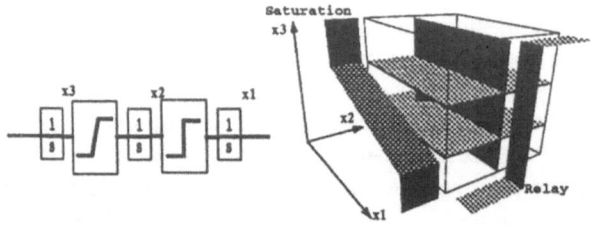

Figure 2. The Geometric Interpretation of a Piecewise Linear System

In figure 2 a block diagram shows a three state system with two PL functions; a saturation followed by a relay. In state space, the system will be in three dimensions. One axis will be split by two planes due to the two breakpoints that appear in the saturation. The other will be split by one plane due to the relay, since although the relay has two breakpoints, they occur at the same instance in the input. As a result, the state space will comprise of six linear regions. The region boundaries relate to the system nonlinearities. In general, a model can be constructed that ensures all regions will be convex (Pettit 1995).

The analysis concept is outlined in Wilson-Jones (1993). It is developed in Pettit (1995) with a good description being given in Pettit and Wellstead (1995b). Details of how an interface to a model can be constructed are described in Besson *et al.* (1994) and an interpreter for the results is described in Pettit *et al.* (1995).

2.1. THE ANALYSIS METHOD

The principal behind the analysis is summarised by figure 3. Given the system is modelled as piecewise linear a geometric model, as illustrated in figure 2, can be constructed (Figure 3(a)). Taking each region separately, the boundaries of each region can be classified according to where dynamics enter a region or leave that region. These partitioned sections of the boundary are mapped to nodes in a graph. For each region, the entrance partitions are projected through the region to find out which exit partitions they can reach. These then appear as connections between nodes for each region (figure 3(b)). All the region graphs are then merged to form a global graph (figure 3(c)). This graph can be analysed using graph theory for patterns that have physical meaning, such as cycles (figure 3(d)). As each node relates uniquely to a system nonlinearity or control logic, the effect of particular logic rules can be inspected and the resulting dynamics traced through the graph and hence the system model.

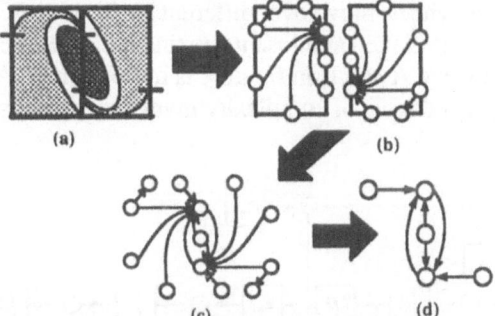

Figure 3. The Analysis Process: (a) The geometric model of the system in state-space, (b) The partitioning of each region and identification of nodes, (c) The creation of a graph capturing the global dynamics of the system, (d) The simplification of the graph.

3. The Anti-Lock Brake System

The aim of an anti-lock braking system is to improve the effectiveness of vehicle brakes by maintaining the tyre braking torque at or near its maximum value. Nowadays commercial ABS controllers are almost all rule-based and have as their control objective the limit cycling of tyre braking torque about its maximum value (Leiber and Czincel 1979). The basic idea can be explained with reference to figure 4 which shows a simplified form of a typical curve relating tyre torque and wheel spin.

The braking torque, or tyre adhesion, is at its highest between wheel spin values marked A and B on the figure. If the wheel spin increases beyond B, the wheel 'locks', tyre adhesion decreases and more importantly the driver loses the ability to steer the vehicle, i.e. the system is considered unstable. The objective of an ABS controller is to keep the wheel spin between A and B. A basic rule-based controller which will do this is:-

```
if wheel spin > B then brake pressure = 0
if wheel spin < A then brake pressure = maximum
```

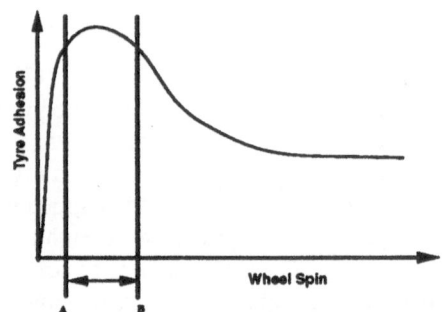

Figure 4. The Tyre Adhesion Curve used in ABS

In reality the tyre torque curve changes rapidly and significantly depending upon the conditions of the road, weather, vehicle and tyres. What's more, sudden 'discrete event' changes will occur if the wheels move over different surface types. For instance an icy road will cause the curve to flatten and lose its distinct peak altogether. This means that selecting thresholds like A and B in figure 4 is not possible. A practical ABS controller will have a high degree of complexity involving perhaps hundreds of rules (Hussain 1986).

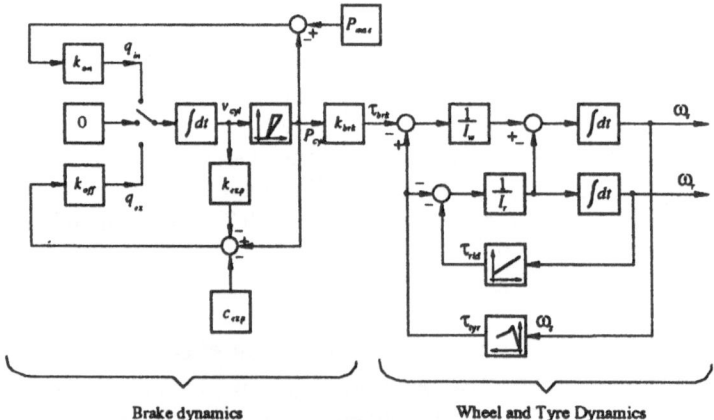

Figure 5. A Simplified Model of an ABS

In order to investigate the effectiveness of the analysis method, a compromise ABS model had to be used. This is given in Figure 5. The main tyre nonlinearity was kept

($\omega_s \rightarrow \tau_{tyr}$ in figure 5). A discrete actuator was modelled that allowed three modes of control: "Pump" (k_{on}) which switched on the brake at some designed rate of increase of brake pressure; "Hold" (0) which held whatever current brake pressure was being applied; and "Dump" (k_{off}) which released the brake pressure at some designed rate. Ten rules were developed, these rules having a total of nine variables that had to be chosen to optimise the brake performance. The design of the rules was done using model knowledge. The variable values were chosen through extensive simulation. This process reflects the current methods often used in rule-based controller design. A more detailed description of the modelling is found in Wilson-Jones (1993).

Simulation showed that the model appeared to perform well for a range of road surfaces (reflected in the model by changing the tyre torque curve). However, the analysis revealed a number of possible behaviour patterns that were unwanted. One example is given in figure 6. Figure 6 shows a detail of the analysis results for a low friction tyre torque curve. This is typical of icy conditions.

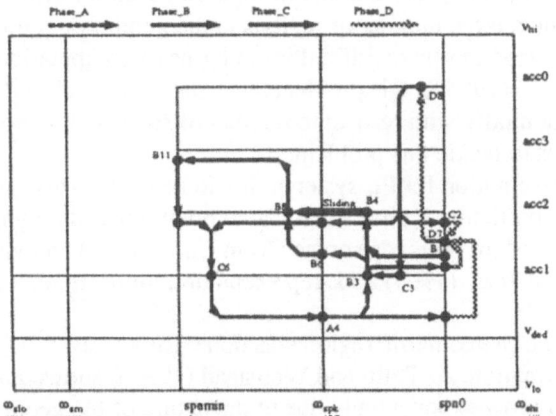

Figure 6. The Analysis Results for a Low friction Surface

Figure 6 revealed a few unexpected patterns. The most worrying is that the only method of reaching the 'dump' (Phase_C or B11→C6) action is for ω_s to increase until spnmin is reached. spnmin represents an emergency rule that should only be fired because for some reason the normal rules have missed a particular dynamic behaviour. For ω_s to reach spnmin means a long time is spent in the unstable part of the tyre torque curve (i.e. the tail of the curve shown in figure 4). This implies the wheel is in its locked position for a long time before unlocking. This is a clear failure in terms of acceptable control action in the ABS, yet one that was not obvious from simulations carried out.

The analysis allowed the problem to be pinpointed as an incorrect scheduling. The rules were redesigned and a final improved controller derived. Overall the analysis revealed around five major weaknesses in the original design (Wilson-Jones 1993).

These were mostly due to incorrect scheduling, but none were spotted during simulation. This was partly because the weaknesses mainly occurred for non-standard initial conditions which were not chosen during simulation. The ABS example showed both the problems with a simulation based design procedure and how the analysis method could add to the design process.

4. Analysing Realistic Systems

Having shown the power of the analysis method, the next phase has been to develop analysis tools that can look at a wide range of more realistic models. The two big problems are:

- In general, nonlinear models of systems will be of high order, i.e. they will contain above ten states. Models can be of several thousand states, but a range of around 10-100 states can be used to approximate a wide range of systems.
- The system can contain a large number of logic rules and/or system nonlinearities. This will generate a large number of regions in the geometric model.

Both these problems lead to severe difficulties in terms of computation time and power. In Pettit and Wellstead (1995a), this problems was shown to lead to algorithms that used memory exponentially with respect to number of states in the model. The paper suggested three steps to tackle this problem.

1. An implicit representation for PL systems should be used. This grows linearly with dimension and a particular region's associated boundaries and dynamics can be extracted when needed. This idea comes from Chua (1977) and van Bokhoven (1991). In Besson *et al.* (1994) a PL representation more suitable to our analysis was developed.
2. The construction of a geometric region was done using convex hull algorithms that used vertex information. As Pettit and Wellstead (1995a) showed, any vertex based algorithm would grow exponentially due to the nature of the geometric regions being manipulated. The solution suggested was to find algorithms that avoided using vertex information. These will be discussed in section 4.1.
3. The final step was to use graph theory to analyse the final system graph. Some of these issues were addressed in Pettit *et al.* (1995).

4.1. VERTEX VERSUS HYPERPLANES

Vertex based convex hull algorithms are detailed in literature (Swart 1986). However, the implicit model returns just the hyperplane constraints bounding a region. The question is, can the region be constructed without resort to using vertices? The answer is yes and algorithms to do this are available (Veres 1996). The proviso is that the adjacency relationships between the hyperplanes that bound the region must be known to identify how dynamics move in and out of the regions. Most algorithms are not adapted to do this, so simple algorithm was developed using linear programming.

The assumption is that a hypercube can be constructed that will contain within it the region to be generated and that all the adjacency relationships of the hypercube are known. This is straightforward and is given below.

$$Gx \leq C, \text{ where } G = \begin{bmatrix} I \\ -I \end{bmatrix}, C = \begin{bmatrix} \max x \\ -\min x \end{bmatrix}, G \in \mathfrak{R}^{2n \times n}, C \in \mathfrak{R}^{2n \times 1}, x \in \mathfrak{R}^{n \times 1}, I \in \mathfrak{R}^{n \times n} \tag{3}$$

I is the identity, n is the number of states and C is a vector of the maximum and minimum extremes for each system state. The adjacency relation between the hyperplanes of the hypercube are given by the matrix below.

$$al = \begin{bmatrix} 0 & 1 & \cdots & 1 \\ 1 & 0 & 1 & \vdots \\ \vdots & 1 & \ddots & 1 \\ 1 & \cdots & 1 & 0 \end{bmatrix}, adj = \begin{bmatrix} al & al \\ al & al \end{bmatrix} \tag{4}$$

where $al \in \mathfrak{R}^{n \times n}$. The row or column position in adj gives the row position for a hyperplane in $\{G,C\}$. That hyperplane's adjacency with others in the hypercube is given by "1" entries in adj. To build a region, the hyperplanes of the region, as found through the implicit representation, are added to the hypercube.

4.1.1. *The Algorithm*

Take one hyperplane $\{\gamma,c\}$ to be added to the hypercube. For each facet F of the hypercube solve:

$$\min_{x \in F} \{\gamma x\}$$

where F is a set of inequality constraints that bound the facet of the hypercube plus the equality constraint of the facet itself. This is solved via a linear program and then repeated to find $\min_{x \in F} \{-\gamma x\}$.

- If $(\gamma x_{\min} < c)$ and $(\gamma x_{\max} < c)$ then facet F not split by $\{\gamma,c\}$ and kept
- If $(\gamma x_{\min} > c)$ and $(\gamma x_{\max} > c)$ then facet F not split by $\{\gamma,c\}$ and rejected
- If $(\gamma x_{\min} > c)$ and $(\gamma x_{\max} < c)$ then facet F is split by $\{\gamma,c\}$
  - $\Rightarrow$ If a facet is split, for each inequality in turn belonging to F, set it to be an equality and repeat the linear program operation. Carry out the same test above. It will not be facets, but facet adjacencies that are tested for a split.

The adjacency matrix is updated as follows:
1. If all facets are not split by the hyperplane, adj is unchanged, otherwise add one more row and column of zeros to adj to represent $\{\gamma,c\}$.
2. If a facet is split, place a "1" in the row and column indicating an adjacency between the split facet and the new hyperplane. Then for every adjacency test that resulted in "not split but rejected", remove the associated "1" in adj.
3. If a facet is not split but kept, adj remains unchanged.
4. If a facet is not split and rejected, its associated row and column in adj is deleted.

This algorithm is repeated for each of the facets in the region. The result will be an adj matrix describing the region in terms of hyperplanes and their adjacencies only. Since adj is symmetric, computation can be saved by using a triangular matrix.

286

Comparison of algorithms The algorithm is compared with a convex hull algorithm from Veres (1996) that uses vertices. Figure 7 shows the results for splitting a hypercube varying from two to ten dimensions with one hyperplane.

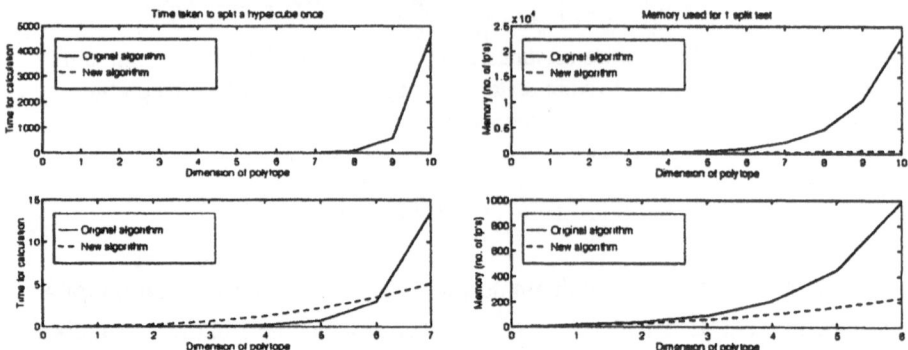

Figure 7. The Time (in seconds) and Memory Usage (in no. floating points stored)

5. Conclusions

An analysis method for nonlinear systems with logic has been outlined. It was then shown to benefit the design process of systems with logic through applying the ideas to a simplified model of an Anti-Lock Braking System. The significant benefit was in identifying logic rules that did not affect the dynamics as predicted by the designer. The problems in extending the tool to deal with a wider range of realistic systems was then discussed and a simple solution to one of these problems given.

6. References

Besson, V., Pettit, N.B.O.L. and Wellstead, P.E. (1994) Representing piecewise linear systems for analysis and simulation, *Proc. 3rd IEEE Conf. on Control App.*, Glasgow, UK, 1815 - 1820.

Chua, L.O. (1977) Section-wise Piecewise Linear Functions: Canonical Representation, Properties and Applications, *Proceedings of the IEEE*, 65, 915 - 929.

Hussain, S.F. (1986) Digital algorithm design for a wheel lock control system, *Proc. SAE*, 860509.

Leiber,H. & Czinczel,A. (1979) Antiskid system for passenger cars with a digital control unit, *Proc. SAE* 790458.

Pettit, N.B.O.L. (1995) *The Analysis of Piecewise Linear Dynamical Systems*, vol. 3 of UMIST Control Systems Centre Series, Research Studies Press, John Wiley and Sons, Taunton, UK.

Pettit, N.B.O.L. and Wellstead, P.E. (1995a) Designing a computation environment for the analysis of piecewise linear systems, *Proc. IFAC NOLCOS'95*, Tahoe City, CA, USA, 947 - 952.

Pettit, N.B.O.L. and Wellstead, P.E. (1995b) Analyzing piecewise linear dynamical systems, *IEEE Control Systems Magazine*, Oct., 43-50.

Pettit, N.B.O.L., Manavis, T. and Wellstead, P.E. (1995) Using graph theory to visualise piecewise linear systems, *Proc. of 3rd European Control Conference*, Rome, 1631-1636.

Swart, G. (1985) Finding the Convex Hull Facet by Facet, *Journal of Algorithms*, 6, 17 - 48.

van Bokhoven, W.M.G. (1981) *Piecewise Linear Modelling and Analysis*, Ph.D. thesis, Eindhoven University of Technology, Eindhoven, The Netherlands.

Veres, S.M. (1996) *Geometric Bounding Toolbox v 2.*, The School of Electronic and Electrical Engineering, The University of Birmingham, Edgbaston, UK.

Wilson-Jones, R. (1993), *A generalised phase portrait for piecewise linear systems*, PhD thesis, Control Systems Centre, UMIST, P.O.Box 88, Manchester, M60 1QD, UK.

Grasping Optimization and Control

Friedrich Pfeiffer

Lehrstuhl B für Mechanik, TU-München
Arcisstr. 21, D-80290 München, Germany

Abstract

Grasping, regrasping are difficult operations requiring optimal coordination and control of the fingers. Paper gives a concept and applies it to a four-fingered hand. All fingers are equal and driven by hydraulic actuators. Comparison of theory and measurements are convincing.

INTRODUCTION

Grasping may be looked at as a process of multiple robots, the fingers, being in contact with some object. Therefore, a description of grasping must include the organization of multiple fingers and in addition the contact phenomena. As grasping by an artificial hand is rather slow we shall neglect in this first approach the dynamical aspects and focus on an optimization of grasping strategies and on the control of a hand with four fingers being modeled kinematically and quasi-statically only.

The first step consists in an optimization of the grasp strategy. From trials with five grasp criteria the best one is evaluated. Best performance is achieved by a minimization of the finger force differences with the additional constraints that force and torque equilibrium is maintained, that contact remains established and that the finger forces are within the friction cone. Starting with this basic optimization problem various additional constraints are included: stability of grasping, relative distances between the fingers, sliding of fingers and changing a finger's contact position. The last operation is the most difficult one including some more constraints which express the necessities that the new contact point can be reached, that the fingers cannot penetrate the object and that no finger has a collision with another finger.

In a second step and on the basis of above results another idea is realized which we call hand planning. It optimises the clearance of motion of each finger and the complete finger arrangement, and it regards additional constraints like finger positioning at the object, penetration aspects, the best finger arrangement and the best orientation and location of the grasping plane. With the tools of the two first steps we are able to establish in a third step a typical manipulation planning, grasp planning and hand planning.

All methods are verified experimentally using a hand with hydraulically driven fingers. This fingers have good positioning accuracy and very sensible force control. Maximum speed is about 0.5 sec for a closing/opening process. The size is near a man's finger size. A kind of damping control has been realized based on a oil model, which works without problems.

D. H. van Campen (ed.), IUTAM Symposium on Interaction between Dynamics and Control in Advanced Mechanical Systems, 287–302.
© 1997 *Kluwer Academic Publishers.*

The first famous artificial hands have been developed in USA and Japan. The UTAH/MIT-Hand [1], the Stanford/JPL-Hand [6] and the WASEDA-Hand are all based on tension-cable-drive-systems, which assure good positioning accuracies and fast motion but not so good force control. In addition cable hands are difficult to design. Up to now direct drives are not small enough with respect to power efficiency, therefore another solution might be a pneumatically or hydraulically driven hand, where hydraulics possesses the advantage of a better density ratio [3]. In the following we shall consider a hydraulic solution.

The hand hardware is one side, the hand software the other one. Grasping, regrasping and manipulation with several fingers require straight and definite strategies which include all physical and geometrical conditions usually connected with processes of that kind. Equilibrium, contact with impacts and friction, questions of reachability, penetration, collision avoidance are some of the essential aspects. In recent years worldwide research focussed on some of these aspects but a comprehensive solution is still missing and, as a matter of fact, still far away of the perfect behavior of the human hand. Strategies of the kind must not only calculate the finger forces necessary to manipulate the object [5], but also locate the fingers on the object in such a way that a stable grasp can be achieved [4]. With a few exceptions [2], the work on grasp planning has focused on one aspect or the other. In this paper, a grasp strategy is demonstrated which accomplishes both tasks. Given the desired external forces on the object and the object geometry, the strategy calculates the grasp points and the finger forces necessary to achieve the desired external wrench on the object.

GRASP STRATEGIES

Finger forces have been decomposed in a first step into components which are normal and tangential to the plane of contact. This deviates from the decomposition into manipulation and internal forces [8], but is more convenient for mechanical reasons. According to Fig. 1 we then write

$$f_{n_i} = f_{n_i} n_i, \quad f_{t_i} = f_{t1_i} e_{t1_i} + f_{t2_i} e_{t2_i}, \quad f_i = f_{n_i} + f_{t_i} = f_{n_i} n_i + f_{t1_i} e_{t1_i} + f_{t2_i} e_{t2_i} \tag{1}$$

The second problem involves an optimization criterion for an evaluation of the finger forces. Five criteria have been investigated [7]: minimum dependence on the friction coefficients, minimum tangential finger forces, minimal sum of all finger force magnitudes, minimum of the maximal finger force, minimum difference of the finger force magnitudes. It turns out that the last criterion gives a best approach for a good distribution of the forces over all fingers. Therefore, for all further considerations finger forces are optimally selected according to the criterion

$$G = \sum_{i=1}^{n} \sum_{\substack{j=1 \\ (j \neq 1)}}^{n} \left(|f_i|^2 - |f_j|^2 \right)^2 \implies \text{min!} \tag{2}$$

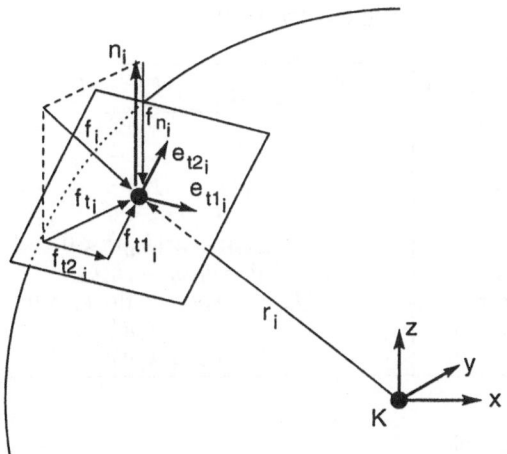

Figure 1: Decomposition of Finger Forces.

Three different optimization processes are considered, normal grasping with stability margins and sufficient finger distances, grasping with controlled sliding and grasping with regrasping. The corresponding optimization processes together with the additional constraints are the following:

- Normal Grasping

Optimization Criterion	$G = \sum_{i=1}^{n} \sum_{j=1(j \neq 1)}^{n} \left(|\boldsymbol{f}_i|^2 - |\boldsymbol{f}_j|^2 \right)^2 \longrightarrow \min$
Necessary Conditions	
Force Equilibrium	$\sum_{i=1}^{n} \left(\boldsymbol{f}_{n_i} + \boldsymbol{f}_{t_i} \right) - \boldsymbol{F}_e = 0$
Moment Equilibrium	$\sum_{i=1}^{n} \tilde{r}_i \left(\boldsymbol{f}_{n_i} + \boldsymbol{f}_{t_i} \right) - \boldsymbol{M}_e = 0$
Contact	$\boldsymbol{f}_{n_i} \cdot \boldsymbol{n}_i < 0$
Friction Cone	$|\boldsymbol{f}_{t_i}|^2 - \mu^2 |\boldsymbol{f}_{n_i}|^2 < 0$
Stability	$|\sum_{i=n}^{n} \boldsymbol{n}_i| \leq S$
Separation	$|\boldsymbol{r}_i - \boldsymbol{r}_j| - \epsilon_{\min} \geq 0 \qquad i \neq j$

- Grasping with Controlled Sliding (see Figure 2)

Optimization Criterion	$G = \sum_{i=1}^{n} \sum_{j=1(j\neq 1)}^{n} \left(|\boldsymbol{f}_i|^2 - |\boldsymbol{f}_j|^2 \right)^2 \longrightarrow \min$
Necessary Conditions	
Force Equilibrium	$\sum_{i=1}^{n} \left(\boldsymbol{f}_{n_i} + \boldsymbol{f}_{t_i} \right) - \boldsymbol{F}_e = 0$
Moment Equilibrium	$\sum_{i=1}^{n} \tilde{r}_i \left(\boldsymbol{f}_{n_i} + \boldsymbol{f}_{t_i} \right) - \boldsymbol{M}_e = 0$
Contact	$\boldsymbol{f}_{n_i} \cdot \boldsymbol{n}_i < 0$
Friction Cone	$|\boldsymbol{f}_{t_i}|^2 - \mu^2 |\boldsymbol{f}_{n_i}|^2 < 0$
Sliding Direction	$d = d_{t1} e_{t1} + d_{t2} e_{t2}$
Sliding Forces	$f_{nr} = -k_r/\mu \quad \text{with} \quad k_r \geq 0$
	$f_{t1_r} = k_r d_{t1}$
	$f_{t2_r} = k_r d_{t2}$

- Grasping with Regrasping

Optimization Criterion	$G = \sum_{i=1}^{n} \sum_{j=1(j\neq 1)}^{n} \left(|\boldsymbol{f}_i|^2 - |\boldsymbol{f}_j|^2 \right)^2 \longrightarrow \min$
Necessary Conditions	
Force Equilibrium	$\sum_{i=1}^{n} \left(\boldsymbol{f}_{n_i} + \boldsymbol{f}_{t_i} \right) - \boldsymbol{F}_e = 0$
Moment Equilibrium	$\sum_{i=1}^{n} \tilde{r}_i \left(\boldsymbol{f}_{n_i} + \boldsymbol{f}_{t_i} \right) - \boldsymbol{M}_e = 0$
Contact	$\boldsymbol{f}_{n_i} \cdot \boldsymbol{n}_i < 0$
Friction Cone	$|\boldsymbol{f}_{t_i}|^2 - \mu^2 |\boldsymbol{f}_{n_i}|^2 < 0$
Regrasping	• Reachability
	• No Penetration
	• No Collision

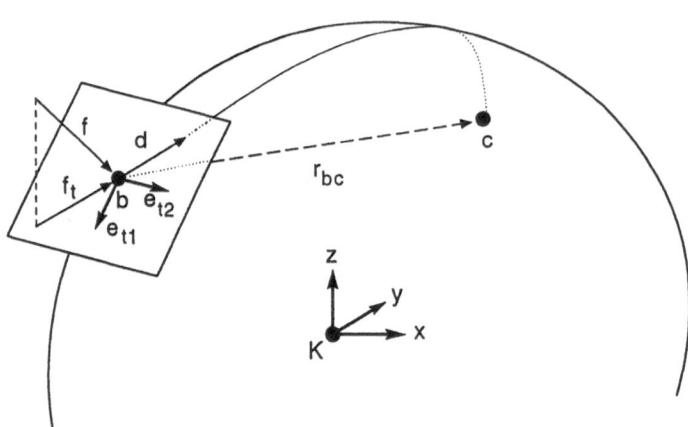

Figure 2: Grasping with Sliding from b to c

The meaning of the various conditions is evident. Neglecting inertia forces the finger forces and the external forces due to gravity must be in static equilibrium. The same is true for the torques ($\bar{a}b = a \times b$ definition of cross product). The contact condition says that the finger forces normal to the contact plane must be negative to assure always pressure forces only. Furtheron the finger forces must be within the friction cone to avoid uncontrolled sliding.

The normal vectors to the object's surface at the grasp points provide a good insight into the stability of the grip: the smaller the sum of the vectors, the more stable the grasp. The grasp is less stable in the direction opposite the resulting sum, which means that it is less capable of resisting disturbances in that direction. This stability writes

$$|\sum_{i=1}^{n} n_i| \leq S ,$$
(3)

where S is the desired stability measure.

The separation condition guarantees that a minimum separation is maintained between the grasp points, so that the fingers do not come too close to one another. For grasping with controlled sliding the sliding direction is given by a direct connection to the target point (point c in Fig. 2). The sliding forces follow the geometry and are controlled by a constant magnitude $k_r \geq 0$.

For regrasping questions of reachability, penetration and collision become important. Normal grasping and grasping with sliding can be performed with three fingers, for regrasping we need at least four fingers. Given the object and the geometry of the fingers we decide geometrically with the help of the fingers' workspaces what points can be reached without violating stability. Furtheron, with known finger geometry we also can evaluate the two problems of penetration and collision. Corresponding formulas and methods are described in [7].

In order to automate the grasping process, a strategy which can orient and locate the hand in such a manner that all fingers can reach their designated grasp points is needed. The object has six degrees of freedom relative to the hand which have to be limited in such a way that the grasp points are reachable. To solve these problems of hand placement a method has been developed which includes several steps: the definition of the grasp-triangle, a rough hand orientation, the finger assignment, and, finally, an optimization of the hand orientation and distance to the object.

Before evaluating these data the following geometric quantities must be known:

- Hand Geometry
  (position and orientation of the fingers on the palm described in hand frames)
- Workspace
  (position and orientation of the robot base described in a robot coordinate frame)
- Path planning
  (position and orientation of the object in a tool frame)
- Grasp Points
  (position of the i-th grasp point in a body-fixed object frame)
- Hand Orientation
  (position and orientation of the robot hand)

With these data known one must check in a first step by applying inverse finger kinematics if the grasp point can be reached without penetrating the object. In a second step position and orientation of the hand are calculated by arranging the palm surface parallel to the grasp triangle and the palm center over the grasp center. Then in a third step the orientation and the distance of the hand are optimized by maximizing the remaining workspace of the fingers.

The last step consists in a planning procedure for a manipulation process which includes all sequences of path planning, grasp planning and hand planning. Figure 3 indicates the corresponding strategy [7].

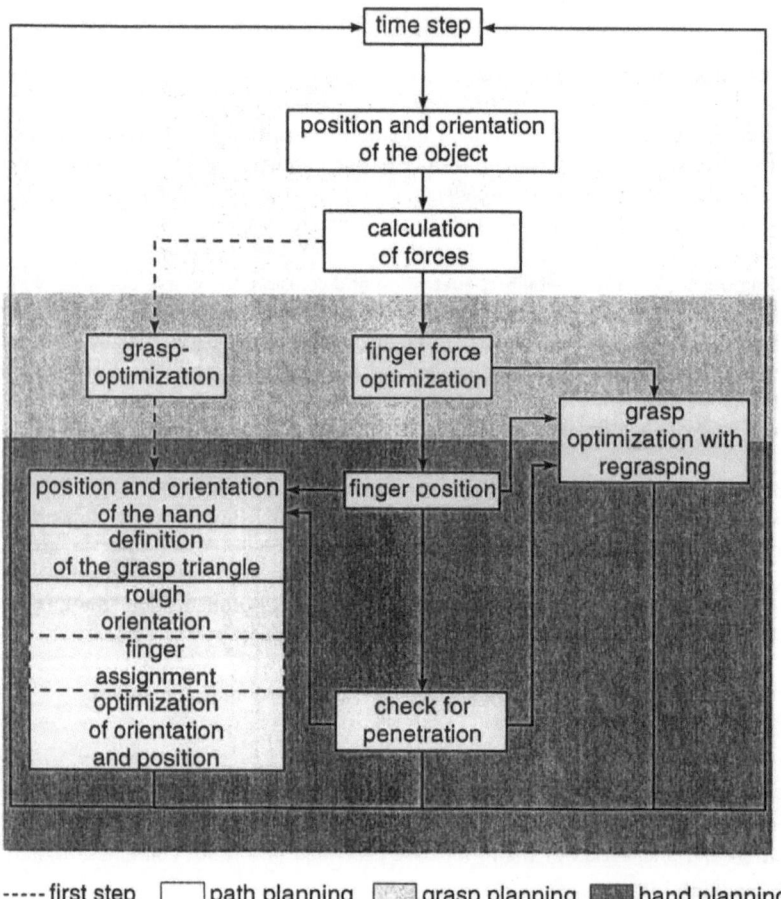

Figure 3: Manipulation Planning

THE TUM-HYDRAULIC HAND

The Design

When starting the development of an artificial hand at the author's institute the following design requirements were established [3]: Size about the human hand, three to four equal fingers which can be exchanged easily, three degrees of freedom per finger, maximum manipulation weight at least 10 N and minimum about 1 N, individual finger force 30 N, one complete grasping motion (open-closed-open) in 0.5 s, sensors to evaluate the fingertip forces with respect to amount, direction and location. A trade-off study with various drive systems (pneumatic, hydraulic, electric, cables) results in a solution with hydraulic drives. They allow excellent force control in a wide range of force magnitudes, on the other hand they have some disadvantages like leakage and difficult calibration. Figure 4 gives an impression of a four-finger arrangement, and Figure 5 shows one finger in more detail [3,7]. The fingers are fixed to the palm by two screws only which allows a quick change of the finger-palm-combination.

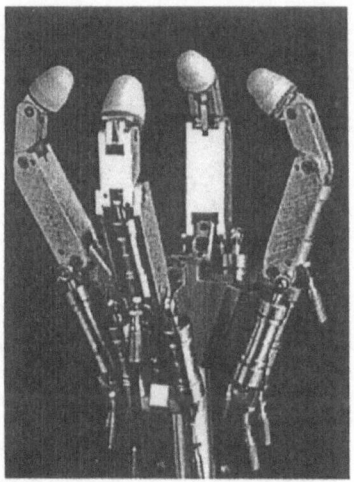

Figure 4: The TUM-Hydraulic Hand

All fingers are equal, and each one possesses three degrees of freedom, one combined degree of freedom for the first two finger joints and additional two degrees of freedom at the finger's root. From this we have realized two DOF in the finger plane and one DOF to allow a motion of the finger plane itself (Fig. 5).

The fingers are driven by hydraulic cylinders which operate in one direction by oil pressure and in the opposite direction by a prestressed spring. The tip and middle links are connected by a simple mechanism combining them to one DOF. The basic joint is driven by two cylinders which can generate two DOF. Altogether this results in three degrees of freedom $\varphi_1, \varphi_2, \varphi_3$. The finger arrangement of Figure 5 has a size like a middle finger of a human hand.

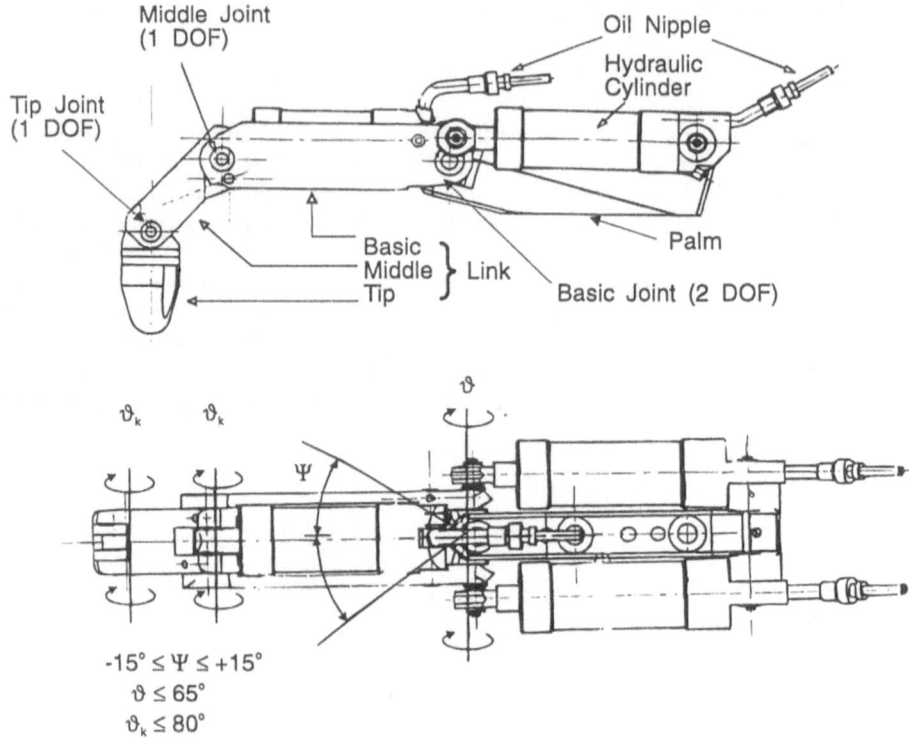

Figure 5: Design of the Hydraulic Finger [3]

Measurement and Control

Measurement and control of the hydraulic finger is realized in the following way, which again represents the outcome of an investigation concerning a large variety of possible solutions.

The piston is driven by oil pressure on one side and by a prestressed spring on the opposite side (Fig. 6). The oil is moved through a 4 m long elastic tube from the hydraulic power station to the piston. The hydraulic power station consists of a motor-gear-combination which drives a gear rack with a piston. This piston moves the oil within a cylinder and from there to the elastic oil tube.

Two measurements are installed. Firstly, an odometer measures the location of the gear rack and with it of the oil piston, which gives an information about the position of the oil column in the cylinder-tube-cylinder combination. Secondly, a pressure sensor measures the oil pressure at the exit of the driving cylinder to the tube. Direct measurements at the finger cylinders are not implemented due to the requirement of having only one connection for each finger cylinder to the ground supported power station.

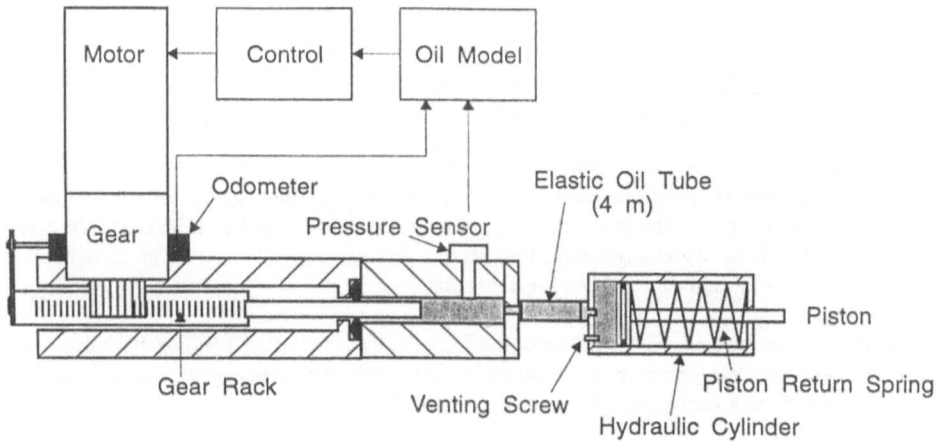

Figure 6: The Hydraulic Finger Control [3]

With these two measurements the motor in Figure 6 cannot be controlled. We need in addition an oil model which takes into account all pressure losses and friction forces from the power station to the finger cylinders. Such a model is used as indicated in Figure 6, therefore it should be as simple as possible. Figure 7 depicts the principal modeling which represents a typical situation for cyclic motion.

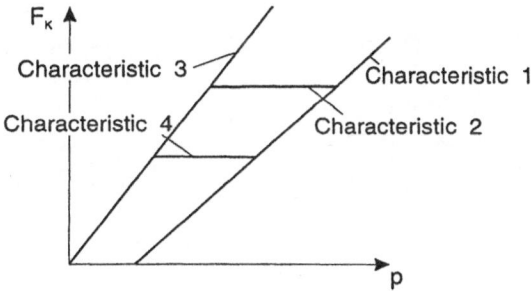

Figure 7: Oil Model

Increasing the pressure by moving the gear rack we walk along characteristic 1. When the pressure time derivative $\dot{p}$ changes sign then the finger piston sticks and its position x_F and its piston force F_K remain constant (characteristic 2). This state is maintained until all external forces like oil pressure force, piston force, spring force are large enough to overcome the stiction state and then to drive the finger piston in the opposite direction. The pressure decreases along the characteristic 3. The piston again sticks when $\dot{p}$ will change sign and x_F, F_K will be constant along characteristic 4. The two characteristics 1, 3 follow the simple equations

$$
\begin{aligned}
F_K &= k_1 x_A + k_2 p + F_r sgn(\dot{x}_F), \\
x_F &= k_3 x_A + k_{4_{1,3}} p, \qquad \text{with } F_r = F_{r_0} + c_r p
\end{aligned}
\tag{4}
$$

where the coefficients are partly determined by experiments [3]. The sign of $\dot{x}_F$ is given with the angular speed of the motor. The four switching points in Fig. 7 can also be evaluated by considering sign $(\dot{x}_F)$. If the velocity $\dot{x}_F$ changes sign, the pressure derivative $\dot{p}$ will change sign as well, at least for the relative slow motion as considered in this case.

For a verification of this oil model we press the finger piston against a bending bar with a strain gauge arrangement. We compare these measurements with the forces recalculated from the oil model. Figure 8 gives a comparison for position x_F and force F_K.

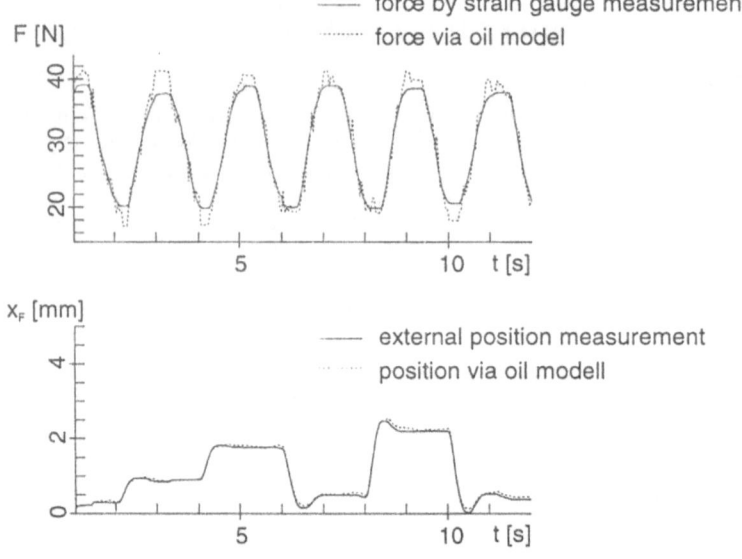

Figure 8: Verification of the Oil Model

The advantages of the solution are obvious. The basic drive is the configuration of Fig. 6, which is the same for all fingers. Each finger possesses three hydraulic drives of that type, and each hand might have any number of equal fingers. The number of connections of the fingers and the ground station is minimized, and all drives are rather simple. Nevertheless any complicated grasping program might be executed by these fingers [3,7].

To execute a complete grasping program we need a supervisory control of each finger cooperating together and performing the grasping sequences, and we need a planning process for manipulating an object with the fingers. Without going into details [3,7], we present two schemes. The first one of Fig. 9 illustrates the hardware of the TUM-hand. All four fingers and all drives of the fingers are connected by a VME-Bus-System which combines a SUN-workstation, a 486 CPU-PC-computer and several AD- and DA-converters. The converters receive the measurement signals and send signals to the finger drives. This set-up allows control of the complete hand.

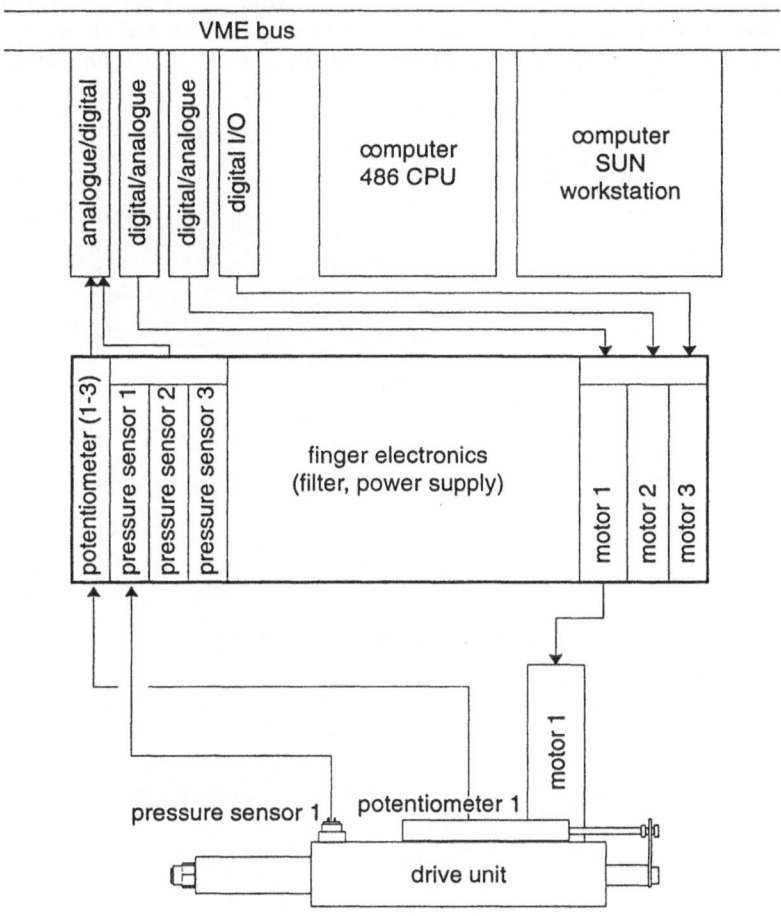

Figure 9: Hardware Scheme of the TUM-Hydraulic Hand

EXAMPLES

Simulation Results

On the basis of the optimizations in the grasping chapter and of the planning procedures (Fig. 3) several simulations have been performed to show the efficiency of the methods in grasping and regrasping [7]. As one typical example we show here the rotation of a sphere by regrasping with a four-fingered hand. A typical grasp pattern as developed in [7] is given with Fig. 10, which is self-explaining. The sequence of finger positions in performing this task is illustrated by the pictures of Fig. 11. We see that the above discussed optimizations generate meaningful sequences of finger operations.

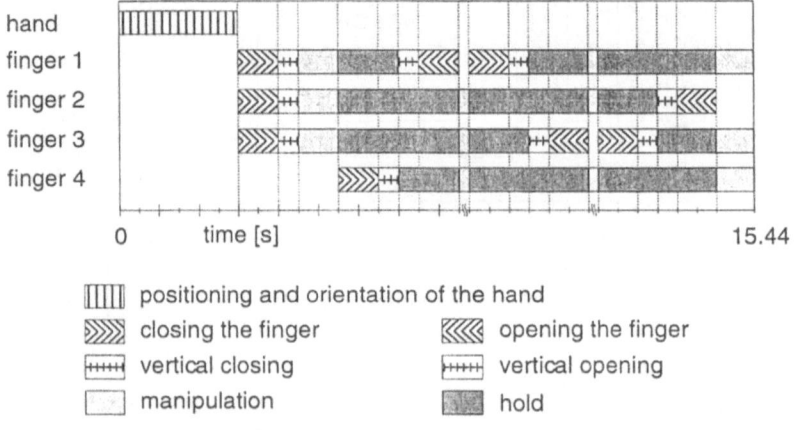

Figure 10: Grasping Pattern [7]

299

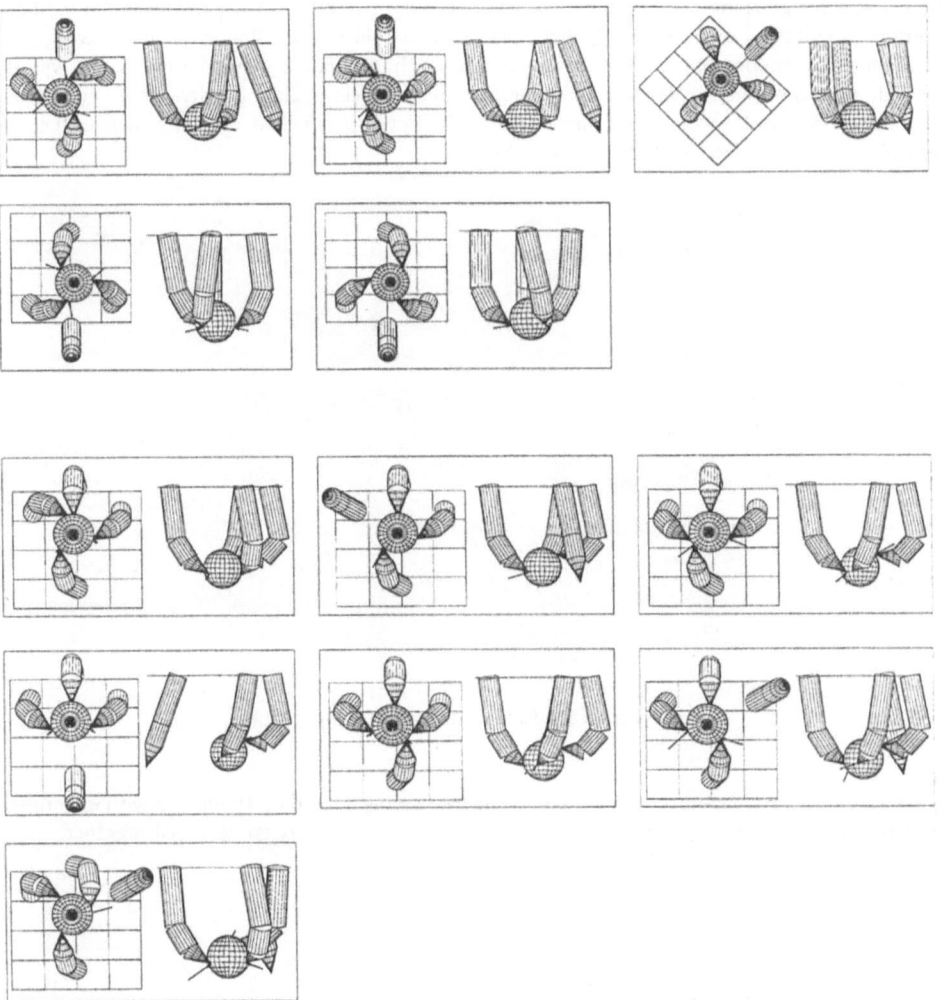

Figure 11: Rotating a Sphere by a Four-Fingerd Hand [7]

The theories for grasping and for the hand, the finger design and the hand-hardware are verified by experiments, rotation of an ellipsoid, regrasping of a cuboid and manipulation of a raw egg. The last mentioned experiment also has been presented at the Hannover Industrial Fair 1994. We show here only the regrasping experiment for a cuboid which is held against gravity. Its weight amounts to 195 g, its size is 15 × 25 × 40 mm. The static friction coefficient amounts to $\mu = 0.4$. Two regrasping steps are performed (see pattern, Fig. 12).

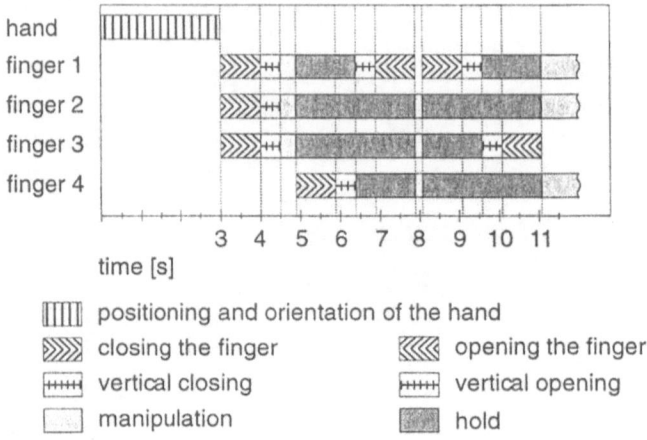

Figure 12: Grasp Pattern for Regrasping a Cuboid

Figure 13 depicts a comparison of theoretical finger planning according to Fig. 12 with experimental measurements of the finger angles (Fig. 6). Theory and experiments go very well together.

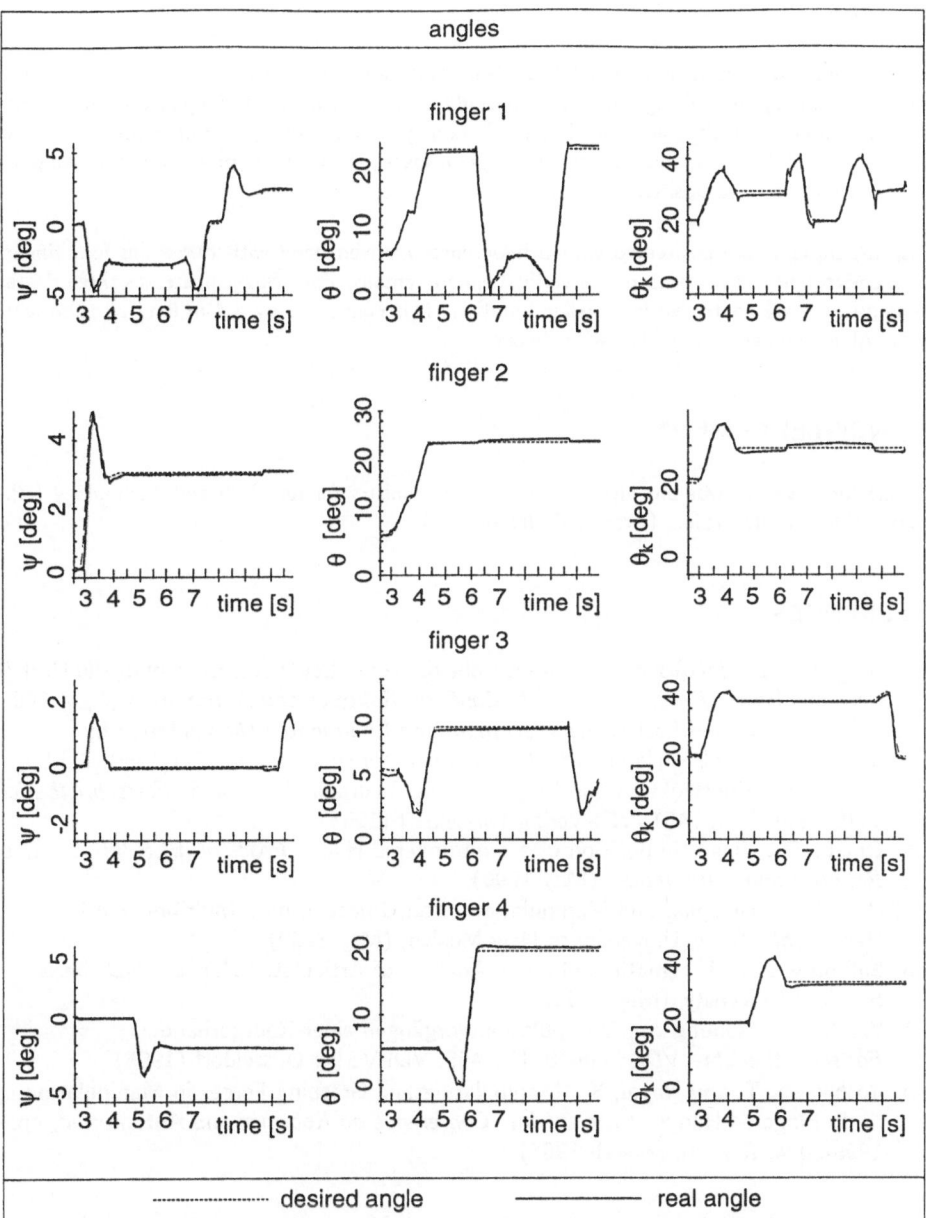

Figure 13: Theory and Measurements for Regrasping a Cuboid,
Angular Motion of the Fingers

SUMMARY

Strategies for cooperating fingers of an artificial hand are considered. Finger force adaptation is carried through by minimizing the finger force differences between the fingers and by taking into account certain constraints like equilibrium, friction, contact, stability, sliding motion, reachability, penetration, collision. A manipulation planning considers path planning of an object, grasp planning of the fingers and hand positioning planning.

Grasp experiments are performed with a hydraulically driven hand with three and four fingers for which design and control concept are verified experimentally. Each finger possesses three degrees of freedom which are controlled by a ground station. It is connected by a 4 m long oil tube with the finger. All experiments agree well with theory.

ACKNOWLEDGEMENT

Funding for this work was provided by the German Ministry for Research and Technology (BMFT) through the SEKON Project (Grant : 01 IN 104 D/7).

REFERENCES

1. Biggers, K.B., Jacobsen, S.C. and Gerpheide, G.E.: Low Level Control of the Utah/MIT Dextrous Hand. *Proc. of IEEE Int. Conf. on Robotics and Automation* (April 1987).
2. Li, Z., Hsu, P. and Sastry, S.S.: Grasping and Coordinated Manipulation by a Multifingered Robot Hand. *Int. Journal of Robotics Research*, 8(4)(1989), 33-50.
3. Menzel, R.: Konstruktion und Regelung einer hydraulischen Hand. *Fortschrittberichte VDI*, Reihe 8, Nr. 451, VDI-Verlag Düsseldorf (1995)
4. Omata, T.: Fingertip position of a multifingered Hand. *Proc. of IEEE Int. Conf. on Robotics and Automation*, (May 1990).
5. Park, Y.C.: Grasping and Manipulation of an Object using a Multifingered Robot Hand. *PhD thesis*, University of New Mexico, (May 1990).
6. Salisbury, J.K.: Kinematic and Force Analysis of Articulated Hands. *PhD thesis*, Stanford University, (May 1982).
7. Woelfl, K.: Planung von Manipulationsvorgängen einer Roboterhand. *Fortschrittberichte VDI*, Reihe 8, Nr. 455, VDI-Verlag Düsseldorf (1995).
8. Yoshikawa, T. and Nagai, K.: Manipulating and Grasping Forces in Manipulation by Multi-Fingered Hands. *Proc. of Int. Conference on Robotics and Automation*, pp. 1998-2004, Raleigh, (March 1987).

EXPERIMENTAL INVESTIGATION OF RANDOM VIBRATION CONTROL THROUGH DRY FRICTION

A. PIRROTTA*   AND   R. A. IBRAHIM**

* Dipartimento di Ingegneria Strutturale e geotecnica
Universita' Di Palermo, Viale delle Scienze 90128, ITALY

** Wayne State University, Department of Mechanical Engineering,
Detroit, MI 48202.

ABSTRACT

The purpose of this experimental investigation is to measure the response statistics in the presence of base friction and other friction sources. The experimental model emulates a one-floor building supported on four leaf springs, subjected to band limited random excitation. Two different types of model base are considered, a friction base and a frictionless base. In both cases friction can also be applied at two sides of the model's main mass against the direction of its motion. Excitation and response transducer signals are processed to estimate excitation and response statistics in the presence and in the absence of top mass friction. Measured statistics include mean squares, autocorrelation functions, power spectra, and probability density functions. The dependence of the mean square response on the excitation level in the presence of friction reveals linear and nonlinear regimes, as well as a drift in the response due to dry friction. Above a certain excitation level the response-excitation relationship displays nonlinearity. A transition from narrow band to wide band response spectra is observed when friction is applied to the system's top mass. The results of this investigation are of interest to civil engineers involved in the structural safety of buildings situated in regions susceptible to earthquakes (Feng, et al, 1993).

1. Experimental Model and Data Acquisition

Figure 1 shows a schematic diagram of the test rig which supports the model and its friction elements. The model consists of a block of mass 1.08 kg. The block is carried on four leaf springs made of vanadium tool steel, each 12 cm long and 2.5 x 0.89-cm in cross-section. The other ends of the leaf springs are clamped to a base block which can move in one direction. Two different types of base contact are used. The first arrangement (Figure 1a) includes a set of four wheels on which the model is carried. The second (Figure 1b) is a set of guided Thomson bearings which have less friction and are constrained to move in one direction. One end of the base block is

D. H. van Campen (ed.), IUTAM Symposium on Interaction between Dynamics and Control in Advanced Mechanical Systems, 303–312.
© 1997 Kluwer Academic Publishers.

linked to the armature of a VTS-600 electromagnetic shaker through a rigid steel bar. The shaker excites the base by a random acceleration $\ddot{U}(t)$. The shaker has a rated output thrust of 2669 N and 0.0254-m peak-to-peak displacement.

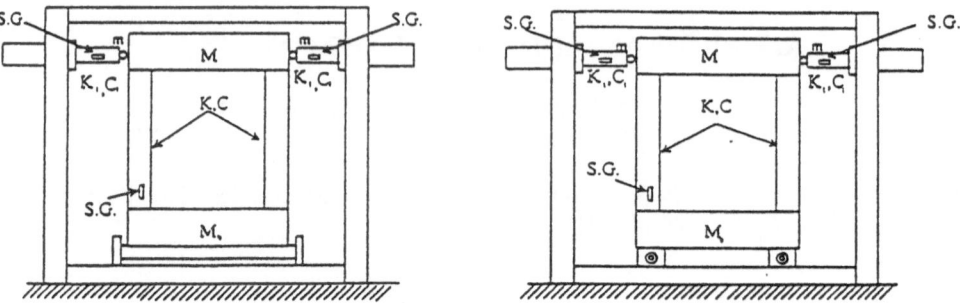

Figure 1a. Model with base wheels   *Figure 1b.* Model with Thomson bearing base

Figure 1. Schematic diagrams of test models (S.G. = strain gage).

Two non-rotating micrometers are installed in the rigid vertical walls of the rig frame facing the two sides of the top mass. The free end of each micrometer is attached to a short leaf spring 3.81 cm long and 1.28 x 0.89 cm in cross-section whose free end holds a friction element. The friction elements are made of a hard steel ball of 0.64 cm diameter. The free end of each ball can slide on each side of the main block along the plane of its motion. An initial normal force can be exerted at the point of contact by rotating the handle of the micrometer without rotating the ball and its attachment. Friction is created at both sides of main mass and is measured using strain gages.

A Data Translation model DT-2828 analog/digital converter board installed in an expansion slot of a 4DX2-66V Gateway 2000 personal computer is used to digitize transducer signals. The signals received by the board are in the form of voltages, and are converted first into binary and then digital numbers. The digitized signals are processed for estimating statistical parameters by using Interactive Laboratory System (ILS) and Data Translation Visual Engineering Environment (DTVEE). The board has four channels with a simultaneous sampling rate of 100,000 samples per second. For this experimental investigation the DTVEE software is used for free vibration tests, and the ILS program is used for forced vibration tests.

2. Test Procedure and Results

The natural frequencies and damping ratios of the model and friction ball elements without contact are measured by performing free vibration tests. The model natural frequency is found to be 9.5 Hz (in the absence of friction elements) while the natural frequencies of the friction elements are 148.4 Hz and 152.3 Hz for the right and left elements, respectively. The average value of the model damping ratio is $\zeta=0.0027$, while the damping ratios of the right and left friction elements are $\zeta_r=0.01$ and $\zeta_\ell=0.007$, respectively. Free vibration tests are also conducted on the model with friction elements in contact with the model mass. The natural frequency of the model is jumped to 16-Hz in the presence of friction under the same normal forces used in forced vibration tests.

The excitation is selected to be a band-limited random process whose center frequency is close to the main system natural frequency with a bandwidth of 3 Hz. The excitation level is decided after several trial tests which define the minimum possible and the safe maximum levels. A series of tests under different excitation levels is conducted to examine the dependence of response statistics on excitation mean square level. The following statistical parameters are estimated from the time history records of each random excitation test in the presence and absence of friction.

2.1 MEAN SQUARES VERSUS TIME

For relatively low excitation levels, the response is characterized by a narrow band random process whose mean square exhibits stationarity after a reasonable transient time. However, for higher excitation levels, the response experiences non-stationarity. The presence of the base friction is found to cause nonstationarity in the response mean square.

When friction is applied to the model and the same tests are repeated, the response mean square is found to be stationary under all excitation levels, associated with a significant drop in the response signal due to friction energy dissipation. Furthermore, the response bandwidth is greater than the corresponding cases without friction. The friction mean squares of both elements are not identical under all excitation levels. This is mainly due to the fact that the two elements do not have exactly the same dynamic characteristics (different natural frequencies and different damping ratios). Another source of friction asymmetry is due to the inevitable lateral motion of the model during large response oscillations; this causes unequal normal loading on each ball. The corresponding time history records with the Thomson bearing base reveal that the base friction with wheels contributes a drift (non-zero mean) to the response time history record.

2.2 AUTOCORRELATION FUNCTIONS

In both base settings, the response is essentially narrow band. There is a significant difference between the autocorrelations of the system response in the absence and in the presence of friction. In the absence of friction, the correlation period is much longer than the corresponding one with friction. The main reason for this is that the response of the system in the presence of friction is coupled with the friction elements, and the friction force is reflected back to the system motion. Furthermore, the system dynamic characteristics, such as natural frequency and damping ratio, are affected by the friction elements.

2.3 POWER SPECTRA

In the absence of friction elements, the response is essentially narrow band, centered at the system natural frequency. However, the response peak with Thomson bearings occurs at a higher frequency (10 Hz) than the peak with wheel bearings (9 Hz). This means that the base friction reduces the system natural frequency.

The power spectra of excitation, system response and friction element signals for wheel base and Thomson bearing base are shown in Figures 2a and 2b, respectively. The power spectra of both response and friction elements are essentially wide band limited and centered at a frequency different from the natural frequency of the system. The reason for this transition to wide band response may be due to abnormal stops (multiple stops per cycle) as reported by Makris and Constaninou (1991) and the friction random variation (Ibrahim, 1994). The excitation spectra always have zero mean-value at zero frequency, while the response and friction spectra have non-zero spectra at zero frequency. It is also seen that the bandwidth of the model response is shifted to the right with a substantial drop in value from the corresponding case without friction. Furthermore, the model response and friction elements have some degree of interaction, as observed by a dip in the spectra of the friction elements at a frequency corresponding to the peak value of the response spectra. The two base settings reflect some differences in the spectra. In the setting with wheels, the response spectrum peak is shifted to the right from the excitation center frequency, while with the Thomson bearing setting, the response peak spectra are shifted to the left. There is also a significant drop in the response peak spectra in the presence of friction as compared to the response spectra in the absence of friction elements.

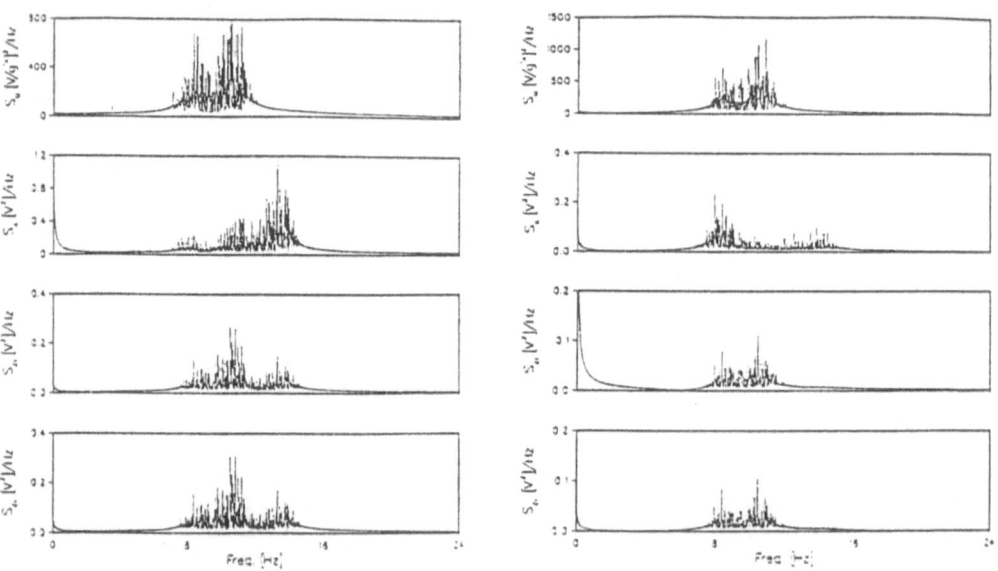

(a) For base wheel mode   (b) For Thomson bearing base model
Figure 2. Power spectra of excitation, response and friction elements

2.4 PROBABILITY DENSITY FUNCTIONS

In an ideal situation, an excitation with Gaussian distribution will result in a normal distributed response, if the system is linear with constant coefficients. The deviation from normality in the response distribution may exist if the system is nonlinear and/or involves dry friction. This deviation can be a function of system parameters, degree of nonlinearity, and excitation level. In order to allow comparison with a normal distribution, the Gaussian distribution is overlaid on each graph by a dotted curve based on the mean and standard deviation of each process. In the absence of friction for both base settings, the probability density functions of the system response are slightly deviated from normality under lower excitation level, and deviate significantly from normality as the excitation level increases. This observed deviation may be due to inherent system nonlinearity and base friction. For this particular system, the main sources of nonlinearity are inertia and curvature nonlinearities. Inertia nonlinearity arises mainly from the vertical drop of the top mass due to the four beams' bending deflection. It is also found that the excitation is not exactly normal, although the input signal to the shaker is Gaussian. It is believed that the deviation from Gaussianity is mainly due interaction of the shaker armature with model dynamics.

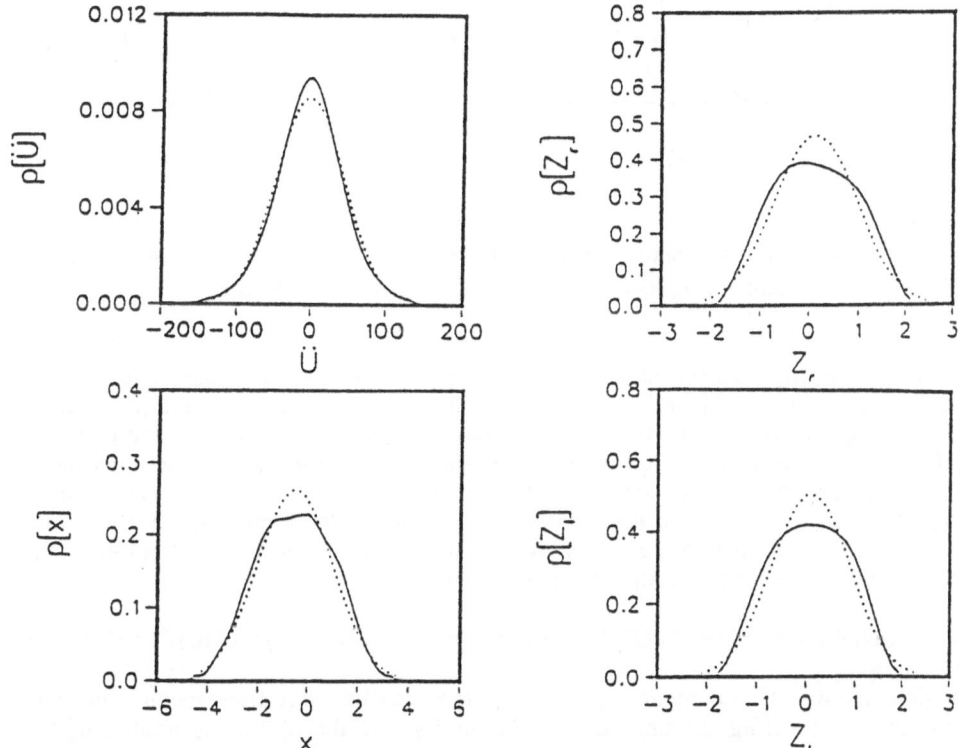

Figure 3a Probability density functions of excitation $\ddot{U}$, response X, and friction elements Z_r and Z_l for the base wheel model

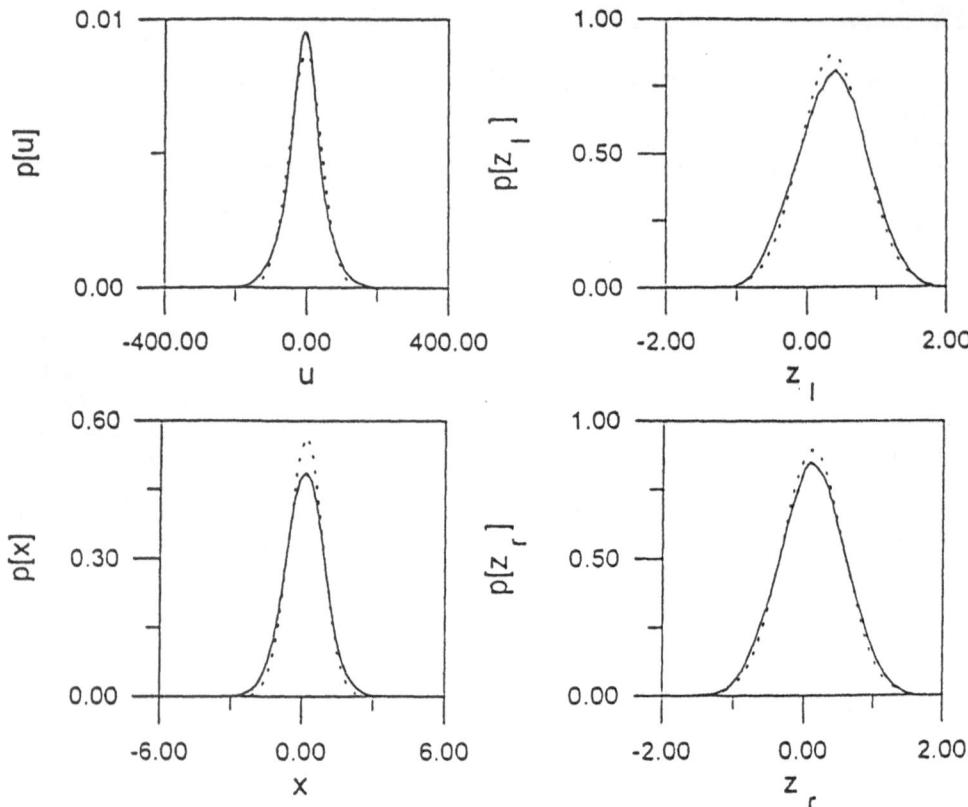

Figure 3b Probability density functions of excitation Ü , response X, and friction
elements Z_r and Z_l for Thomson bearing base model

The probability density plots for the excitation, model response, and friction
elements are shown in Figures 3a and 3b for the two base settings in the presence of
friction, respectively. It is seen that the base friction of the setting with wheels
results in non-Gaussianity in the response, as shown in Figure 3a. On the other hand,
when Thomson bearings are used the system response exhibits slight deviation from
normality. As the excitation level increases, the deviation from normality becomes
more obvious. Here the friction adds another source of nonlinearity. In addition, the
friction probability density is strongly non-Gaussian.

2.5 DEPENDENCE OF RESPONSE MEAN SQUARE ON EXCITATION LEVEL

The dependence of the mean square response on the excitation mean square is
important in revealing the degree of nonlinearity. If the system is nonlinear, the

relationship between response and excitation becomes nonlinear. By measuring the mean square responses versus the excitation mean squares for different excitation levels, one can establish excitation level ranges for linear and nonlinear response regimes. In the absence of friction elements, Figures 4a and 4b show the dependence of mean square response on the excitation mean square for the two base settings, respectively. The tests were conducted twice, and each figure shows two curves. The solid curve belongs to a series of tests which start with a relatively higher excitation level, while the dotted curve is obtained when the excitation level is increasing. With wheel base setting, Figure 4a, the relationship is almost linear up to excitation level $E[\ddot{U}^2]=1,200$, above which it is nonlinear due the inherent system nonlinearity and base friction. However, this relationship experiences an off-set due to the base friction. With the Thomson bearings, the base friction is almost eliminated, and the linear relationship emanates from the origin and remains linear up to excitation level $E[\ddot{U}^2]=800$. One may also observe a lack of repeatability in Figure 4a due to the base friction, while it is almost preserved with Thomson bearings, as shown in Figure 4b.

In the presence of friction elements, Figures 5a and 5b show the same type of tests for the two settings. By comparing Figures 4a and 5a and Figures 4b and 5b, one may observe a significant drop in the response mean squares due to friction elements, by a factor exceeding 100 under the same excitation levels. On the other hand drift response can be inferred by extrapolating the data which intersect the vertical axis at RMS close to 1.1 for the case of the wheel base where the base friction is significant. The dotted curve is much lower than the solid curve, which implies that the friction force associated with the dotted curve is much higher than the one for the solid curve, mainly due to the enlargement of the contact area resulting from surface wear. This drift is almost negligible for the case of Thomson bearings, where base friction is almost negligible.

3. Conclusions

Experimental measurements of response statistics of two different single-degree-of-freedom oscillators with different base isolations are obtained under random base acceleration. Both base arrangements were tested in the presence and in the absence of friction applied at the oscillator mass. Statistical measurements included mean squares, autocorrelation functions, power spectra, and probability density functions. In the absence of friction elements, the response was found to be narrow band, with a spike located at the system's natural frequency. On the other hand, in the presence of friction elements, the response spectra are essentially wide band random processes. The dependence of the response mean squares on the excitation level exhibits linearity for low excitation levels and becomes nonlinear for relatively higher excitation levels. In the presence of friction, the relationship does not pass through the origin, revealing a drift caused by friction.

The measured response statistical parameters can be used to identify the friction law curve and system inherent nonlinearity. It is clear that the measured response statistics exhibit nonlinearity. In this case one should include nonlinear inertia due

310

to axial drop of the four supporting beams of the main mass, and the curvature nonlinearity of the leaf springs. The friction elements of this model form another degree of freedom which interacts with the model where the contact friction establishes the coupling of the two systems. The analytical modeling and response, analysis together with Monte Carlo simulation, are currently under investigation by the authors.

Acknowledgment: This research was supported by NSF grant INT-9311774. The first author would like to acknowledge the Italian scholarship awarded to her during the period of her stay at Wayne State University.

4. References

Feng, M. Q., Shinozuka, M. and Fujii (1993) Friction Controllable Sliding Isolation System, *ASCE Journal of Engineering Mechanics* 119(9), 1845-1964.

Ibrahim, R. A. (1994) Friction-Induced Vibration, Chatter, Squeal, and Chaos, Part: Mechanics of Contact and Friction, and Part II: Dynamics and Modeling, *ASME Applied Mechanics Reviews* 47(7), 209-226& 227-243.

Makris, N. and Constantinou, M. C. (1991) Analysis of Motion Resisted by Friction. I: Constant Coulomb and Linear Friction, *Mech. Struct. & Mech.* 19(4), 477-500.

311

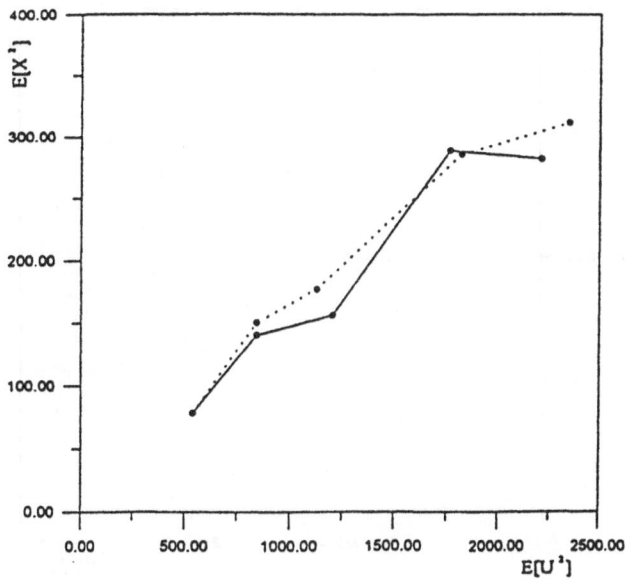

Figure 4a. Base wheel model

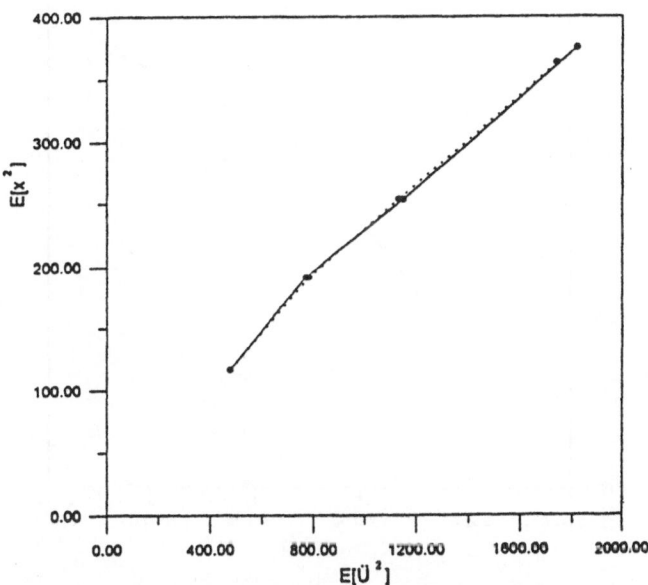

Figure 4b Thomson bearing base model.
Figure 4. Dependence of mean square response on excitation level in the absence of
friction elements
————— excitation level is decreased,   – – – – excitation level increases.

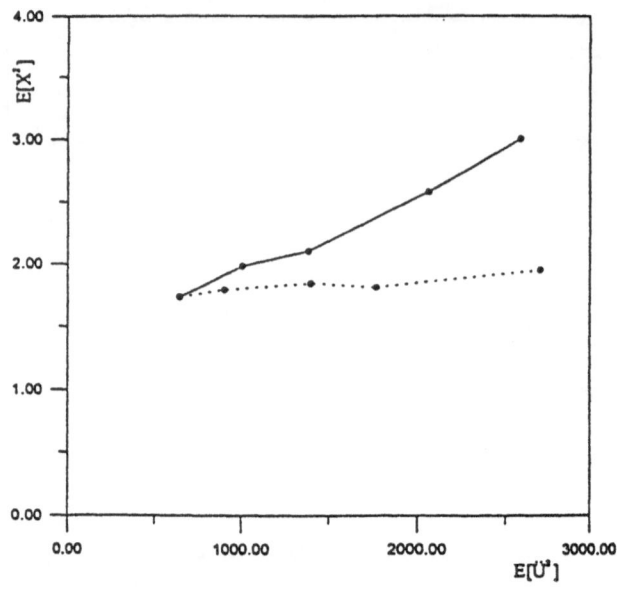

Figure 5a. Base wheel model

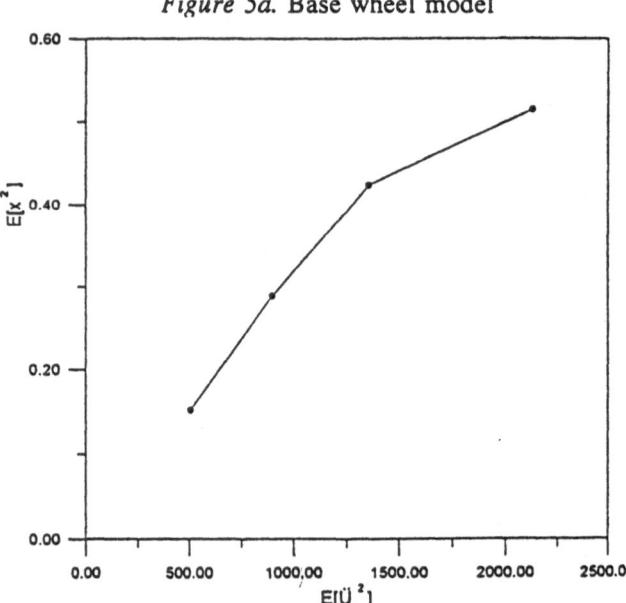

Figure 5b Thomson bearing base model.
Figure 5. Dependence of mean square response on excitation level in the presence of friction elements

————— excitation level is decreased   – – – – – excitation level increases.

SOME DEVELOPMENTS IN COMPUTATIONAL TECHNIQUES IN MODELING ADVANCED MECHANICAL SYSTEMS

D. POGORELOV
Dept. of Applied Mechanics, Bryansk State Technical University
bulv. 50 let Oktyabrya 7, 241035 Bryansk, Russia

Some methods and algorithms for optimal computer-aided modeling of multibody systems are considered. Special approaches to the computerized symbolic generation of motion equations are discussed. The methods described are realized in the program package Universal Mechanism (UM). Their application to modeling technical objects such as a six-legged walking mechanism and spatial cable systems are presented.

1. Equations of Motion

Computerized generation of symbolic motion equations of multibody systems provides an efficient and reliable basis for their dynamics and kinematics. In comparison with numerical generation of the equations, it allows to reduce in many cases the number of floating-point operations by motion simulation and in addition to this to use computer algebra methods for analytical study of the equations. Many recently developed software packages use special symbolic manipulation modules, which are more efficient for derivation multibody system motion equations than general-purpose systems [1,2].

Consider a system of rigid bodies connected by hinges of various types. If the system has closed kinematic loops, a minimal set of hinges has to be cut to open the closed loops [3]. The system obtained has a tree structure and the kinematic relations can be derived in the following form:

$$r_i^0 = r_i^0(q,t), \quad A_{0i} = A_{0i}(q,t). \tag{1}$$

The vector r_i^0 and rotation matrix A_{0i} determine the position of body-fixed frame i relative to the inertial coordinate system. The upper index corresponds to the reference frame. The vector q denotes the set of object coordinates. The calculation of these quantities is based on the formulas of a relative orientation of a pair of adjacent bodies i,j,

$$r_{ij}^i = r_{ij}^i(q_{ij},t), \quad A_{ij} = A_{ij}(q_{ij},t), \tag{2}$$

with q_{ij} as relative coordinates in a connecting hinge. Thus the coordinate vector q is made up from local coordinates q_{ij} corresponding to the tree of the system.

313

D. H. van Campen (ed.), IUTAM Symposium on Interaction between Dynamics and Control in Advanced Mechanical Systems, 313–320.
© 1997 *Kluwer Academic Publishers.*

The relations (1) can be derived from Eqs.(2) by using a simple recurrence procedure starting from the inertial frame.

By differentiating Eqs. (1), (2), we obtain the velocities and accelerations

$$v_i^0 = D_i^0(q,t)\dot{q} + v_i'^0(q,t), \quad \omega_i^0 = B_i^0(q,t)\dot{q} + \omega_i'^0(q,t), \tag{3}$$

$$a_i^0 = D_i^0(q,t)\ddot{q} + a_i'^0(q,\dot{q},t), \quad \varepsilon_i^0 = B_i^0(q,t)\ddot{q} + \varepsilon_i'^0(q,\dot{q},t),$$

for global and

$$v_{ij}^i = B_{ij}^i(q_{ij},t)\dot{q} + v_{ij}'^i(q_{ij},t), \quad \omega_{ij}^i = B_{ij}^i(q_{ij},t)\dot{q} + \omega_{ij}'^i(q_{ij},t), \tag{4}$$

$$a_{ij}^i = D_{ij}^i(q_{ij},t)\ddot{q} + a_{ij}'^i(q_{ij},\dot{q}_{ij}t), \quad \varepsilon_{ij}^i = B_{ij}^i(q_{ij},t)\ddot{q} + \varepsilon_{ij}'^i(q_{ij},\dot{q}_{ij}t),$$

for relative kinematic quantities.

For systems having closed kinematic loops, constraint equations have to be added. They have the form of nonlinear algebraic equations,

$$h(q,p,t) = 0, \tag{5}$$

and depend both on the coordinates q and on local coordinates p in cut hinges [4]. Differentiation of Eq. (5) yields the differential constraint equations,

$$H_q\dot{q} + H_p\dot{p} + h' = 0, \quad H_q\ddot{q} + H_p\ddot{p} + h'' = 0. \tag{6}$$

For example, if the cut hinge connecting the bodies i,j has an arbitrary number of rotational and translational degrees of freedom, the corresponding closing conditions can be found in the form of the following equations:

$$r_i^0(q,t) + A_{0i}(q,t)r_{ij}^i(p_{ij},t) - r_j^0(q,t) = 0, \tag{7}$$

$$\frac{1}{2}\sum_{k=1}^{3}\tilde{e}_k^0 A_{0i}(q,t)A_{ij}(p_{ij},t)A_{j0}(q,t)e_k^0 = 0,$$

where e_k^0 are unit vectors of the inertial frame, $\tilde{e}_k^0$ are the corresponding antisymmetric matrices, p_{ij} are the local coordinates. Eqs.(7) are in general independent and give a minimal set of 6 scalar conditions. The first equation is well-known, but the second has a very simple expression for the derivative,

$$\omega_i^0 + \omega_{ij}^0 - \omega_j^0 = 0.$$

The constraint equations for a contact joint used for modeling the rolling of bodies have an analogous form. Let

$$\rho_i^i = \rho_i^i(p_i), \quad \rho_j^j = \rho_j^j(p_j)$$

be the parametrization of contact manifolds in the body-fixed frames. Thus, there are six types of contact joints from (0,0) up to (2,2) according to manifold dimensions. The constraint equations include the condition

$$r_i^0(q,t) + A_{0i}(q,t)\rho_i^i(p_i) - A_{0j}(q,t)\rho_j^j(p_j) - r_j^0(q,t) = 0, \tag{8}$$

as well as one or two additional equations

$$\tau_{i,k}^i(p_i)A_{i0}(q,t)A_{0j}(q,t)n_j^j(p_j) = 0, \ k=1 \text{ or } k=1,2,$$

for joints of the types (1,2) and (2,2). The last equation means that the normal n_j^j to the second manifold is perpendicular to the tangent $\tau_{i,k}^i$ to the first manifold. The nonholonomic rolling without sliding satisfies the equation

$$A_{0i}\rho_i'^i p_i - A_{0j}\rho_j'^j p_j = 0.$$

Generalization of contact joints leads to one-sided constraints. In this case, Eq. (8) should be replaced by

$$r_i^0(q,t) + A_{0i}(q,t)\rho_i^i(p_i) - \mu A_{0j}(q,t)n_j^j(p_j) - A_{0j}(q,t)\rho_j^j(p_j) - r_j^0(q,t) = 0,$$

where the parameter μ is the distance between the manifolds and must be included in the set of variables.

In general, the linear nonholonomic constraints are described by the equations

$$S_q q^{\cdot} + S_p p^{\cdot} + s = 0. \tag{9}$$

The multibody system equations of motion can be obtained using the Newton-Euler formalism [2,5]. For systems having closed kinematic loops, they lead to

$$M(q,t)q^{\cdot\cdot} + k(q,q^{\cdot},t) = f(q,q^{\cdot},p,p^{\cdot},t) + G(q,t)^T \lambda, \tag{10}$$

where the inertia matrix M and the vectors k, f of generalized inertia and active forces are represented by sums of terms corresponding to every body of the object,

$$M = \sum M_i = \sum \left(m_i D_i^{0T} D_0^0 + B_i^{iT} I_i^i B_i^i \right), \tag{11}$$

$$k = \sum k_i = \sum \left(m_i D_i^{0T} a_i'^0 + B_i^{iT} (I_i^i \varepsilon^{i} + \widetilde{\omega}_i^i I_i^i \omega_i^i) \right),$$

$$f = \sum f_i = \sum \left(D_i^{0T} F_i^0 + B_i^{iT} M_i^i \right),$$

where m_i, I_i^i are the mass and inertia tensor of body i, and F_i^0, M_i^i are the active force and torque. The Lagrange multipliers λ correspond to reaction forces in the cut joints and the nonholonomic constraints. The matrix G can be determined after the elimination of the auxiliary variables p from the differential and the nonholonomic constraint equations (6), (9),

$$G(q,t)q^{\cdot} + g'(q,t) = 0.$$

The closed equation system contains Eqs. (5), (6), (9) and (10).

2. Computerized Generation of Symbolic Motion Equations

Two facts are important for the computerized symbolic generation of the equations above. First, if all hinges in the system tree have only translational and rotational degrees of freedom, i.e. contact joints are cut, expressions corresponding to the kinematical relations above and the constraint equations are polynomial. The elements of the matrices and vectors in Eqs. (11), except matrix G, are polynomial as well. In general, explicit expressions for the matrix G can be found only for cut joints having an arbitrary number of rotational and translational degrees of freedom. Elimination of the auxiliary variables p_i, p_j for contact joints is connected with solving transcendental algebraic equations, such as (8), and can be executed numerically.

Secondly, exponents of polynomial variables, such as inertia parameters, identifiers of length parameters, sine and cosine of angular coordinates, active forces, time functions and so on, are positive integer numbers which do not exceed 2. This fact was proved for systems with a chain structure [6] but it is also true for systems having closed loops [4]. Thus, a very efficient coding of symbolic expressions is possible as dynamic ordered chains containing codes of polynomial terms [7]. The code is an integer number presenting the polynomial term exponents in numerical systems with the base 3. The ordering of chains eliminates the problem of the grouping like terms.

The number of floating-point operations in the symbolic equations can be reduced drastically using an optimization routine. The optimization consists in a recursive procedure of taking some variables out of brackets and building a dynamic chain of elementary expressions.

Two alternative generation methods are realized in the package UM. The first method is based on coding of expressions which is different from the one mentioned above. The expressions are regarded as trigonometric sums with polynomial coefficients. To avoid operations of trigonometric simplification, Euler's representation of sine and cosine functions is applied. For coding trigonometric exponents, the numerical system with base 5 is used. The symbolic expression is regarded as a two-level ordered tree. The chain of the first level corresponds to the trigonometric sum, the chains of the second level are attached to each element of the first chain and present the polynomial coefficients. All elements of the matrices included in the equations of motion are generated in a full symbolic form, i.e. they have an explicit form. Then, a special optimizing module reduces the number of operations in the whole set of equations and converts it into an implicit recurrent form.

TABLE 1. Multibody pendulum. The first method of the motion equation generation. Processor P5, 90 MHz.

Number of bodies	CPU time of generation, s	CPU time of optimization, s	Number of operations before optimization	Number of operations after optimization	Coefficient of optimization
2	0.52	0.58	211	75	2.8
4	1.51	3.7	4070	575	7.1
6	8.1	34.1	40991	1973	20.8
8	69	320	312084	4685	66.6

Table 1 presents some results of the application of the method to a multibody pendulum. All characteristics, except the number of floating-point operations after the optimization, show exponential growth of the computational burden. This means that the described approach to the equation generation can be used for multibody systems without long kinematic chains having rotational degrees of freedom.

The second method applies the optimization procedure to symbolic expressions at several breakpoints of the generation process. In this case the equations have an implicit recurrent form, but as a rule, the number of operations as well as the time of generation are much less than those obtaind by the first method. The breakpoints of optimization can be set in different parts of process. The following scheme could be suggested as an efficient one. The breakpoints are set after generation of:

   a) local kinematic expressions (2),(4) for each hinge;
   b) global kinematic expressions (1),(3) for each body;
   c) active forces and torques for each body;
   d) constraint equations (5),(6),(9) for each cut hinge;
   d) mass matrix and generalized inertia and active forces (11).

Table 2. Multibody pendulum. The second method of the motion equation generation

Number of bodies	CPU time of generation, s	Number of operations	CPU time of a simulation step, s
8	4.78	1173	0.0024
16	33.9	5213	0.007
32	657	29933	0.04

Table 2 shows that the second approach is much more efficient for long kinematic chains, but the number of rotational degrees of freedom in the chain is limited as well.

3. Technique of Subsystems

The examples in the previous section show that the symbolic generation of motion equations both in explicit and implicit forms, leads to an increase of the computational burden. It becomes impossible to obtain directly the equations of motion for systems having some hundreds or even thousands of bodies. In this case, the approach based on breaking up the system into separate substructures (subsystems) via cutting some hinges is more efficient. The equations of motion for each subsystem

$$M_i(q_i,t)\ddot{q}_i + k_i(q_i,\dot{q}_i,t) = f_i(q,\dot{q},p,\dot{p},t) + G_i(q_i,t)^T \lambda_i + \sum G_{ij}(q_i,q_j,t)^T \lambda_{ij} \,, \quad (12)$$

$$h_i(q_i,p_i,t) = 0, \quad H_{iq}\dot{q}_i + H_{ip}\dot{p}_i + h_i' = 0, \quad H_{iq}\ddot{q}_i + H_{ip}\ddot{p}_i + h_i'' = 0,$$

should be complemented by closing conditions for hinges connecting different subsystems

$$h_{ij}(q_i,p_i,q_j,p_j,t) = 0, \quad (13)$$

and their derivatives.

The equation of motion (12) as well as the main parts of the constraint conditions (13) can be obtained for each subsystem separately. Moreover, most technical objects can be partitioned into subsystems some of which are kinematically identical. The equations for all identical subsystems are generated by UM only once. An example of a spatial cable modeling using the subsystem technique is considered below.

4. Simulation of a Cable Dynamics[1]

Consider a spatial cable model as a chain of identical elements connected by hinges with two rotational degrees of freedom. Each element is presented by a rigid rod. To simulate the underwater motion of the cable, some further development of the formalism for the motion equation generation is necessary.

As a simplest model, consider an ideal and incompressible fluid. Equations of a single body motion can be obtained using the following expression for the kinetic energy of the body and fluid [8]:

$$2T = v^{0T} M_{11}^0 v^0 + 2v^{0T} M_{12}^0 \omega^0 + \omega^{0T} M_{22}^0 \omega^0 ,$$

where the matrices M_{11}^0, M_{12}^0, M_{22}^0 present inertia properties of the system. The equations of the motion are derived using Lagrange's equations and have the form suitable for a symbolic generation,

$$J^T MJ\ddot{q} + J^T (VMu + Mw) = f , \qquad (14)$$

where

$$J = \begin{bmatrix} D \\ B \end{bmatrix}, \quad V = \begin{bmatrix} \tilde{\omega} & 0 \\ \tilde{v} & \tilde{\omega} \end{bmatrix}, \quad M = \begin{bmatrix} M_{11} & M_{12} \\ M_{21} & M_{22} \end{bmatrix}, \quad u = \begin{bmatrix} v \\ \omega \end{bmatrix}, \quad w = \begin{bmatrix} a' - \tilde{\omega}v \\ \varepsilon \end{bmatrix},$$

the vector f corresponds to generalized active forces. The superscripts are omitted for the tensors and vectors can be presented in any coordinate system.

Generalizing Eq. (14) for systems with a tree structure results in the following equation:

$$\sum J_i^T M_i J_i \ddot{q} + \sum J_i^T (V_i M_i u_i + M_i w_i) = \sum f_i ,$$

where i is the number of a body.

In the case of a viscous fluid, the cable motion equations can by used in a simplified form [9]. By this approach, the body-fluid interaction is taken into account via additional terms in the mass matrix and damping forces.

A suitable cable model should contain a large number of bodies. Therefore, the direct generation of the symbolic motion equations is impossible. In this case the subsystem technique is very efficient. The cable is presented by a chain of kinematically identical subsystems.

Some characteristics of different variants of the cable models are shown in table 3. The models contain about 100 bodies and 200 degrees of freedom. According to these results, the subsystems for the optimal model consist of seven bodies.

---

[1] Research is carried out in a co-operation with the Technical University Hamburg-Harburg

Table 3. Generation of the cable equations. Simulations characteristics

Number of bodies in subsystem	CPU time of generation, s	Number of operations	Number of subsystems in the cable	CPU time of a simulation step, s
5	51	3957	20	0.273
7	122	6817	14	0.246
9	257	10523	11	0.272
11	497	15159	9	0.306

5. Model of a 6-Legged Walking Mechanism[2]

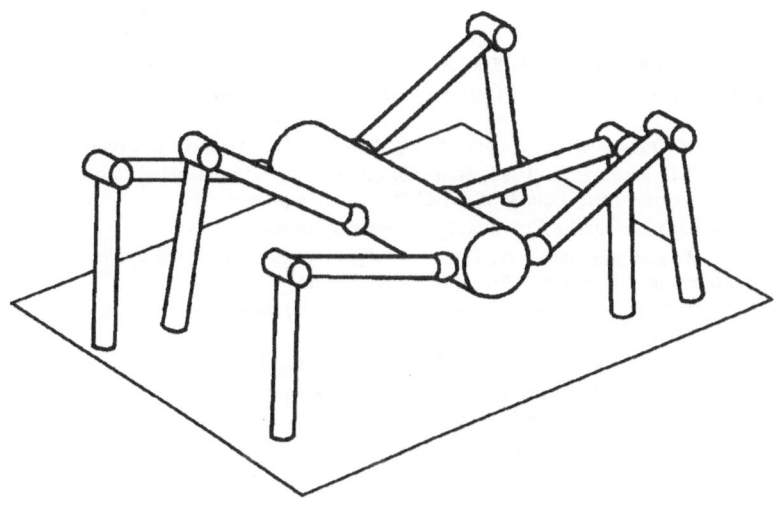

Figure 1.

The model of a walking mechanism is shown in the figure 1. It consists of 13 bodies and 13 hinges. One of the hinges is a fictitious one and determines a central body position in space. The legs are connected with the central body by hinges with two rotational degrees of freedom. The knee hinges have one degree of freedom. For a normal force acting on the leg terminus, a linear viscous-elastic model is used. The mechanism has 24 degrees of freedom. The alternative model has 6 degrees of freedom and all angles in the leg hinges are prescribed functions of time. Some characteristics of the model equations are shown in table 4.

---

[2] Research is carried out in a co-operation with the Keldysh Institute of Applied Mathematics RAS.

320

Table 4. A comparison of the mechanism models

D.o.f.	CPU time of generation, s	Number of operations	CPU time of a single step, s
24	79	7458	0.013
6	56	5927	0.0073

6. Conclusion

The examples above show the high efficiency of the symbolic generation of the equations of motion of a mechanical system. In fact, the subsystem technique allows to model multibody systems up to some thousands degrees of freedom.

7. References

1. Schiehlen, W. (ed.) (1990) *Multibody Systems Handbook*, Springer-Verlag`, Berlin.

2. Kreuzer, E. (ed.) (1994) *Computerized Symbolic Manipulations in Mechanics*, Springer-Verlag, Wien - New York.

3. Wittenburg, J. (1977) *Dynamics of Systems of Rigid Bodies*, B.G. Teubner, Stuttgart.

4. Efimov, G.B. and Pogorelov, D. (1993) Some algorithms for computer-aided generation of multibody system equations, *Preprint Inst. Appl. Math. RAS* **80**, Moscow (in Russian).

5. Schiehlen, W. (1986) *Technische Dynamik*, B.G. Teubner, Stuttgart.

6. Vukibratovic, M. and Kircanski, N. (1985) *Real-time Dynamics of Manipulation Robots*. Series: Scientific Fundamentals of Robotics **4**, Springer-Verlag, Berlin.

7. Pogorelov, D. (1993) On coding the symbolic expressions by generation of multibody system motion equations, *Techn. Cybern* **6**, 209-213 (in Russian).

8. Sedov, L.I. (1984) *Mechanics of Continuum* **2**, Moscow, Nauka (in Russian).

9. Kleczka, W. (1994) *Symbolmanipulationsmethoden zur Analyse nichtlinearer dynamischer Systeme am Beispiel Fluid-gekoppelter Strukturen*, VDI-Verlag, Duesseldorf.

OPTIMISATION OF NON-LINEAR INTER-VEHICLE ACTIVE SUSPENSION CONTROL LAWS APPLIED TO RAILWAY TRAINS.

I. PRATT AND R. M. GOODALL
Department of Electronic & Electrical Engineering
Loughborough University, LE11 3TU, UK.

Abstract

Inter-vehicle railway dampers are utilised on high speed railway vehicles to improve passenger ride comfort. Extensive theoretical and analytical assessment of the performance benefits offered by inter-vehicle dampers has not previously been attempted, partly due to the lack of adequate modelling software and the computational demands involved in analysing the dynamic behaviour of long trains, but also because suspension design in the railway industry has progressed in a rather heuristic and experimental manner. This paper analyses the behaviour of trains containing between three and seven vehicles. It utilises a non-linear simplex search method to evaluate the inter-vehicle spring and damper rates a train should possess for optimal ride quality. It also investigates the effects of non-linear active inter-vehicle connections and comments on the contrasting performance.

1. Introduction

Inter-vehicle dampers are currently used to improve passenger ride quality, and are installed on the British Rail Mk IV, French TGV, and the Japanese Shinkansen trainsets (Fujimoto and Miyamoto, 1995) to cite just a few examples. They are primarily used as lateral dampers to improve the lateral ride quality, this being the most demanding direction for ride improvement. The use of inter-vehicle dampers and other forms of of connections is not restricted to the lateral direction: they may also be fitted vertically. The focus of this paper is the optimisation of vertical inter-vehicle connections, but the ideas espoused here may readily be extended to the lateral direction. The lateral behaviour is not dealt with here purely for reasons of simplicity; the lateral direction possesses many more coupled degrees of freedom than the vertical. It is the authors' opinion that vertical inter-vehicle connections also can offer significant advantages.

The structure of the paper is divided as follows: modelling, performance assessment, description of search optimisation methods, inter-vehicle connection control methods, and finally the conclusion of the work.

The modelling section describes the techniques and outlines the software used to simulate the sideview vertical behaviour of a railway train. The performance assessment section describes the methodology by which a train ride quality index and other important performance indices may be evaluated. The Nelder-Mead non-linear simplex search method is the optimisation method used throughout to locate the inter-connection parameters required

D. H. van Campen (ed.), IUTAM Symposium on Interaction between Dynamics and Control in Advanced Mechanical Systems, 321–328.
© 1997 *Kluwer Academic Publishers.*

for optimal ride performance. An overview of this search method is given in section 4. A variety of inter-connection structures will be addressed in the next section, which describes the concept of *square-root damping,* and *maximum force damping*, and the rationale for utilising a different inter-connection from the more traditional *linear* damper.

2. Modelling Sideview Train Dynamics

Figure 1 shows the sideview structure of a railway train and the elements influencing its vertical dynamic behaviour. The primary suspension is defined as the suspension below the axles and the bogie frame, usually formed through a parallel coil spring and damper, but rubber primary suspension elements are also common. The secondary suspension of a high speed train is usually an airbag located between the body and the bogie. A detailed description of individual vehicle dynamics is given in section 2.1, and the amalgamation of these vehicles into a train model is given in section 2.2.

2.1 Vehicle sideview dynamics

The secondary suspension parameters are tuned to give good ride quality. The model posseses six degrees of freedom, two associated with the leading and trailing bogie bounce and pitch modes. A typical optimised high speed suspension will have a body bounce mode of 0.8 Hz with 15% damping and a body pitch mode of 0.9 Hz with 20% damping. The bogie bounce and pitch modes have a somewhat higher frequency, 6Hz with 60% damping, and 10 Hz with 30% damping respectively. The airbag is a crucial element in the ride performance of a vehicle, and a combination of linear springs and dampers is used to describe its behaviour as shown in Figure 1. Although the true behaviour is non-linear this model is accepted to be a reasonably accurate description (Williams, 1986). In the model, considering vehicle 'm', $\dot{Z}_{WM1}$ to $\dot{Z}_{WM4}$ represent the vertical wheel/axle velocities, F_{M1} and F_{M2} represent the inter-vehicle forces acting on the body. The model outputs of interest are the body accelerations at various measurement points, and the secondary suspension deflection; also absolute position and velocity measurements at the vehicle ends because these variables are used to connect vehicles. These variables are contained within the output vector Y. The state-state format of the model is given by equations (2.1.1) and (2.1.2)

$$\dot{X} = AX + W \begin{bmatrix} \dot{Z}_{WM1} \\ \dot{Z}_{WM2} \\ \dot{Z}_{WM3} \\ \dot{Z}_{WM4} \end{bmatrix} + B \begin{bmatrix} F_{M1} \\ F_{M2} \end{bmatrix} \qquad (2.1.1)$$

$$Y = CX + W \begin{bmatrix} \dot{Z}_{WM1} \\ \dot{Z}_{WM2} \\ \dot{Z}_{WM3} \\ \dot{Z}_{WM4} \end{bmatrix} + B \begin{bmatrix} F_{M1} \\ F_{M2} \end{bmatrix} \qquad (2.1.2)$$

2.2 Simulink modelling of train dynamics

Matlab® and Simulink® are the primary software tools used to simulate train behaviour. The graphical nature of Simulink readily permits inter-connection of the single vehicle models developed in section 2.1.

3. Performance Assessment

Train behaviour is evaluated through computer simulation on a time history of a typical good quality high speed track. The test track used throughout being a 2 km section of the English East Coast mainline near North Allerton, Yorkshire, traversed at 200 km/h (55ms⁻¹). The length of the section must be sufficient to ensure accurate ride assessment. In high speed railway vehicles the vertical dynamic modes occur just below 1Hz, which implies that track wavelengths of 55m will be the dominant excitation at 55ms⁻¹. Approximately 35 cycles are experienced during a 2km test run, which is suffiicient to ensure accurate ride assessment.

For each vehicle the ride quality is taken to be the average of the r.m.s. vertical acceleration experienced at three points on the vehicle: one at the centre of gravity ($\ddot{z}_B$), one above the leading bogie ($\ddot{z}_{BOG1}$), and one above the trailing bogie ($\ddot{z}_{BOG1}$). The ride index for a train is simply taken to be the average ride quality experienced by the individual vehicles. The definition of ride quality for a train is given more formally in equations (3.1) to (3.3).

$$
\begin{bmatrix}
\ddot{z}_{BOG11} \\
\ddot{z}_{B1} \\
\ddot{z}_{BOG12} \\
\cdot \\
\cdot \\
\ddot{z}_{BOGn1} \\
\ddot{z}_{Bn} \\
\ddot{z}_{BOGn2}
\end{bmatrix} = C_{ACC}\, \underset{\sim}{Y}
\tag{3.1}
$$

$$
\underset{\sim}{y}(t) = \int_{\tau=0}^{t} C_{ACC} C e^{A(t-\tau)} W \underset{\sim}{\zeta}(\tau)\, d\tau
\tag{3.2}
$$

$$
PI_{RIDE} = \frac{\sum_{i\,=\,1}^{3n} \left[\int_{t=0}^{T} tr\{\underset{\sim}{y}(t)\underset{\sim}{y}^{T}(t)\} dt \right]_i}{3\,n}
\tag{3.3}
$$

The system response to random track ζ is evaluated and the averaged r.m.s value obtained.

An important performance indicator for the design of secondary suspensions is the suspension deflection in response to deterministic inputs. This is strictly speaking a constraint, and its value must not exceed a particular level. This is defined to be the maximum displacement between body and bogie for any bogie in the train when negotiating a vertical transition of 0.4%g sustained vertical acceleration. This definition is given more formally in equations (3.4) and (3.5).

$$
\begin{bmatrix}
Z_{B1} + lt\,\phi_{B1} - Z_{BOG11} \\
Z_{B1} - lt\,\phi_{B1} - Z_{BOG12} \\
\cdot \\
\cdot \\
Z_{Bn} + lt\,\phi_{Bn} - Z_{BOGn1} \\
Z_{Bn} - lt\,\phi_{Bn} - Z_{BOGn2}
\end{bmatrix} = C_{DEFL}\ \underset{\sim}{Y} \tag{3.4}
$$

$$
PI_{DEFL} = \| \int_{\tau=0}^{t} C_{DEFL}\, C\, e^{A(t-\tau)}\, W\, \underset{\sim}{V}(\tau)\ d\tau \|_{\infty} \tag{3.5}
$$
$$
\forall\ 0 \le t \le T
$$

These are used in a similar manner to the ride assessment method, in this instance V represent the deterministic track feature. It is important to calculate this index even in the design of inter-vehicle connections since it is possible to violate the secondary suspension deflection constraint with a particular class of inter-vehicle connections. Violation of the suspension deflection constraint will result in high accelerations and reduced ride quality due to contact with the bump stops. This will be demonstrated in subsequent sections.

4. Non-Linear Simplex Search

The non-linear simplex search method of Nelder and Mead (William, 1992) is used to locate optimal parameters for the various inter-vehicle connections described and analysed in subsequent sections. With reference to Figure 2 a procedural description may be given for the search for a pair of parameters. From an initial guess X^0 a simplex is formed with vertices: X^1, X^2, and X^W. The search involves taking the largest functional value of any of the vertices and reflecting this point. Assuming $f(X^W)$ in this instance to be the largest value, a new simplex (X^1, X^2, X^W) is formed . This procedure is repeated, reflecting the worst point of the simplex to form a new simplex. The value of the function at the reflected point should continue to decrease as the procedure continues. If it does not, or if a boundary condition is violated, the second worst point of the simplex is reflected. If this point again fails in the same way, the simplex is contracted. The contraction continues until the reflected point is reduced and the boundary conditions are met.

Figure 3 shows the worst secondary suspension deflection experienced by each vehicle in a 3 vehicle train under the conditions described in section 3. It is important to keep this value constrained, resulting in a boundary condition on the maximum stiffness and damping of the inter-vehicle connection.

5. Inter-Vehicle Connection

The dynamic properties of the inter-vehicle connection are comprise both design and parasitic effects. A typical high speed train will possess both a coupler and a gangway connection between vehicles. The coupler is essentially a joint constraint, permitting the adjacent vehicles to move freely in yaw and pitch motions, and the gangway provides a sheltered passage between vehicles; both posses significant stiffness and damping properties. The dynamic behaviour of the train is assumed to be modelled as shown in Figure 1, and the inter-connector structure to be that of a parallel spring-damper combination. It is accepted that several elements (secondary airbags, gangways, and couplers) will in reality deviate from this model, but not significantly. In the subsequent sections 5.1 to 5.3 the desired inter-vehicle characteristic must ultimately be formed from a hybrid of the inherent gangway and coupler characteristics, and additional controlled dampers and stiffnesses, the net effect being to obtain the characteristics described in sections 5.1 to 5.3.

5.1 Linear spring-damper

This section is dedicated to searching for the optimal parallel linear spring-damper combination between vehicles. Figure 4 shows the characteristics of a linear damper.

The surface contour of Figure 5 illustrates the effect on ride performance of varying the inter-vehicle damping and stiffness. This enables the optimum for three vehicles to be found. This result is for a three vehicle train, the non-linear simplex search routine is used to find the optimum for trains with higher numbers of vehicles, which counteracts increased computational requirements imposed by larger trainsets by requiring fewer simulations to find the optimum. The results for larger trainsets are given in section 6.

5.2 "Square-root" damping and linear stiffness

It is often desirable for dampers to have non-linear characteristics; for example, in motorcycles, progressive damping is used to avoid excessive "diving" under heavy braking conditions (Dixon J.C, 1991). In fact it requires design effort in order for a damper to behave in a linear fashion because the nature of fluid flow giving rise to an inherent non-linearity. However, the advent of controllable dampers and electro-rheological devices implies that an infinite variety of characteristics is possible. In this section the performance of a square-root damper in parallel with a linear spring are assessed. The damper characteristic is shown in Figure 4. The hypothesis behind proposing the use of a "square-root" characteristics the increased power dissipation per cycle of such a device (as shown by the area between the linear damper and the "square-root" damper). It is surmised that this will improve ride quality by dissipating more vibrational energy in contrast to its linear counterpart. Figure 6 illustrates the ride quality for a variety of stiffness and damping rates on a 3 vehicle train. The results for longer trainsets are given in section 6.

5.3 "Maximum force" inter-connector

According to our energy dissipation hypothesis, the ultimate form of inter-connection would be of the type shown in Figure 7. This gives a considerable dissipation due to the generation of a very large force for small inter-vehicle velocities. The ride qualities predicted with this control strategy are given in section 6 for a variety of trainset lengths.

326

6. Results

Parameter optimisation of the control strategies outlined in section 5 was performed using the Nelder-Mead search method of section 4. Table 1 presents the percentage ride quality improvement over an un-optimised high speed trainset with the basic inter-vehicle connection.

Number of vehicles	Parallel linear spring /damper	Square-root damping /stiffness	Maximal force inter-connector
3	11	14	-5
4	12	15	-7
5	13	17	-8
6	13	18	-9
7	13	18	-10

Table 1 - Percentage improvement in ride quality with different control strategies and varying train lengths

7. Conclusions

The results indicate that the original concept of using a non-linear damping characteristic to increase vibrational energy dissipation and improve ride quality is a valid one. However, the "maximal force" damping control law, which was predicted to have the best performance benefits in fact deteriorated the ride quality. The reason behind this being the lack of an equilibrium point with the "maximal force" damping control law.

8. References

Dixon J.C., (1991) Tyres, suspension and handling, *Cambridge University Press*.

Fujimoto H, Miyamoto M, (1995) Lateral vibration and its decreasing measure on a Shinkansen train, *Proc of 14th IAVSD Symposium on dynamics of vehicles on roads and tracks*, Ann Arbor, Michigan, USA, 40-42.

William H, (1992) Numerical recipes in FORTRAN: the art of scientific computing, *Cambridge University Press*, 2nd Edition.

Williams R.A., (1986) A comparison of classical and optimal active suspension control systems, *Loughborough University Ph.D. thesis*, Chap 3, 56-64.

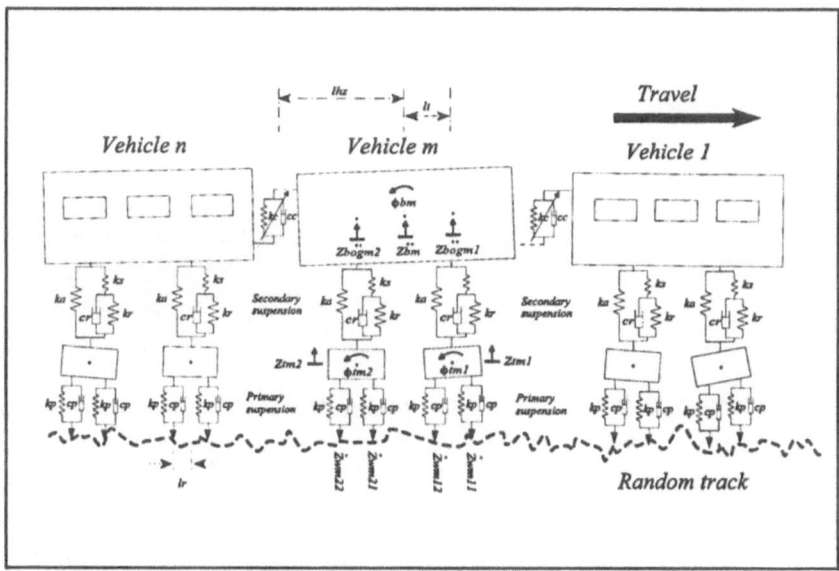

Figure 1 - Train of 3 vehicles

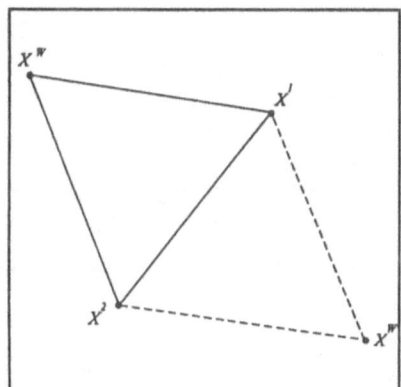

Figure 2 - Simplex search method
(2-dimensional)

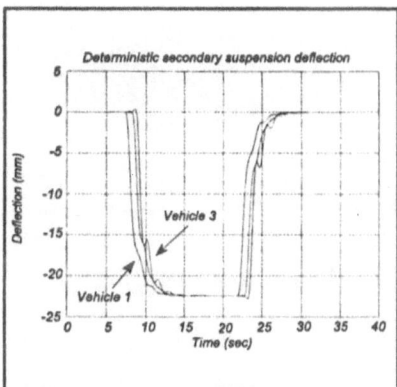

Figure 3 - Secondary deflection
(low IV damping)

328

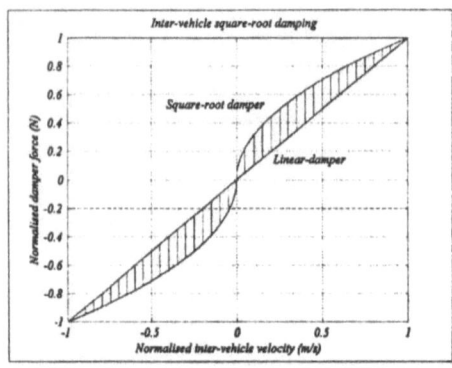

Figure 4 - "Square-root" damping

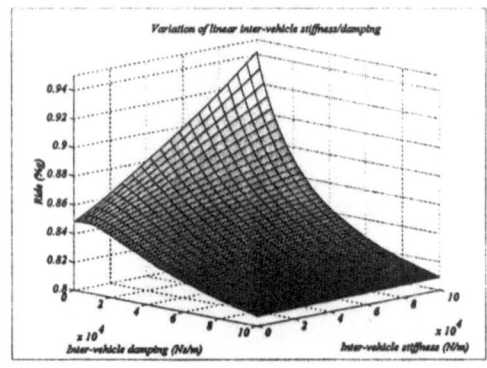

Figure 5 - Passive linear inter-connection

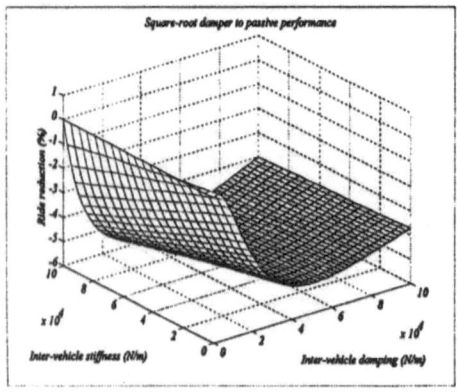

Figure 6 - "Square-root" damping

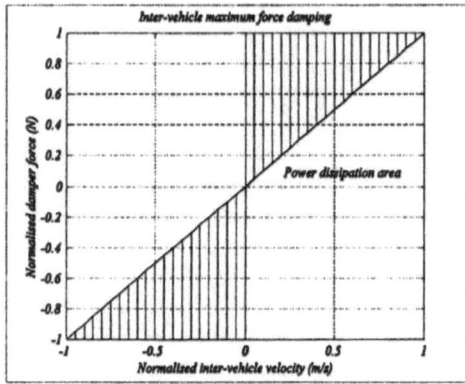

Figure 7 - "Maximum force" damping

ACTUATOR AND SENSOR LOCATION FOR STRUCTURAL IDENTIFICATION AND CONTROL

M. RADEŞ
University POLITEHNICA of Bucharest
Strength of Materials Section
RO-79590 Bucureşti, Romania

Abstract

A procedure is presented for location of sensors and exciters in structural testing, based on data from an *a priori* finite element model. In a first step, a sub-set of sensors is placed on a candidate set of response measurement locations such that the linear independence of target mode shapes is maximized. In the second step, excitation degrees of freedom are ranked according to their contribution to the frequency response information. A criterion based on the multivariate mode indicator function is used to establish the minimum number of exciters and to assess the quality of the chosen set. Numerical simulations illustrate application of the procedure.

1. Introduction

The problem of optimal sensor and exciter placement is analyzed from a structural dynamics point of view, where modal parameters extracted from test data are used to validate or update a finite element model (FEM). Finite bandwidth dynamics, correlation between sensor and exciter noise sources, as well as their interaction with the test structure are neglected.

During the pre-test analysis phase, the FEM, which contains more degrees of freedom (DOF) than can be surveyed, is reduced to a condensed test/analysis model (TAM). Assuming that the original FEM is an accurate representation of the physical structure, the aim is to choose a reduced order model which represents the FEM as closely as possible. But while the purpose of a TAM is to describe the FEM characteristics at all low frequencies, in control dynamics and health monitoring applications the objective is to track only a preselected sub-set of target modes, discarding some low-frequency, non-observable or non-important modes. This makes modal reduction best for both sensor placement and model condensation. During the test phase, the quality of collected data is heavily influenced by the selection of exciter and sensor locations.

D. H. van Campen (ed.), IUTAM Symposium on Interaction between Dynamics and Control in Advanced Mechanical Systems, 329–336.

The exciter placement is dependent on the damping distribution in the actual structure, which is only approximately added to the modal FEM. Use of synthesized frequency response function (FRF) data is useful in the selection of exciter locations and can be extended to the general problem of test/analysis model correlation.

This paper reports experience in using a location procedure which combines the effective independence (EfI) method (Kammer, 1991) with a frequency selective ranking based on the dyadic expansion of the FRF matrix (Radeş, 1995).

2. Selecting Response Measurement Locations

2.1. EFFECTIVE INDEPENDENCE METHOD

Test/analysis correlation techniques require linearly independent test modes; otherwise the cross-orthogonality matrix check fails. The requirement of absolute identifiability is more stringent than that of observability demanded by most identification and control techniques. Sensors must be located so that the resulting modal vectors can be spatially differentiated. In the EfI method, candidate sensor locations are ranked according to their contribution to the linear independence of the FEM target modes.

Starting with the full modal matrix $[\Phi]$ of the FEM, the first step is to remove all coordinates which cannot be measured (e.g.: rotations) or which are considered of little significance. Next, the target modes are determined using a modal selection procedure (Blelloch et al., 1989) which orders the modes in terms of their contribution to the input/output dynamics of the model. Let $[\Phi_r]$ be the reduced matrix of N_m target modes truncated to the N_o candidate sensor locations.

The cross-product matrix $[A_r]$ is then formed

$$[A_r] = [\Phi_r]^T [\Phi_r] = \sum_{j=1}^{N_0} \lfloor \Phi_j \rfloor^T \lfloor \Phi_j \rfloor \tag{1}$$

which will be referred to as the Fisher information matrix (FIM).

The problem is to search the best set of N_s sensor locations from N_o candidate locations so that $\det[A_r] = \det(\text{FIM})$ is maximized.

The orthogonal projector onto the column space of $[\Phi_r]$ is the matrix

$$\left[P_\Phi \right] = [\Phi_r][\Phi_r]^+ \tag{2}$$

(where + denotes the pseudo-inverse), whose trace is equal to its rank and to the number of target modes N_m (idempotent matrix). Each diagonal element P_{jj} represents the fractional contribution of the j-th sensor to the rank of $[P_\Phi]$, and hence to the independence of the target modes.

Elements P_{jj} are sorted based on magnitude. The sensor location with minimum contribution to the rank of $[P_\Phi]$, indicated by the smallest element, is removed from the candidate set. The matrix $[P_\Phi]$ is then recomputed, and the process is repeated, sensors being deleted one at a time, until the number N_s is attained or until eliminating one additional sensor creates a rank deficiency. It is useful to track the P_{ij} values. If several DOFs have the same P_{ij} value, then they are deleted simultaneously.

2.2. QUALITY OF CHOSEN MEASUREMENT LOCATIONS

Several criteria have been investigated (Penny et al., 1994) to assess the quality of any set of response measurement locations. The 2-norm condition number of the eigenvector matrix, defined as the ratio of the largest singular value to the smallest, was considered the best. The value of det(FIM) varies with the number of sensors, so it cannot give an absolute assessment of the quality of a set of locations. However, plots of det(FIM) versus the number of retained sensors are useful to determine both the quality of the elimination and a lower limit when the FIM becomes rank deficient.

Another useful information is obtained tracking the elimination process on the diagram of cond(FIM) vs. the number of retained sensors. The smallest the condition number, the best the choice of locations. Minima on these plots show optimal values for the number of retained sensors, while a sudden increase denote the lower limit set by the rank deficiency of the FIM matrix.

3. Selecting Excitation Locations

Optimal selection of excitation locations using the EfI method has been considered by Lim (1993). Though the basic assumption - linear independence of the forcing vectors is of questionable physical significance, use of the method in a first stage of selection may be helpful. Considerations regarding the energy input from the exciter locations yield similar results.

3.1. PRESELECTION BY EfI METHOD

Consider that the N_s response measurement locations are chosen as potential locations for the N_f exciters (N_f<N_s). The problem of finding the optimum exciter configuration for the N_m target modes was defined as one of combinatorial optimisation (Niedbal, 1990, Lallement et al., 1991).

Introducing proportional damping as a good first approximation, a matrix of forcing vectors is computed premultiplying the modal matrix $[\Phi_s]$ by either the corresponding reduced stiffness or the mass matrix. Assume that

$$[F_s] = [K_s][\Phi_s].$$

The EfI method is then applied to the cross-product matrix

$$[C_s] = [F_s]^T [F_s].$$

The number of rows N_s is reduced to N_m when the process is stopped to avoid rank deficiency of the force matrix.

3.2. USE OF SYNTHESIZED FRF DATA

Adding damping to the FEM modal data set, we create a proportionally damped system, from which the FRF matrix $[H]$ can be calculated either for the frequency range spanned by the target modes or only at the undamped natural frequencies (UNF). Writing the covariance of the $(N_o \times N_i)$ FRF matrix as a sum of dyadic products

$$[H][H]^H = \sum_{l=1}^{N_i} \{H_l\}\{H_l\}^H = \sum_{l=1}^{N_i} [G_l] \tag{3}$$

where $\{H_l\}$ is the l-th column of $[H]$, we find that the information from the l-th exciter is given by the $[G_l]$ matrix. While $[H]$ is complex, $[H][H]^H$ is hermitian, hence with real diagonal elements. The trace of the rank one matrix $[G_l]$ can be used as a measure of the l-th exciter contribution to the FRF information. Plots of $\mathrm{Tr}[G_l]$ versus frequency help to rank exciter locations.

3.3. MMIF-BASED SELECTION

The final selection of exciters is based on the analysis of the multivariate mode indicator function (MMIF) calculated from synthesized FRF matrices. MMIF curves are plotted using one, two and three excitation DOFs (columns in the FRF matrix) in turn, and N_s response measurement DOFs (rows in the FRF matrix). The minimum number of exciters is determined by the pseudo-multiplicity of eigenfrequencies and is increased if supplementary important local modes have to be excited. Controllability and local impedance characteristics (to avoid large non-linearities) have to be taken into account, as well as coupling of excitation DOFs by skew mounting of exciters.

A measure of how effective the exciter configuration is at exciting the target modes is obtained by calculating the average mode purity index

$$\Delta_m = \frac{1}{N_m} \sum_{j=1}^{N_m} \left[1 - MMIF\left(\omega_j\right)\right]$$

where $MMIF(\omega_j)$ is calculated at the j-th UNF located by the MMIF. Values $\Delta_m > 0.95$ indicate good location.

4. Examples of Sensor/Exciter Selection

4.1. 11-DOF SYSTEM

Consider the 11-DOF system with structural damping from Fig.1. Physical parameters and UNFs are given by Radeş (1993). Due to mass and stiffness symmetry, the system has pairs of modes with close natural frequencies. The right hand branch (masses 6 to 11) has higher damping values than the left one, hence the system exhibits non-symmetrical complex mode shapes.

Assuming that the first two modes have to be identified using four sensors, the EfI method selects locations 1,2,10,11. If we want to identify modes 3,4,5, and 6, EfI locates four sensors at 1,5,8,11. If the target modes are the first six modes, the EfI method locates four sensors at 1,4,8 and 11. However plots of det(FIM) and cond(FIM) show rank deficiency for a number of sensors less than the number of target modes (Fig.2). Location of six sensors results in 1,2,4,8,10,11.

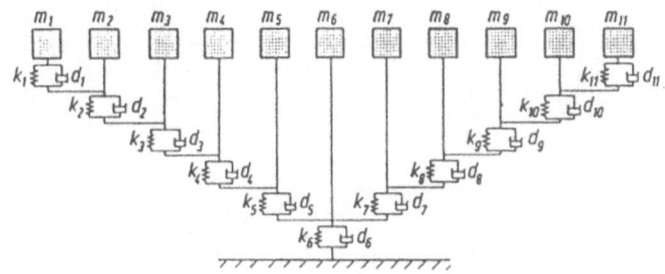

Figure 1. 11-DOF system

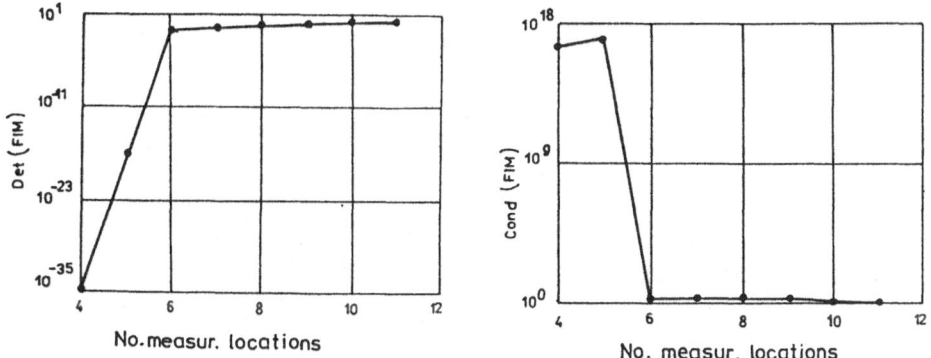

Figure 2. Det(FIM) and Cond(FIM) plots for the 11-DOF system

For exciter location by EfI, the matrix of undamped eigenvectors is replaced by the matrix of forces, calculated by premultiplying the modal matrix by the hysteretic damping matrix reduced to the coordinates selected as sensor locations. If the first two modes are target modes, the extension of EfI method places two exciters at locations 3 and 11. If the target modes are 3,4,5,6, trying to locate only two exciters, EfI yields locations 8 and 11. This is wrong, because a minimum of four exciters have to be located to avoid rank degeneracy, and both exciters are on the same half of the system.

If a 6x11 FRF matrix is set up for coordinates 1,2,4,8,10,11 selected by EfI for sensors, and the traces of the eleven dyadic product components are calculated and plotted against frequency, two exciters are located at 6 and 7. Figure 3a shows the

334

MMIF for a 6x2 FRF matrix calculated with forces at 6,7, and responses at 1,2,4,8,10,11. Figure 3b shows the MMIF for forces at 6,7 and responses at 1,3,5,7,9,11, where the undamped mode shapes have large values. The second location performs better, hence the sub-optimal character of the suggested procedure. Figure 3c shows the MMIF for forces at 3, 11 located by the extended EfI method.

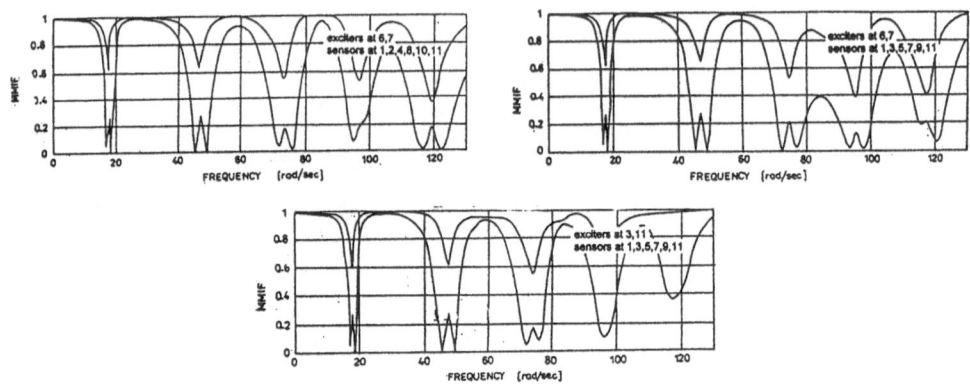

Figure 3. MMIF plots for the 11-DOF system

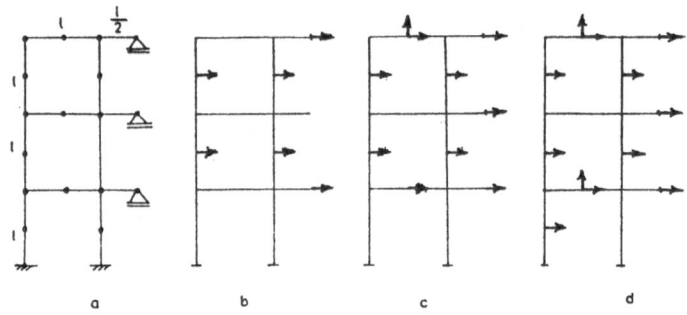

E=210 GPa, ρ=7810 kg/m³, l=0.2032 m, A=80.6 mm², I=271 mm⁴

Figure 4. Plane frame and EfI location of sensors

4.2. PLANE FRAME

Figure 4*a* shows half of a rigid-jointed plane frame used to investigate planar antisymmetrical vibrations.. The frame was modelled with 51 DOFs, using 21 beam elements, neglecting shear deformations and rotatory inertia.

The model was first reduced eliminating 12 axial displacements among the vertical members and 18 rotations. The remaining 21 translations have been used to select a reduced number of sensor locations. If the first six modes of vibration are selected as target modes, the location of 6, 9 and 12 sensor positions chosen by EfI and K/M methods is shown in Figs.4*b* to *d*.

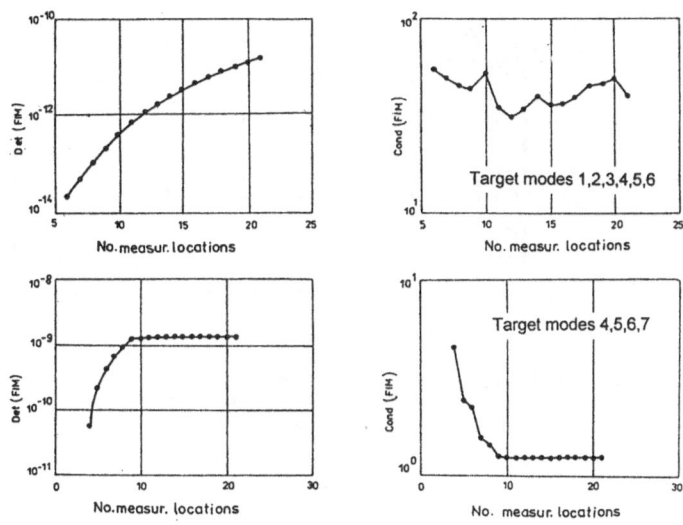

Figure 5. Det(FIM) and Cond(FIM) for the plane frame model

Figure 5a shows the evolution of det(FIM) and cond(FIM) for six sensors. Because the target modes are the lowest 6 modes there is no rank deficiency. Det(FIM) decreases monotonically, but cond(FIM) is lowest for 12 sensors. Figure 5b shows the same plots when the four target modes are the modes 4,5,6,7. If four sensors are selected, the plots indicate rank deficiency for values lower than 9. The same tendency was seen for three target modes 9,10,11, and six selected sensors. The minimum number of 9 sensors corresponds to the nine identical lowest k_{ii}/m_{ii} values. Further research is needed to explain this behaviour.

4.3. TEST RIG

Figure 6*a* shows a laboratory supporting and loading frame. The cross section properties are as for rolled steel shapes W24x492. For the study of planar vibrations, the frame was modelled with 72 DOFs, using 26 beam elements. After elimination of 24 rotations and 10 axial displacements in vertical members, the remaining 38 candidate locations have been reduced to 5 by EfI, considering the lowest 5 target modes.

Location of four exciters by the FRF method was used to place two skew exciters as in Fig. 6*b*. Guyan reduction method yielded six valid eigenfrequencies for ten sensors,

336

but only four (71.45, 241.94, 318.42, 334.42 Hz) with excellent accuracy. Eigenvector expansion is necessary to improve the spatial resolution of deflected shapes.

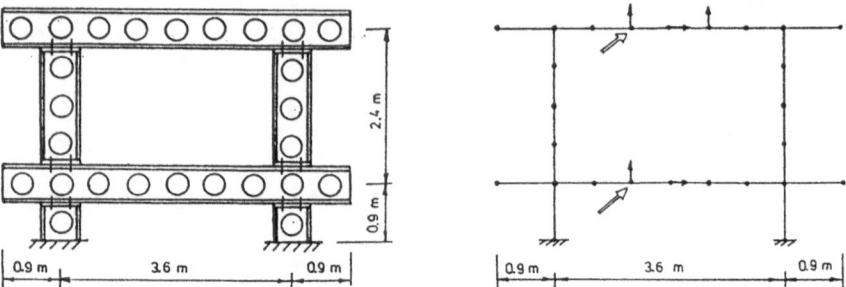

Figure 6. Test rig with 2 exciters and 5 sensors

5. Conclusions

A systematic and sub-optimal selection procedure has been presented in which decisions have to be made about: a) the number and the index of the target modes retained in the modal matrix; b) the number of sensor locations required to describe, with acceptable accuracy, the target modes; c) where to place the sensors, once their number is established; d) the number of exciter locations required; e) where and on what direction to apply the excitation; f) what is the best exciter/sensor combination. An evaluation criterion based on the MMIF has been used to assess the accuracy of the selected sensor/exciter set. The robustness of the procedure has to be tested based on prescribed errors in the finite element model.

6. References

Blelloch, P.A. and Carney, K.S. (1989) Modal selection in structural dynamics, *Proc. 7th I.M.A.C.*, 742-749.
Kammer, D.C. (1991) Sensor placement for on-orbit modal identification and correlation of large space structures, *J. Guidance, Control, and Dynamics* 14, 251-259.
Lallement, G., Cogan S., and Andriambololona, H. (1991) Optimal selection of the measured degrees of freedom and application to a method of parametric correction, *Proc. 9th I.M.A.C.*, 369-375.
Lim, T.W. (1993) Actuator/sensor placement for modal parameter identification of flexible structures, *Modal Analysis* 8, 1-14.
Niedbal, N. and Klusowski, E. (1990) Optimal exciter placement and force vector tuning required for experimental modal analysis, *Proc. 15th I.S.M.A.*, Leuven, 1195-1222.
Penny, J.E.T., Friswell, M.I., and Garvey, S.D. (1994) Automatic choice of measurement locations for dynamic testing, *AIAA Journal* 32, 407-414.
Radeş, M. (1993) A comparison of some mode indicator functions, *Mech. Syst. and Signal Proc* 8, 459-474.
Radeş, M. (1995) *Modal Testing Using Multiple Sinusoidal Excitation*, UNICAMP, Campinas.

EXPERIMENTAL AND THEORETICAL INVESTIGATION OF NONLINEAR DYNAMICS IN AN ELASTIC SYSTEM WITH INITIAL CURVATURE

G. REGA[1] , F. BENEDETTINI [2] and R. ALAGGIO [2]
[1] *Dipartimento di Ingegneria Strutturale e Geotecnica, Università di Roma "La Sapienza", via A. Gramsci 53, 00197, Roma, Italy*
[2] *Dipartimento di Ingegneria delle Strutture, Acque e Terreno, Università dell'Aquila, Monteluco Roio, 67040, L'Aquila, Italy*

1. Introduction

Nonlinear phenomena play an important role in the finite dynamics of elastic structural systems with initial curvature, such as suspended cables and arches. In this paper, we document their richness and analyze some of them with reference to two different multidegree-of-freedom models of an elastic cable, a system which can easily exhibit multiple internal resonance conditions with ensuing accretion of the nonlinear and chaotic coupling phenomena: an experimental cable/mass system and a discrete theoretical model of a continuous system. Attention is devoted to multimodal interaction, coexistence and competition of various response classes, transition mechanisms, nonstationary motions, reconstruction of attractors in pseudo-phase space, and quantitative characterization of chaotic responses.

2. Experimental cable/mass system

An experimental system made by a small sag nylon wire carrying eight equally spaced concentrated masses and hanging at supports which are given in-phase or out-of-phase vertical motions, is analysed. Based on the linearized dynamics features of the underlying continuous elastic cable, which are substantially reproduced - though at lower values of the λ^2 cable control parameter - in the lower modes of a symmetric cable/mass suspension (Cheng and Perkins, 1992), the experimental system naturally exhibits nearly simultaneous 1:2:2 internal resonances involving the first nonplanar symmetric mode (H1) and the first planar (V2) and nonplanar (H2) antisymmetric modes. By properly adjusting its mechanical and geometrical parameters, we can realize the crossover fundamental condition of the in-plane frequencies, where the first symmetric mode (V1) has the same natural frequency as the corresponding antisymmetric one, and is thus involved in the resonance condition, too. Systematic response measurements have been made for an experimental system close, but not exactly realizing, this latter situation, which would further increase the intricateness of the finite response due to the basic unavoidable resonances. By considering different

D. H. van Campen (ed.), IUTAM Symposium on Interaction between Dynamics and Control in Advanced Mechanical Systems, 337–344.
© 1997 *Kluwer Academic Publishers.*

338

amplitudes of supports motion in a large frequency range including meaningful
external resonance conditions, such as the fundamental and principal resonances of the
H1 mode, and the principal resonance of the V2 and H2 modes, we have obtained and
discussed overall response charts in the excitation control parameter plane (Benedettini
and Rega, 1996).

2.1 MULTIMODAL INTERACTION, COMPETITION BETWEEN REGULAR CLASSES OF MOTION, TRANSITION MECHANISMS

A meaningful portion of one chart showing the more robust classes of regular response
occurring in the closed neighbourhood of the 1/2-subharmonic (fundamental)
resonance of antisymmetric modes under in-phase supports motion is shown in Figure
1. Various regions are distinguished through labels referring to the prevailing modal
components contributing to the system response, which however exhibits also further
(secondary) components not shown in the figure (see Rega *et al.*, 1996, for a detailed
analysis). Apart from the simple vertical drag motion occurring in the chart
background, various fundamental *multimodal response* classes are observed, namely
the prevailing planar antisymmetric V2, the antisymmetric ballooning V2H2, the
nonplanar antisymmetric H2, and the mixed nonplanar H1H2. By determining the
maximum extent of each region through up or down sweep of the control parameter,
we have detected meaningful zones of *coexistence* of different attractors onto which
the response settles down, depending on initial and environmental conditions,
sometimes after fairly long transients. No irregular motions are observed in Figure 1,
which however refers to a range of very low values of excitation amplitudes. By
properly extending the window of experimental investigation in the excitation
parameters plane, we can observe several meaningful regions of both quasiperiodic
and chaotic motions in practically all the examined frequency ranges.

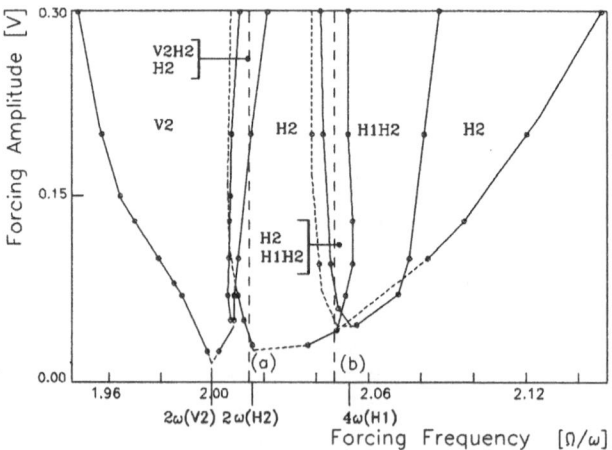

Figure1. Chart of regular responses at 1/2-subharmonic resonance.

Specific forcing amplitude-response pictures (Figure 2) obtained in zones of overlapping regions with different response highlight the *role* of various modal components of different amplitude and periodicity: the contribution from *higher* components (e.g., the second planar V4 and nonplanar H4 antisymmetric modes), the different *bifurcation paths* followed by the system response when increasing or decreasing the control parameter, the *competition* between response classes, their onset or disappearance through several smooth or sudden *transition mechanisms*. Some of them referring to the passage between orbits of prevalently unimodal or bimodal regular motions with increasing frequency are illustrated in Figure 3. The transition is represented in terms of transient Poincarè points moving in the reported frequency interval between the steady points corresponding to each orbit.

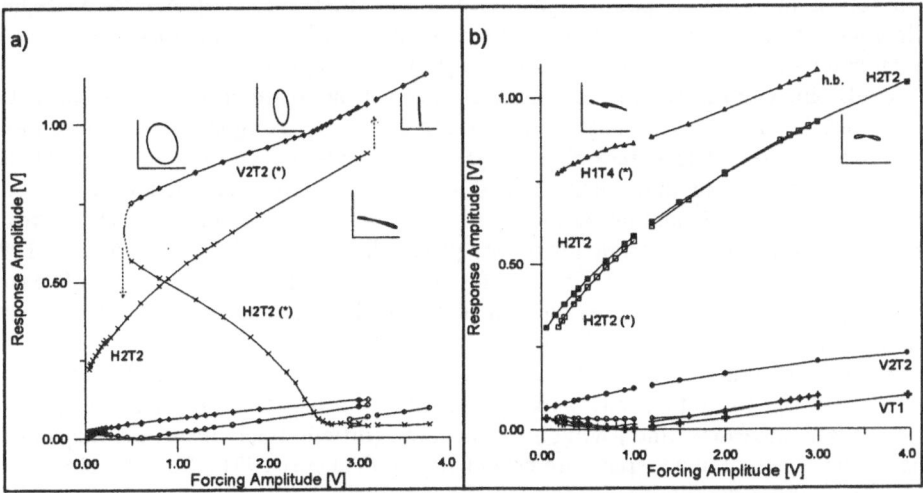

Figure 2. Forcing amplitude-response diagrams in two zones ((a) and (b)) of overlapping responses.

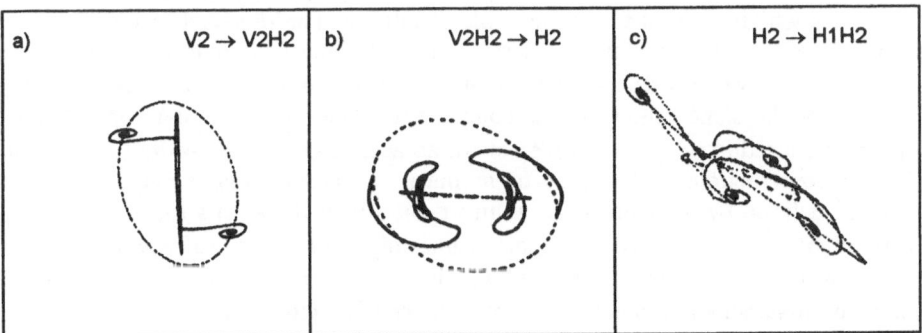

Figure 3. Transitions between couples of classes of motion with increasing frequency.

2.2 QUASIPERIODICITY AND CHAOS

In regular dynamic regimes the experimental investigation furnishes precious hints in obtaining minimal analytical models able to catch, in terms of modal components, the actual behavior of the system. In a strongly developed irregular regime (QP or chaotic), it can help us even more in calibrating those models on the basis of synthetic measures of strangeness and/or chaoticity: in fact, it is important to check whether these properties pertain rather to the performed analytical modelling (discretization, approximation, truncation) than to the real behaviour of the system.

Unfortunately, when testing multidegree-of-freedom or continuous systems, we are often in a position to measure only one or few points, obtaining a coordinate subset of the real state space. Such a situation is mainly due to the high cost of simultaneous monitoring of different points. Therefore, it is necessary to appeal to procedures allowing us to reconstruct the global properties of the attractors necessary to understand and classify the irregular regimes, from a single (or few) time series.

Hints about the possibility of characterizing various periodic regimes exhibited by the system have been given in the previous section by reporting some overall and local response pictures. The experimental system also exhibits various zones of chaotic response (Benedettini and Rega, 1996), whose preliminary classification has been made on the base of qualitative measures, such as time laws, phase space portraits, power spectra, autocorrelation functions, probability density functions (Moon, 1992; Nayfeh and Balachandran, 1995). Going deeper into the understanding of system global dynamics requires the evaluation of the likely different levels of "strangeness" and/or "chaoticity".

When a quantitative classification is needed, the delay embedding procedure has to be applied: the theoretical bases for reconstructing a pseudo-phase space using only one (or few) measures, and giving an embedding (differentiable invertible map) of the inaccessible, actual phase space are established in (Takens, 1981). Such an embedding preserves, under certain conditions, all the topological properties of the manifold $\mathfrak{m}$ containing the attractor of the real system. Among them, in particular, it preserves the possible fractal dimension of the attractor, as well as the divergence features of nearby trajectories which account for the possible chaoticity. The practical technique is quite simple: starting with a scalar time series constituted of N discrete measures acquired with step Δt, we build m-dimensional evolution pseudovectors having delayed measures of the same time-series as components. The reconstruction depends on the delay τ measured in integer multiples of Δt, as well as on the embedding dimension m of the reconstruction. The values of m must be chosen large enough so that the mapping defined by the embedding is invertible. Even if Taken's theorem states that the biunivocity is guaranteed for generic values of τ, and for a "sufficient" embedding dimension m, the unknown values of d, the physical nature of noisy data, and the truncation in acquisition, make the correct choice of m and τ mandatory.

Based on such measures, we can derive a quantitative comparison between different chaotic states and a possible classification criterion accounting for the fractal measure of the attractors and the velocity of diverging trajectories.

A qualitative and quantitative comparison of two different chaotic states obtained in corresponding regions of the excitation parameters plane, is shown in Figures 4 and 5,

341

respectively. They refer to two different cables, the slacker one already discussed in the previous section as to regular motions, whose internal resonance condition involves only the three fundamental modes, and a cable at first crossover point exhibiting internal resonance of higher multiplicity. For each cable, Figure 5 shows: i) the classical log-log curves, for different embedding dimensions, necessary for calculating the correlation dimension according to the (Grassberger and Procaccia, 1983)

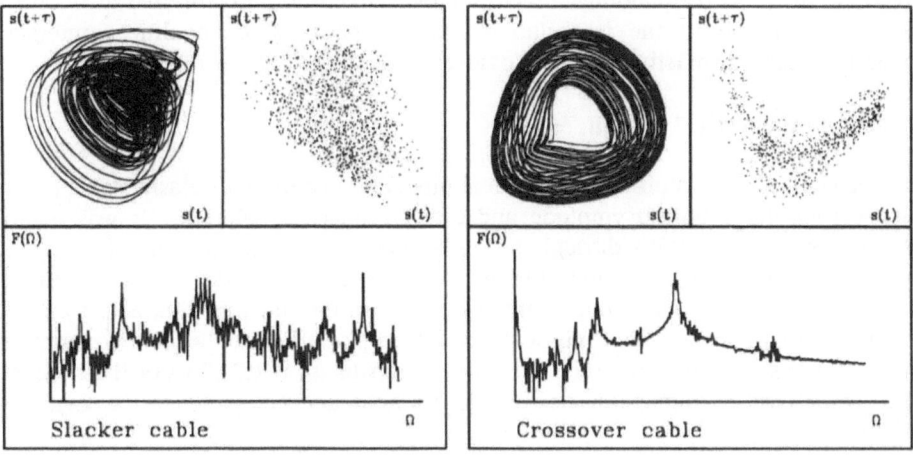

Figure 4. Qualitative characterization of chaotic response for slacker (a) and crossover (b) cable: reconstructed phase space section and Poincarè section, power spectrum.

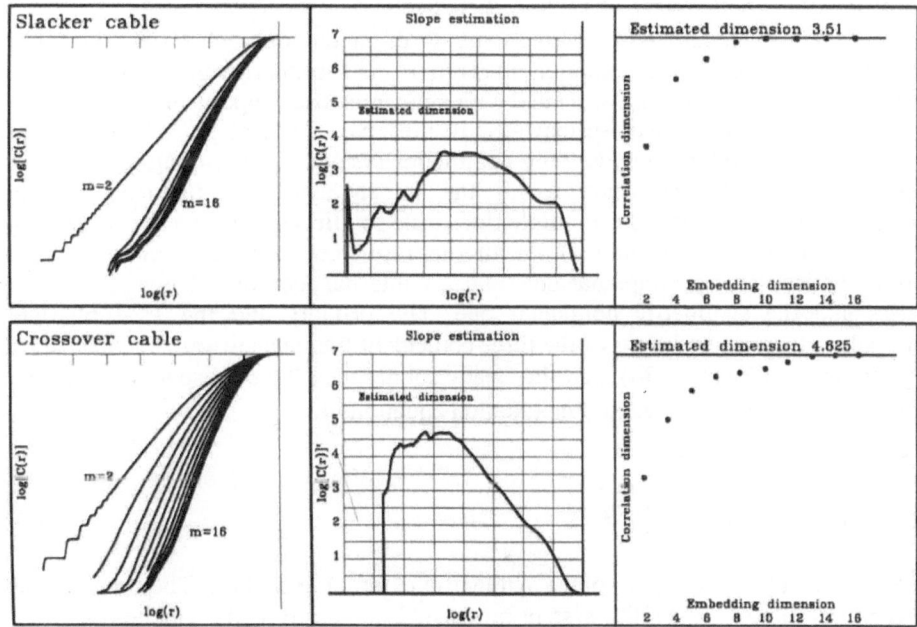

Figure 5. Quantitative characterization of chaotic response for slacker (a) and crossover (b) cable: correlation and embedding dimensions.

342

algorithm (left), ii) an example of slope plot, necessary to identify the plateau for estimating that dimension (middle), and iii) the saturation results for evaluating the embedding dimension (right). In agreement with the pictures in Figure 4 in which the slacker cable presents a less established broad-band spectrum, the relevant saturation value for the fractal dimension (D_c=3.51) is smaller than the corresponding one for the cable at crossover (D_c=4.62), thus highlighting smaller strangeness and less developed chaoticity of the corresponding attractor. This is likely to be connected also with the increase occurring in the irregular regime of the stronger nonlinear interaction phenomena already existing for the crossover cable in the regular regime.

3. Discrete theoretical system

The second system is a discrete theoretical model of a continuous elastic cable of small sag, susceptible to both asymptotic and direct numerical solutions. It was obtained (Benedettini et al., 1995) through discretization of the relevant partial differential equations of nonplanar motion through a classical four-mode technique: the two fundamental symmetric and antisymmetric modes of the linearized parabolic cable are accounted for both the planar vertical and out-of-plane horizontal displacement. The ensuing ODEs system in the relevant time unknowns q_i, i=1,2 (3,4) of the symmetric and, respectively, antisymmetric planar (nonplanar) components, is written in functional form:

$$\ddot{q}_i + \mu_i\dot{q}_i + \lambda_i^2 q_i + N_i(q_1,q_2,q_3,q_4) + P_i(q_1,q_2,q_3,q_4;q_{oa}(t)) = E_i(p_i(t);q_o(t)) \qquad i=1,4 \qquad (1)$$

where N_i are even and odd nonlinear terms of modal interaction, E_i are external excitations due to directly applied loads $p_i(t)$ and supports motion $q_o(t)$, P_i are parametric excitation terms associated with antisymmetric supports motion $q_{oa}(t)$.

This discrete theoretical system strictly preserves the features of the linearized dynamics exhibited by the underlying continuous system when varying its mechanical and geometrical characteristics, mostly the already mentioned frequency crossover phenomenon. A second-order multiple time scale solution to system (1) was developed in such a situation, which entails 1:1:1 internal resonance involving both planar modes and the antisymmetric nonplanar one, and 2:1 internal resonance between the former modes and the symmetric nonplanar one. The primary and the 1/2-subharmonic external resonance conditions of the three coincident frequencies were studied (see also Benedettini and Rega, 1994), and the steady solutions of the corresponding systems of amplitudes a_i (t) and phases $\gamma_i(t)$ modulation equations:

$$\dot{a}_i = f_i(a_1,a_2,a_3,a_4;\gamma,\vartheta_2,\vartheta_3,\vartheta_4)$$
$$a_i\dot{\gamma}_i = g_i(a_1,a_2,a_3,a_4;\gamma,\vartheta_2,\vartheta_3,\vartheta_4) \qquad i=1,4 \qquad (2)$$

were analyzed in detail for in-plane symmetric external excitation. They correspond to periodic solutions to the ODE system produced by either directly applied (distributed) harmonic loads or in-phase harmonic motion at the supports. Strong features of *multimodal interaction*, and *coexistence* and *competition* of different classes of steady motion, as well as richness of *local bifurcation mechanisms*, occur for the discrete system, too. Notwithstanding some important differences existing between the previous experimental cable/mass system and the present theoretical one, there is general qualitative agreement between the relevant main classes of regular response in terms of spatial shape, amplitude and periodicity of the participating modal components, and external frequency ranges where they are more easily activated (Benedettini and Rega, 1996).

In the context of the asymptotic approach, irregular responses of the theoretical system can be deduced from the analysis of the nonstationary motions $a_i(t)$, $\gamma_i(t)$ of equations (2), which correspond to periodically or chaotically amplitude modulated motions $q_i(t)$ in the original ODEs (1) (Rega and Benedettini, 1995). Here, they are obtained through direct numerical integrations of the ODE system itself.

3.1 QUASIPERIODIC AND CHAOTIC MOTIONS IN THE ODEs

Very rich and varied scenarios of bifurcations to quasiperiodic and chaotic motions are seen to occur, associated with the eight-dimensional state space, with coexistence of several regular and irregular attractors onto which the response settles depending on the set of initial conditions, and with rather large number of control parameters.

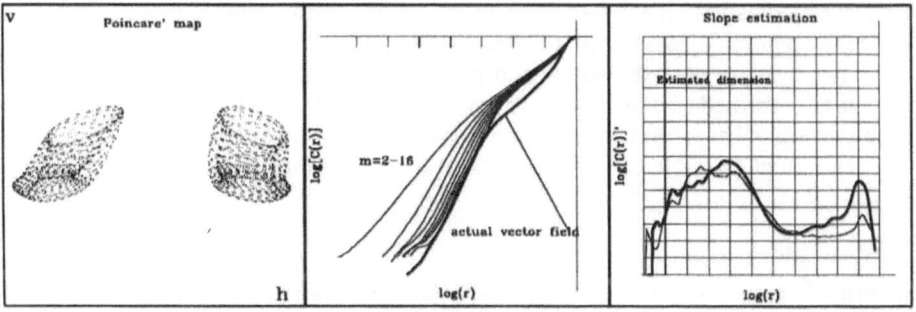

Figure 6. A seemingly chaotic attractor (a) and its quantitative characterization (b,c) with direct and embedding procedures.

Figure 6 refers to a seemingly chaotic motion derived, via a complex sequence of bifurcations with decreasing excitation amplitude, from a four-mode periodic solution occurring in the neighbourhood of primary resonance of the antisymmetric modes. A Poincarè section of the attractor in the (horizontal-vertical) configuration plane of the system is shown in Figure 6a. The quantitative characterization (Figs. 6b,c) has been made by calculating the correlation dimension with two different procedures, either directly operating on the actual vector field of eight first order equations equivalent to system (1) (thick line in the figures), or reconstructing an embedding phase-space from the numerical scalar time series of a singledegree-of-freedom (thin lines). Apart from the right part of the log-log curve in Figure 6a - which however has little significance in the dimension estimation for referring to too large distances between points on the attractor -, overall satisfactory agreement between the two results is obtained. This is a check on the reconstruction procedure. The two plateaux for estimating the dimension

344

in Figure 6c are fairly close, and correspond respectively to a value of ≈ 3.4 with the direct procedure (which looks more reliable) and of ≈ 3.0 with the embedding procedure (whose result is however seen to be slightly dependent on the particular choice of the time series used in the reconstruction). Of course, in all reconstruction problems, the actual meaning of the estimated correlation dimension - which, for qualitatively comparable responses, can be related to even notably different values of the embedding necessary for saturation - has to be understood clearly.

4. Concluding remarks

Experimental results showed coexistence and competion of several classes of regular motions in main resonance zones, involving strong multimodal interaction and a rich picture of periodic-to periodic bifurcations. Based on these results, it seems impracticable to account for actual nonlinear behaviour of elastic cables in different regions by using too simple models; at least four degrees-of-freedom are necessary in a discrete theoretical system to describe coupling among fundamental components. The two considered (experimental and theoretical) systems, however, concordantly highlighted robustness of some basic regular steady solutions involving only few modes. These solutions can possibly be described by simpler and more handable theoretical models with a selected number of degrees-of-freedom, to be used in specific zones of the control parameter plane of excitation.

The two systems also showed a rich picture of quasiperiodic and chaotic motions. Their more refined comparison to be made through systematic quantitative characterizations based on the use of the synthetic measures previously discussed, can allow to establish possible classification criteria of irregular regimes.

5. References

Benedettini, F. and Rega, G. (1996) Experimental Investigation of the Nonlinear Response of a Hanging Cable. Part II: Response Charts, *Nonlinear Dynamics* (submitted).

Benedettini, F. and Rega, G. (1994) Analysis of Finite Oscillations of Elastic Cables Under Internal/External Resonance Conditions, in *Nonlinear and Stochastic Dynamics ASME*, **AMD-192**, 39-46.

Benedettini, F., Rega, G., and Alaggio, R. (1995) Nonlinear Oscillations of a Four-Degree-of-Freedom Model of a Suspended Cable under Multiple Internal Resonance Conditions, *Journal of Sound Vibration* **182**, 775-798.

Cheng, S.P. and Perkins, N.C. (1992) Closed-Form Vibration Analysis of Sagged Cable/Mass Suspensions, *Journal of Applied Mechanics* 92-WA/APM-7.

Grassberger, P. and Procaccia, I. (1983) Characterization of Strange Attractors, *Physical Review Letters* **50**, 346-349.

Moon, F. 1992. *Chaotic and Fractal Dynamics*. Wiley, New York.

Nayfeh, A.H. and Balachandran, B. 1995. *Applied Nonlinear Dynamics*. Wiley Series in Nonlinear Science.

Ott, E., Sauer, T., and Yorke, J.A. 1994. *Coping with Chaos*. Wiley Series in Nonlinear Science.

Rega, G., Alaggio, R., and Benedettini, F. (1996) Experimental Investigation of the Nonlinear Response of a Hanging Cable. Part I: Classes of Motion and Modal Interaction, *Nonlinear Dynamics* (submitted).

Rega, G. and Benedettini, F. (1995) Nonlinear Interaction, Bifurcation and Chaos in Multidegree-of-freedom Cable Models, *International Symposium on Cable Dynamics*, Liège, Belgium, 141-148.

Takens, F. (1981) Detecting Strange Attractors in Turbulence, in *Dynamical Systems and Turbulence*, eds. D.A. Rand and L.S. Young, Springer Lecture Notes in Mathematics **898**, New York, 266-281

CONTROL CONCEPTS FOR LATERAL MOTION OF ROAD VEHICLES IN CONVOY

W. O. SCHIEHLEN and U. N. PETERSEN
Institute B of Mechanics, University of Stuttgart
Pfaffenwaldring 9, 70550 Stuttgart, Germany
e–mail: wos@mechb.uni–stuttgart.de, pu@mechb.uni–stuttgart.de

Abstract

This contribution deals with the lateral dynamics and control of two vehicles in convoy. Both vehicles are connected by a tow–bar where the second vehicle is to follow the leading one autonomously. Main attention is focused on the derivation and analysis of controllers on the basis of a linear single track model of the convoy. In addition to a mechanically motivated controller which proves to be unstable at higher velocities, a standard PDT_1 controller and an $H\infty$ approach based on a normalized left coprime factorization of the plant are discussed. Both, the standard and the $H\infty$ controller provide good performance where the latter is more robust.

1. Introduction

There are two main concepts for the lateral guidance of a convoy of vehicles along a road. Firstly, each vehicle of the convoy is guided laterally by some kind of inertial reference system such as an inductive wire or magnetic markers, e.g. Peng and Tomizuka [8]. Secondly, the leading vehicle of the convoy follows some inertial reference and all other vehicles of the convoy use their preceding vehicle as reference, see Narendran and Hedrick [6] or Fujioka and Suzuki [2]. Following the latter approach, this contribution deals with the lateral dynamics and control of two vehicles in convoy. The leading vehicle is controlled by a driver and the second vehicle is steered by an actuator to follow autonomously. Considering lateral dynamics only, both vehicles are connected by a rigid tow–bar, which serves as an ideal longitudinal controller maintaining constant distance between the vehicles all the time.

2. Vehicle Models and Control Structure

The theoretical analysis is carried out using two vehicle models of different complexity, Petersen, Rükgauer and Schiehlen [9]. On the one hand, a linearized single track model of the vehicle convoy considering lateral and yaw motion is used for linear system analysis and control design. On the other hand, a nonlinear multibody system model serves as

345

D. H. van Campen (ed.), IUTAM Symposium on Interaction between Dynamics and Control in Advanced Mechanical Systems, 345–354.
© 1997 *Kluwer Academic Publishers.*

346

more realistic representation for simulation studies. The equations of motion of both vehicle models are derived using the multibody system formalism NEWEUL, e.g. Kortüm and Sharp [4]. The data used in this research are based on two experimental vehicles, a BMW 325i Touring as leading vehicle and a BMW 518i as following vehicle.

The nonlinear complex model of the convoy, is composed of two single vehicle models, see Glora [3]. For each vehicle the six degrees of freedom of the car body and the motion of the four suspensions are taken into account. Since one degree of freedom is lost by connecting both vehicles with the tow–bar the resulting convoy system has 19 degrees of freedom. Another important difference to the linear vehicle model is the use of the nonlinear Magic Formula tire model, see Pacejka and Bakker [7], which considers tire saturation.

The simple vehicle model of the convoy is implemented on the basis of two single track models connected by a rigid bar, as shown in Figure 1. Simplistic assumptions for this model are constant longitudinal velocity and plane motion, i.e. lateral and yaw motion are considered but heave, roll and pitch are neglected. In addition, both wheels of each axis are represented by one tire in the middle of the axis. The tire forces are implemented using a linear tire model. The state space equations read as

$$\dot{x} = Ax + Bu \qquad (1)$$

with the state and control vector

$$x = \begin{bmatrix} \mu_1 & \mu_2 & v_{y1} & \omega_{z1} & v_{y2} & \omega_{z2} \end{bmatrix}^T, \qquad u = \begin{bmatrix} \delta_1 & \delta_2 \end{bmatrix}^T, \qquad (2)$$

see Petersen, Rükgauer and Schiehlen [9]. As system states the lateral velocity v_{y1} of the leading vehicle with respect to the given path, the yaw rate ω_{z1} of the leading vehicle, the corresponding coordinates v_{y2} and ω_{z2} of the second vehicle, and the angles μ_1 and μ_2 between the tow–bar and the vehicles are used. The input vector is composed of the steering angle δ_1 of the leading vehicle and the steering angle δ_2 of the following vehicle. The longitudinal platoon velocity v_x is considered as a system parameter, not as a degree of freedom.

Possible sensor information are the tow–bar angles μ_1 and μ_2, as well as the longitudinal velocity v_x of the convoy. In this paper only the tow–bar angle μ_2 will be used which is considered as a characteristic variable to be zero. Transforming equations (1) and (2) with μ_2 as system output to the Laplace Space one gets

$$\mu_2(s) = \mu_{21}(s) + \mu_{22}(s) = G_1(s)\delta_1(s) + G_2'(s)\delta_2(s) \qquad (3)$$

where $G_1(s)$ and $G_2'(s)$ denote the system's transfer functions from δ_1 and δ_2, to μ_{21} and μ_{22}, respectively, which add up to the tow–bar angle μ_2. The Laplace transform variable is given by s. Up to this point, the steering dynamics has been considered to be ideal, but

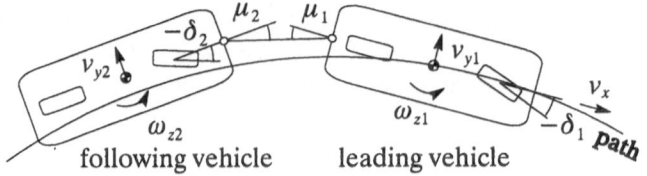

Fig. 1. Single track vehicle model.

of course this is not the case in the real world. The dynamics of the steering system includ-ing the actuator and its control which sets the actual steering angle to the desired value shows some delay. Since the real steering system has not been modelled in detail yet it is represented by a simple first order system with a bandwidth of 2 Hz, in the following denoted as G_{steer}. This might be a conservatively low bandwith but it allows to observe more clearly possible effects to the overall vehicle dynamics. The structure of the control problem with a controller transfer function $K(s)$, the extended plant transfer function $G_2 = G_{steer} G_2'$, input disturbances d' and d, measurement noise η and reference input r is displayed as a block diagram in Figure 2. It is obvious that the leading vehicle acts as a disturbance to the system via δ_1.

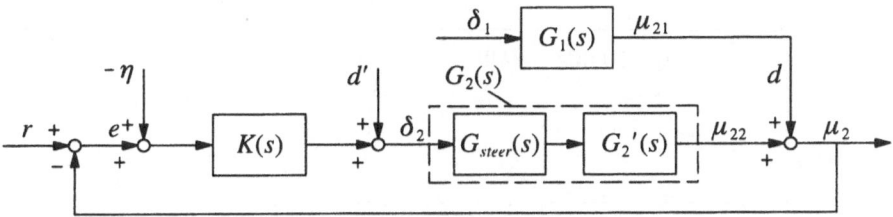

Fig. 2. Block diagram of control structure

3. Control Principles

Three different approaches for controlling the vehicle convoy are presented. The first one results from a vehicle dynamics engineers point of view, then a standard PDT_1 controller and an $H\infty$ controller are designed.

3.1 TRAILER PRINCIPLE

Motivated by the steering mechanism of a tractor trailer system, this principle feeds back the tow–bar angle μ_2 into a proportional controller with $K_{tr}(s) = 1$ and $r = 0$. Physi-cally, this means the front wheels of the following vehicle will always be aligned parallel to the tow–bar. To examine the stability of this control concept, a Bode plot of the open loop transfer function $G_2 K_{tr}$ is displayed in Figure 3. The different lines indicate the longi-tudinal velocities 1, 10, 20, 30, 40 and 50 m/s. The system is stable up to almost 20 m/s but instable for higher velocities. Thus, applying the trailer principle as controller for higher velocities is not recommended. This crucial result could not necessarily be ex-pected since the trailer principle works well for commercial tractor trailer systems even at speeds higher than 20 m/s. An investigation using the plant G_2' without the dynamics of the steering system showed that stability is given for all velocities considered here, but the phase margin decreases with increasing velocity. Hence the instability observed is due to the steering system dynamics.

3.2 STANDARD CONTROLLER

In order to overcome the stability problem, a standard PDT_1 controller is now designed. With the general structure of the transfer function

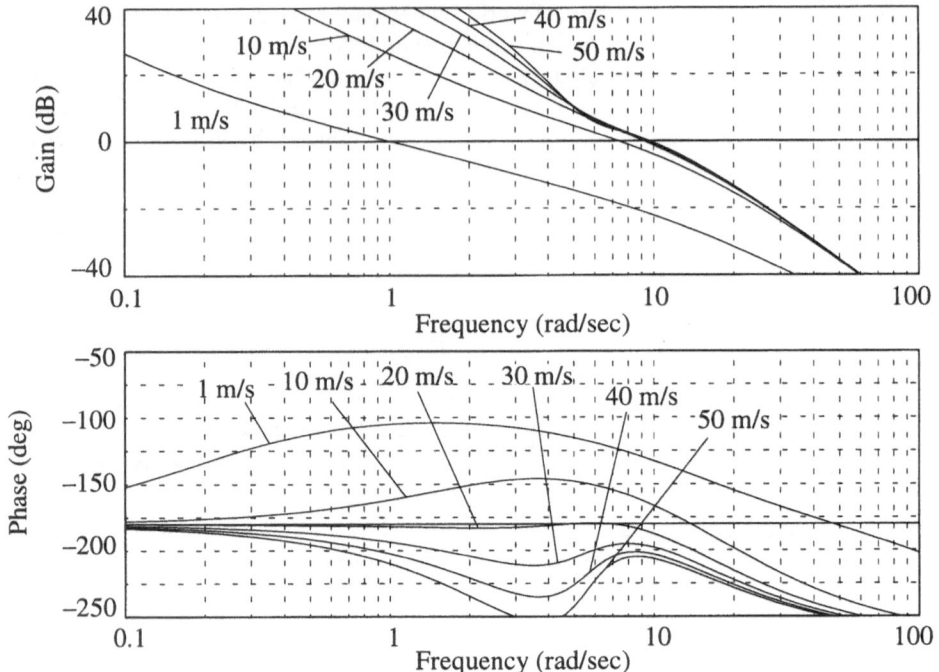

Fig. 3. Bode plot of open loop transfer function $G_2 K_{tr}$

$$K_{st}(s) = K_{stat} \frac{1 + T_d s}{1 + T_1 s} . \tag{4}$$

there are three parameters to be set. The stationary gain K_{stat} determines the position of the following vehicle relative to the leading vehicle when driving stationary curves, see Petersen, Rükgauer and Schiehlen [9]. A satisfactory behavior is obtained for low and medium velocities for $K_{stat} = 1$ as known from the trailer principle. Figure 3 indicates that additional phase has to be provided by the controller which results in the choice of a lead–element represented by the DT_1–part. After several iterations, aiming on phase and gain margin, a good set was found in $T_d = 0.25$ and $T_1 = 0.005$, see Figure 4 for the open loop Bode plot of $G_2 K_{st}$. The phase margin ranges from 89 degrees at 1 m/s to 23 degrees at 50 m/s. Lowering the gain for higher velocities would increase the phase margin a little but decrease the upper gain margin. Again, the phase margin decreases with increasing velocity but the system remains stable up to 50 m/s.

An interesting measure for the overall performance of the vehicle convoy is given by the transfer function $K_{st} S_{st} G_1$ from the steering angle δ_1 of the leading vehicle to the steering angle δ_2 of the following vehicle, where S_{st} denotes the sensitivity function $S_{st} = (I + G_2 K_{st})^{-1}$, see Figure 5. For velocities of more than 20 m/s a peak occurs in the frequency range of 4 to 8 rad/s and a second smaller peak at 20 rad/s. Considering that a driver is well capable to steer in the frequency range of the first peak, the steering input to the leading vehicle will cause steering actions of the following vehicle with higher amplitude. Regarding the linearized vehicle model, this causes no trouble. But with respect

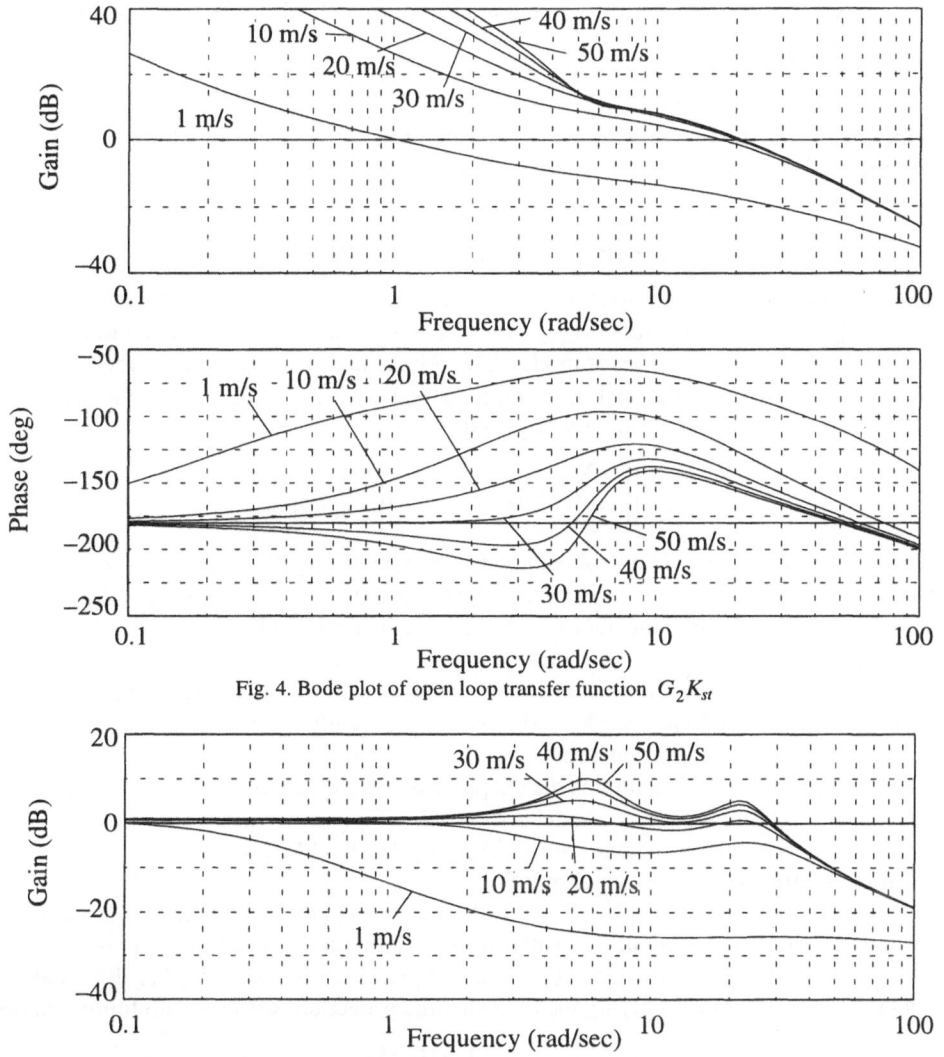

Fig. 4. Bode plot of open loop transfer function $G_2 K_{st}$

Fig. 5. Bode plot of transfer function $K_{st} S_{st} G_1$

to the real (nonlinear) vehicle convoy where increased steering amplitude means getting closer to the limits of tire adhesion this also means that the following vehicle can reach the physical limit to instability even if the driver of the leading vehicle steers within reasonable bounds.

3.3 ROBUST $H\infty$ CONTROLLER

In this paper, an $H\infty$ controller is developed based on a normalized left coprime factorization (NLCF) of the plant, see McFarlane and Glover [5] or Raisch [10]. The calculations are carried out using MATLAB, Chiang and Safonov [1]. The NLCF approach uses the decomposition

$$G = M^{-1}N \tag{5}$$

of the plant transfer function where M and N are stable transfer functions representing a normalized left coprime factorization of G. The supposed real plant with model errors is given by

$$G_r = M_r^{-1}N_r = (M + \Delta_M)^{-1}(N + \Delta_N). \tag{6}$$

Δ_M and Δ_N are stable transfer functions representing the class of factorized errors $\mathfrak{D}_{MN} := \{[\Delta_M \ \Delta_N] | \bar{\sigma}[\Delta_M(j\omega) \ \Delta_N(j\omega)] < l_{MN}(\omega)\}$, where $l_{MN}(\omega)$ is a bound for the error and thus determines the error model. The robust stabilization problem then is to find a feedback controller which stabilizes G_r for all $[\Delta_M \ \Delta_N] \in \mathfrak{D}_{MN}$. This is the case if the controller stabilizes the nominal plant G and

$$\left\| \begin{bmatrix} I \\ K \end{bmatrix} (I + GK)^{-1}M^{-1} \right\|_\infty \leq \frac{1}{\varrho}, \tag{7}$$

where $\varrho = \sup_\omega l_{MN}(\omega)$. Thus, robustness is given regarding the class of factorized model errors $\mathfrak{D}_{MN}$. An advantage of this approach is that the optimal bound $\varrho_{\max}$, the measure for robustness regarding the factorized errors, can be calculated in advance without choosing an error model $l_{MN}(\omega)$ explicitly. In order to introduce quantitative requirements, the singular values (an extension of the Bode diagram to MIMO systems), of the plant G can be formed via open loop shaping using weighting matrices $W_i(s)$ and $W_o(s)$ for the plant inputs and outputs, respectively. The $H\infty$ algorithm then tries to calculate a controller $K'_{H\infty}(s)$ that maintains the shape demanded by the user with the best possible robustness regarding factorized model errors. The weighting matrices are finally incorporated into the controller as $K_{H\infty} = W_i K'_{H\infty} W_o$. In the SISO case instead of the two weighting matrices only one weighting function $W(s)$ is needed.

For the vehicle convoy a NLCF controller has been calculated without applying loop shaping as a first step. A stable controller is obtained with the phase margin ranging from 76 degrees at 1 m/s to 31 degrees at 50 m/s and gain margins being higher than with the standard controller, too. Applying the loop shaping procedure with the weighting function

$$W(s) = \frac{1 + 0.25s}{1 + 0.005s} \tag{8}$$

to increase the slope of the magnitude of the open loop transfer function at the gain cross-over frequency, these margins are increased further, see Figure 6. The $H\infty$ controller is more robust than the standard controller. Phase margin now ranges from 93 degrees at 1 m/s to 53 degrees at 50 m/s. Note that a unique controller has been calculated for each longitudinal velocity ($v_x = 1, 10, 20, 30, 40, 50$ m/s) in order to examine this control scheme in greater detail. Figure 7 displays the Bode plot of the transfer function $K_{H\infty} S_{H\infty} G_1$ from δ_1 to δ_2, with the sensitivity function $S_{H\infty} = (I + G_2 K_{H\infty})^{-1}$, for such an adaptive controller. The main magnitude peak could not be diminished but compared to the standard controller the second, smaller peak is not present any more. Further inves-

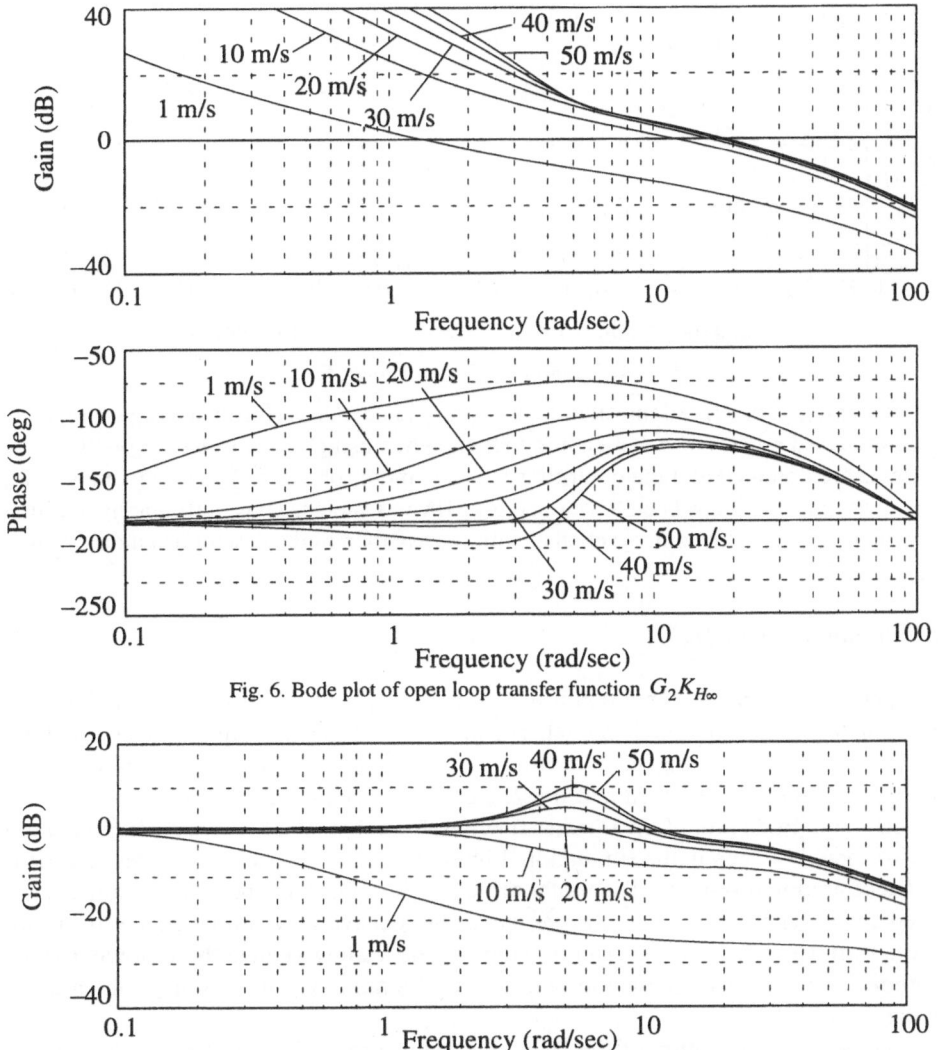

Fig. 6. Bode plot of open loop transfer function $G_2 K_{H\infty}$

Fig. 7. Bode plot of Transfer Function $K_{H\infty} S_{H\infty} G_1$

tigations showed that the main peak cannot be reduced by a linear standard or a NLCF controller. This seems to be due to the use of the single sensor signal μ_2 to control the convoy.

The Bode plot of the pure $H\infty$ controller part $K'_{H\infty}$ is shown in Figure 8. Obviously, with increasing velocity, additional phase is added by the $H\infty$ approach which improves stability as shown above. Another interesting point is the stationary gain of the controller. At 1 m/s the stationary controller gain is 0.98 (−0.17 dB) but it decreases with increasing velocity to 0.49 (−6.2 dB) at 50 m/s which improves stability even more. At higher velocities the stationary tracking performance is less sensitive to the stationary gain of the controller.

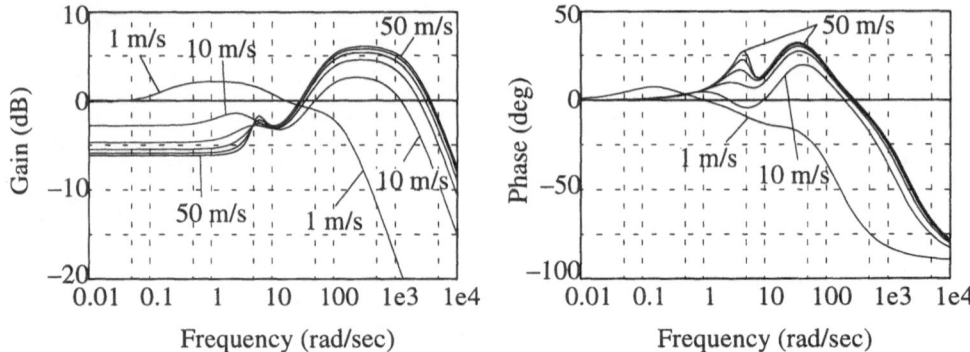

Fig. 8. Bode plot of $H\infty$ controller $K'_{H\infty}$

In summary, the $H\infty$ controller does not provide better performance regarding distur-bances caused by steering the leading vehicle when using μ_2 as sensor signal but better margins are obtained and the controller is hence more robust. A further advantage of the $H\infty$ control is that it has been developed for MIMO systems. Thus, introducing additional sensor signals into the control scheme is more easy than in the case of a standard control-ler.

4. Simulation Results

The performance of the three controllers is now investigated through simulations using the complex nonlinear vehicle model. The maneuver chosen is a sinusoidal steering input by the driver of the leading vehicle with a frequency of 6 rad/s and a longitudinal velocity of 30 m/s. Both, the standard and the $H\infty$ controller show the same peak magnitude of 4.8 dB in the Bode plot of the transfer function from δ_1 to δ_2, hence a corresponding be-havior should be seen in the simulation. Figure 9 displays the steering angles and the lat-eral acceleration responses. The trailer principle shows instable behavior as already indi-cated above. The standard and the $H\infty$ controller behave very similar except for a slight phase shift. Compared to the steering angle of the leading vehicle the following vehicle operates with higher amplitude in both cases. The ratio of 1.84 (5.3 dB) compares well to the Bode plots.

The lateral acceleration of the following vehicle exceeds that of the leading vehicle as expected, too, and hence the following vehicle is closer to the limits. This is also indi-cated by the amplitude of the lateral deviation from the center line. The simulations show 4 cm for the leading vehicle and 6 cm for the following vehicle. If the steering amplitude is increased, the following vehicle reaches the physical limits first and starts to skid, espe-cially if the steering frequency is reduced. Nevertheless, the results of the linear system analysis are confirmed well.

5. Experimental Results

An experimental convoy has been set up using two passenger cars connected by a rigid tow bar where the following vehicle is equipped with a steering actuator, see Petersen,

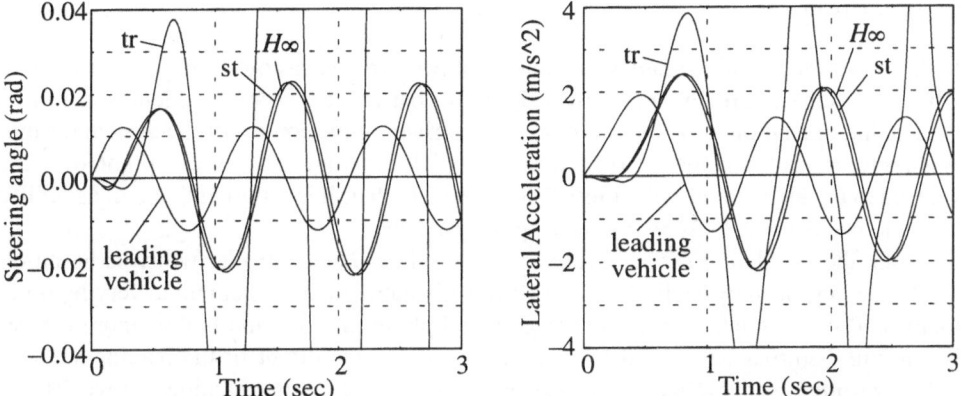

Fig. 9. Simulation results with complex vehicle model, sinusoidal steering input of leading vehicle

Rükgauer and Schiehlen [9]. The experiments are still at an early stage using the standard controller. Due to safety considerations the velocity is restricted to low speed at the time being. As a first step attention is focused on the limit where the vehicle convoy reaches instability. Therefore, the driver of the leading vehicle was given the task to follow a sla-lom–path. The following vehicle was steered using a purely proportional controller ($T_d = 0$, $T_1 = 0$) where the gain was increased with every run. Figure 10 (left) displays the resulting steering wheel angle δ_{sw} and tow–bar angle μ_2 for a controller gain of 10 where the system gets unstable at walking speed already. Almost immediately after the longitudinal motion started the steering wheel starts to swing up until, after some few peri-ods, the steering wheel stop is reached. Similarly the tow–bar angle μ_2 increases, too. A simulation using the complex vehicle model with a controller gain of $K_{stat} = 10$ is stable. If the gain is increased to 20 a behavior similar to the experiment can be seen, Figure 10 (right), but the rise of the oscillation amplitude is slower.

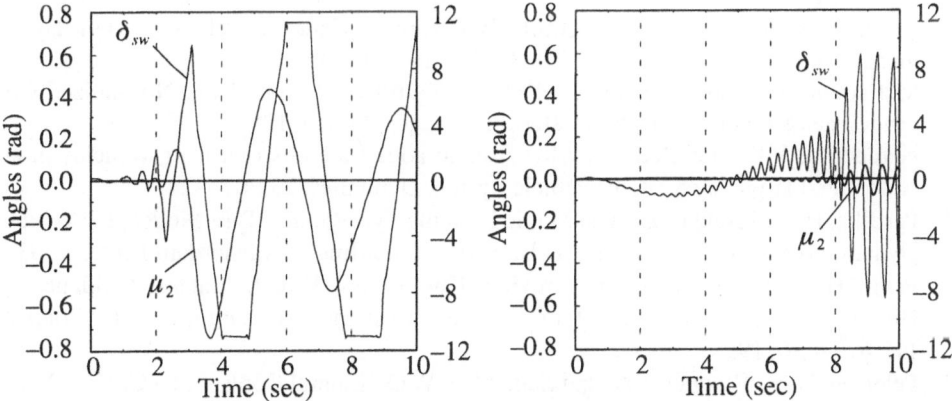

Fig. 10. Comparison of experiment (left plot) and simulation (right plot): low speed with instable controllers, left scale: steering wheel angle μ_2 , right scale: tow–bar angle δ_{sw}

6. Conclusions

In this paper the lateral dynamics and control of two vehicles in convoy have been studied, where the vehicles are connected by a rigid tow–bar and the second vehicle is to follow the leading one autonomously. Main focus was given to the control of the system design. A simple control scheme derived from a vehicle dynamics engineers point of view showed to be unstable for velocities of 20 m/s and higher. It is reported that the instability is due to phase lag caused by the steering mechanism.

Both, the standard PDT_1 controller and the robust $H\infty$ controller which is based on the NLCF approach by McFarlane and Glover [5] stabilize the system up to very high velocities. The $H\infty$ controller has better gain and phase margins and is thus more robust. Regarding disturbances caused by steering actions of the driver of the leading vehicle, both controllers show a peak in a relevant frequency range for velocities above 20 m/s which seems to be due to the use of the tow–bar angle μ_2 as the sensor signal only. Due to this, the following vehicle may reach the physical limit even if the driver of the leading vehicle steers within resonable bounds. Simulations confirm this result. Furthermore, experiments have been carried out using two specially equipped passenger cars. Investigations on the border to instabilty for low velocities correspond well with simulation results, where the real vehicle convoy behaves more vigorously than in simulations.

References

1. Chiang, R.Y. and Safonov, M.G.: Robust Control Toolbox for Use with MATLAB – User's Guide. Natick, MA: The Math Works Inc., 1992.
2. Fujioka, T. and Suzuki, K.: Control of Longitudinal and Lateral Platoon Using Sliding Control. Vehicle System Dynamics, 23:647–664, 1994.
3. Glora, M.: Dynamik von Fahrzeuggespannen. Stuttgart: University, Institute B of Mechanics, 1994.
4. Kortüm, W. and Sharp, R.S.: Multibody Computer Codes in Vehicle System Dynamics. Swets & Zeitlinger, Lisse, The Netherlands, 1993.
5. McFarlane, D.C. and Glover, K.: Robust Controller Design Using Normalized Coprime Factor Plant Descriptions. Berlin: Springer–Verlag, 1989.
6. Narendran, V.K. and Hedrick, J.K.: Autonomous Lateral Control of Vehicles in an Automated Highway System. Vehicle System Dynamics, 23:307–324, 1994.
7. Pacejka, H. and Bakker E., The Magic Formula Tyre Model, Tyre Models for Vehicle Dynamic Analysis. Ed. Pacejka, H.B., Swets & Zeitlinger, Amsterdam/Lisse, 1993.
8. Peng, H. and Tomizuka, M.: Preview Control for Vehicle Lateral Guidance in Highway Automation. Journal of Dynamic Systems, Measurement and Control, 115:679–686, 1993.
9. Petersen, U.N., Rükgauer, A. and Schiehlen, W.O.: Lateral Control of a Convoy Vehicle System. In: The Dynamics of Vehicles on Roads and Tracks. Ed.: L. Segel. Swets & Zeitlinger, Lisse, Netherlands, 1996.
10. Raisch, J.: Mehrgrößenregelung im Frequenzbereich. München; Wien: Oldenbourg Verlag, 1994.

CONTROL OF NONLINEAR BEAM VIBRATIONS BY MULTIPLE PIEZOELECTRIC LAYERS

K. SCHLACHER
Institute of Automatic Control and Electrical Drives
Johannes Kepler University of Linz, A-4040 Linz, Austria

H. IRSCHIK
Institute of Technical Mechanics and Foundations of Machine Design
Johannes Kepler University of Linz, A-4040 Linz, Austria

AND

A. KUGI
Institute of Automatic Control and Electrical Drives
Johannes Kepler University of Linz, A-4040 Linz, Austria

1. Introduction

In the first part of this contribution the mechanical model of the piezoelectric beam with multiple layers is derived. Since some results of the control loop design by differential geometric methods are restricted to ordinary differential equations, the partial differential equation of the beam is approximated by a finite set of ordinary differential equations following Galerkin's method. No damping is assumed, therefore the mechanical model is a Hamiltonian one. The approximating equations are of a special type, called Hamiltonian AI–systems.

The second part of this contribution presents the design of the controller. After a short introduction to Hamiltonian AI–systems, the main results of the input output linearization for this special type are summarized. In contrast to general AI–systems, only the knowledge of Poisson brackets is needed to do all calculations, and the stability test is based on the Hamiltonian function only. Finally, numerical simulations show the excellent behavior of the controlled beam.

2. The Piezoelectric Beam

A straight composite beam under the action of lateral loadings is considered. The case of plane flexural vibrations taking place in the (x, z)–plane is studied, x denoting the axial coordinate with $0 \leq x \leq L$. In order to control the beam vibrations, piezoelectric layers are used as actuators, where the piezoelectric layers and the substrate are perfectly bonded. Linear constitutive equations for an elastic isotropic material are taken into account, ε

D. H. van Campen (ed.), IUTAM Symposium on Interaction between Dynamics and Control in Advanced Mechanical Systems, 355–362.

denoting the strains

$$\begin{bmatrix} 1 & -\nu & -\nu \\ -\nu & 1 & -\nu \\ -\nu & -\nu & 1 \end{bmatrix} \begin{bmatrix} \sigma_{xx} \\ \sigma_{yy} \\ \sigma_{zz} \end{bmatrix} = E \left(\begin{bmatrix} \varepsilon_{xx} \\ \varepsilon_{yy} \\ \varepsilon_{zz} \end{bmatrix} - \begin{bmatrix} c_{xxz} \\ c_{yyz} \\ c_{zzz} \end{bmatrix} E_z \right) , \tag{1}$$

where it is assumed, that the electrical field densities E_x and E_y vanish (Nowacki, 1975 and Tzou et al, 1992). The electrical flux density D_z is related to the stresses σ and the electrical field density E_z by

$$D_z = \begin{bmatrix} c_{xxz} & c_{yyz} & c_{zzz} \end{bmatrix} \begin{bmatrix} \sigma_{xx} \\ \sigma_{yy} \\ \sigma_{zz} \end{bmatrix} + \epsilon_{zz} E_z , \tag{2}$$

with Young's modulus E and Poisson's ratio ν. The voltage U and flux density D_z in a single piezoelectric layer with thickness h satisfy the eq.

$$U = \frac{h}{\epsilon_{zz}} D_z . \tag{3}$$

Under the assumptions $\sigma_{yy} = \sigma_{zz} = 0$, eqs. 1, 2 and 3 lead to

$$\sigma_{xx} = E^* (\varepsilon_{xx} - \Lambda) , \qquad E^* = E \frac{\epsilon_{zz}}{\epsilon_{zz} - c_{xxz}^2 E} , \qquad \Lambda = \frac{c_{xxz}}{h} U \tag{4}$$

where Λ denotes the eigenstrain induced by the electrical field. The longitudinal strain ε_{xx} is expressed by means of the Bernoulli–Euler assumption in the non-linear form (Ziegler, 1995), w and u denoting the deflection and and the axial deformation, respectively

$$\varepsilon_{xx}(z) = \frac{\partial u}{\partial x} + \frac{1}{2}\left(\frac{\partial w}{\partial x}\right)^2 - z\frac{\partial^2 w}{\partial x^2} . \tag{5}$$

Inserting this relation into eq. 4 and integrating with respect to the cross-section $A = \sum_{i=0}^m A_i$ gives the stress resultants

$$M = \sum_{i=0}^m \int_{A_i} \sigma_{xx} z \mathrm{d}A_i = \sum_{i=0}^m \int_{A_i} E^* (\varepsilon_{xx} - \Lambda_i) z \mathrm{d}A_i = -B\left(\frac{\partial^2 w}{\partial x^2} + \sum_{i=1}^m \kappa_i\right) ,$$

$$N = \sum_{i=0}^m \int_{A_i} \sigma_{xx} \mathrm{d}A_i = \sum_{i=0}^m \int_{A_i} E^* (\varepsilon_{xx} - \Lambda_i) \mathrm{d}A_i = D(\varepsilon_{xx}(0) - \bar{\varepsilon}_{xx}) , \tag{6}$$

where the voltage U_0 vanishes in the substrate $i = 0$ and the voltages U_i in the areas $i = 1, \ldots, m$ serve as plant inputs. In eq. 6, the beam axis $z = 0$ is chosen such that $\int_A E^* z \mathrm{d}A = 0$. The effective stiffness parameters are $B = \int_A E^* z^2 \mathrm{d}A$ and $D = \int_A E^* \mathrm{d}A$. The induced curvature κ_i is given by

$$\kappa_i = \frac{1}{B} \int_{A_i} E^* \Lambda_i z \mathrm{d}A_i . \tag{7}$$

For beams built up symmetrically with respect to $z = 0$, with a skew symmetric distribution of all Λ_i, the mean induced strain $\bar{\varepsilon}_{xx} = \frac{1}{D} \sum_{i=1}^m \int_{A_i} E^* \Lambda_i \mathrm{d}A_i$ in eq. 6 vanishes. Thus

$$\bar{\varepsilon}_{xx} = 0 \rightarrow N = D\varepsilon_{xx}(0) . \tag{8}$$

Neglecting longitudinal and rotational inertia, we use the principle of D'Alembert to derive the equations of motion in the form

$$\frac{\partial N}{\partial x} = 0 , \qquad \frac{\partial}{\partial x}(N\frac{\partial w}{\partial x} + \frac{\partial M}{\partial x}) + q_z = \mu\frac{\partial^2 w}{\partial t^2} . \qquad (9)$$

The lateral loading is denoted by q_z, and $\mu = \int_A \rho dA$, with the mass density ρ. Simply supported flexural boundary conditions and fixed longitudinal end conditions are studied subsequently

$$x \in \{0, L\} : \qquad w = 0 , \qquad M = -B\left(\frac{\partial^2 w}{\partial x^2} + \sum_{i=1}^{m}\kappa_i\right) = 0, \qquad u = 0 . \qquad (10)$$

Taking $B = const.$, $D = const.$, we derive the following nonlinear initial–boundary–value problem for the deflection w from the eqs. above in a straightforward manner

$$\mu\frac{\partial^2 w}{\partial t^2} + B\frac{\partial^4 w}{\partial x^4} - \frac{D}{L}\left(\frac{1}{2}\int_0^L(\frac{\partial w}{\partial x})^2 dx\right)\frac{\partial^2 w}{\partial x^2} = q_z - B\sum_{i=1}^{m}\frac{\partial^2}{\partial x^2}\kappa_i \qquad (11)$$

for $0 \le x \le L$ and

$$x \in \{0, L\} : \qquad w = 0, \qquad \frac{\partial^2 w}{\partial x^2} = -\sum_{i=1}^{m}\kappa_i . \qquad (12)$$

Proper initial conditions have to be prescribed for a unique solution of this problem. In the following, the two lateral loadings $q_1 + q_2 = q_z$ are assumed to be space-wise constant, $q_1(x, t) = F_1(t)$, and linear, $q_2(x, t) = (x/L) F_2(t)$, respectively. Accordingly, the space–wise distributions of the induced curvatures are taken in the form

$$\kappa_1(x, t) = U_{c,1}(t)\left(1 - \frac{x}{L}\right)x , \qquad \kappa_2(x, t) = U_{c,2}(t)\left(1 - \left(\frac{x}{L}\right)^2\right)\frac{x}{6} , \qquad (13)$$

where $U_{c,1}(t)$ and $U_{c,2}(t)$ formally serve as the plant inputs at the overall beam level. Using the following problem–oriented scaling with respect to the total height of the beam $H = \tilde{H}L$

$$x = \tilde{x}L , \qquad w = \tilde{w}H , \qquad F_1 = \frac{BH}{L^4}\tilde{F}_1 , \qquad U_{c,1} = \frac{H}{2L^3}\tilde{U}_{c,1} ,$$

$$t = \tilde{t}\sqrt{\frac{\mu L^4}{B}} , \qquad \tilde{k} = \frac{DL^2}{B} , \qquad F_2 = \frac{BH}{L^4}\tilde{F}_2 , \qquad U_{c,2} = \frac{H}{L^3}\tilde{U}_{c,2} , \qquad (14)$$

where a tilde refers to a non–dimensional quantity, we find that the problem defined in eq. 11 and eq. 12 becomes

$$\frac{\partial^2 \tilde{w}}{\partial \tilde{t}^2} + \frac{\partial^4 \tilde{w}}{\partial \tilde{x}^4} - \tilde{k}\left(\frac{\tilde{H}^2}{2}\int_0^1(\frac{\partial \tilde{w}}{\partial \tilde{x}})^2 d\tilde{x}\right)\frac{\partial^2 \tilde{w}}{\partial \tilde{x}^2} = \tilde{F}_1(\tilde{t}) + \tilde{U}_{c,1}(\tilde{t}) + \tilde{x}\left(\tilde{F}_2(\tilde{t}) + \tilde{U}_{c,2}(\tilde{t})\right) \qquad (15)$$

for $0 \le \tilde{x} \le 1$ and

$$\tilde{x} \in \{0, 1\} : \qquad \tilde{w} = 0 , \qquad \frac{\partial^2 \tilde{w}}{\partial \tilde{x}^2} = 0 . \qquad (16)$$

It is a special feature of the actuator characteristic presented in eq. 13 that the effect of the piezoelectric actuator layers is the same as the influence of the loadings.

The deflection now is approximated using the finite series

$$\tilde{w}^*(\tilde{x}, \tilde{t}) = \sum_{i=1}^{l} X_i(\tilde{t}) \sin(i\pi\tilde{x}) , \quad 0 < l < \infty , \tag{17}$$

Running through Galerkin's procedure (Ziegler, 1995), we obtain the following set of non-linear ordinary differential equations

$$\ddot{X} + (i\pi)^2 X_i \left((i\pi)^2 + \frac{\tilde{k}\tilde{H}^2}{4} \sum_{j=1}^{l} (j\pi)^2 X_j^2 \right) = \tag{18}$$

$$\frac{2}{\pi i} (1 - \cos i\pi) \left(\tilde{F}_1(\tilde{t}) + \tilde{U}_{c,1}(\tilde{t}) \right) - \frac{2}{\pi i} \cos i\pi \left(\tilde{F}_2(\tilde{t}) + \tilde{U}_{c,2}(\tilde{t}) \right)$$

for $i = 1, \ldots, l$.

With the state vector $x \in R^n$, $n = l$, eq. 18 can be written as

$$\frac{d}{dt} x = a(x) + b(x)u + g(x)d , \tag{19}$$

$$x = \begin{bmatrix} x_1 \\ x_2 \end{bmatrix} \quad \text{and} \quad \mathbf{x}_1^T = [X_1, \ldots, X_l] , \quad \mathbf{x}_2^T = \left[\dot{X}_1, \ldots, \dot{X}_l \right]$$

with the plant input $u = \begin{bmatrix} \tilde{U}_{c,1} & \tilde{U}_{c,2} \end{bmatrix}^T$, the disturbance $d = \begin{bmatrix} \tilde{F}_1 & \tilde{F}_2 \end{bmatrix}^T$ and the vectors $a(x)$, $b(x)$ and $g(x)$ of real valued functions following from eq. 18. Note that for $u = 0$ and $d = 0$, eq. 19 possesses only one stationary point, that is $x_s = 0$. Note too that the finite state space model eq. 19 is a Hamiltonian AI–system with the Hamiltonian function

$$H(q, p) = \sum_{i=1}^{l} \frac{2q_i}{\pi i} ((1 - \cos(i\pi)) (u_1 + d_1) - \cos(i\pi) (u_2 + d_2)) + \tag{20}$$

$$\frac{1}{2} \sum_{i=1}^{l} p_i^2 + \sum_{i=1}^{l} \frac{(i\pi)^2}{2} \left((i\pi)^2 + \frac{\tilde{k}\tilde{H}^2}{8} \sum_{j=1}^{l} (j\pi)^2 q_j^2 \right) q_i^2$$

with $X_i = q_i$ and $\dot{X}_i = p_i$. Therefore it is natural to pose a question: What can control theory offer for such systems.

3. Hamiltonian AI–Systems

Let $\mathcal{M}$ be the configuration manifold with cotangent bundle $T^*\mathcal{M}$ of a Hamiltonian AI–system with input $u \in R^m$ and Hamiltonian function $H : T^*\mathcal{M} \times R^m \to R$

$$H = H_0 - \sum_{i=1}^{m} H_i u_i . \tag{21}$$

The equations of motion are given locally by

$$\frac{d}{dt} q_i = \frac{\partial}{\partial p_i} H_0(q, p) - \sum_{i=1}^{m} u_i \frac{\partial}{\partial p_i} H_i(q, p)$$

$$\frac{d}{dt} p_i = -\frac{\partial}{\partial q_i} H_0(q, p) + \sum_{i=1}^{m} u_i \frac{\partial}{\partial q_i} H_i(q, p) , \tag{22}$$

where $(q_i, p_i) \in D \subset T^*R^n$ are canonical coordinates and H_0 is the Hamiltonian of the free system (Nijmeijer *et al*, 1989). The cotangent bundle T^*R^n possesses the non degenerate two form $\Omega = \sum_{i=1}^n dq_i \wedge dp_i \in \wedge^2(T^*R^n)$, which is called symplectic. With the abbreviation $x^T = \left[q^T, p^T\right]$, eq. 22 takes the form $\frac{d}{dt}x = X_H$, with the vector field $X_H \in TT^*R^n$, $X_H \rfloor \Omega = dH$, where $X_H \rfloor \Omega$ denotes the interior product of the vector $X_H \in TT^*R^n$ and the two form Ω. dH is the exterior derivative of H. These considerations can be easily extended from T^*R^n to $T^*\mathcal{M}$, since $T^*\mathcal{M}$ possesses a non degenerate symplectic form Ω (Choquet–Bruhat *et al*, 1982).

The rate of change of a function $f : T^*\mathcal{M} \to R$ as a result of the evolution of the system is measured by the Poisson bracket (Nijmeijer *et al*, 1989)

$$\{H, f\} = X_H \rfloor df = X_H \rfloor (X_f \rfloor \Omega) \ , \tag{23}$$

and the eqs. of motion can be recovered from the Hamiltonian function by the relations

$$\frac{d}{dt}x_i = \{H, x_i\} \ . \tag{24}$$

The rate of change of the total energy H_0 of the system eq. 24

$$
\begin{aligned}
\{H, H_0\} &= \left\{H, H + \sum_{i=1}^m H_i u_i\right\} \\
&= \sum_{i=1}^m \{H, H_i\} u_i
\end{aligned}
$$

leads to the natural outputs $y_i = H_i$, $i = 1, \ldots, m$ of the Hamilton control system, since $\dot{y}_i u_i = \{H, H_i\} u_i$ can be interpreted as the flow of power into the system caused by the input u_i (Nijmeijer *et al*, 1989).

4. Input Output Linearization

The design of the controller for the piezoelectric beam is based on the well known input output linearization (Isidori, 1989). There are some simplifications of this approach for Hamiltonian AI–systems eq. 24 with natural output y_j, $j = 1, \ldots, m$, (Nijmeijer *et al*, 1989)

$$
\begin{aligned}
\dot{x} &= X_{H_0} - \sum_{j=1}^m X_{H_j} u_j \\
y_j &= H_j \ .
\end{aligned}
\tag{25}
$$

With the definition of the k times repeated Poisson bracket eq. 23

$$\mathrm{ad}^0_{H_0} H_i = H_i \ , \qquad \mathrm{ad}^{k+1}_{H_0} H_i = \left\{H_0, \mathrm{ad}^k_{H_0} H_i\right\}$$

the ρ_i times differentiation of the output y_i leads to

$$y_i^{\rho_i} = \mathrm{ad}^{\rho_i}_{H_0} H_i + \sum_{j=1}^m \left\{\mathrm{ad}^{\rho_i-1}_{H_0} H_i, H_j\right\} u_j$$

where ρ_i is the smallest integer such $\left\{ \mathrm{ad}_{H_0}^{\rho_i-1} H_i, H_j \right\} \neq 0$ for some j. If the decoupling matrix

$$A_{ij} = \left\{ \mathrm{ad}_{H_0}^{\rho_i-1} H_i, H_j \right\} \tag{26}$$

is regular, which is assumed from now on, then the plant input transform

$$u_j = \sum_{k=1}^{m} B_{jk} \left(v_k - \sum_{i=0}^{\rho_k} \alpha_{k,i} \mathrm{ad}_{H_0}^i H_k \right) \qquad \text{with} \qquad \alpha_{k,\rho_k} = 1, \tag{27}$$

where B denotes the inverse of A yields a linear and decoupled closed loop system. This system is characterized by the transfer functions matrix

$$G(s) = \mathrm{diag} \left(\frac{1}{d_j(s)} \right) \quad , \quad d_j(s) = \sum_{i=0}^{\rho_j} \alpha_{j,i} s^i . \tag{28}$$

Clearly a further controller design for the linear and decoupled system eq. 28 is rather easy.

This method is applicable only, if the zero dynamics is stable. It is a well known result, that the closed loop is stable in the neighborhood of a stationary point x_0, iff the system eq. 25 restricted to the output nulling manifold

$$\mathcal{N} = \left\{ x \in R^{2n} \mid H_i = \ldots = \mathrm{ad}_{H_0}^{\rho_i-1} H_i = 0 , \quad i = 1, \ldots, m \right\} \tag{29}$$

is stable around $x_0 \in \mathcal{N}$ (Isidori, 1989 and Nijmeijer et al, 1989). The dimension $n_{\mathcal{N}} = 2n - \sum_{i=1}^{m} \rho_i > 0$ of $\mathcal{N}$ is assumed, otherwise there is no zero dynamics. If i denotes the inclusion map $i : \mathcal{N} \to \mathcal{M}$, then the pull back of H is given by i^*H and the pull back of Ω by $i^*\Omega$ respectively. It can be shown, that $\mathcal{N}$ is a symplectic submanifold (the form $i^*\Omega$ is non degenerate), iff $\mathcal{N}$ is not empty and the decoupling matrix A eq. 26 is regular. Therefore the zero dynamics of a Hamiltonian AI–systems has a special structure, it is a Hamiltonian system, too. The following theorem summarizes the important results for the closed loop design (Nijmeijer et al, 1989).

Theorem 1 $Consider$ the $Hamiltonian$ AI–$system$ $eq.$ 25 on a $symplectic$ $manifold$ $\mathcal{M}$ $with$ $X_{H_0}(x_0) = 0$, $H_j(x_0) = 0$, $j = 1, \ldots, m$ $such$ $that$ A $eq.$ 26 is $regular$ for all $x \in \mathcal{M}$. $Suppose$ $that$ the $pull$ $back$ i^*H_0 of H_0 to the $output$ $nulling$ $manifold$ $\mathcal{N}$ $eq.$ 29 has a $strict$ $local$ $minimum$ in $x_0 \in \mathcal{N}$. $Then$ $there$ $exists$ a $decoupling$ $regular$ $static$ $feedback$ $u = \alpha(x) + \beta(x)v$ $such$ $that$ the $closed$ $loop$ is $locally$ $stable$ $around$ x_0 for $v = 0$.

There are some important impacts on the application of this method. Only the knowledge of the Hamiltonian function eq. 21 and the Poisson bracket eq. 23 is necessary for doing all calculations. If the decoupling matrix A is regular, then $\mathcal{N}$ is a symplectic submanifold. The stability test uses only the pull back i^*H_0 of the Hamiltonian function H_0 to $\mathcal{N}$. The equilibrium point x_0 of the closed loop is stable but never asymptotically stable.

5. Controller Design and Numerical Simulations

The controller design is based on the input output linearization of Hamiltonian AI–systems, where the natural output functions are taken. All symbolic calculations are done with the computer algebra system MACSYMA and the package $nlinear.fas$ for AI–systems

(Kugi, 1995), all numerical calculations are done with MATLAB. Since there can be handled only a finite number of differential equations, here the choice is $l = 6$ in eq. 18. During all presented simulations the values for the parameters of the mechanical model are

$$\tilde{k} = 1.2 \cdot 10^5 , \qquad \tilde{H} = 0.01$$

and the following initial conditions

$$X_i(0) = 0, i \geq 1 , \qquad \dot{X}_i(0) = 0, i \geq 1$$

are used. The choice of the parameters of the controller is $\alpha_{1,0} = \alpha_{2,0} = 100$ and $\alpha_{1,1} = \alpha_{2,1} = 10$. In order to guarantee that the steady state error equals zero, two additional PI–controllers

$$R_1(s) = R_2(s) = 100\frac{s + 10}{s}$$

are used for the decoupled linear system eq. 28. Applying theorem 1 to the Hamiltonian function eq. 20, we see that the zero dynamics is stable.

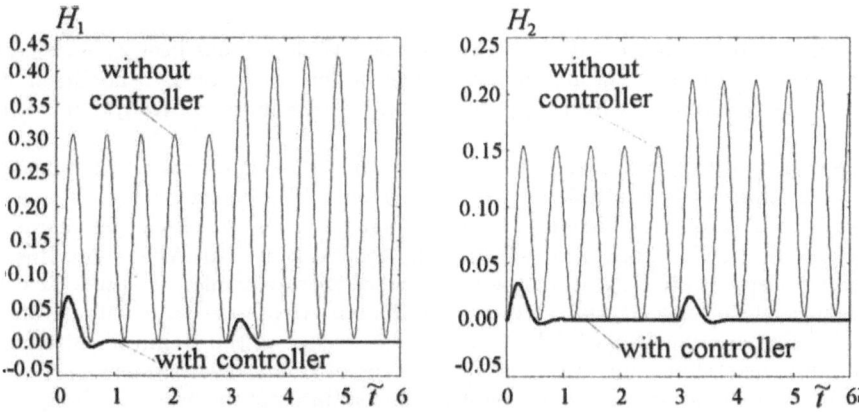

Figure 1. Simulations of the natural outputs for the open and closed loop.

Fig. 1 presents the behavior of the mechanical model for the two natural outputs $H_1 = \frac{4}{\pi}X_1(\tilde{t})$ and $H_2 = \frac{1}{\pi}(2X_1(\tilde{t}) - X_2(\tilde{t}))$ of the open loop (without controller) and the closed loop (with controller) for the disturbances $\tilde{F}_1(\tilde{t}) = 10\sigma(\tilde{t})$ and $\tilde{F}_2(\tilde{t}) = 10\sigma(\tilde{t} - 3)$ with the unit step $\sigma(\tilde{t})$. The first two modes are taken into consideration, because higher modes contribute less to the deflection $\tilde{w}^*$ of eq. 17. The deflection $\tilde{w}^*(\tilde{t}, 0.5)$ is approximately given by $\tilde{w}^*(\tilde{t}, 0.5) \approx X_1(\tilde{t}) \sin\frac{\pi}{2}$ and therefore $\lim_{\tilde{t} \to \infty} \tilde{w}^*(\tilde{t}, 0.5) \approx 0$ holds.

The simulations of the plant inputs $\tilde{U}_{c,1}(\tilde{t})$ and $\tilde{U}_{c,2}(\tilde{t})$ are shown in fig. 2. One can see the excellent performance of the closed loop system. By a simple change of the parameters of the control law it is possible to achieve a faster or slower behavior of the control system, depending on the allowed quantity of the plant inputs $\tilde{U}_{c,1}(\tilde{t})$ and $\tilde{U}_{c,2}(\tilde{t})$.

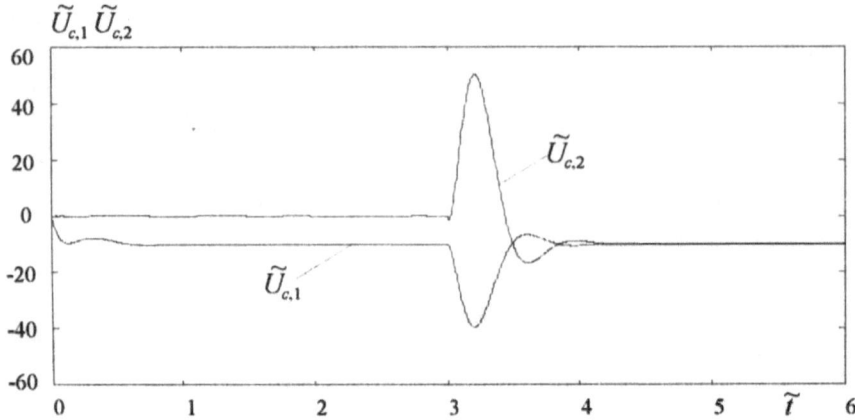

Figure 2. Simulations of the plant inputs.

6. Conclusion

A straight composite beam with multiple piezoelectric layers has been considered under the action of two loads. The mechanical model of the deflection is a nonlinear initial–boundary–value problem, which is approximated by a Hamiltonian AI-system with finite state.

The design of the controller is based on the theory of input output linearization for MIMO (multiple input multiple output) AI–systems. The control law obtained in this way leads to a linear and decoupled input–output behavior, which can be characterized by a diagonal transfer function matrix. For the general case of an AI–system, this design procedure is applicable only, if a subsystem, called zero dynamics, is stable. It can also be shown that the zero dynamics of a Hamiltonian AI–system with natural output is a Hamiltonian system. In particular, only the knowledge of the Hamiltonian function is required for the stability investigations of this type of Hamiltonian system.

Finally, numerical simulation results show the feasibility of the proposed design method for intelligent structures as well as the excellent performance of the proposed control loop.

References

Choquet–Bruhat, Y. and D_E Witt–Morette, C. (1982) *Analysis Manifolds and Physics*, North–Holland.
Isidori, A. (1989) *Nonlinear Control Systems*, Springer Verlag.
Kugi, A. (1995) *Nonlinear Control of Electrical Systems*, Ph.D. thesis, Johannes Kepler University of Linz, Austria.
Nijmeijer, H. and Van der Schaft, A.J. (1989) *Nonlinear Dynamical Control Systems*, Springer Verlag.
Nowacki, W. (1975) *Dynamic Problems of Thermoelasticity*, Noordhoff International Publishing, Leyden.
Tzou, H.S. and Anderson, G.L. (1992) *Intelligent Structural Systems*, Kluwer, Dordrecht.
Ziegler, F. (1995) *Mechanics of Solids and Fluids*, Springer Verlag.

DECOUPLING IN AUTOMOTIVE ACTIVE SUSPENSION SYSTEM DESIGN

R. S. Sharp
School of Mechanical Engineering
Cranfield University

Abstract

The problem considered is that of combining results and ideas obtained from simplified models of actively suspended vehicles to obtain a complete scheme for a real vehicle. It is argued that, if the road is considered to be isotropic and the vehicle laterally symmetrical, the symmetric motions of a vehicle in response to a cylindrical road profile and the anti-symmetric motions in response to a pure cross-level input can be superposed (assuming linearity) to give the total motion. The symmetric problem has been considered extensively in previous work but the anti-symmetric problem has not been separated out from the total and studied on its own before. The paper contains such a study. A mathematical model of a car with a limited bandwidth active suspension system and roll freedoms is excited by a cross-level input deriving from white noise. Suspension control laws are obtained by application of a form of linear quadratic optimal control theory and cost function values and frequency responses are used to evaluate performance. It is found that the active suspension is able to make only very modest contributions to improving the performance, in this anti-symmetric mode, of an adequately damped passive suspension.

Notation

ap, bp, cp	Padé filter matrices
C_s	suspension damping coefficient
k_s, k_t	suspension spring and tyre vertical stiffnesses
I_a, I_b	axle and body roll inertias
K	control gain matrix with elements k_{11} ...... k_{24}
l	vehicle wheelbase
q_1 ...... q_6	cost function weighting parameters
s, w	spring/damper and vehicle half track widths
t, T	time and time period
u	vehicle speed
u_c	control signal
z	state vector for Padé filter

363

D. H. van Campen (ed.), IUTAM Symposium on Interaction between Dynamics and Control in Advanced Mechanical Systems, 363–374.
© 1997 *Kluwer Academic Publishers.*

α	distance constant of road input filter
ϕ_a, ϕ_b, ϕ_i	axle, body and road input angles
ϕ_c	actuator angle relative to vehicle body
ξ_1, ξ_2	white noise processes
ζ	damping factor for actuator dynamics
ω_n	natural circular frequency for actuator dynamics

subscripts: f, front; r,rear

1. Introduction

Over the last 20 years, a large volume of research has been done into active suspension systems for road vehicles of various types. Mostly, the research has been notionally associated with more expensive cars although much of what has been discovered is of a generic nature and it is widely applicable. Several reviews have been written [1,2,3,4,5] and a comprehensive classified bibliography has been published [6]. Most active suspension studies can be characterised as separating out from the whole of the vehicle operating spectrum a particular segment, leaving in the background the matter of converting any narrowly conceived design into a useful form on an actual vehicle. A number of papers, on the other hand, relate to complete working systems [7,8,9,10] but mostly they do not describe the control schemes in detail. [10] is an exception to this rule. In that case, some simplification of the control system design was obtained from combining measurands into bounce, pitch, roll and warp components, each motion being treated by classical methods.

With respect to generic studies, the segmentation of the vehicle usage spectrum is useful insofar as it keeps the studies simple, it often allows straightforward generation of control laws and it allows easily understood results to be obtained [11,12,13,14,15]. The most common separation is into ride comfort issues (mainly vertical motions) and handling qualities issues (mainly horizontal motions but with a vital contribution from body roll). Most studies on ride comfort and tyre/road contact control have been conducted with quarter car models, in fact, so that there is then only one system disturbance input, there is nothing in the model to represent body pitching and there is no recognition of cross-level excitations. Also, most studies have included the actuator as an ideal force producer, without bandwidth limitations. The major results from these linear quarter car models are that (1) only one of the frequency response functions relating body acceleration, working space and tyre load variation to road vertical velocity input is independent and there are two invariant points in these responses [14]; (2) passive damping forces are necessary to maintain good tyre to ground contact; (3) actuator forces proportional to absolute body vertical velocity damp the body resonance condition very effectively without spoiling the high frequency vibration isolation properties; and (4) actuator forces to cancel the spring (if one is included to relieve the active system from supporting the body weight all the time) help to reduce body acceleration, enabling lower levels of damping force to be adequate.

Quarter car studies have been extended to half car ones, including front and rear, with the idea that pitching motion can be studied, the effect of the rear input being a time delayed version of that at the front can be included and advantage may be gained, in setting the control scheme, from the "wheelbase" preview afforded by that system [16,17]. Another small group of ride comfort studies is significant. In this group, control schemes for a full car have been devised. Linear optimal control theory was used in [12,18,19], without yielding much advance over the simpler cases, while in [20], a controller design was obtained by a modal decoupling procedure, following which the modes were dealt with independently. Very recently, a new method for achieving a complete system design has been devised [21,22]. The main idea here is to employ a high level controller to supervise the operation of lower level ones, which can be designed separately without prejudicing performance or robustness. The preferred strategy for devising a complete active suspension control scheme for a road vehicle remains an issue, and it is with contributing to progress in this area that the present paper is concerned.

The main ideas are that combining an isotropic roadway with a laterally symmetric vehicle will allow the suspension design problem to be broken down into two separate, simpler problems. The first of these is the bounce/pitch problem, which has been treated extensively already, while the second is the roll problem, which has not previously been separated out from the remainder of the problem. The paper proceeds to explain the basis for the separation into sub-problems and develops a control scheme for the anti-symmetric case, to illustrate the possibilities. A starting model for the study of the anti-symmetric problem contains front and rear axles with vertically flexible tyres, front and rear suspension systems, each with a roll degree of freedom, the cross-level road excitation described above and the time delay relating to the wheelbase travel time, this being the lag between the excitation arriving at the front and then the rear axle. This system has some obvious similarities to quarter-car and half-car systems but is significantly different from them, so that some new difficulties arise in the devising of control laws for it. The performance properties are largely unknown too, so that it is of interest to follow up on the control design by carrying out performance calculations.

The design envisaged and the method involved do not accommodate the handling aspect of the vehicle operating spectrum but this is only weakly coupled to the ride motions, depending on the mass centre height and some suspension system details, and the plan is to deal with that as a separate issue. The strong idea in this approach is that the suspension controls should be designed to give roll-free steering responses and that the system should be driven either by feedforward of steering wheel and road speed signals [23] or by feedback of lateral acceleration [24,25]. The paper will be concerned with the setting up of appropriate disturbance input and vehicle models for the anti-symmetric case described, with the techniques by which control laws for the system can be derived and their relationship to previously developed techniques, and with the performance

366

properties of the controlled systems resulting. The results will lead to a relatively simple way (in view of what is already known and understood) of envisaging and realising a complete control scheme for ensuring the ride and handling qualities of a two track road vehicle.

2. Anti-symmetric road and suspension models

2.1 THE ROAD SURFACE MODEL

The treatment of road surfaces as isotropic Gaussian random processes has been developed in [26,27,28] and applied to vehicles in [29]. The ideas were applied with clarity to active suspension research in [30], results from which will be used directly. An isotropic surface profile can be generated by combining two independent elevation signals, each derived by filtering white noise. One of the filtered signals represents the road elevation along the centreline of the vehicle path, relative to some zero mean level, directly, while the other represents an angular displacement at the centreline which has to be multiplied by the vehicle half track dimension to give the elevation contribution at the port or starboard wheels. The process is illustrated in Figure 1.

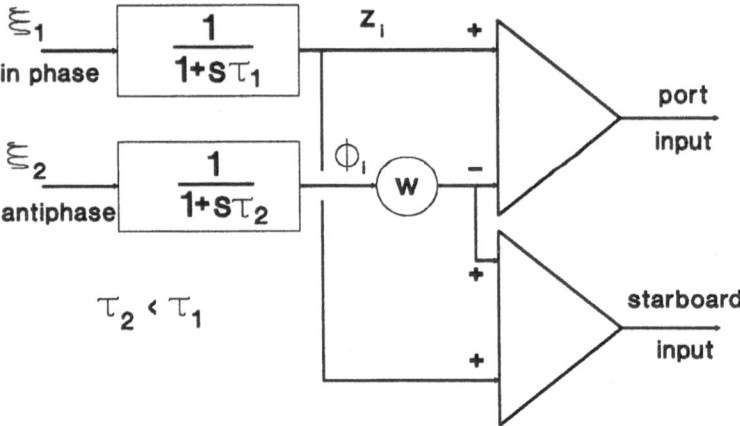

Figure 1 Generation of two tracks of an isotropic road surface from independent random signals.

Venhovens [30] established that employing a first order low pass filter with a distance constant of 0.833 m/rad for processing the white noise, cross-level, input component gives a reasonable representation of real roads. For a vehicle travelling with speed, u, the required filter time constant is $(0.833/u)$ s. For the anti-symmetric suspension study in focus, only this part of the road input is needed. The same input is taken to be applied to the front and rear axles of the vehicle, with the wheelbase travel time delay between the two.

2.2   THE VEHICLE MODEL

The front end of the vehicle is shown diagrammatically in Figure 2. The rear end corresponds precisely to the front and is not shown. The tyres are taken to be vertically compliant and to be completely free to move laterally without producing side forces. A rolling tyre relaxes after a short distance, to conform to this model [31]. Each axle and the body are taken to roll about their respective mass centres, without applying any mutual lateral constraints. Alternative treatments involving a common roll centre for axle and body would be somewhat more complex and almost imperceptibly different in result.

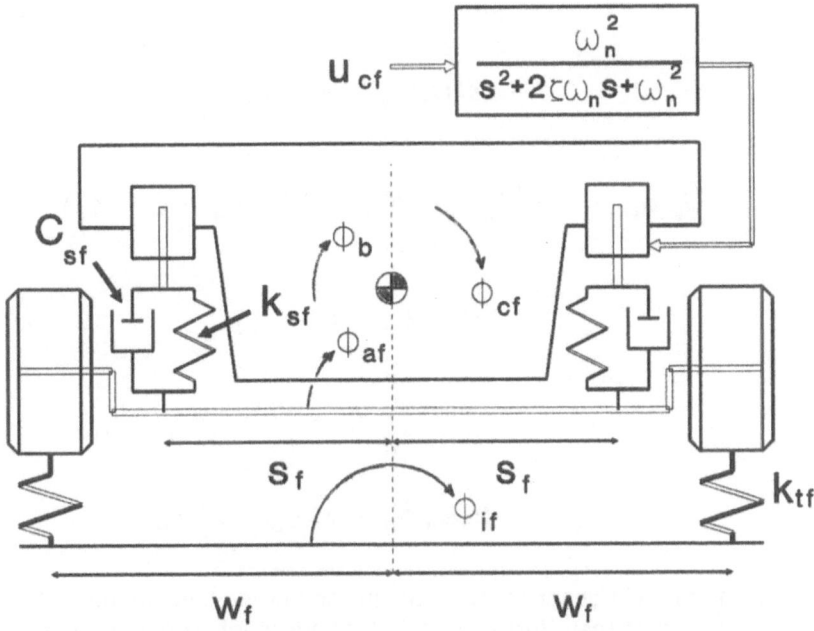

Figure 2 Diagrammatic representation of front end suspension system.

The suspension actuators are considered to be displacement controllers each in series with a parallel spring/damper conventional suspension unit, this being a mechanical analogue of those active suspensions which have been made commercially [7,8], although the latter are somewhat non-linear. Each controller has a bandwidth limitation, set by $\omega_n = 12\pi$ and $\zeta = 0.7$, to represent the idea that the system is slow-active as opposed to full authority. If the front and rear road inputs are initially considered to be uncorrelated, the equations of motion of the vehicle model can be written down by inspection of the system as:

$$\dot{\phi}_{if} = -2\pi U\alpha \; \phi_{if} + \xi_1 \qquad\qquad (1)$$

$$\dot{\phi}_{lr} = -2\pi u\alpha \; \phi_{lr} + \xi_2 \tag{2}$$

$$I_b\ddot{\phi}_b = 2s_f^2 k_s(\phi_{af} - \phi_b - \phi_{cf}) + 2s_f^2 C_s(\dot{\phi}_{af} - \dot{\phi}_b - \dot{\phi}_{cf})$$
$$+ 2s_r^2 k_s(\phi_{ar} - \phi_b - \phi_{cr}) + 2s_r^2 C_s(\dot{\phi}_{ar} - \dot{\phi}_b - \dot{\phi}_{cr}) \tag{3}$$

$$I_a\ddot{\phi}_{af} = 2w_f^2 k_t(\phi_{lf} - \phi_{af}) - 2s_f^2 k_s(\phi_{af} - \phi_b - \phi_{cf})$$
$$- 2s_f^2 C_s(\dot{\phi}_{af} - \dot{\phi}_b - \dot{\phi}_{cf}) \tag{4}$$

$$I_{ar}\ddot{\phi}_{ar} = 2w_r^2 k_t(\phi_{lr} - \phi_{ar}) - 2s_r^2 k_s(\phi_{ar} - \phi_b - \phi_{cr})$$
$$-2s_r^2 C_s(\dot{\phi}_{ar} - \dot{\phi}_b - \dot{\phi}_{cr}) \tag{5}$$

$$\ddot{\phi}_{cf} = -2\xi\omega_n \dot{\phi}_{cf} - \omega_n^2 \phi_{cf} + \omega_n^2 u_{cf} \tag{6}$$

$$\ddot{\phi}_{cr} = -2\xi\omega_n \dot{\phi}_{cr} - \omega_n^2 \phi_{cr} + \omega_n^2 u_{cr} \tag{7}$$

Equations (1) and (2) describe the road surface cross-level inputs to the axles, on the assumption that they derive from independent white noise processes. If, on the other hand, the rear input is to be regarded as a time delayed version of that for the front, a Padé approximation can be used in the structure of Figure 3, so that both axle disturbance signals derive from a single white noise process. A Padé filter, in state space form, to generate the rear input, ϕ_{lr}, from that at the front, ϕ_{lf}, is:

$$\dot{z} = [ap] \; z + [bp] \; \phi_{lf} \tag{8}$$

$$\phi_{lr} = [cp] \; z + \phi_{lf} \tag{9}$$

and the elements of [ap], [bp] and [cp] are generated automatically in MATLAB from the specification of the filter order and the time delay. To incorporate the

Padé filter into the former system, equation (2) is replaced by equations (8) and (9). The system order is raised by one less than the order of the Padé filter.

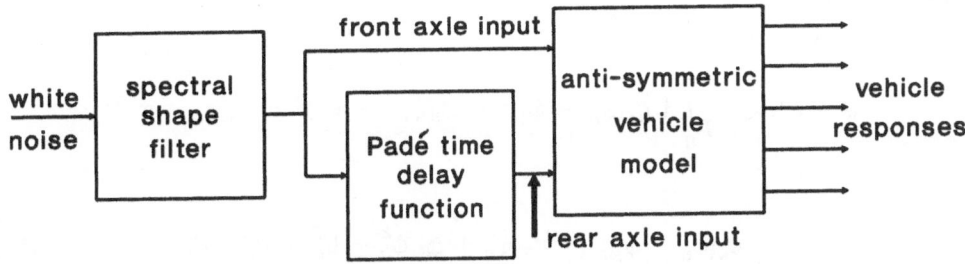

Figure 3   Use of a Padé approximation for the time delay between front and rear wheel inputs.

The equation sets described, (1-7) or (1,3-9) can be used to derive control laws for the actuators, as will be described below. When control laws relating u_{cf} and u_{cr} to the system states have been decided, the performance of the closed loop system can be examined through calculation of frequency responses. This is conveniently done in four stages, using equations (1-7) and the state feedback equations for u_{cf} and u_{cr}. Firstly, the front input (replacing ξ_1) is made a unit amplitude sinusoid while the rear input (replacing ξ_2) is made zero, and the frequency responses of interest are calculated and stored. Secondly, the front input is made zero and the rear input is made a unit amplitude sinusoid, and the responses are again calculated and stored. Thirdly, the responses to the rear input are each phase lagged by $180\,\ell\omega/(u\pi)$ degrees to account for the wheelbase travel time delay and fourthly, the responses to front and rear inputs are superposed.

2.3   CONTROL LAW DETERMINATION

Following much previous work in active suspension research, it is natural to think of applying linear optimal control theory to this anti-symmetric system problem. Limited state, output, feedback control is considered appropriate to the task, with measurements of body roll velocity and roll angle, and of front and rear axle roll angles relative to the body being assumed feasible. The control law then takes the form:

$$
\begin{vmatrix} u_{cf} \\ u_{cr} \end{vmatrix} = \begin{vmatrix} k_{11} & k_{12} & k_{13} & k_{14} \\ k_{21} & k_{22} & k_{23} & k_{24} \end{vmatrix} \begin{vmatrix} \dot{\phi}_b \\ \phi_b \\ \phi_b - \phi_{af} \\ \phi_b - \phi_{ar} \end{vmatrix} \tag{10}
$$

The objective function to be minimised must be of quadratic form and it has been chosen to include terms representing tyre load variation, suspension working space, body roll velocity and body roll angle, in addition to the control signals. Specifically,

$$J = \lim_{T \to \infty} \frac{1}{T} \int_0^T \{ q_1(\phi_{af} - \phi_{tf})^2 + q_2(\phi_{ar} - \phi_{tr})^2 + q_3(\phi_b - \phi_{af})^2$$

$$+ q_4(\phi_6 - \phi_{ar})^2 + q_5 \dot{\phi}_b^2 + q_6 \phi_b^2 + u_{cf}^2 + u_{cr}^2 \} \, dt \tag{11}$$

The optimisation process is an elaboration of that described in [32]. It involves iterative improvement through gradient descent with stability checking at each iteration. Routines from [33] translated into MATLAB have been used. Control laws generated through equations (1-7), assuming no front/rear input correlations can be expected to have symmetry properties, corresponding to the idea that front states driving the rear actuator will be mirrored by rear states driving the front actuator. This property provides something of a check on the operation of the optimiser. The control gains for no input correlations also may provide a useful starting point for the more refined calculations including the Padé filter. Choosing the weighting constants, q_1 - q_6, provides a means of steering the control system design towards the various objectives. Initially, estimates of values which would make each contribution to the cost of about equal size were made. There were then refined by observing changes in the cost function value resulting from changes to the weights, to ensure that each term was making a significant contribution.

3.   Results

Results concern the optimal control, the cost reduction and the frequency response properties of the actively controlled systems in comparison with a conventional passive one, referred to as p_h, and an underdamped passive case, referred to as p_l. A baseline parameter set, representative of a European family car, has been adopted as follows: $C_{sf} = C_{sr}$ = 1280 Ns/m; $l_{af} = l_{ar}$ = 50 kg m²; l_b = 160 kg m²; $k_{sf} = k_{sr}$ = 31250 N/m; $k_{tf} = k_{tr}$ = 192000 N/m; l = 2.7 m; $s_f = s_r$ = 0.6 m; u = 20 m/s; $w_f = w_r$ = 0.75 m; α = 0.191 cycles/m; ξ = 0.7; ω_n = 12π rad/s.

Many cases involving different weighting constants, q_1 .... q_6, in the cost function have been considered. Invariably, the cost value for the underdamped passive system, p-l, is relatively high and sometimes the active control applied to the underdamped suspension is not able to reduce the cost as much as the additional damping does. However, for any given passive arrangement and cost definition, the active control will always bring a cost reduction, which is reassuring. Some examples are shown in Figure 4, in which system A has its priority on roll angle reduction, system B is biassed towards wheel load control and system C is a balanced design.

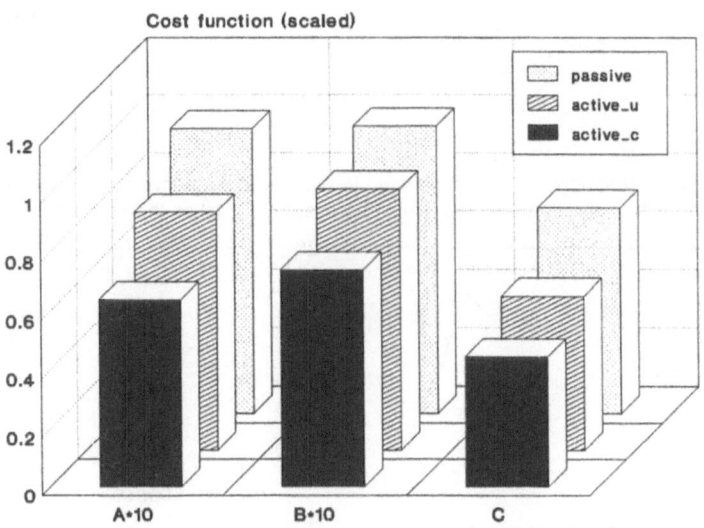

Figure 4   Cost reductions by active systems: J _u, not accounting for input correlations; J_c, accounting for them via the Padé filter.

An exemplary set of frequency responses is shown for the case of system C in Figures 5 and 6, the first of which shows the roll angle response magnitudes to a unit amplitude displacement into the cross-level spectral shaping filter while the second shows front and rear suspension working space usage and front and rear dynamic tyre compressions. In the roll angle response plots, the front/rear symmetric properties of the passive systems and the "uncorrelated" active systems makes for extremely effective wheelbase filtering when the vehicle wheelbase is ½ or 1½ wavelengths, as can be observed.

These results show that the active control is able to accentuate one quality of the suspension performance at the expense of others. The responses of the active systems overall do not appear significantly better than those of the baseline passive system. This is not an exhaustive study of the possibilities, since other vehicle speeds, other bases for the control system design, gain scheduling adaptation according to load and speed etc. can all be considered. Each of these additional features may bring some reward. However, the results in [19] suggested that little advantage may be gained from accommodating the left/right track correlations.

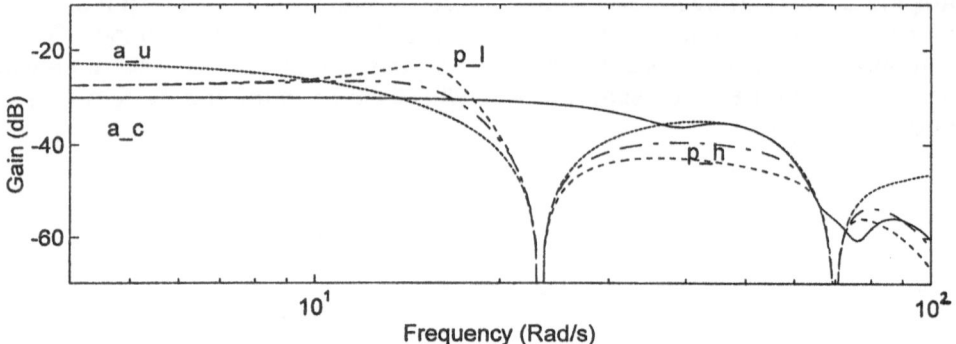

Figure 5   Roll angle responses to unit amplitude input; a_u is the active system with its control optimised for no front/rear input correlations while the law for a_c takes the correlation into account.

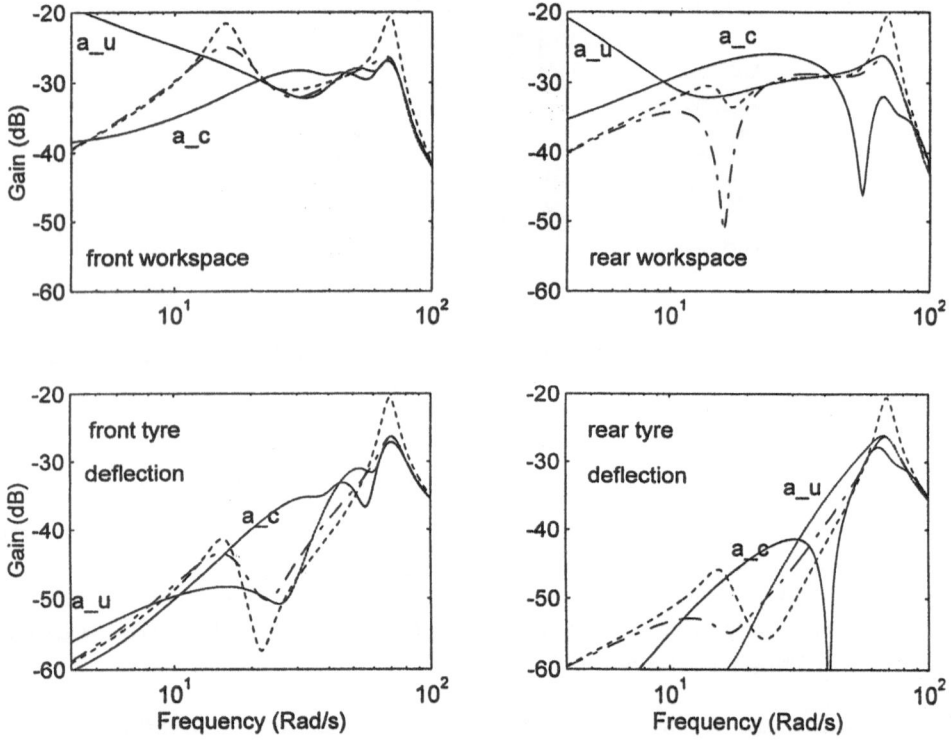

Figure 6   Working space and dynamic tyre compression responses to unit amplitude input.

4.   Conclusions

Using the ideas that the road is isotropic and the vehicle has lateral symmetry, ride control of an actively suspended car can be separated into two largely independent sub-problems. One of these, involving symmetric motions and a cylindrical road input profile has been studied extensively in previous work. The other has been in focus here and it has been shown how linear optimal control theory can be applied to the derivation of control laws. Cost reductions through the use of active control have been demonstrated but these have been modest and improving one aspect of performance significantly has necessitated making another one worse. Although the study conducted is by no means exhaustive, the results suggest that the part of an active suspension control scheme designed to provide vibration isolation enhancement in respect of cross-level excitation will make only very modest contributions to enhancing performance. This is in contrast with the corresponding situation for cylindrical road excitation.

Acknowledgement: The author is pleased to acknowledge the contribution of D J Cole of Cambridge University Engineering Department to this work, through the provision of translations of the optimisation routines BRENT, FRPRMN, LINMIN and MNBRAK from [33] into MATLAB.

REFERENCES

[1] Hedrick J K and Wormley D N, Active suspension for ground support vehicles - a state of the art review, ASME-AMD, 15, 21-40, 1975.
[2] Morman K N and Giannopoulos F, Recent advances in the analytical and computational aspects of modelling active and passive vehicle suspensions, ASME-AMD, 50, 75-115, 1982.
[3] Goodall R M and Kortüm W, Active controls in ground transportation - a review of the state of the art and future potential, Vehicle System Dynamics, 12, 225-257, 1983.
[4] Sharp R S and Crolla D A, Road vehicle suspension design - a review, Vehicle System Dynamics, 16, 167-192, 1987.
[5] Wallentowitz H and Konik D, Actively influenced suspension systems - survey of actual patent literature, EAEC 3rd International Conference, Vehicle Dynamics and Power Train Engineering, Strasbourg, 1991.
[6] ElBeheiry E M, Karnopp D, ElAraby M E and Abdelraaouf A M, Advanced ground vehicle suspension systems - a classified bibliography, Vehicle System Dynamics, 24, 231-258, 1995.
[7] Aoyama Y, Kawabata K, Hasegawa S, Kaobari Y and Sato M, Development of full active suspension by Nissan, SAE 910747, 1990.
[8] Yokoya Y, Ryotiei K, Kawaguchi H, Ohashi K and Otino H, Integrated control system between active control suspension on four wheel steering for the 1989 Celica, SAE 910748, 1990.
[9] Acker B, Darenberg W and Gall H, Active suspensions for passenger cars, The Dynamics of Vehicles on Roads and Tracks, R J Anderson (ed..) Vehicle System Dynamics Supplement to 18, 15-26, 1990.
[10] Williams R A, Best A and Crawford I L, Refined low frequency active suspension, Vehicle Ride and Handling, MEP, Bury St. Edmunds, 285-300, 1993.
[11] Chalasani R M, Ride performance potential of active suspension systems - part 1: Simplified analysis based on a quarter-car model, ASME-AMD, 80, 1987-202, 1986.
[12] Chalasani R M, Ride performance potential of active suspension systems - part 11: Comprehensive analysis based on a full-car model, ASME-AMD, 80, 205-234, 1986.
[13] Sharp R S and Hassen S A, The relative performance capabilities of passive, active and semi-active car suspension systems, Proc. I.Mech.E., Journal of Automobile Engineering, 200(D3), 219-228, 1986.
[14] Hedrick J K and Butsuen T, Invariant properties of automotive suspensions, Advanced Suspensions, MEP, Bury St. Edmunds, 35-42, 1988.
[15] Yue C, Hedrick J K and Butsuen T, Alternative control laws for automotive active suspensions, Trans. ASME, Jour. Dyn. Syst. Meas. and Control, 111, 286-290, 1989.
[16] Louam N, Wilson D A and Sharp R S, Optimal control of a vehicle suspension incorporating the time delay between front and rear wheel inputs, Vehicle System Dynamics, 17, 317-336, 1988.

[17] Pilbeam C and Sharp R S, On the preview control of limited bandwidth vehicle suspensions, Proc. I.Mech.E., Journal of Automobile Engineering, 207(D3), 185-194, 1993.

[18] Fruehauf F, Kasper R and Luckel J, Design of an active suspension for a passenger vehicle model using input processes with time delays, The Dynamics of Vehicles on Roads and Tracks, O Nordstrom (ed..) Vehicle System Dynamics Supplement to 15, 126-138, 1986.

[19] Crolla D A and Abdel Hady M B A, Active suspension control; performance comparisons using control laws applied to a full vehicle model, Vehicle System Dynamics, 20, 107-120, 1991.

[20] Malek K M and Hedrick J K, Decoupled active suspension for improved automotive ride quality and handling performance, The Dynamics of Vehicles on Roads and Tracks, O Nordstrom (ed..) Vehicle System Dynamics Supplement to 15, 383-398, 1986.

[21] Gordon T J, An integrated strategy for the control of complex mechanical systems based on sub-system optimality criteria, IUTAM Symposium on Optimisation of Mechanical Systems, D Bestle and W Schiehlen (eds), Kluwer, Dordrecht, 97-104, 1996.

[22] Gordon T J, An integrated strategy for the control of a full vehicle active suspension system, Proc. IAVSD Symposium on The Dynamics of Vehicles on Roads and on Tracks, L Segel (ed..) Swets and Zeitlinger, Lisse, in printing.

[23] Lang R and Walz U, Active roll reduction, EAEC 3rd International Conference on Vehicle Dynamics and Powertrain Engineering, EAEC, Strasbourg, 88-92, 1991.

[24] Sharp R S and Pan D, On active roll control for automobiles, The Dynamics of Vehicles on Roads and Tracks, G Sauvage (ed.) Vehicle System Dynamics Supplement to 20, 566-583, 1992.

[25] Sharp R S and Pan D, On the design of an active roll control system for a luxury car, Proc. I. Mech.E., Journal of Automobile Engineering, 207(D4), 275-284, 1993.

[26] Dodds C J and Robson J D, The description of road surface roughness, J. Sound and Vibration, 31, 175-183, 1973.

[27] Kamash K M A and Robson J D, Implications of isotropy in random surfaces, J. Sound and Vibration, 54, 1-15, 1977.

[28] Robson J D, Road surface description and vehicle response, Int. J. Vehicle Design, 1, 25-35, 1979.

[29] Rill G, The influence of correlated random excitation processes on dynamics of vehicles, The Dynamics of Vehicles on Roads and Tracks, J K Hedrick (ed..) Swets and Zeitlinger, Lisse, 449-459, 1984.

[30] Venhovens P J Th, Optimal control of vehicle suspensions, Doctoral Dissertation, Delft University of Technology, 1993.

[31] Clark S K (ed..) Mechanics of pneumatic tires - 2nd ed., NBS Monograph 122, Washington D.C., 1982.

[32] Wilson D A, Sharp R S and Hassan S A, The application of linear optimal control theory to the design of active automotive suspensions, Vehicle System Dynamics, 15, 105-118, 1986.

[33] Press W H, Teukolsky S A, Vetterling W T and Flannery B P, Numerical recipes in FORTRAN; the art of scientific computing - 2nd ed., Cambridge University Press, 1992.

DYNAMICS OF A SINGLE-AXLE STEERED BOGIE

EVA SLIVSGAARD* AND HANS TRUE**
*Danish State Railways Consult, Pilestraede 58,
DK-1112 Copenhagen K, Denmark
**Department of Mathematical Modelling, The Technical University of
Denmark, DK-2800 and ES-Consult, Staktoften 20,
DK-2950 Vedbaek, Denmark

1. Abstract

The drive towards energy savings in the railway industry have forced the railway
companies and manufacturers to examine unconventional vehicle designs. The Danish
State Railways together with Professor Frederich from RWTH in Aachen and the
company Linke-Hoffmann-Busch have developed a steered single-axle bogie, which will
save weight compared with the conventional two-axle bogie. The single-axle bogie has
been built under an S-train car replacing a conventional two-axle bogie. Tests have
shown that the wheels on the steered single-axle bogie wear less than the wheels on the
conventional bogies, so we save both energy and material.

The purpose of the steering is mainly to secure an optimal radial position of the
wheelset in curves. The running stability of the vehicle is governed by the steering in
combination with the suspension once the wheel-rail geometry is given. Therefore, the
suspension and the steering must be optimized with respect to both curving performance
and vehicle stability. In this paper the effect of the stiffness of the steering on the stability
is reported.

2. Problem Description

2.1. THE TRAIN

The train is a two-car electric train set used on the Copenhagen 1500 V DC suburban
railway system. It consists of two motor cars, which are connected with a coupling rod.
The train set is shown on figure 1. One of the motor cars is a driving trailer on which the
front two-axle bogie has been replaced by the steered single-axle driven bogie. The other
motor car runs on two conventional two-axle bogies with motors on three of the axles.
The single-axle bogie is driven in order to simulate the operating conditions of the
production model in the best possible way. The mechanical steering consists of

*D. H. van Campen (ed.), IUTAM Symposium on Interaction between Dynamics and Control in Advanced
Mechanical Systems, 375–382.*

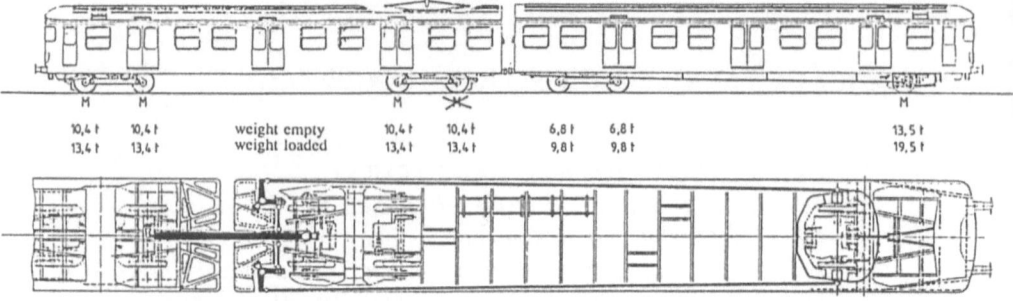

Figure 1. The S-train set with the steered single-axle bogie to the far right is seen on the upper part. "M" indicates a driven axle, and the axle loads of the individual axles are shown. On the lower part the arrangement of the mechanical steering system is shown. Reproduced with permission of LHB.

the coupling rod between the two cars and two angle rods connected to the single-axle bogie by long bars. It is shown on figure 1. The connections between the angle and coupling rods are very stiff. The angle rods are attached to the car body by links, such that their centers of gravity are fixed relative to the car body. The links transfer the coupling forces between the cars to the car body. The attachments are also very stiff. The angle of deflection between the two car bodies is used as a passive control signal. When one of the cars enters a curve the coupling rod will rotate around the vertical and force the angle rods to yaw. Hereby the single-axle bogie is steered into a radial position in the curve. The long bars introduce an important elasticity in the control system. Their total longitudinal stiffness c_s including the flexibility of the connections is termed the stiffness of the steering system.

The bearings of the wheelsets of all the bogies support bogie frames on a primary suspension. The bogie frames carry the car bodies on a secondary suspension. The primary suspensions on the two-axle bogies consist of two springs acting in all three space dimensions and two vertical dampers. The secondary suspensions consist of two springs, which also act in all three space dimensions, two vertical dampers and one lateral damper. The primary suspension of the single-axle bogie consists of four springs acting in all three directions and four vertical dampers. The secondary suspension consists of two springs acting in all three directions, two longitudinal springs to control pitch, one lateral and two vertical dampers and two yaw dampers with serial stiffness. The steering bars are attached to the frame of the single-axle bogie. For further construction details we refer to Slivsgaard (1995).

2.1. THE MATHEMATICAL MODEL

The car bodies, the bogie frames, the wheelsets and the rods are all assumed to be perfectly rigid bodies. The suspension elements are all assumed to have linear characteristics and the longitudinal steering stiffness c_s is constant. The model of the steering system is shown on figure 2. The track is also assumed to be rigid. The nonlinearities enter the problem through the contact mechanics in the rail-wheel contact areas. The kinematic conditions that express the geometrical contact between a wheel and the rail as well as the creepage-creep force relations in rolling contact are nonlinear. We use the program RSGEO by W. Kik to compute the kinematic conditions, the approximation by Sauvage and Pascal (1990), for the calculation of the normal forces in the idealized contact point and the formula by Shen, Hedrick and Elkins (1983), for the calculation of the creep forces.

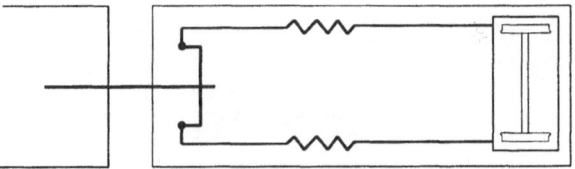

Figure 2. The steering system of the single-axle bogie.

We use Newton's law of motion to formulate the dynamical system. In the resulting equations of motion the dependent variables are denoted q_i, i=1,2,... . When we make the substitutions: $x_{2i-1}=q_i$, i=1,2,.., and $x_{2j}=dq_j/dt$, j=1,2..., in the second order equations we obtain a nonlinear dynamic system of 121 first order differential equations for determination of $x_n(t)$; n=1,2,..,121. The dynamical system is analysed numerically. The program used is a railway vehicle simulation program developed by Slivsgaard and Jensen. It is an extension of the program described in Jensen (1995). For the time integration a Runge-Kutta method with adaptive step control is used.

The stiffness of the steering system must ensure the asymptotic stability of the point of operation in phase space, which is an equilibrium point in an appropriate moving frame. See Slivsgaard (1995) for details. We shall therefore examine the running stability of the train set in dependence on the speed and the stiffness of the steering system of the single-axle bogie for the combinations of rail-wheel contact geometries that may be found on the Copenhagen S-train system. We investigate the stability for 0<V<55 m/s (198 km/h), where V is the speed of the trainset, and $1<c_s<10^5$ kN/m. The upper limit for the speed is well above the highest permitted speed on the S-train lines (120 km/h) in Denmark. The wheel profile is DSB82-1 and it must be combined with the four track systems:

UIC60 rails with an inclination of 1/20 and 1/40

DSB45 rails with an inclination of 1/20 and 1/40

The gauge is the standard gauge of 1435 mm.

3. The Results

First we make a linear stability analysis. The maximum real part of the eigenvalues of the linearized system are calculated for the different track configurations. The calculations indicate two main areas of instability: one for high speeds and low steering stiffnesses and another for low speeds and high steering stiffnesses. Furthermore the low speed range with low stiffnesses seems to have a low damping rate. The worst track configuration with respect to vehicle stability is DSB45 rails inclined at 1/20, while the

UIC60 rails inclined at 1/40 seems to be best. The results of the linear stability analysis for the DSB45 rail are shown on figure 3.

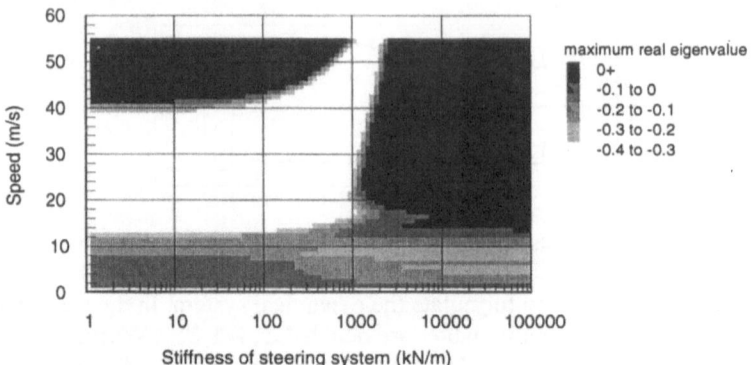

Figure 3. The maximum real part of the eigenvalue as a function of the stiffness of the steering system c_s (in logarithmic scale) and the vehicle speed V. The track is DSB45 rails inclined at 1/20.

When we examine the eigenvectors of the problem, we find that the single-axle bogie is the source of the instability. When the steering stiffness is low and the speed high the mode is a combined lateral and yaw oscillation of the single-axle bogie and its wheelset. The oscillations have a very small effect on the car body. When the steering stiffness is high and the speed low both the single-axle bogie and the body of the front car oscillate with a frequency, which is one third of the frequency of the oscillations at the low stiffness. The two different kinds of oscillations are shown on figures 4 and 5. Both kinds of instability are well-known in railway vehicle dynamics see Wickens (1965).

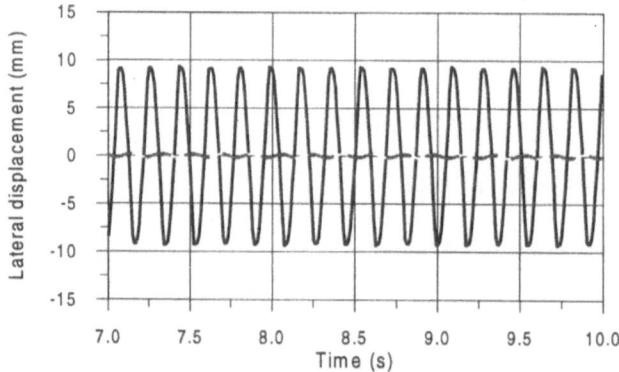

Figure 4. The lateral displacement of the wheelset on the single-axle bogie (solid line) and of the car body (dotted line) versus time. The track is DSB45 rails inclined at 1/20, c_s=100 kN/m and V=50 m/s.

Next we perform an analysis of the transients of the full nonlinear problem three seconds after the front bogie is displaced 4 mm laterally. The result for the wheelset is shown on the figure 6 and for the car body on figure 7.

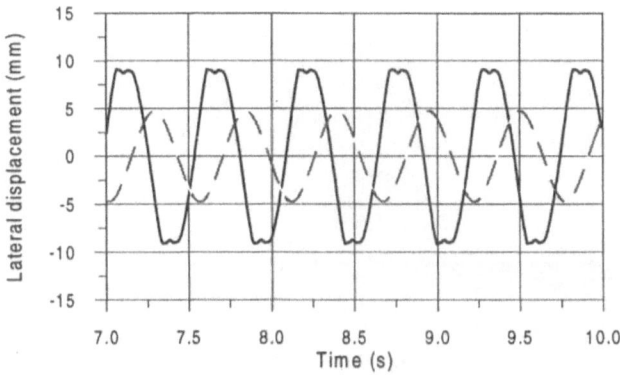

Figure 5. The lateral displacement of the wheelset on the single-axle bogie (solid line) and of the car body (dotted line) versus time. The track is DSB45 rails inclined at 1/20, $c_s=10^4$ kN/m and V=30 m/s.

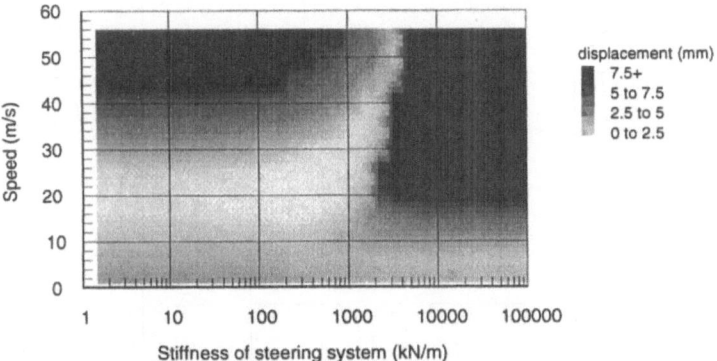

Figure 6. The maximum lateral displacement of the wheelset on the single-axle bogie after 3 secs as a function of the stiffness of the steering system c_s (in a logarithmic scale) and the vehicle speed V. The track is DSB45 rails inclined at 1/20.

When we compare the wheelset displacements on figure 6 with the results of the linear stability analysis, we see that the periodic hunting starts out at lower speeds than the linear critical speed. That indicates that the loss of stability is through a subcritical Hopf bifurcation The car body instability at low speeds and high steering stiffnesses is clearly seen on figure 7. We find that a disturbance of the wheelset is transmitted almost undamped to the car body even though the amplitude of the wheelset oscillation is decreasing.

In order to obtain a more accurate value for the critical speed V_C of the train set we calculate a bifurcation diagram figure 8 for the most critical track configuration namely DSB45 rails inclined at 1/20. The figure shows the maximum lateral displacement of the wheelset on the single-axle bogie versus the vehicle speed.

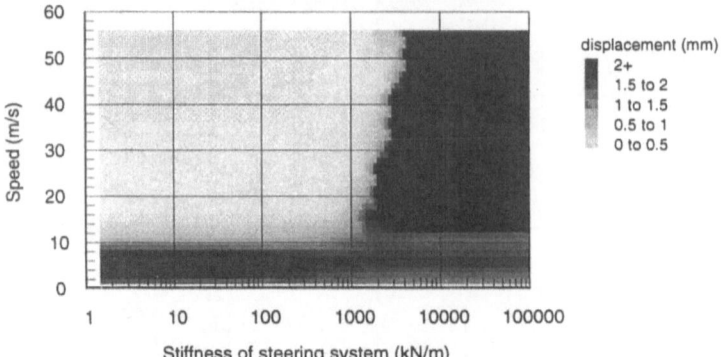

Figure 7. The maximum lateral displacement of the car body after 3 secs as a function of the stiffness of the steering system c_s (in a logarithmic scale) and the vehicle speed V.
The track is DSB45 rails inclined at 1/20.

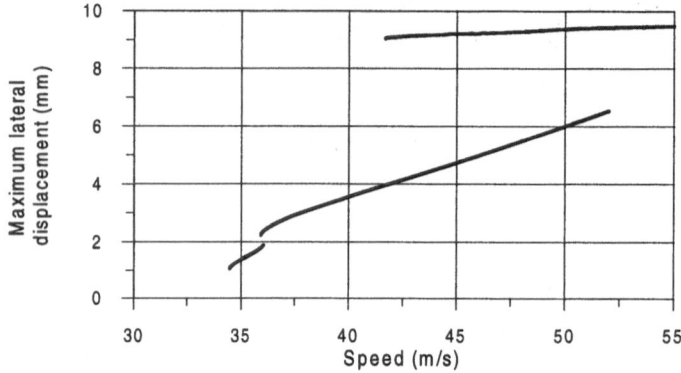

Figure 8. Bifurcation diagram showing the maximum lateral displacement of the wheelset on the single-axle bogie versus the vehicle speed for c_s=500 kN/m and DSB45 rails inclined at 1/20.

Only the asymptotically stable solutions are shown on figure 8. We find that a periodic solution with flange contact exists in the speed range 41.7 m/s <V<55 m/s. V=41.7 m/s is a saddle-node bifurcation point. Another saddle-node at V=52 m/s creates a stable periodic solution, which disappears in a new saddle-node bifucation at V=35.9 m/s. Finally a low amplitude periodic solution exists for 34.4 m/s <V<36.1 m/s. The lowest speed for which a stable periodic solution exists is found to be 34.4 m/s (=124 km/h), and the critical speed V_C for the train set for c_s=500 kN/m is therefore V_C=34.4 m/s. On figure 9 we show the saddle-node bifurcation that determines the critical speed in dependence on the steering stiffness c_s for all four track configurations. We see that the critical speed V_C increases with the steering stiffness.

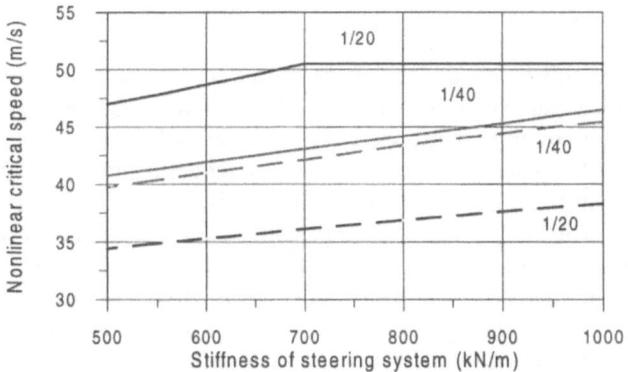

Figure 9. The critical speed V_C versus the stiffness of the steering system c_s. The dashed lines are for DSB45 rails and the solid lines are for UIC60 rails.

For the UIC60 rail inclined at 1/20 there is a kink in the curve at c_s=700 kN/m. The kink is caused by a qualitative change in the periodic oscillation that is created in the saddle-node bifurcation. For c_s<700 kN/m as well as for UIC60 rails inclined at 1/40 the oscillations have flange contact. For UIC60 rails inclined at 1/20 and c_s>700 kN/m as well as for DSB45 rails the oscillations have no flange contact. When we compare the results of the bifurcation analysis with the results of the transient analysis, we find that V_C is lower than the critical speed found in the transient analysis. The reason is that the basin of attraction for the oscillations with flange contact is small near the critical speed, so near V_C we do not enter the basin with a lateral displacement of only 4 mm.

On figure 10 we show the difference between the linear critical speed and V_C versus the stiffness of the steering system c_s for DSB45 and UIC60 rails both inclined at 1/20. The difference is significant and underlines the danger of making decisions on the basis of linear analyses in railway vehicle dynamic problems.

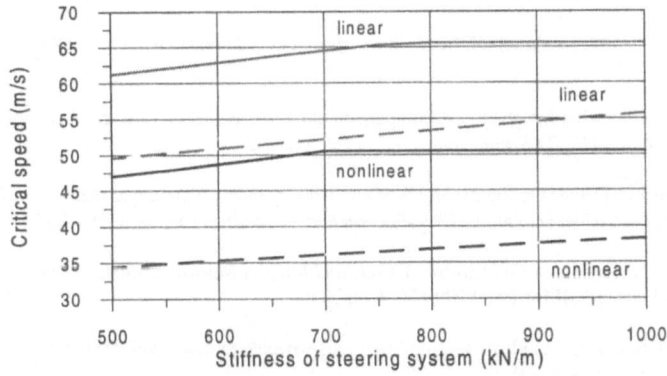

Figure 10. The linear and nonlinear critical speeds versus the stiffness of the steering system c_s. The dashed lines are for DSB45 rails and the solid lines are for UIC60 rails both inclined at 1/20.

Slivsgaard and Jensen have analysed the stability of the train set, when the direction of travel is reversed so the single-axle bogie runs at the rear end (Slivsgaard *et al.*, 1994). When the single-axle bogie is at the rear end the critical speeds increase a little.

4. Conclusions

It is a surprising result of this investigation that the stiffness of the steering system must be chosen within a narrow range of values in order to secure the global asymptotic stability of the equilibrium point in phase space that represents the steady motion of the train set in a sufficiently large speed range. For the practical application it is not sufficient to guarantee the stability up to a speed that is 10 % higher than the maximum speed on the S-train network. The analyses have been performed for ideal and new wheel and rail profiles, and since the critical speed generally has a tendency to decrease with increasing wear a sufficiently large speed margin above the theoretical critical speed must exist to guarantee the smooth running of the train set even when the wheelsets are worn.

It is the changes in the wheel-rail contact geometry that influence the critical speed, and fortunately the contact geometry in reality only changes very little, when the single-axle wheelset is worn. The single-axle wheelset can therefore run several hundred thousands of kilometers in daily service, before it becomes necessary to reprofile the wheels in order to prevent hunting. An important goal for the development has thus been achieved.

The successful tests with the train set encouraged DSB to order a new type of S-train sets with steered single-axle bogies only. Linke-Hoffmann-Busch has delivered the first two train sets and in extensive trials they live up to the expectations. The production train sets differ from the test train in having a hydraulic steering of the single-axle bogie, which has the same stiffness as the mechanical steering on the test train. Ahrens (1994) made a linear stability analysis of the new S-train sets.

5. References

Ahrens, R. (1994) S-bahn Kopenhagen Fahrzeugdynamik eines Gliederzuges mit kurvengesteuerten Einzelradsatz-Fahrwerken, *Proceedings of Systemdynamik der Eisenbahn*, Hennigsdorf, 109-118.

Jensen J.C. (1995) *Teoretiske og eksperimentelle dynamiske undersøgelser af jernbanekøretøjer*, Ph.D. Thesis, Technical University of Denmark, IMM-PHD 1995-9, Lyngby.

Sauvage G. and Pascal J-P. (1990) Solution of the multiple wheel and rail contact dynamic problem, *Vehicle System Dynamics* **19**, 257-272.

Shen Z.Y., Hedrick J.K. and Elkins J.A. (1983) A comparison of alternative creep force models for rail vehicle dynamics analysis, *Proc of 8th IAVSD-Symposium*, Cambridge, 591-605.

Slivsgaard E. and Jensen J.C. (1994) On the dynamics of a railway vehicle with a single-axle bogie, *Proc of 4th Mini Conference on Vehicle System Dynamics, Identification and Anomalies*, Budapest, 197-207.

Slivsgaard E. C. (1995) *On The Interaction between Wheels and Rails in Railway Dynamics*, Ph.D. Thesis, Technical University of Denmark, IMM-PHD 1995-20, Lyngby.

Wickens A.H. (1965) The dynamic stability of a simplified four-wheelded railway vehicle having profiled wheels, *Int J. Solid Structures* **1**, 385-406.

CONTROL DESIGN FOR A PICK-AND-PLACE MECHANISM

M. STEINBUCH AND M.J. VERVOORDELDONK
Philips Research/Philips CFT
P.O.Box 218
5600 MD Eindhoven, The Netherlands

1. Introduction

The pick-and-place system consists of a four-beam mechanism of which the driving motor rotates over 180 degrees, see Fig.1.

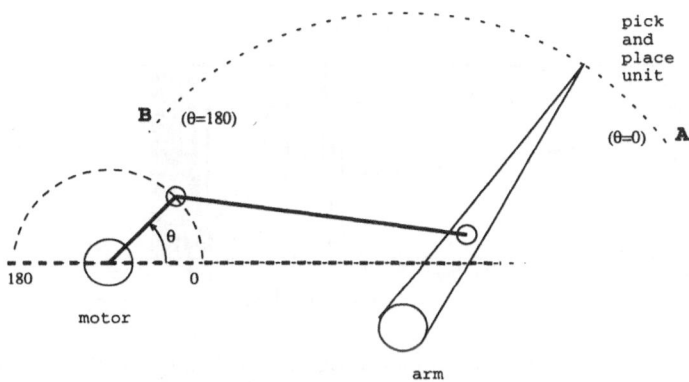

Figure 1. Mechanism - top view

The measured variable is θ and the input to the system is the motor current I. Primary goal of the active position control system is to realize a cycle-time as short as possible. The current performance is a movement from A to B within only 100 ms, with an accuracy of a few encoder increments ($\frac{\pi}{1000}$ rad) as final positioning error.

In order to further reduce the cycle time, a study is done to investigate whether it is possible to improve the performance by making use of optimization based feedback design and learning control for feedforward design. The main problem is that the system dynamics strongly depend on posi-

D. H. van Campen (ed.), IUTAM Symposium on Interaction between Dynamics and Control in Advanced Mechanical Systems, 383–390.

384

tion, while the current implementation environment limits the controller to be linear time-invariant.

2. Parametric System Identification

Frequency responses of the mechanism, from motor current I to angle θ, are collected in five operating points: $\theta = 0, \pi/4, \pi/2, 3\pi/4, \pi$ rad. In Fig.2 the measurement is shown in $\theta = 0$ rad and in Fig.3 for $\theta = \pi/2$ rad.

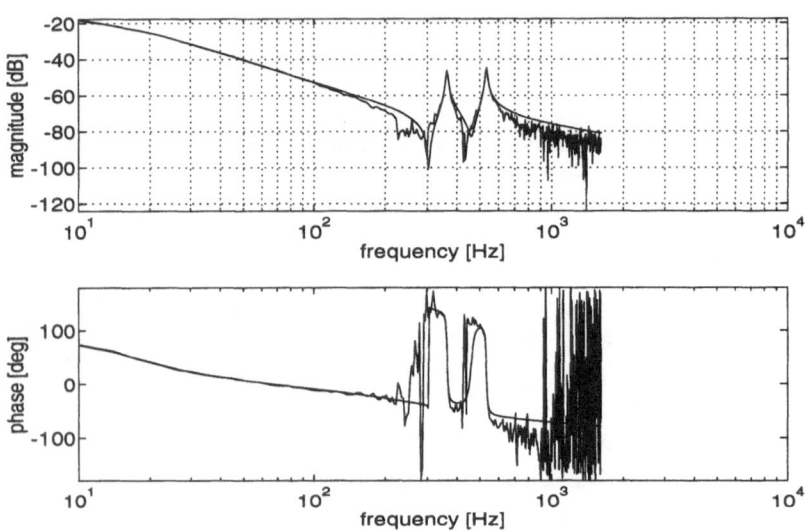

Figure 2. Measured frequency response of the mechanism in $\theta = 0$ rad, and of the identified 7th order model.

At low frequencies, the rigid body mode shows some damping, which is due to cables and friction effects. At high frequencies resonances appear, which depend on the operating point. Since the desired bandwidth is approximately 100 Hz (in $\theta = 0, \pi$), these parasitic dynamics have to be accounted for in the control system design. In addition, the gain varies with a factor 10, due to the fact that the inertia strongly depends on position as seen from the motor.

The rigid body dynamics can be described by the equation of motion:

$$J(\theta)\ddot{\theta} = -\dot{J}(\theta)\dot{\theta} + kI \tag{1}$$

in which the inertia $J(\theta)$ (defined at the motor side) depends on the angle θ, and where the motor is modelled as static gain k. The equation shows that when the angular speed is high, and recalling the fact that the

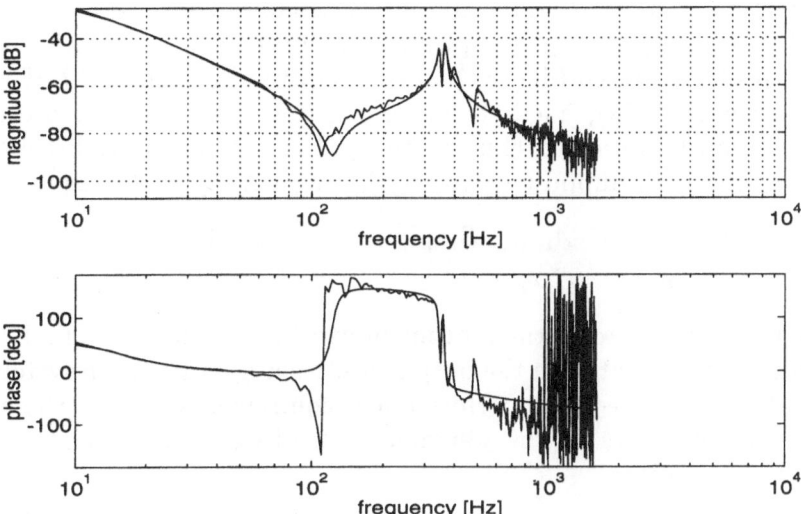

Figure 3. Measured frequency response of the mechanism in $\theta = \pi/2$ rad, and of the identified 7th order model.

inertia changes rapidly with a factor 10, the additional nonlinear term adds significant damping (either positive or negative) to the system dynamics.

Linearizing the system in each operating point ($\dot{\bar{\theta}} = 0$, $\bar{I} = 0$), yields $J(\bar{\theta})\ddot{\theta} = kI$, in which $\bar{\theta}$ denotes the operating point, and where θ and I now denote small perturbations around the operating point. This shows that the rigid body linear model has a double integrator structure, of which the gain is a function of the operating point.

Instead of using the (nonlinear) rigid body model for control design, (linear) system identification is used because the parasitic dynamics play a crucial role and have to be captured accurate by the control design model. With frequency-domain based system identification the measured frequency responses have been fitted using a least-squares error criterion. The frequency responses of the resulting 7th order models are also shown in Fig.2-3.

3. Feedback Control Design

The control problem can be stated as finding a stabilizing, high performance feedback controller with a fixed controller structure, for a set of (linearized) models of the original nonlinear system. This problem is also known as multi-model fixed-structure control.

One method appropriate for the problem is Linear Quadratic Optimal Control in which the performance objective is specified as an integral of squared deviations of the states and input variables (Anderson and Moore, 1971). By directly posing the optimization problem as a (static) output feedback problem, called Linear Quadric Output Feedback (LQOF) (Levine and Athans, 1970), fixed structure controllers as well as multi-model controllers can be designed (Looze, 1983). Here we will only briefly summarize the method. Suppose we have r linear state space models:

$$
\begin{aligned}
\dot{x}_i &= A_i x_i + B_i u_i & i = 1, ..., r & \quad (2) \\
y_i &= C_i x_i
\end{aligned}
$$

The multi-model robustness problem can be formulated as finding one controller F, such that with the output feedback $u_i = F y_i$ all r systems (2) are stabilized and meet the performance requirements. Define the overall performance objective as the weighted sum of quadratic performance in each operating point:

$$
J_{tot} = \sum_{i=1}^{r} w_i \int_0^\infty (x_i^T Q_i x_i + u_i^T R_i u_i) dt, \quad (3)
$$

in which Q_i and R_i are weighting matrices for each operating point $i = 1, ..., r$. The weighting scalars $w_i > 0$ can be used to weight some models more than others, expressing their relative importance. The Multi-Model LQOF problem is to find a feedback F such that J_{tot} is minimized. To solve this optimization problem a numerical algorithm is used.

A dynamic feedback law can be obtained by augmenting the open-loop system with integrators and to define an optimization problem for the enlarged system (Levine and Athans, 1970). In this way it is also possible to optimize PD (lead/lag) controllers with respect to a quadratic performance criterion.

Using the LQOF method a design has been made for the five models obtained, see Fig.4. For comparison a controller is also designed using loop-shaping techniques, i.e. combining PID filters and a notch. This loopshaping design is denoted PID[+].

Experimental results are obtained by digital implementation at a sampling frequency of 4 kHz. The open-loop frequency responses of the controlled system are shown in Fig. 5. Clearly, the bandwidth of the PID[+] controlled system (—) is in $\theta = 0$ rad (left) about 60 Hz, whereas the LQOF (- - -) has a 100 Hz bandwidth. In $\theta = \pi/2$ rad (right) both controllers give a low bandwidth due to the reduced proces gain.

4. Setpoint and Feedforward Design

In this section results are presented concerning the tracking performance. Major issue is how to determine the best setpoint and feedforward signals.

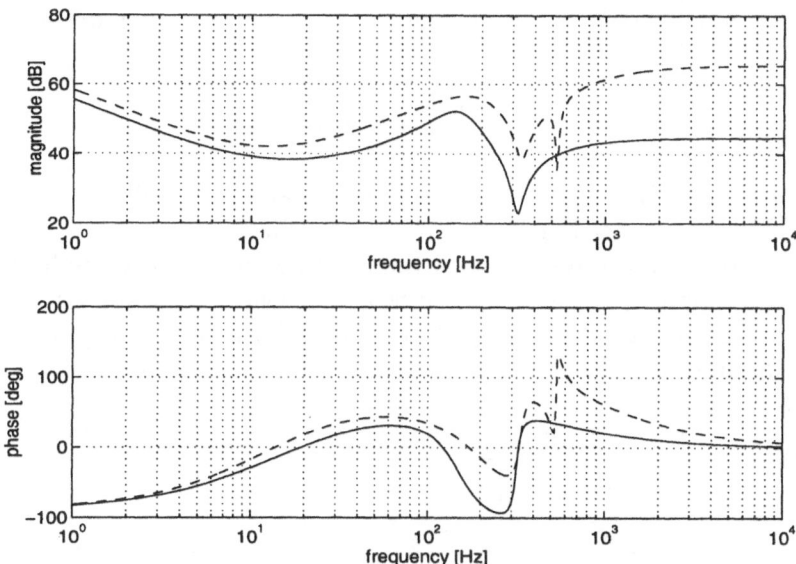

Figure 4. Frequency responses of the PID⁺ (—), and of the LQOF control (- - -) .

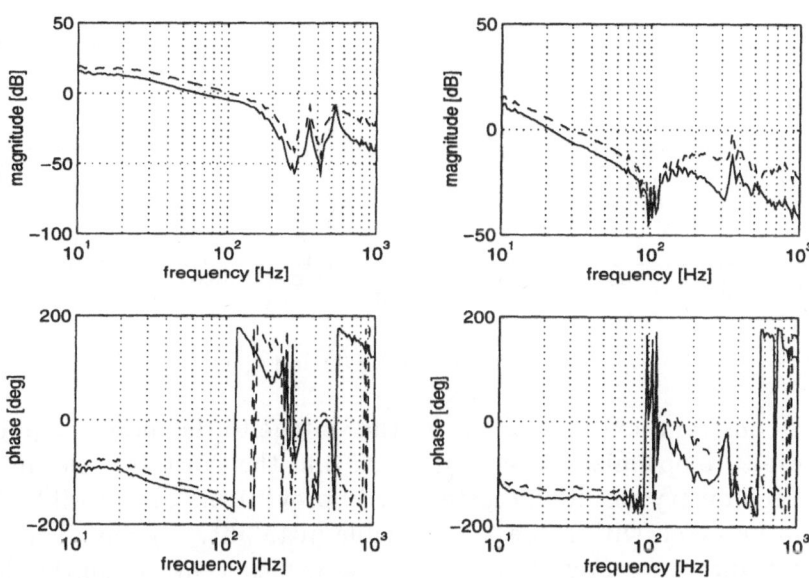

Figure 5. Measured open-loop frequency responses of the mechanism in $\theta = 0$ rad (left) $\theta = \pi/2$ rad (right) with the PID⁺ (—) and with the LQOF control (- - -).

From the model for the rigid body dynamics of the mechanism (1), it is easy to see how to calculate the force (i.e. current) feedforward signal.

Given a setpoint profile $\theta(t)$ and hence $\dot{\theta}(t)$ and $\ddot{\theta}(t)$ the required current is:

$$I = \frac{J(\theta)}{k}\ddot{\theta} + \frac{1}{k}J'(\theta)\dot{\theta}^2 \qquad (4)$$

Notice that for linear systems the feedforward is calculated as $I = \frac{J}{k}\ddot{\theta}$ in which J/k is the acceleration feedforward gain.

In order to compare results with the existing situation, we will adopt a third order setpoint profile, i.e. the jerk is piecewise constant. In Fig.6 the desired values for $\theta(t)$, $\dot{\theta}(t)$ and $\ddot{\theta}(t)$ are shown for a step of 180 degrees (π radians) forth and back.

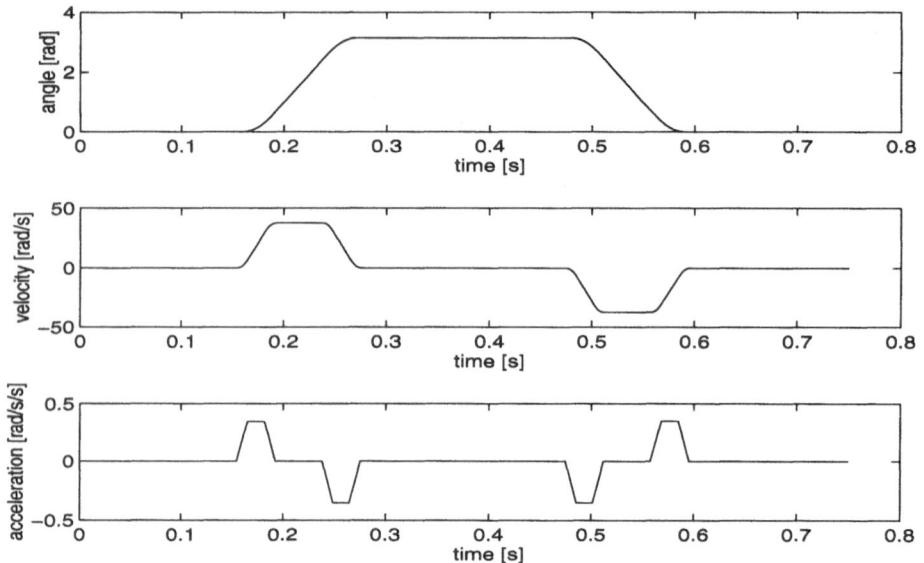

Figure 6. Setpoints $\theta(t)$, $\dot{\theta}(t)$ and $\ddot{\theta}(t)$

Since a constraint exists of $\pm 10V$ for the voltage to the motor amplifier, it is important to optimize the setpoint such that the shortest cycle-time is realized. Preliminary results indicate that improvements are possible, but care has to be taken with respect to possible instability regions. This will be subject of future research. In this paper we will concentrate on the third order setpoint profile.

Using the setpoints from Fig.6 various designs for the feedforward signal has been made: (i) direct substitution into equation (4), yielding errors up to 70 increments during experiments, (ii) using (4) but including a tuning parameter for the centrifugal term ($dm = \frac{\partial J(\theta)}{\partial\theta}\dot{\theta}^2$), with which the maximum error could be lowered to 25 incr., for 0.4 dm as best value, (iii) the best from (ii) but now including a viscous friction compensation (denoted fv), tuned using experiments, and showing results with maximum error approximately 15 incr. Finally, *iterative learning control (ILC)* was used. We will elaborate on ILC before showing results.

Iterative learning control (Arimoto et al., 1984; Tomizuka et al., 1989) is useful when systems have to perform repetitive tasks or are subject to repetitive disturbances. A (feedforward) signal is generated which cancels most of the repetitive error signals. A block-diagram of a system with ILC is shown in Fig.7.

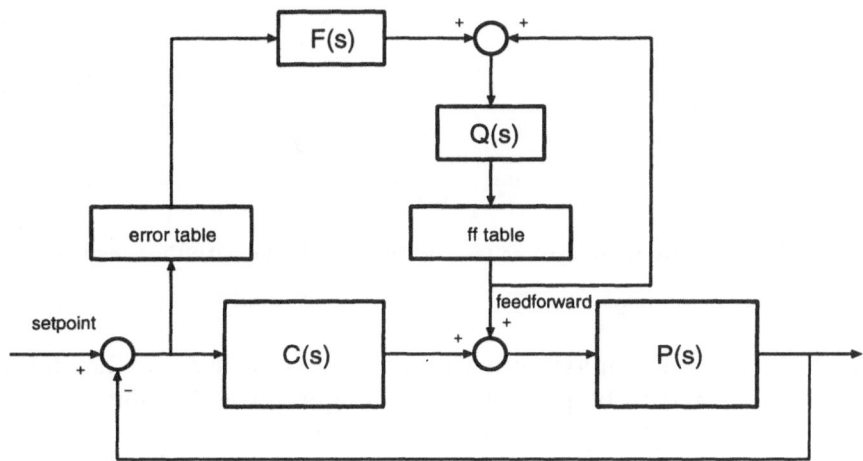

Figure 7.   Control system block diagram including Learning feedforward

At each training cycle, the error is logged into the error table and the feedforward is read from the feedforward table. They are processed off-line to compute the feedforward signal that will be applied the next cycle. With the right choice of filters F and Q convergence can be proven, either by simple small-gain arguments for linear systems (Moore et al., 1992) or by Lyapunov analysis for nonlinear systems (Guglielmo, Sadegh, 1996).

For the pick-and-place mechanism a learning feedforward was used up to 50 Hz, with the experimental results shown in Fig.8, compared to the best 'hand-tuned' design. The results show an important reduction of the error down to 5 increments.

5.   Conclusions and future work

The pick-and-place mechanism shows significant non-linear behaviour. A high gain optimal controller has been designed such that in all operating points stable behaviour is obtained with good error reduction capabilities. Using learning feedforward control the error was successfully reduced to 5

390

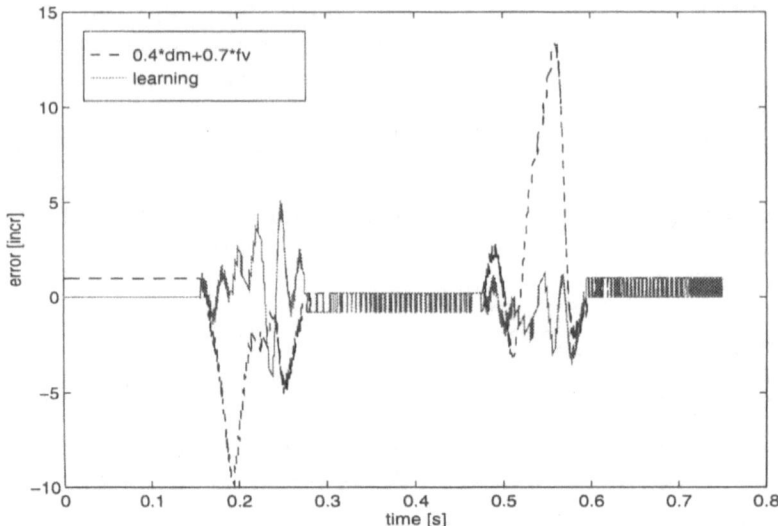

Figure 8. Position error with learning feedforward (solid) and with the tuned feedforward (dashed)

increments. Since the maximum voltage is not reached, further improvements are expected to be possible, by redesigning the setpoint signal.

References

Anderson, B.D.O., and Moore, J.B. (1971) *Linear optimal control*, Prentice-Hall, Englewood Cliffs, N.J.

Arimoto, S., Kawamura, S. and Miyazaki, F. (1984) Bettering operation of robots by learning, *Journal of Robotic Systems*, pp.123-140.

Guglielmo, K. and Sadegh, N. (1996) Theory and implementation of a repetitive robot controller with cartesian trajectory description, *Journal of Dynamic Systems, Measurement, and Control*, 118(3), pp.15-21.

Levine, W.S., and Athans, M. (1970) On the determination of the optimal constant output feedback gains for linear multivariable systems, *IEEE Trans. on Aut. Control*, AC-15, pp.44-48.

Looze, D.P. (1983) A dual optimization procedure for Linear Quadratic robust control problems, *Automatica*, 19, pp. 299-302.

Moore, K.L., Dahleh, M. and Bhattacharyya, S.P. (1992) Iterative learning control: a survey and new results, *Journal of Robotic Systems*, 9(5), pp.563-594.

Tomizuka, M., Tsao, T.C. and Chew, K.K. (1989) Analysis and synthesis of discrete-time repetitive controllers, *Journal of Dynamic Systems, Measurement, and Control*, 111(3), pp.353-358.

DYNAMICS OF DIGITALLY CONTROLLED MACHINES

G. STÉPÁN
Department of Applied Mechanics
Technical University of Budapest, H-1521 Budapest, Hungary

E. ENIKOV
Department of Mechanical Engineering
University of Illinois at Chicago, Chicago, IL 60607, U.S.A.

G. HALLER
Division of Applied Mathematics
Brown University, Providence, RI 02912, U.S.A.

1. Introduction

In this paper we model and explain the existence of small, irregular oscillations that are frequently observed near the desired equilibrium of digitally controlled, linearly unstable mechanical systems. A one-dimensional (1D) map is derived that captures exactly the dynamics of the continuous system. Using this micro-chaos map, the existence of a hyperbolic strange attractor can be proved for a large set of parameter values.

This mapping and its multi-dimensional analogs have a central role in describing the local dynamics of digitally controlled systems. When a processor is used to stabilize the unstable equilibrium of a mechanical system, the sampling delay and the roundoff errors at the analog-digital converters frequently result in small amplitude stochastic vibrations around the desired equilibrium. Such problems were considered by Ushio and Hsu (1987), who studied the dynamics of a corresponding two-dimensional map. Delchamps (1990) formulated the general control problem of an n-dimensional (nD), discrete linear system and analyzed the $n = 1$ case in more detail for a measure zero, nowhere-dense set of the parameters. Stépán (1994) studied experimentally the small amplitude stochastic motion of an inverted pendulum attached to a moving cart. In that example digital control was used to stabilize the upright position of the pendulum. In linear approximation the corresponding discrete control problem can be described by a three dimensional (3D) micro-chaos map.

Although most of these problems are multi-dimensional, even the study of the general one dimensional digital control problem has been missing in the literature. Our goal in this paper is to provide a characterization of the one dimensional case described by this map for a certain parameter domain. We also present estimates

391

D. H. van Campen (ed.), IUTAM Symposium on Interaction between Dynamics and Control in Advanced Mechanical Systems, 391–398.

for some characteristic quantities like amplitude and frequency range, entropy, and fractal dimension.

We prove the existence of an invariant and globally attractive set and the sensitive dependence on initial conditions within this set for the case of the digital stabilization of an inverted pendulum. Numerical simulation results are also presented which confirm the analytical and experimental results.

2. 1D Micro-Chaos Map

Consider the near-equilibrium motion of a one-degree-of-freedom mechanical system under the effect of velocity dependent forces. In particular, assume that the system is subject to some negative velocity-dependent dissipation which is linear in the velocity. Then, in non-dimensionalized form, the velocity v satisfies the linear differential equation

$$\dot{v} - kv = 0$$

with $k > 0$. We want to counteract the effect of the force kv by introducing a computer-controlled dissipation term which is linear in the velocity. Ideally, such a force would change the above equation to

$$\dot{v} - kv = -pv,$$

where $p > k$ is the damping coefficient. However, the control we use is assumed to have two deficiencies. First, the computer samples the velocity only at discrete time instances with sampling time $\Delta t > 0$. As a result, the dissipative force applied by the control system would be $-pv(j\Delta t)$ throughout the time interval $[j\Delta t, (j+1)\Delta t)$, where j is a positive integer. Second, the velocity measurement has a finite resolution, i.e., velocity is measured by the system in terms of the multiples of some small velocity unit $h > 0$. This implies that for $t \in [j\Delta t, (j+1)\Delta t)$ the actual force applied by the control system will be $-ph \operatorname{int}(v(j\Delta t)/h)$. Introducing the notation $t_j = j\Delta t$, we arrive at the following equation for the velocity v:

$$\dot{v}(t) - kv(t) = -ph \operatorname{int} \frac{v(t_j)}{h}, \quad t \in [t_j, t_{j+1}). \tag{1}$$

This equation arises, e.g., in the study of stick-and-slip motion of certain machine tool parts (see, e.g., Tobias (1965). For these systems digital control is used to achieve a stable, small feed rate for the tool. The corresponding mechanical model consists of a block sliding on a surface near some prescribed velocity v_0 under the action of an electric motor (see Figure 1).

At low speed the combined dry and viscous friction force C acting on the block is locally decreasing as the velocity v increases. The electric motor introduces a dissipative force described by the torque-speed $(T-\omega)$ characteristics of the motor, but the system may still be unstable at v_0. In that case, an additional control force provided by the electric motor (with input voltage U) is used to keep the velocity v_0 stable. This introduces artificial dissipation which increases linearly with the velocity.

393

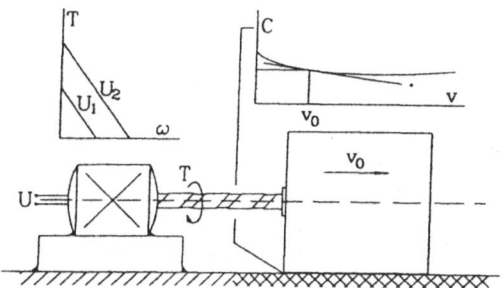

Figure 1. Control of stick-and-slip motion.

One finds that for $t \in [t_j, t_{j+1})$, (1) admits the solution

$$v(t) = (v_j - \frac{ph}{k} \operatorname{int} \frac{v_j}{h})e^{k(t-t_j)} + \frac{ph}{k} \operatorname{int} \frac{v_j}{h}, \qquad (2)$$

where $v_j = v(t_j)$. From (2) we directly obtain that

$$v_{j+1} = \lim_{t \to t_{j+1}} v(t) = (v_j - \frac{ph}{k} \operatorname{int} \frac{v_j}{h})e^{k\Delta t} + \frac{ph}{k} \operatorname{int} \frac{v_j}{h}, \qquad (3)$$

or

$$v_{j+1} = v_j e^{k\Delta t} - \frac{p}{k}(e^{k\Delta t} - 1)h \operatorname{int} \frac{v_j}{h}. \qquad (4)$$

Let us introduce the parameters

$$b = e^{k\Delta t} > 1, \quad c = \frac{p}{k}(e^{k\Delta t} - 1) = \frac{p}{k}(b - 1) > b - 1.$$

Equation (4) shows that the velocity values at the time instances $j\Delta t$ can be obtained by iterating the one dimensional mapping

$$x \mapsto bx - \frac{c}{q} \operatorname{int}(qx), \qquad (5)$$

where $q = 1/h$ and the initial value for the iteration is $x_0 = v_0$, the velocity at some time t_0.

3. Characterization of the 1D map

For convenience, we introduce the rescaling $x \to qx$ to obtain the equivalent map

$$F: [0, m] \to I = [0, m],$$
$$x \mapsto ax - b \operatorname{int} x, \qquad (6)$$

which we shall refer to as the micro-chaos map. Note that F is a piecewise linear, monotone, upper semicontinuous map with discontinuities at the integers $i = 1, 2, ..., m$. For these parameter values

$$b = 2.5, \quad c = 2.0 \qquad (7)$$

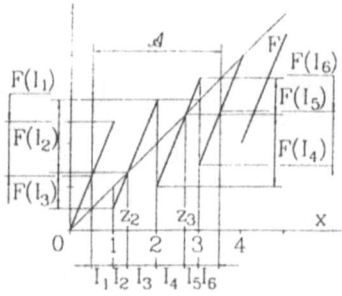

Figure 2. Micro-chaos map with $b = 5/2$, $c = 2$.

the graph of F is shown in Figure 2.

Simple numerical calculations show that the 3 fixed points of this map are located at

$$z_1 = 0, \quad z_2 = \frac{4}{3}, \quad z_3 = \frac{8}{3},$$

and the set

$$\mathcal{A} = [\frac{1}{2}, \frac{7}{2}]$$

is invariant and attractive with respect to F. Since F has no fixed points in $I - \mathcal{A} \cup \{0\}$, the domain of attraction for $\mathcal{A}$ is $I - \{0\}$, i.e. all trajectories starting away from the origin end up in the attractor.

Let us construct a symbolic dynamics on $\mathcal{A}$ with the help of the 6 intervals

$$I_1 = [\frac{1}{2}, 1], \quad I_2 = [1, \frac{4}{3}], \quad I_3 = [\frac{4}{3}, 2],$$
$$I_4 = [2, \frac{8}{3}], \quad I_5 = [\frac{8}{3}, 3], \quad I_6 = [3, \frac{7}{2}],$$

also shown in Figure 2. This provides a Markov-partition of $\mathcal{A}$. The symbolic dynamics is described by the matrix $A \in \mathbb{R}^{6 \times 6}$ defined as usual (see Guckemheimer and Holmes (1986)):

$$a_{ij} = \begin{cases} 1 & \text{if } \overline{F(I_i)} \supset I_j \\ 0 & \text{otherwise.} \end{cases} \tag{8}$$

Then the transition matrix takes the form

$$A = \begin{pmatrix} 0 & 0 & 1 & 0 & 0 & 0 \\ 1 & 1 & 0 & 0 & 0 & 0 \\ 0 & 0 & 1 & 1 & 1 & 0 \\ 0 & 1 & 1 & 1 & 0 & 0 \\ 0 & 0 & 0 & 0 & 1 & 1 \\ 0 & 0 & 0 & 1 & 0 & 0 \end{pmatrix},$$

as it can also be checked in Figure 2. The size and the structure of this transition matrix may vary, of course, for parameter values different from (7), but it can easily be shown that for a measure non-zero set of the parameters it remains the same. For example, A is the same as above if

$$b = 5/2 \quad \text{and} \quad c \in [\frac{75}{38}, \frac{225}{112}]$$

which is satisfied by (7), of course. If b is perturbed, the interval for c changes somewhat, but its length still remains non-zero.

It is proved by Haller and Stépán (1996) that $\mathcal{A}$ is a hyperbolic strange attractor and there exists a hyperbolic invariant Cantor set Λ for the map F, on which F is topologically semi-conjugate to a one-sided subshift with the transition matrix A. The topological entropy ν_F of the map F restricted to the Cantor set in $\mathcal{A}$ can be computed as

$$\nu_F = \log |\lambda|_{max} \approx \log 2.32 \approx 0.842$$

where $|\lambda|_{max} \approx 2.32$ is the dominant eigenvalue of the transition matrix A (see Mané (1987)). The other important characteristic quantity of this map is its Ljapunov exponent μ_F which is simply

$$\mu_F = \log a = \log 2.5 \approx 0.916 .$$

The Haussdorff- and fractal dimensions of the hyperbolic Cantor set $\Lambda \subset \mathcal{A}$ obey the estimate

$$HD(\Lambda) = C(\lambda) = \frac{\nu_F}{\mu_F} \approx \frac{\log 2.32}{\log 2.5} \approx 0.92$$

as it can be concluded from Ledrappier (1981).

We believe that by proving the existence of chaotic attractor for large sets of parameter values, constructing instability charts, and characterizing the strange attractor with its domain of attraction, our study lays the groundwork for the development of more advanced design principles for digitally controlled, one dimensional systems. An important direction for future research is the extension of the one-dimensional results to the higher dimensional digital control problems listed in the Introduction.

4. Mechanical Example for 3D Micro-Chaos Map

The mechanical system in question consists of a pendulum hinged to a mobile cart (see Figure 3). The cart and the pendulum are constrained to move in the same vertical plane. The cart consists of an electric motor with a gear-box connected to a pair of wheels by a chain. If the wheels roll on the ground without slipping, our system has two degrees of freedom and its motion can be described by two generalized coordinates: the angle θ of the pendulum and the horizontal position q of the cart.

The equations of motion assume the form

$$\begin{pmatrix} \frac{1}{3}m_1 l^2 & \frac{1}{2}m_1 l \cos\theta \\ \frac{1}{2}m_1 l \cos\theta & m_1 + m_2 \end{pmatrix} \begin{pmatrix} \ddot{\theta} \\ \ddot{q} \end{pmatrix} - \begin{pmatrix} m_1 g \frac{l}{2} \sin\theta \\ m_1 \frac{l}{2} \dot{\theta}^2 \sin\theta \end{pmatrix} = \begin{pmatrix} 0 \\ Q \end{pmatrix} \tag{9}$$

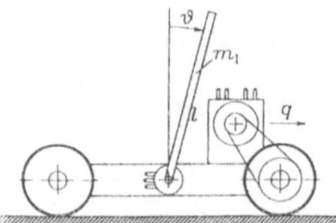

Figure 3. Inverted pendulum on a cart.

Here, m_2 includes the mass of the whole cart, the mass moments of inertia of its wheels, and the rotor of the electric motor, while m_1 and l are the mass and the length of the homogeneous pendulum, respectively. In the horizontal direction, a viscous damping force is considered with damping factor b which also describes the linear torque-speed characteristic of the DC motor. The bar is attached to the cart with a ball bearing, so the viscous friction is negligible along θ.

The control force Q is applied via the electric motor and the Coulomb friction at the contact points of the wheels and the ground. It is considered in the form of a simple PD controller:

$$Q = k_1\theta + k_2\dot{\theta} \tag{10}$$

where the gains are $k_{1,2} \in \mathbb{R}$. When quantization both in time and space are modelled, the control force has the form

$$Q(t) = h\operatorname{int}\frac{k_1\theta(t - r(t)) + k_2\dot{\theta}(t - r(t))}{h}, \quad r(t) = \tau + t - \tau\operatorname{int}(t/\tau)$$

where h denotes the value of one digit converted into control force, while the sampling effect is described by the time delay $r(t)$ where τ stands for the sampling time. After eliminating the cyclic coordinate q and linearization with respect to θ, we obtain the following simple ordinary ordinary differential equation:

$$\ddot{\theta}(t) - \beta^2\theta(t) = u_j, \quad t \in [j\tau, (j+1)\tau), \quad j = 1, 2, \ldots \tag{11}$$

with the piece-wise constant control given by

$$u_j = -\gamma h\operatorname{int}\frac{k_1\theta((j-1)\tau) + k_2\dot{\theta}((j-1)\tau)}{h}$$

and with the new parameters introduced as

$$\beta = \sqrt{\frac{6(m_1 + m_2)g}{(m_1 + 4m_2)l}}, \quad \gamma = \frac{6}{(m_1 + 4m_2)l}.$$

By integrating (11) for each sampling time interval $[j\tau, (j+1)\tau)$, and using the Poincaré sections of the trajectories at the time instants $j\tau$, $j = 1, 2, \ldots$, we construct a three dimensional linear map with respect to the new variable $\mathbf{x}_j = \operatorname{col}(\theta(j\tau)\ \dot{\theta}(j\tau)\ u_j) \in \mathbb{R}^3$ in the form:

$$\mathbf{x}_{j+1} = \mathbf{B} \cdot \mathbf{x}_j + \mathbf{b}\operatorname{int}(\mathbf{c} \cdot \mathbf{x}_j), \quad j = 0, 1, \ldots, \tag{12}$$

$$\mathbf{B} = \begin{pmatrix} \text{ch}(\beta\tau) & \frac{\text{sh}(\beta\tau)}{\beta} & \frac{\text{ch}(\beta\tau)-1}{\beta^2} \\ \beta\,\text{sh}(\beta\tau) & \text{ch}(\beta\tau) & \frac{\text{sh}(\beta\tau)}{\beta} \\ 0 & 0 & 0 \end{pmatrix}, \quad \mathbf{b} = \begin{pmatrix} 0 \\ 0 \\ -h\gamma \end{pmatrix}, \quad \mathbf{c} = \begin{pmatrix} \frac{k_1}{h} & \frac{k_2}{h} & 0 \end{pmatrix}.$$

The control parameters are chosen in a way that the matrix $\mathbf{A} = \mathbf{B} + \mathbf{b} \circ \mathbf{c}$ has eigenvalues within the unit circle of the complex plane, that is, the upper vertical position of the pendulum would be asymptotically stable if there were no quantization effect. However, the solution is unstable in (12) since the eigenvalues of the linear part are $\lambda_{1,2} = e^{\pm\beta\tau}$, $\lambda_3 = 0$, that is one of them has an absolute value greater than 1.

In the case of this 3D micro-chaos map, an invariant and globally attractive set can be found in the form

$$\mathcal{A} = \left\{ \mathbf{x} \in \mathbb{R}^n \mid \|\mathbf{x}\| \leq \frac{\|\mathbf{b}\|}{1-\rho} \right\} \tag{13}$$

where we suppose that

$$\rho = \sigma(\mathbf{B} + \mathbf{b} \circ \mathbf{c}) < 1 \tag{14}$$

is true for the spectral radius of $\mathbf{A}$. This can be proved if we re-write the map (12) as

$$\mathbf{x}_j = (\mathbf{B} + \mathbf{b} \circ \mathbf{c}) \cdot \mathbf{x}_{j-1} - \mathbf{b}(\mathbf{c} \cdot \mathbf{x}_{j-1} - \text{int}(\mathbf{c} \cdot \mathbf{x}_{j-1})) = (\mathbf{B} + \mathbf{b} \circ \mathbf{c}) \cdot \mathbf{x}_{j-1} - \mathbf{b}\chi_{j-1},$$

where $\chi_j \leq 1$ for any $j = 0, 1, \ldots$. Repeating this j times we obtain:

$$\mathbf{x}_j = (\mathbf{B} + \mathbf{b} \circ \mathbf{c})^j \cdot \mathbf{x}_0 - \sum_{i=0}^{j-1} (\mathbf{B} + \mathbf{b} \circ \mathbf{c})^i \mathbf{b}\chi_i.$$

Thus, the norm of $\mathbf{x}_j$ can be estimated as:

$$\|\mathbf{x}_j\| \leq \rho^j \|\mathbf{x}_0\| + \|\mathbf{b}\| \frac{1-\rho^j}{1-\rho}, \tag{15}$$

where $\mathbf{x}_0$ is the initial state. Now, if $\mathbf{x}_0 \in \mathcal{A}$ then for any consecutive iterate we will have $\mathbf{x}_j \in \mathcal{A}$, so $\mathcal{A}$ is invariant under (12). Further, if we let $j \to \infty$, then the right hand side of (15) approaches $\|\mathbf{b}\|/(1-\rho)$, thus, a solution starting from an arbitrary point $\mathbf{x}_0$ will approach the sphere $\mathcal{A}$ asymptotically.

With similar mathematical tools, we can also prove that the map (12) has sensitive dependence on initial conditions, and it has countable number of periodic orbits all contained in $\mathcal{A}$ if (14) fulfills.

Figure 4 shows a numerical simulation of the 3D micro-chaos map (12) when (14) is satisfied by the parameters of the real experimental device. Note that the state space is built of parallel planes defined by $x_3 = jh\gamma$, $j = 0, \pm1, \pm2 \ldots$, referring to the quantization in the control signal. The cloud-like pattern formed by the iterated points follow the pieces of hyperbolas (also emphasized by the

398

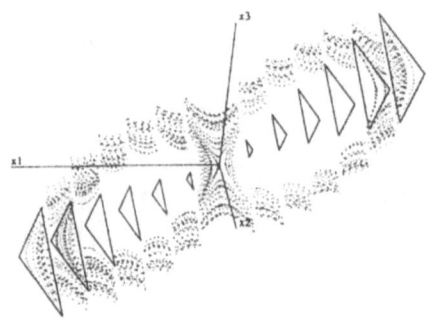

Figure 4. Numerical simulation of non-linear map (12).
triangles) prescribed by the saddle-like dynamics of the piece-wise linearization, and the cloud, which may represent the forecasted chaotic attractor, is within the attractive set $\mathcal{A}$ as predicted in (13).

5. Conclusions

When digital control is used to stabilize otherwise unstable mechanical systems, small-amplitude stochastic vibrations will typically arise around the desired equilibrium or stationary motion of the system. This stochastic oscillation can be described by a simple deterministic model which has a chaotic attractor. We believe that the results including also the characterization of the chaotic attractor are of substantial practical importance and may be used to improve the design of digitally controlled systems.

Acknowledgments

This research was partially supported by the Hungarian Scientific Research Foundation OTKA Grant No. 4-041, the US-Hungarian Science and Technology Program Grant No. 336, and the NSF Grant DMS-95011239.

References

Delchamps, F. D. (1990) Stabilizing a linear system with quantized state feedback, *IEEE Trans. Autom. Contr.* **35** 916-924.

Guckenheimer, J. and Holmes, P. (1986) *Nonlinear Oscillations, Dynamical Systems, and Bifurcations of Vector Fields*, Springer, N.Y.

Haller, G. and Stépán, G. (1996) Micro-chaos in digital control, accepted in *Int. J. of Nonlinear Science*.

Ledrappier, F. (1981) Some relations between dimensions and Ljapunov exponents, *Comm. Math. Phys.* **81** 229-241.

Mané, R. (1987) *Ergodic Theory and Differentiable Dynamics*, Springer-Verlag, N.Y.

Stépán, G. (1994) μ-chaos in digitally controlled mechanical systems, Thompson, J. M. and Bishop, S. R. (eds.), *Nonlinearity and Chaos in Engineering Dynamics*, Wiley, Chichester (UK), pp. 143-151.

Tobias, S. A. (1965) *Machine Tool Vibration*, Blackie, London.

Ushio, T. and Hsu, S. (1987) Chaotic rounding error in digital control systems, *IEEE Trans. Circuits Syst.* **34**, 133-139.

LOAD ADAPTION OF A LINEAR MAGLEV GUIDE FOR MACHINE TOOLS

K.-D. TIESTE AND K. POPP
Institute of Mechanics
University of Hannover
Appelstr. 11
D-30167 Hannover
Germany

1. Introduction

Machine tools require linear guides for high slide velocities with high stiffness and damping as well as very high position accuracy.

The aim of our research consists in answering the following two questions: Firstly, can the principle of active magnetic levitation (maglev) be used for a linear support unit in a machine tool? Secondly, what design potential offers this new technology?

Thus, a maglev guide has been developed for the experimental investigation of control techniques for the levitation and guidance task. In the first part of this paper the "decoupled cascade control" is shown as a control technique which allows an efficient adaption of the load. In the second part a parameter identification method is shown using a pseudo random binary signal (PRBS) as test signal and the COR-LS method, which has been applied to parameter identification and parameter adaption of the maglev guide.

2. The Maglev Guide

The maglev guide consists of a guide block with six pairs of electromagnets in differential arrangement, and a guideway of three rails, cf. Fig. 1. The guide block has a mass of about 30kg, is able to stand forces up to 2kN and is designed to carry workpieces up to 100kg. Automatic adaption of the mass parameters is necassary due to changing mass of the workpieces.

In a first step, the so called "degree-of-freedom control" has been developed, which is based on the decoupling of the five degrees of freedom of the

D. H. van Campen (ed.), IUTAM Symposium on Interaction between Dynamics and Control in Advanced Mechanical Systems, 399–406.

maglev guide, cp. (Tieste, Popp, 1994), (Popp, Tieste, 1995b). A parameter identification method using the minimization of the model error is found in (Popp, Tieste, 1995a). The "degree-of-freedom control" works properly, but has the disadvantage of poor adaptability to different workpieces. The "decoupled cascade control" presented in this paper tries to avoid this disadvantage.

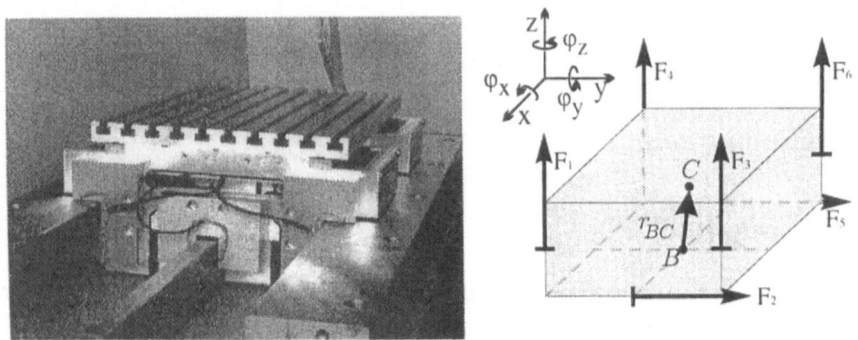

Figure 1. The maglev guide and free-body diagram of the maglev guide

3. System Description

The mechanical system is described as a rigid body based on the free-body diagram, cf. Fig. 1. The model of the entire maglev guide is shown in Fig. 2 by means of a block diagram.

The electrical behaviour of the six electromagnets is described using the diagonal matrices $\mathbf{L}^{-1} = \mathrm{diag}\,(L_1^{-1} \ldots L_6^{-1})$ and $\mathbf{T}^{-1} = \mathrm{diag}\,(\tau_1^{-1} \ldots \tau_6^{-1})$. The actuator forces $\mathbf{f}^{(A)}$ are gained from the six magnet currents using the force-current matrix $\mathbf{K_i} = \mathrm{diag}\,(k_{i_1} \ldots k_{i_6})$. The six actuator forces are applied to the mechanical system shown in the free body diagram,

ELECTROMAGNETS   **MECHANICAL SYSTEM**

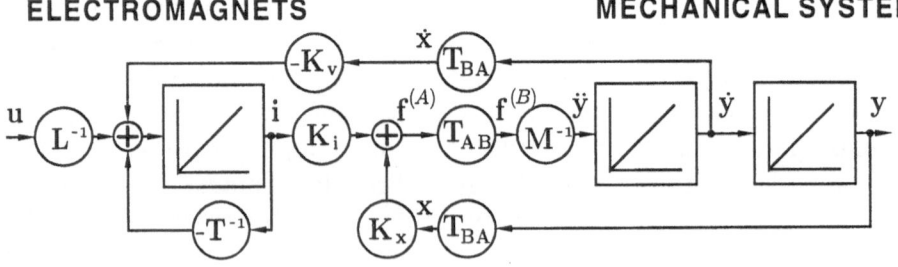

Figure 2. Block diagram of the maglev guide

cf. Fig. 1. The equations of motion are written with respect to the reference point B using the state vectors $\mathbf{y}^{(B)} = (x, y, z, \varphi_x, \varphi_y, \varphi_z)^T$ and $\dot{\mathbf{y}}^{(B)} = (\dot{x}, \dot{y}, \dot{z}, \dot{\varphi}_x, \dot{\varphi}_y, \dot{\varphi}_z)^T$. The selection of this symmetry point as reference point is leading to exceptionally simple equations for the connection between the actuator forces and the mechanical system. Due to the small angles of rotation, the force vector $\mathbf{f}^{(A)} = (F_1 \ldots F_6)^T$ of the six electromagnets is transformed into the generalized force vector $\mathbf{f}^{(B)} = (F_x, F_y, F_z, M_x, M_y, M_z)^T$ using the Jacobi matrix $\mathbf{T}_{AB}$, which is *independent* from the load.

The F_x-component of the generalized force vector $\mathbf{f}^{(B)}$ is zero due to the arrangement of the electromagnets in the experimental setup, cf. Fig. 1. The actual research aims at the development of control strategies for the support and guidance task of the maglev guide. The development of a linear drive in x-direction will be the subject of further research. Nevertheless, the equations of motion are written using all six degrees of freedom including the uncontrolled x-direction: $\mathbf{M}^{(B)}\, \ddot{\mathbf{y}}^{(B)} = \mathbf{f}^{(B)}$,

$$
\begin{pmatrix}
m & 0 & 0 & 0 & m\,r_z & -m\,r_y \\
0 & m & 0 & -m\,r_z & 0 & m\,r_x \\
0 & 0 & m & m\,r_y & -m\,r_x & 0 \\
0 & -m\,r_z & m\,r_y & J_{xx} & J_{xy} & J_{xz} \\
m\,r_z & 0 & -m\,r_x & J_{yx} & J_{yy} & J_{yz} \\
-m\,r_y & m\,r_x & 0 & J_{zx} & J_{zy} & J_{zz}
\end{pmatrix}^{(B)}
\begin{pmatrix}
\ddot{x} \\ \ddot{y} \\ \ddot{z} \\ \ddot{\varphi}_x \\ \ddot{\varphi}_y \\ \ddot{\varphi}_z
\end{pmatrix}^{(B)}
=
\begin{pmatrix}
F_x \\ F_y \\ F_z \\ M_x \\ M_y \\ M_z
\end{pmatrix}^{(B)}.
\tag{1}
$$

The mass matrix $\mathbf{M}^{(B)}$ depends on the mass m, the six components of the symmetric moment of inertia tensor $\mathbf{J}$, and three coordinates of the position vector $\mathbf{r}_{BC}$. The negative stiffness $\mathbf{K}_x\mathbf{T}_{BA}$ of the maglev guide is responsible for the instability of the uncontrolled system. The internal feedback $-\mathbf{K}_v\mathbf{T}_{BA}$ characterizes the relation between mechanical and electrical power of the electromagnets.

4. The Decoupled Cascade Control System

The components of the displacement vector $\mathbf{y}^{(B)}$ – except x – are gained from six eddy current displacement sensors which are placed next to the electromagnets. Furthermore, the magnet currents $\mathbf{i} = (i_1 \ldots i_6)^T$ are measured by Hall-effect current sensors inside the power amplifiers.

The levitation of the system is stabilized in five degrees of freedom using a multi-DOF controller, which has 5 inputs and 6 outputs. The idea of the decoupled cascade control consists of a control technique using a moderately fast PID controller or an equivalent state feedback controller for levitation control, and a secondary very fast current controller, which is implemented as an analogous controller inside the power amplifiers, cf. Fig. 3. This control technique is known from rotational magnetic bearings. Here, it must be extended to variable loads of the linear maglev guide.

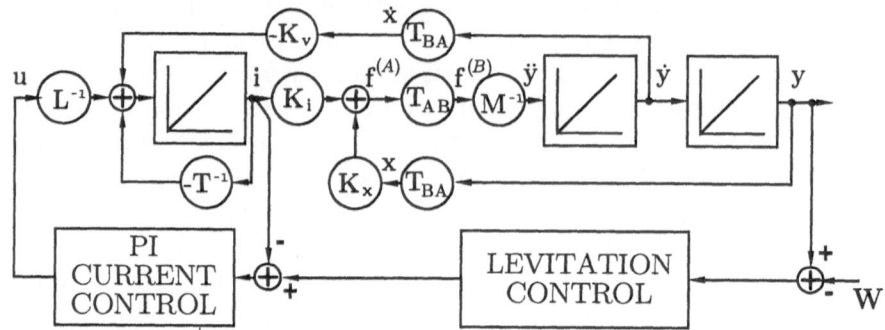

Figure 3. Block diagram of the maglev guide with cascade control

The levitation control has to achieve high stiffness and high damping of the maglev guide. Moreover, the changing load of the maglev guide requires an adaption of the changing mass parameters in the control system. The very fast dynamics of the electromagnets with current controllers can be neclected compared to the dynamics of the levitation control system. So, the levitation control of the mechanical system can be separated from the dynamics of the electromagnets.

The levitation control is assembled in two parts: i) The compensation of the negative stiffness using the feedback $-\mathbf{k}_x \mathbf{T}_{BA}$ and ii) the gap control of the compensated system, cf. Fig. 4. For each degree of freedom, a PID feedback with the acceleration output $\ddot{\mathbf{y}}_s^{(B)}$ is used. The feedback parameters are independent of the changing load due to the normalization of the PID controllers. The acceleration output of the PID controllers is transformed into the required force vector $\mathbf{f}_s^{(B)}$ for reference point B by the mass matrix $\mathbf{M}$, which depends on the load of the maglev guide. The required actuator

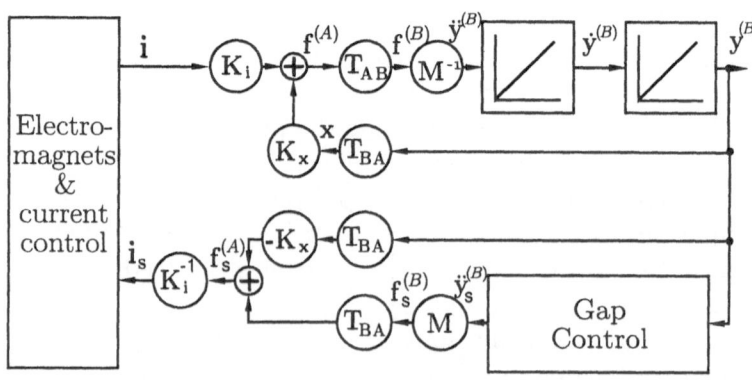

Figure 4. Block diagram of the levitation control

forces are evaluated using the Jacobi matrix $\mathbf{T}_{BA}$.

The advantage of the "decoupled cascade control", beside of the decoupling of current- and levitation control mentioned above, is the physical sense of each parameter. Nearly all matrices and parameters are independent of the load of the maglev guide. Changing workpieces only change the mass matrix $\mathbf{M}$, which varies in a wide range. Thus, parameter adaption of the mass matrix is required.

5. Parameter Identification

If all parameter matrices except the elements of the mass matrix are well known, the aim of the parameter identification is the determination of the changing mass matrix $\mathbf{M}$. In the present case of parameter adaption, the parameter identification has to be done under the following conditions:

- The identification procedure must take only few seconds, including measurements and parameter evaluation.
- Closed loop parameter identification is required because of the instability of the uncontrolled maglev guide.
- The identification method should profit from the structural knowledge about the maglev guide.
- An efficient test signal and a reliable identification method has to be used.

In a first step, the signals $\mathbf{y}^{(\mathbf{B})}(n)$, $\mathbf{i}(n)$ and $\xi(n)$ are recorded using the sampled data of the levitation controller, which operates with a sample rate of $T = 300\mu s$, cf. Fig. 5.

For direct identification of the inverse mass matrix $\mathbf{M}^{-1}$ (cf. Fig. 2) the acceleration vector $\ddot{\mathbf{y}}^{(B)}$, which is gained by filtering and differentiating

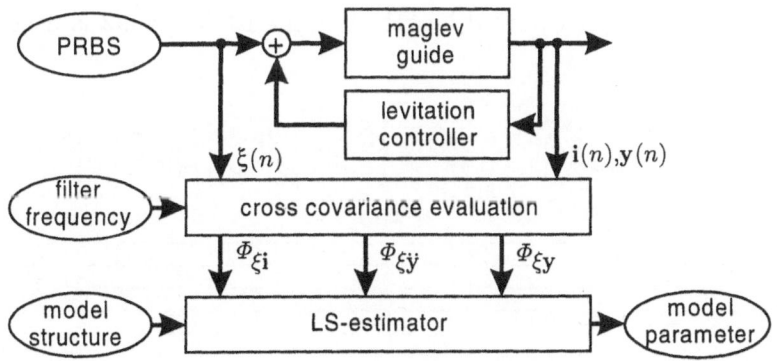

Figure 5. Signal processing structure for parameter identification

$\mathbf{y}^{(B)}$, and the force vector

$$\mathbf{f}^{(B)} = \mathbf{T_{AB}} \ \mathbf{K_x} \ \mathbf{T_{BA}} \ \mathbf{y}^{(B)} + \mathbf{T_{AB}} \ \mathbf{K_i} \ \mathbf{i} \qquad (2)$$

have to be calculated. Then, the mass matrix can be derived from the equation

$$\ddot{\mathbf{y}}^{(B)} = \mathbf{M}^{-1} \ \mathbf{f}^{(B)}. \qquad (3)$$

5.1. THE TEST SIGNAL

Some required properties of the test signal are:

- For simultaneous measurements of the MIMO-system of the maglev guide, five orthogonal test signals are required.
- The test signal must excite frequencies in the range of $1 \ldots 200$Hz.
- Improvements of the parameter identification quality shall simply be done by extending the measurement time.

A pseudo random binary signal (PRBS) has been used as test signal, cp. (Isermann, 1992). It combines advantages of periodic signals (averaging over several periods) and stochastic signals (excitation in a wide frequency range). Furthermore, the spectral properties of the PRBS can be adapted to the identification problem.

The measurements of the MIMO-system of the maglev guide can be done in two ways: i) Five test signals can be applied sequentially to each input of the mass matrix, leading to independent measurements. ii) Five orthogonal test signals can be applied to all inputs simultaneously.

The generation of orthogonal test signals can be done using a PRBS with period length N sufficiently long. Five orthogonal signals are generated using a time shift of $m \sim N/5$ samples between each input signal, where m is the nearest integer value of the fraction, cf. Fig. 6. For the maglev guide, a PRBS with the period length $N = 4 * 1023 = 4092$ is used, which

Figure 6. Generation of orthogonal test signals by time shifting

is subsampled by the factor 4. The sample rate of the levitation control amounts to $T = 300\mu s$, giving a period length of about 1.24 seconds for the test signal. Due to the transient response of the controlled maglev guide, which goes to zero within less than 0.15 seconds, a separation of the coupling within the MIMO-system is possible.

5.2. THE CORRELATION METHOD

During measurements, the five time shifted PRBS $\xi_i(n)$ are added to the output values of the levitation controller. After waiting for fading the transients, the six currents $i(n)$ as well as the five output displacements $y(n)$ one records during one period or averages them over several periods of the test signal, cf. Fig. 5, see also (Isermann, 1992).

For separation of the responses to the 5 centered test signals the cross correlation vectors $\Phi_{\xi i}(\tau) = \frac{1}{N} \sum_{k=1}^{N} \xi(k+\tau)\, i(k)$ and $\Phi_{\xi y}$ are evaluated. The cross correlation vector $\Phi_{\xi \ddot{y}}$ must be evaluated using time differentiation of y. This can either be done using a low pass filter of at least second order with subsequent differentiation, and the inevitable amplification of the measurement noise. The best way is to use the interchangeability of derivation with respect to the variables,

$$\Phi_{\xi \ddot{y}}(\tau) = \frac{d^2}{dt^2}\Phi_{\xi y}(\tau) = \Phi_{\ddot{\xi} y}(\tau). \tag{4}$$

In this way $\Phi_{\xi \ddot{y}}$ can be evaluated using the filtered and derivated PRBS, which is free from measurement noise.

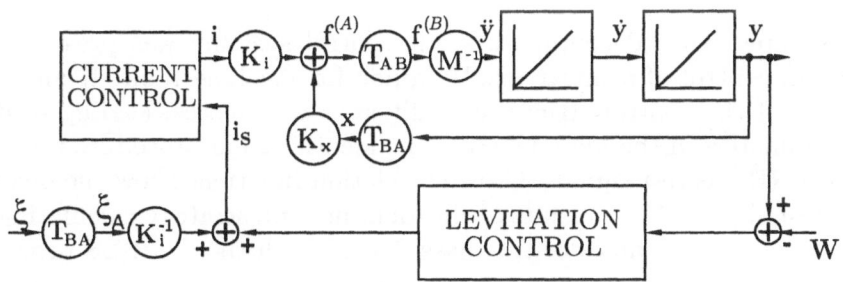

Figure 7. Generating the test signal

Each of the five components of the force vector $\mathbf{f}^{(B)}$ should be excited independently. But, only the magnet currents enable the generation of forces. So, the vector ξ of the 5 test signals is first transformed into the test vector ξ_A for the magnet currents using the equation

$$\xi_A = \mathbf{T_{AB}}\, \xi. \tag{5}$$

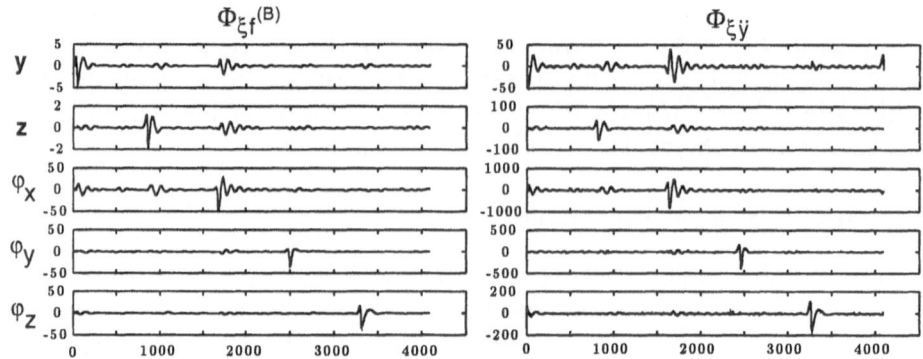

Figure 8. Cross correlation functions

Then, this test vector is added to the magnet currents, cf. Fig. 7. The amplitude of the test signal is set as high as possible. After recording the time signals, we calculate the cross correlation functions $\Phi_{\xi_1 f(B)}$ and $\Phi_{\xi_1 \ddot{y}}$ shown in Fig. 8. The five time shifted test signals lead to a shifting of the cross correlation functions in the τ-domain. Each peak in a cross correlation function shows the response of each measured signal to the test signal. The parts of the cross correlation functions, which contain most information (e.g. $\tau = 0 \ldots 200$, $\tau = 800 \ldots 1000 \ldots$) can be used for evaluating the coefficients of the inverse mass matrix using a Gauss estimator.

6. Conclusion

In this paper, the "decoupled cascade control system" was presented for levitation control of a linear maglev guide for machine tools. It allows an adaption of the control structure to different workpieces, leading to different parameters of the mass matrix. The identification procedure uses time shifted PRBS as test signals. Cross correlation functions show the qualities of the test signals. First results of the identification strategy calculating the 10 load depending values of the mass matrix with the COR-LS procedure are promising, but further research is necessary.

References

Tieste,K.-D.; Popp,K.: Dynamical Behaviour of a Linear Maglev Support Unit for Fast Tooling Machines, 4th *International Symp. on Magnetic Bearings*, ETH Zürich 1994.

Popp,K.; Tieste,K.-D.: Optimization Methods for Parameter Adaption in Mechatronic Systems, *IUTAM Symp. on Optimization of Mechanical Systems*, Stuttgart 1995.

Popp, K.; Tieste, K.-D.: A Linear Maglev Guide for Machine Tools, 14th *International Conference on Magnetically Levitated Systems*, Bremen 1995.

Isermann R.: Identifikation dynamischer Systeme, Berlin-Heidelberg 1992.

ACTIVE DECOUPLING OF DYNAMIC STRUCTURE-FOUNDATION INTERACTIONS

H. ULBRICH, J. BORMANN *
K.-H. STENVERS, U. MUTZBERG **

*   University of Essen,
         Faculty of Mechanical Engineering - Chair of Mechanics
         Schützenbahn 70, D-45117 Essen, Germany
**   VERTEX ANTENNENTECHNIK GmbH
         Baumstr. 47, D-47198 Duisburg, Germany

1.   Introduction

Telescopes are used to observe far-away objects. This requires a very high positioning accuracy. When installed on the ground, sometimes even the rotation of the earth must be accounted for to achieve a sufficiently high resolution.

The VERTEX ANTENNENTECHNIK GmbH at Duisburg developed a universal light-weight telescope that can be used to realize slow high-precision tracking movements (Fig. 1.1). For this purpose, the telescope is mounted on a hexapod system being known from the field of robot techniques (parallel robots). By varying the length of the individual legs, we can control the six spatial rigid body degrees of freedom with iterative reverse kinematics.

In order to become independent from stationary systems, a mobile carrier system for a telescope could be considered for future applications - e.g. a train, a ship or an airplane - which would enable one to choose the most favorable place for the observation.

Fig 1.1: Hexapod telescope

D. H. van Campen (ed.), IUTAM Symposium on Interaction between Dynamics and Control in Advanced Mechanical Systems, 407–416.
© 1997 *Kluwer Academic Publishers.*

408

While today's terrestrial systems have to realize only „slow" tracking movements with large amplitudes to position the antenna, telescopes installed in mobile carrier systems can be disturbed by „quick" interfering movements which are caused by structural vibrations of the carrier system and can be transmitted to the telescope. These structural vibrations can stem from the outside, i.e. from movements of the carrier system itself, or from system noise (machines, gears, etc.). The consequence is that the required directing accuracy can no longer be realized with common constructive means, and the telescope must be dynamically decoupled from its carrier system in the frequency range of the parasitic oscillations.

This paper is based on a case study which was carried out to find whether it is possible to prevent the transmission of these „quick" interfering movements.

1.1   PROBLEM

The case study deals with the installation of a telescope in an airplane. The telescope itself is mounted on the hexapod system (Fig. 1.2) which in his turn is installed in the airplane.

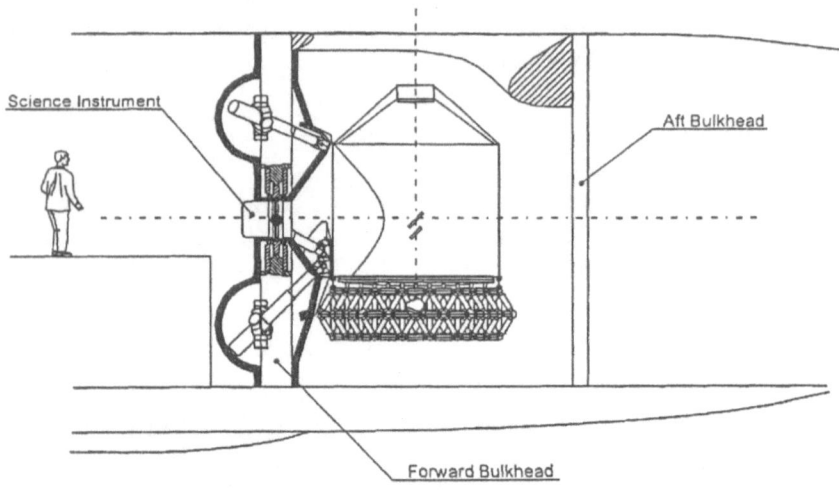

Fig. 1.2: Installation of a light-weight telescope in an airplane

The hexapod consists of six identical length-adjustable precision spindles. This spindle system enables slow movements in large ranges. It can thus be used to realize tracking movements and also to compensate slow disturbances of the base (i.e. slow changes of course of the airplane). The spindle system has a cut-off frequency of around 1Hz, so small interfering movements can still occur in the amplitudes (in a frequency range of 1 to 50Hz caused by structural vibrations of the airplane). These must be barred from transmission to the telescope. The active vibration isolation, therefore, aims at dynamically decoupling the telescope from these quick interfering movements of the airplane.

In the present case, hydraulic actuators are particularly suitable because of the high quasi-static forces to be supported and the rather low regulating frequencies to be realized [Ulbrich, 1]. The high quasi-static forces are caused by the weight of the antenna and can vary significantly for each leg depending on the operating position of the telescope; up to ± 15kN per leg. In contrast to magnetic and piezo-electric acutators, hydraulic actuators have the advantage that they do not generate any electromagnetic waves which might disturb the navigation of the plane or the operation of the telescope (if it is a radio-telescope). Their general usability has already been proven in different applications in the field of rotor dynamics, both theoretically and practically [Althaus, 2].

1.2   STRATEGY TO SOLVE THE PROBLEM

When working on this problem, we assumed that each hexapod leg could be controlled separately in order to achieve a dynamical decoupling of the complete system; i.e. if we succeeded in dynamically balancing each individual leg, the complete system would also be dynamically balanced. In this way, six independent controller circuits would be created. This assumption is allowed because the six legs of the hexapod are kinematically independent of each other. Based on these modelling ideas, this study is limited to the general design of the actuator and the subsequent development of the controller for only **one** hexapod leg.

Section 2 presents the actuator developed for this application. It deals with the design of the actuator type according to the regulating distances and forces to be realized, and also the integration of the actuator into the existing hexapod leg.

Section 3 deals with the modelling of the system to be controlled (i.e. one hexapod leg with an equivalent telescope mass), the simulation of its dynamic behaviour as well as the design and selection of the controller. The results are graphically displayed.

2.   Hydraulic Actuator

As already explained in the introduction, a hydraulic actuator system is especially suitable in this case to realize the regulating action. The reasons lie in the special combination of the requirements:

– Regulating distances up to ± 0.2mm (depending on the frequency),
– regulating frequencies between 1 and 50Hz,
– high static loads in the regulating direction (max. ± 15kN).

Fig. 2.1 shows the working principle of the actuator. It consists, essentially, of a lower and an upper body elastically connected to each other via two membranes.

410

These membranes form, together with the lower body, two opposite regulating chambers filled with oil, the pressure of which can be controlled by a commercial

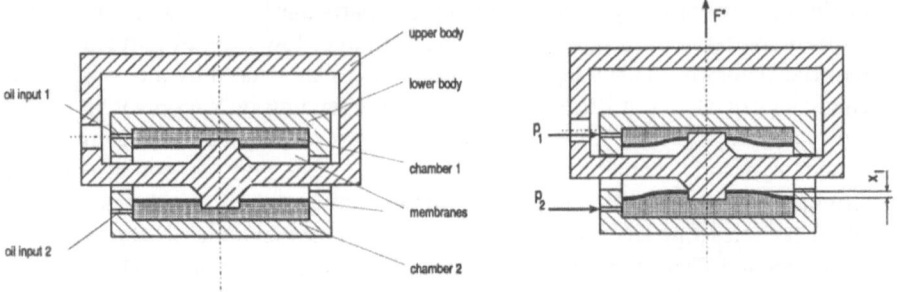

Fig. 2.1: Sketch showing the principle of the hydraulic actuator

servo-valve. The membranes give the system the required flexibility in the axial direction together with a high radial stiffness. This has the advantage of no friction and no sealing problems, in contrast to, e.g., hydraulic cylinders. Since there are no friction losses, an excellent dynamic behaviour can be achieved.

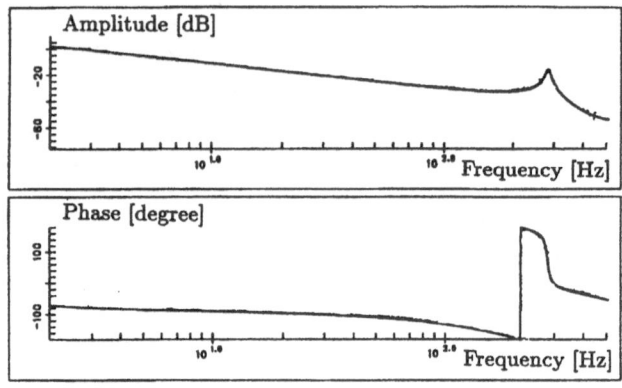

Fig. 2.2: Typical transfer function of a hydraulic actuator
(Source: Althaus [3])

The typical transfer function of a hydraulic actuator is shown in Fig. 2.2. It can be seen that the actuator has an integrating behaviour. This is revealed when looking at the phase which has a value of approx. -90° at frequency zero.

The generated control force can be described by the following equation:

$$F^*(t) = A^* \cdot \Delta p_K(t) - c_1 \cdot x_1(t) \tag{2.1}$$

where F^* is the control force of the actuator, A^* is the characteristic membrane area, Δp_K is the differential pressure between the chambers ($\Delta p_K(t) = p_2(t) - p_1(t)$) and, therefore, the control pressure of the actuator, c_1 is the effective membrane stiffness and x_l is the regulating distance in force direction.

The control pressure Δp_K depends on the fluid-mechanical losses in the hydraulic system and the dynamics of the servo-valve. Thus, the realizable control frequencies and the regulating distances are essentially determined by the cut-off frequency of the servo-valve and the fluid dynamical influences (oil pipes, hydraulic oil, etc.). The relation between generated control pressure $\Delta p_K(t)$, regulating distance $x_1(t)$ and the output of the servo-valve $Q_u(t)$ can be described by the following differential equation [Althaus, 3]:

$$K^* \Delta \dot{p}_K(t) + K_{pq} \Delta p_K(t) + K_R c_t \ddot{x}_1(t) + K_{pq} c_t \ddot{x}_1(t) + A^* \dot{x}_1(t) = \sqrt{\frac{\Delta p_S}{\Delta p_N}} Q_u(t), \qquad (2.2)$$

with K^* as the effective compressibility, K_{pq} the leakage value; K_R considers the compressibility of the oil in the pipes; c_t considers the inertia of the oil; Δp_S and Δp_N are pressure values being related to the characteristic of the servo-valve. The coordinate $Q_u(t)$ represents the flow output of the servo-valve. It depends on the dynamics of the servo-valve, and should be described as a PT_2-element, as indicated by the manufacturer [Fa. Moog, 4]. In the frequency range of up to 50Hz, the servo-valve shows a nearly proportional transfer characteristic, so that the coordinate $Q_u(t)$ can be considered proportional to the control voltage of the servo-valve in this case.

Equations (2.1) and (2.2) describe the dynamical behaviour of the actuator. They must be implemented later into the model of the leg to be decoupled in order to represent the closed-loop control circuit.

2.1   DESIGN OF THE ACTUATOR

During the design of the hydraulic actuator, optimal values must be found for the inner radius and the thickness of the membranes. The control forces and regulating distances were specified. Limiting values were the upper stress limit of the membranes, their outer radius (due to requirements on the size of the actuator) and the maximum pressure of the oil supply. The design task could only be solved iteratively.

Figure 2.3 shows plots of the stress at the inner and outer edge of the membranes of the chambers 1 and 2 versus the inner radius of the membranes. The diagrams have been calculated with the formulae of the theory of thin shells [Timoshenko, 5] and served for the iterative design of the actuator. A verification of the results was obtained with finite element simulations.

2.2 INTEGRATION OF THE HYDRAULIC ACTUATOR SYSTEM INTO THE HEXAPOD LEG

An important topic when using actuators is the integration into existing designs, in this case the integration into the hexapod leg. Fig. 2.4 shows one hexapod leg with integrated actuator system.

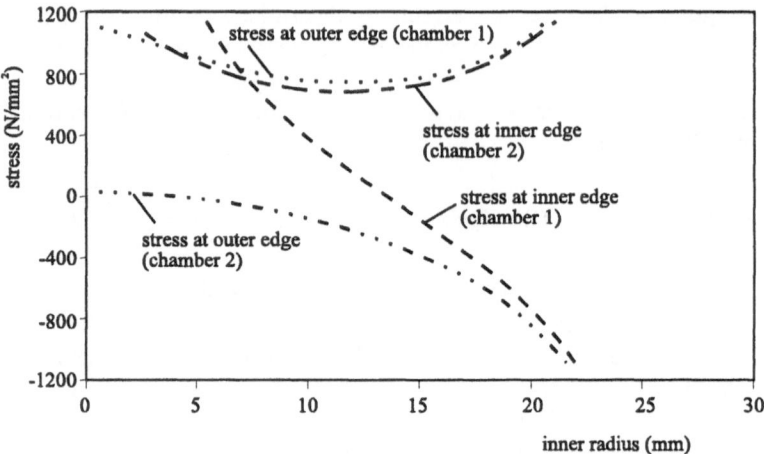

Fig. 2.3: Membrane stress as a function of the inner radius
(point load 3kN, supply pressure p_0 = 30bar, deflection of membrane x = 0.2mm)

Since the design of one single actuator resulted in a max. regulating force of 3kN at a regulating distance of 0.2mm, an annular arrangement of 6 actuators around the hexapod leg was chosen. This leads to a maximum regulating force of 18kN (as opposed to 15kN required), which gives a safety against failure of one actuator.

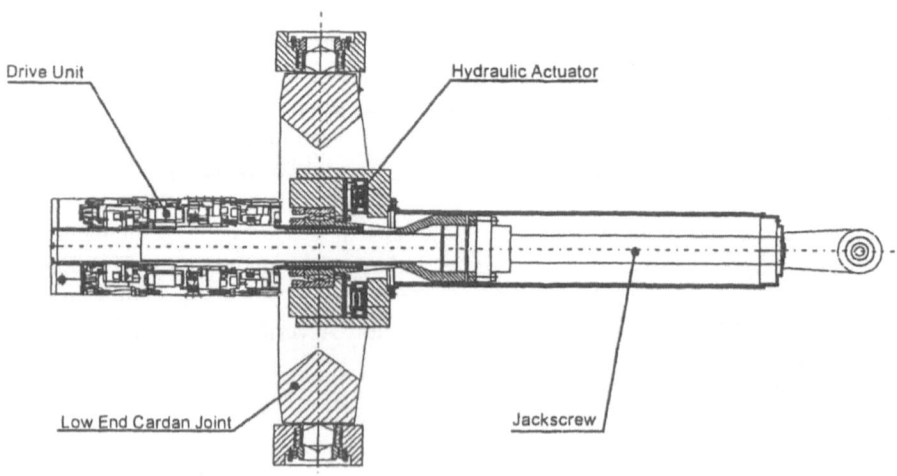

Fig. 2.4: Hexapod leg with integrated actuator system

3.   Modeling, Simulation and Design of the Controller

In the introduction we stated that the six legs of the hexapod are kinematically de-coupled and that it is therefore sufficient to design the active vibration isolation for only **one** leg.

In this section, the equivalent model of a hexapod leg is presented and the corre-sponding equations of motion are formulated. Subsequently, the developed control cirquit is explained and the results are presented.

In the telescope, the control of the antenna position and the vibration isolation control are realized by two separate controllers. The design principle of the two control circuits is shown in Figure 3.1.

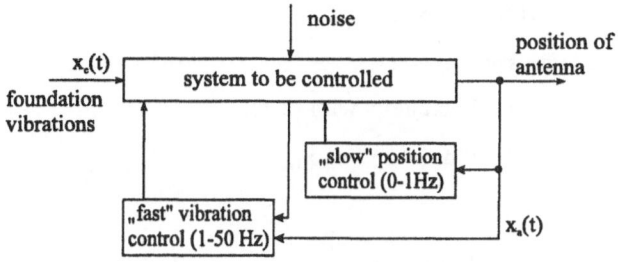

Fig. 3.1: Principles of the two antenna control circuits

To realize the position control of the antenna, a laser gyroscope is applied as a sensor that serves to determine the absolute position of the antenna. The position is then controlled by modifying the length of the individual legs in a „slow" control circuit (< 1Hz).

The system values for the active vibration isolation will be determined at both the actuator and the antenna dish ($\hat{=}$ equivalent telescope mass in the model) by distan-ce and acceleration sensors.

3.1   EQUIVALENT MODEL

The system to be investigated can be described by the equi-valent model shown in Figure 3.2. It is assumed that the hexapod leg can be modelled as an elastic spring. The model considers the possibility of an additional disturbance force F(t) acting upon the equivalent telescope mass m_a which can be, e.g., wind-excited forces.

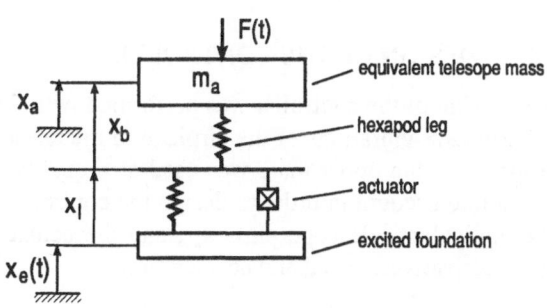

Fig. 3.2: Equivalent model of active hexapod leg

414

The actuator is described according to equation (2.1).

The simulation of the total system is based on 5 equations which are derived from the model assumptions as well as from the model of the hydraulic actuator according to section 2.

1) Force equilibrium at the upper mass:

$$m_a \cdot \ddot{x}_a = F^* - F(t),$$ (3.1)

with m_a the equivalent telescope mass and $F(t)$ the disturbance force (e.g. noise, etc.).

2) Modelling of the hexapod leg as a spring:

$$F^* = -c_b \cdot x_b,$$ (3.2)

with c_b the stiffness of a hexapod leg.

3) Equilibrium at the 6 actuators with (2.1):

$$F^* = 6(\Delta p_K \cdot A^* - c_l \cdot x_l).$$ (3.3)

4) Kinematics:

$$x_a = x_e + x_l + x_b.$$ (3.4)

5) Differential equation of the hydraulic actuator according to (2.2).

From these 5 equations, a **differential equation for the description of the whole system** can be obtained:

$$\dddot{x}_a \cdot A^* \cdot K_R \cdot c_t \cdot \frac{m_a}{c_b} + \dddot{x}_a \cdot A^* \cdot K_{pq} \cdot c_t \cdot \frac{m_a}{c_b} + \ddot{x}_a \cdot \left[K^* \cdot m_a \cdot \left(\frac{1}{6} + \frac{c_l}{c_b} \right) + A^{*2} \cdot \frac{m_a}{c_b} + A^* \cdot K_R \cdot c_t \right] +$$
$$\ddot{x}_a \cdot \left[K_{pq} \cdot m_a \cdot \left(\frac{1}{6} + \frac{c_l}{c_b} \right) + A^* \cdot K_{pq} \cdot c_t \right] + \dot{x}_a \cdot \left[K^* \cdot c_l + A^{*2} \right] + x_a \cdot K_{pq} \cdot c_l =$$
$$\dddot{x}_e \cdot A^* \cdot K_R \cdot c_t + \ddot{x}_e \cdot A^* \cdot K_{pq} \cdot c_t + \dot{x}_e \cdot \left[K^* \cdot c_l + A^{*2} \right] + x_e \cdot K_{pq} \cdot c_l + A^* \cdot \sqrt{\frac{\Delta p_S}{\Delta p_N}} \cdot \frac{Q_N}{u_N} \cdot u_V.$$

(3.5)

3.2   DESIGN OF THE CONTROLLER

The design of the controller was performed with the goal to prevent the transmission of structural vibrations of the airplane to the telescope. The amplitude of the system response at the upper mass was used as a quality criterion. Six system values were taken into account in order to design the control law. These are the displacements of the equivalent telescope mass x_a, and the actuator output $x_{al} = x_e + x_l$, as well as their derivatives, speed and acceleration.

When we introduce new variables x_i, $i = 1, ..., 6$, with $x_1 = x_a$, $x_2 = \dot{x}_a$, $x_3 = \ddot{x}_a$, $x_4 = x_{al}$, $x_5 = \dot{x}_{al}$, $x_6 = \ddot{x}_{al}$, the control vector can be written as

$$u_v = \sum_{i=1}^{6} k_i\, y_i\,; y_i = \begin{cases} x_i\,; & x_i \le (x_m)_i \\ (x_m)_i\,; & x_i > (x_m)_i \end{cases}, \tag{3.6}$$

where $(x_m)_i$ are the maximum admissible amplitudes of the variables x_i.

A quadratic integral criterion was chosen for the design:

$$J = \int_0^\infty \sum_{i=1}^{3} a_i\, x_i^2\, dt \rightarrow \min, \tag{3.7}$$

with a_i as weighting factors. The critical point in order to optimize the controller is to find suitable weighting factors a_i. Using the criterion (3.7), we systematically optimized the parameters k_i. In order to check the stability of the overall system, simulations were carried out for the initial conditions $x_i\,(t = 0) = 2\,(x_m)_i$.

The calculated results, which have been obtained using the optimized controller, are shown in Fig. 3.3. As system input, a harmonic sum excitation $x_e\,(t)$ between 1 and 50Hz has been applied. The figure clearly shows the decreased system response after the control was switched on (at $t \approx 0.7s$). The desired reduction of the vibrations at the antenna ($x_a \le 1\mu m$) is reached within a period of about 0.2s.

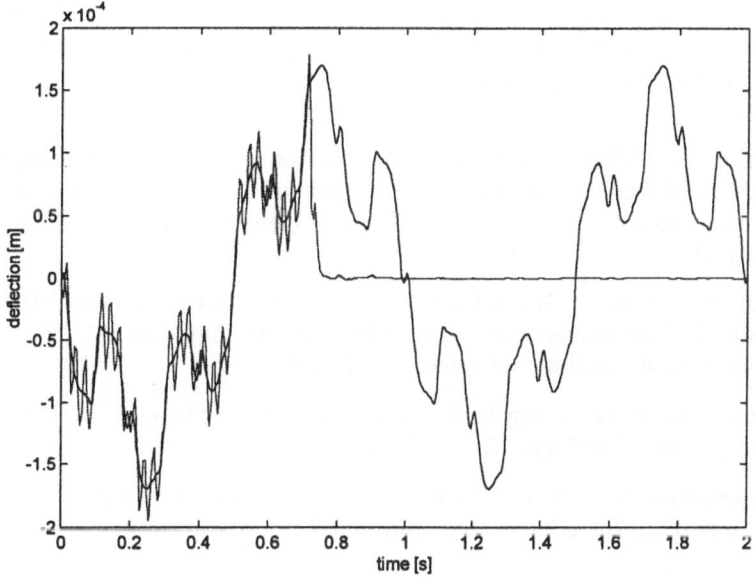

Fig. 3.3: System response $x_a(t)$ for a sum excitation $x_e(t)$

416

4. Discussion of the Results and Final Comments

The design of the actuator (section 2) shows that it is possible to develop a hydraulic actuator which allows one to realize regulating distances of up to 0.2mm for loads up to 3kN (push or pull). The diameter of the membranes used for this purpose was 70mm, which leads to a total diameter of the actuator of approx. 90mm. When using 6 actuators (annular arrangement) and taking as safety the possible failure of one actuator, we can apply a load of 15kN to each hexapod leg.

The results of the simulation of the dynamic behaviour, as well as the design of the robust controller, show that the desired control quality of 1μm, i.e. residual system movements at the telescope mass, can be achieved under the assumed conditions in the range of 1 to 50Hz. The investigations are still in progress. The results already obtained show that more sophisticated control strategies would enable a further increase of the positioning accuracy of the telescope system if necessary. However, this should be done with caution and only after a more precise determination of the required control quality, since the robustness of the closed-loop system naturally decreases with increasing gain factors.

5. References

[1]   Ulbrich, H.; Wang, Y.-X.; Bormann, J.: **Design of Actuators for Mechanism Control.** Proceedings of the Mechanical Engineering Publications Limited, IUTAM Symposium, University of Bath, ISBN 08 52 98 91 64, 1994.

[2]   Althaus, J.; Stelter, P.; Feldkamp, B.: **An Active Hydraulic Bearing for a Solid-Bowl Screw Centrifuge.** Proceedings of the Mechanical Engineering Publications Limited, IUTAM Symposium, University of Bath, ISBN 08 52 98 91 64, 1994.

[3]   Althaus, J.; Ulbrich, H.: **A Fast Hydraulic Actuator for Active Vibration Control.** Proceedings of the Inst. of Mechanical Engineers, Int. Conference „Vibrations in Rotating Machinery", Bath, 1992.

[4]   Firma Moog: **Servoventile der Baureihe 769.** Catalogue D 769-09.92, Moog GmbH, Böblingen, Deutschland.

[5]   Timoshenko, S.; Woinowsky-Krieger, S.: **Theory of Plates and Shells.** McGraw-Hill, 2nd edition, New York, 1959.

UNIVERSAL BIFURCATIONS IN IMPACT OSCILLATORS

JOHN DE WEGER[1], DOUG BINKS[1], JAAP MOLENAAR[2] AND
WILLEM VAN DE WATER[1]
[1] *Physics Department,* [2] *Mathematics Department,*
Eindhoven University of Technology,
P.O.Box 513, 5600 MB, Eindhoven, the Netherlands

Abstract. We give experimental evidence for a new bifurcation structure
that arises when smooth dynamical systems cross a boundary. Our experiment concerns a driven impacting leaf-spring oscillator with a very precise
control of the driving amplitude. The results are in surprisingly good agreement with the predictions of a simple nonlinear mapping that is valid near
grazing impact (*i.e.* impact with zero velocity). The agreement is surprising because a multitude of vibration modes of the spring is excited upon
impact whereas the mapping is two-dimensional. Additionally, we consider
the case where the impact is not instantaneous due to collisions with a
nonrigid stop. This case can be captured by a mapping and we find strong
support for the universality of impact phenomena, even in systems that
have nonidealities.

Impact oscillators have an oscillating mass that impacts with a boundary when its amplitude is large enough. In between impacts the dynamics
is smooth and often linear, but it derives a strong nonlinearity from the
presence of the impact. These systems are known to exhibit a richness of
bifurcation phenomena [1, 2] and a plethora of periodic and chaotic states
has been found [3]. The existence and stability of these states depends in
a very complicated way on the system parameters, and a strong guiding
principle has, so far, been lacking. Impact oscillators are prototypical for
nonlinear phenomena in engineering systems that are often designed with
loose fittings. A timely application of impact oscillators is in atomic force
microscopy [4].

A grazing impact is a boundary impact with zero velocity. In order to

*D. H. van Campen (ed.), IUTAM Symposium on Interaction between Dynamics and Control in Advanced
Mechanical Systems, 417–424.*
© 1997 *Kluwer Academic Publishers.*

418

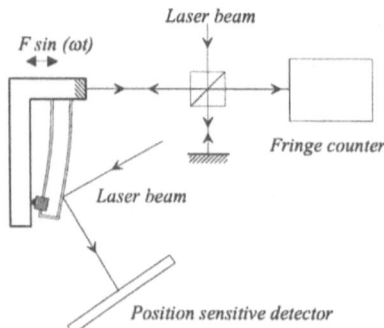

Figure 1. A U-shaped leaf spring is brought into oscillation by horizontally oscillating its support. At a large enough forcing amplitude F, the attached mass impacts with a stop. Collisions take place between a hard ceramic ball and a hardened steel plate. The amplitude of the exciter is measured interferometrically. The deflection of the leaf spring is registered by reflecting a laser beam off the spring onto a position sensitive diode.

describe the events that occur when an orbit evolves to grazing impact and beyond, let us define ρ as the bifurcation parameter, such that ρ increases with the driving amplitude. When the driving amplitude is increased smoothly from zero, impacts first occur at $\rho = 0$. These impacts may be with a relatively large velocity and the transition that has taken place may be hysteretic, i.e. impacts may remain when ρ is subsequently smoothly decreased and may only vanish at a negative value of ρ. When the impacts are about to disappear, the orbit is closest to grazing. In a recent paper Nordmark [5] has reduced the dynamics of impact oscillators to a simple nonlinear mapping that is valid for orbits close to grazing ones. A striking phenomenon in this mapping is the emergence of a square root singularity.

An extensive bifurcation analysis of this mapping has recently been reported by Wai Chin *et al.* [6]. Among the found characteristic phenomena are infinite series of inverse period addings for overdamped oscillators and an infinite series of hysteretic period 1 (p_1) to period m (p_m), $m = 3, 4, \ldots$ transitions for the underdamped case. These phenomena are a consequence of the square root singularity of the mapping and are independent of the details of the dynamics in between collisions. They are, therefore, predicted to be *universal.* This situation is strongly reminiscent of the series of bifurcations that is observed in mappings with a quadratic nonlinearity, where quantitative universality is the consequence of the quadratic nature of the nonlinearity.

The experiment is sketched in Fig. 1. The oscillator consists of a 13 cm long brass leaf spring mounted on a large electromagnetic exciter which oscillates horizontally. The beam is weakly damped by strips of damping material which are glued to it. It is U-shaped to suppress torsional modes but the excitation of higher-order bending modes is unavoidable. Collisions occur between a ceramic ball that is attached near the end of the beam and a hardened steel plate on the exciter. The motion of the beam is registered by reflecting a laser beam off the beam onto a position sensitive photodiode. The measurement and control of the excitation amplitude is a crucial aspect of the experiment. The excitation amplitude (typically 2 mm) is measured using a laser interferometer and fringe counter and is digitally controlled with a long-term stability of 0.5 μm [7].

The experiment is approximately described by the differential equation (overdots denote time derivatives)

$$\ddot{\xi} + \mu\dot{\xi} + \Omega^2\xi = F\sin(\omega t), \tag{1}$$

where Ω is the eigen frequency of the free oscillator, μ represents the damping and where the excitation force has amplitude F and frequency ω. When the position $\xi(t)$ comes to the boundary at, say, t_c, further motion is determined by the properties of the stop. For a rigid stop we assume an instantaneous reflection of the velocity as $\dot{\xi}(t_{c+}) = -\tau\dot{\xi}(t_{c-})$, where τ is the restitution coefficient. However, the impact process for an elastic stop is more complicated, and will be discussed later on. Equation (1) is approximate in that small oscillation amplitudes are assumed and, more important, that higher-order modes of the beam are ignored.

For near-grazing orbits in a broad class of second-order dynamical systems (one of which would be Eq. (1)) Nordmark derived the following mapping [5]

$$\begin{cases} x_{n+1} &= \alpha x_n + y_n + \rho \\ y_{n+1} &= -\gamma x_n \end{cases} \text{ for } x_n \leq 0 \tag{2}$$

$$\begin{cases} x_{n+1} &= -\text{sign}(\beta)\sqrt{x_n} + y_n + \rho \\ y_{n+1} &= -\gamma\tau^2 x_n \end{cases} \text{ for } x_n > 0 \tag{3}$$

where (x_n, y_n) are transformed coordinates of the $(\xi, \dot{\xi})$ space at stroboscopic times $t_n = n2\pi/\omega$, and where ρ is the bifurcation parameter that measures the distance to the point of grazing impact. If no collision occurs between t_n and t_{n+1}, the linear map Eq. (2) applies, whereas Eq. (3) describes the dynamics in the case that an impact will occur on $[t_n, t_{n+1}]$. For oscillators described by Eq. (1) the parameters of the mapping can be expressed explicitly in those of the differential equation, $\gamma = \exp(-\mu/f)$, $\beta = \gamma^{1/2}\sin(2\pi/fT)/(2\pi/fT)$, $\alpha = 2\gamma^{1/2}\cos(2\pi/fT)$, and $\rho = (F/F_g - 1) \times [1 - \alpha + \gamma] / [8\pi^2\beta^2(1+\tau)^2]$; where $f = \omega/2\pi$ is the excitation frequency and F_g is the excitation amplitude at grazing impact. For underdamped oscillators $T = 2\pi/[\Omega^2 - (\mu/2)^2]^{1/2}$ is the period of the free damped oscillations, whereas for the overdamped case the same formulas, but with imaginary T, apply. However, we emphasize again that Eqs. (2,3)

420

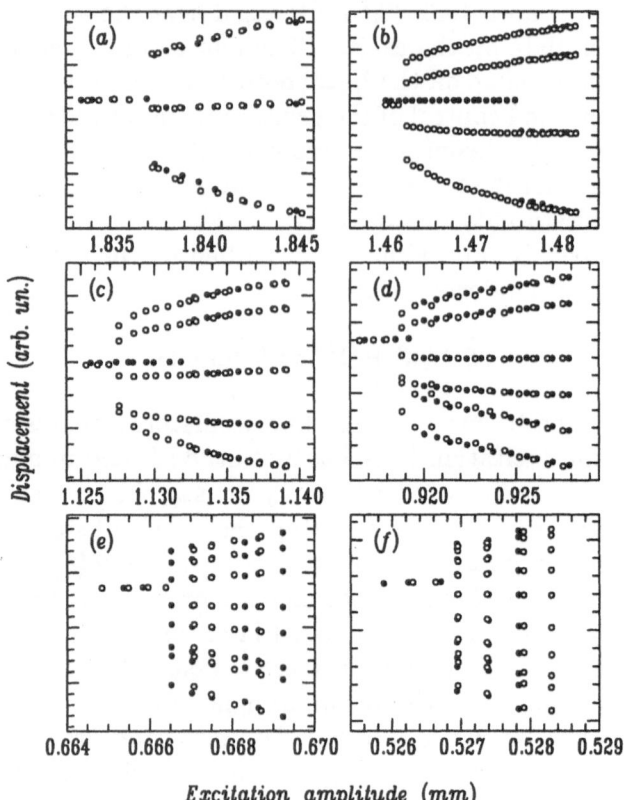

Excitation amplitude (mm)

Figure 2. Experimental bifurcation diagrams of $p_1 \leftrightarrow p_m$ transitions with $m = 3, 4, 5, 6, 8, 10$ for figures (a) through (f), respectively. The displacement of the beam is measured at stroboscopic times $t_n = n2\pi/\omega$. The closed dots are for the upward scan of the excitation amplitude; the open circles are for the downward scan. The period of the free swinging beam is $T = 40.80 \pm 0.02$ ms. The excitation frequencies are $f =$20.90 Hz, 21.50 Hz, 22.20 Hz, 22.60 Hz, 23.05 Hz, and 23.40 Hz, for figures (a) through (f), respectively. In all cases the observed periodicity agrees with the prediction from the Nordmark map (Eqs. (2,3)).

apply to more general impacting oscillators than those described by Eq. (1).

The presence of the square root in Eq. (3) is a key aspect of the mapping; it causes the Jacobian to be singular [8] along the line $x = 0$ and it gives rise to the grazing-bifurcations reported here. The square root can simply be explained by the relation between elapsed time and travelled distance in presence of constant acceleration together with a discontinuity in the velocity, but the precise derivation of the mapping is highly nontrivial. In general, the non-differentiability of a continuous mapping along some curve in phase space gives rise to the characteristic *border-collision bifurcations* [9], of which grazing-bifurcations are an important class.

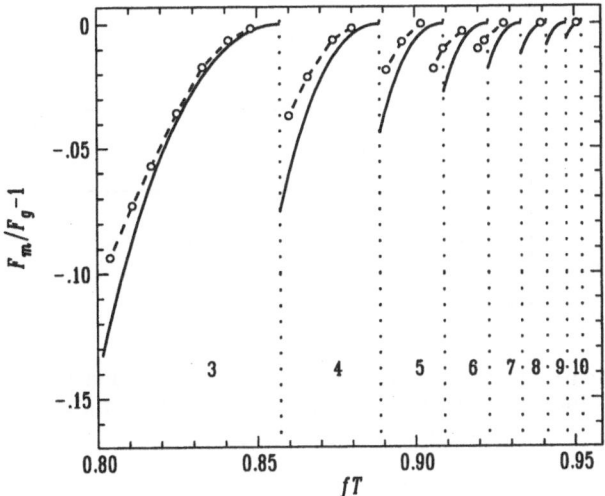

Figure 3. Hysteresis of "maximal" p_m orbits as a function of excitation frequency. Full lines: as computed from the Nordmark map Eqs. (2,3). Dotted vertical lines: stability boundaries of the p_m orbits as predicted by the Nordmark map; the corresponding periods m are indicated. Open circles connected by dashed lines: experimental results.

Our experiment corresponds to the underdamped case, $\Omega^2 > (\mu/2)^2$, and an analysis of the mapping predicts an infinite series of $p_m(m = 3,\ldots,\infty)$ saddle-node bifurcations at $\rho = \rho_m \leq 0$ as the parameter α is varied from 0 to $2\sqrt{\gamma}$ [6]. The transition $p_1 \to p_2$ occurs at $-2\sqrt{\gamma} < \alpha < 0$, a case that was not considered in [6]. Our mapping differs from that analyzed in [6] in the emergence of the factor $\text{sign}(\beta)$ that can be both positive and negative. The transition $p_1 \to p_1$, that is the bifurcation to a (grazing) impact orbit with period 1, occurs for negative $\text{sign}(\beta)$. The stable period m orbit has a single impact per period and is called a *maximal* periodic orbit [6]. At $\rho = 0$ the unstable m-saddle collides with the p_1 orbit. Because the p_m orbit is born away from the p_1 orbit, the transition $p_1 \leftrightarrow p_m$ *appears* to be hysteretic: whereas the transition $p_1 \to p_m$ takes place at $\rho = 0$, the transition $p_m \to p_1$ in a smooth downward scan of ρ takes place at $\rho_m \leq 0$.

A measured series of transitions $p_1 \leftrightarrow p_m$, for $m = 3$ to $m = 10$ is shown in Fig. 2. In our experiments the period 1 orbit is the trivial periodic motion of the non-colliding beam. The highest observed period (p_{10}) is at the limit of our experimental resolution and stability. The shown series of bifurcations accurately follows the prediction of the mapping.

The $p_1 \leftrightarrow p_m$ transitions occur in bands in the (α, γ) parameter plane of the mapping [6] that are crossed approximately transversely by scanning the excitation frequency in our experiment. At the upper boundary of the m-region, the p_m saddle node is born exactly at $\rho = 0$ with no hysteresis. The hysteresis increases towards the upper boundary of the lower lying

422

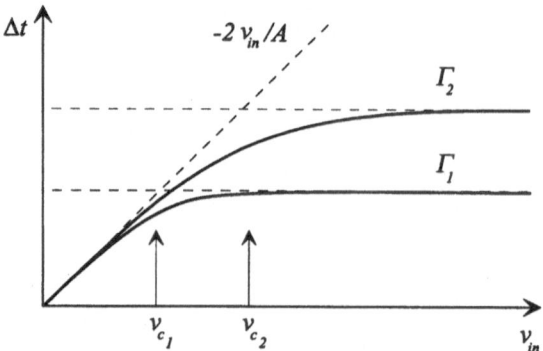

Figure 4. Schematic behavior of the time delay after collision with a yielding stop as a function of the impact velocity. The two values of the spring ratio are $\Gamma_1 > \Gamma_2$; the corresponding cross-over velocities are v_{c_1} and v_{c_2}.

$(m-1)$ boundary. Figure 3 compares the predicted hysteresis $F_m/F_g - 1 = \rho_m \left[8\pi^2 \beta^2 (1+\tau)^2/(1-\alpha+\gamma) \right]$ with that found experimentally, where F_m is the excitation amplitude at which the period m orbit vanishes in a smooth downward scan of F.

The excitation of higher-order modes upon impact is a characteristic feature of impacts in continuous systems. The energy in these modes is quickly dissipated, for example by the radiation of sound. As a consequence, the collision is highly inelastic. This is accounted for by a non-unity restitution coefficient τ. It is important to realize that the mapping is derived for orbits that have a collision that is near grazing. As was explained by us elsewhere [10], the energy loss in collisions helps to stay close to these orbits.

In practical situations, such as described by Eq. (1), secondary bifurcations may be seen for $\rho_m < \rho < 0$ that complicate the observed bifurcation scenario. The mapping Eqs. (2, 3) does *not* predict these secondary bifurcations on $\rho_m < \rho < 0$. However, we surmise that this is observable in the dynamics of mappings that has been extended with higher order terms as compared with Eqs. (2, 3) [11].

The observed bifurcation phenomena are a consequence of the square root singularity of the mapping, which may be smoothed by nonidealities. Here, we consider the nonideality of the excitation of higher-order modes at impact. As a model [12], we image the impacts to occur with a non-rigid stop such that at impact the spring stiffness changes by a factor $\Gamma \gg 1$. In fact, grazing impacts with a yielding boundary are typically encountered in atomic force microscopy of soft condensed matter.

The most important consequence of having a nonrigid stop is that, when impact takes place, the time Δt of contact between the oscillator and the stop has a *finite* value. It is found that Δt depends on the velocity immediately after impact v_{in} and the stiffness ratio Γ in a characteristic manner,

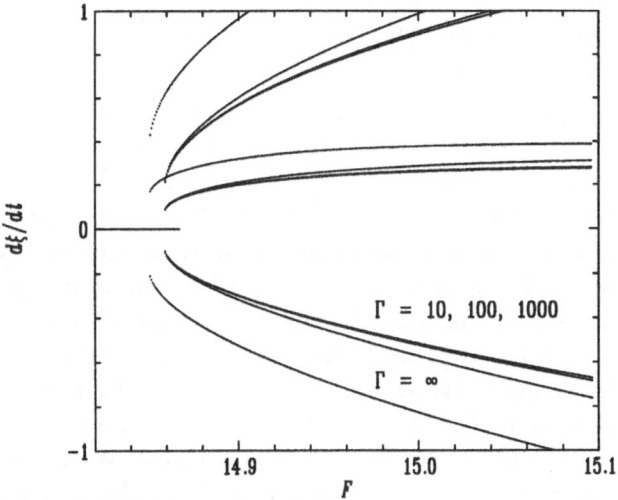

Figure 5. The influence of a yielding stop on grazing impact bifurcations. The bifurcation diagram for a $p_1 \rightarrow p_3$ transition was computed from Eq. (1). The spring-stifness ratios are $\Gamma = 10, 100, 1000$ and $\Gamma = \infty$, respectively. There is an abrupt change of the hysteresis when $\Gamma \rightarrow \infty$. The parameters are $\Omega^2 = 45.30$, $\mu = 0.1962$, $\omega = 2\pi$, $\tau = 0.1$, and the stop is located at $\xi = -1$.

which is illustrated schematically in Fig. 4. Here, we model the relation between v_{in} and the velocity at impact v_{imp} by $v_{in} = \tau v_{imp}$, where τ accounts for energy loss at first contact. A simple analysis shows that there exists a cross-over velocity v_c, where $v_c \sim \Gamma^{-1/2}$, that separates two asymptotic cases. For $v_{in} \ll v_c$, the time delay increases linearly $\Delta t = v_{in}/2A$, where A is the acceleration experienced at grazing, whereas for $v_{in} \gg v_c$, $\Delta t = \pi/\Gamma^{1/2}\Omega$, which is equal to half of the period corresponding to the free vibration of the oscillator in contact with the stop. Our analysis shows that for *each* asymptotic case, a mapping similar to Eq. (3) can be derived, but with different parameter values [11].

The softness of the stop has an important effect on the dynamics which can be understood by considering the time delay upon impact. This is illustrated by Fig. 5, where bifurcation diagrams are shown of a $p_1 \rightarrow p_3$ transition for $\Gamma = 10, 100, 1000$ and $\Gamma = \infty$. The results are obtained by numerically simulating the system of Eq. (1) [13]. The differences between the behaviour for $\Gamma \leq 1000$ on the one hand, and for a rigid stop on the other hand, follow from the impact velocity v_s at the end of the hysteresis. For $\Gamma \leq 1000$ the cross-over velocity v_c is *above* v_s, so that Δt and consequently the dynamics do *not* depend on Γ, whereas for $\Gamma = \infty$ we have $\Delta t = 0$ and the dynamics correspond to the ideal case. Apparently, the change of the cross-over velocity v_c from above v_s to below v_s occurs at a value of $\Gamma > 1000$ (the dependence of v_s on Γ is very weak).

424

The nature of the bifurcations becomes completely different when Γ has a finite value and when the collisions are perfectly elastic ($\tau = 1$). For example, in the case of Fig. 5 a $p_1 \rightarrow p_3$ transition is still observed, but it has become supercritical. This phenomenon can be understood from our analysis, which shows that for a finite Γ and $v_{in} \ll v_c$ the square root singularity of the mapping vanishes when τ exactly equals 1.

In conclusion, we have given experimental evidence for a new class of universal bifurcations that arises when smooth dynamical systems cross a boundary. These bifurcations are predicted by a mapping that reduces the dynamics to its essentials: namely the occurrence of a square root singularity. This singularity appears to be robust and is present even in systems that involve a soft stop. We therefore expect that these phenomena should be observed in a wide class of boundary crossing systems.

We thank Jan Niessen for technical assistance. We gratefully acknowledge financial support by the "Nederlandse Organisatie voor Wetenschappelijk Onderzoek (NWO)".

References

1. S.W. Shaw and P. Holmes, Phys. Rev. Lett. **51**, 623 (1983); S.W. Shaw and P.J. Holmes, J. Sound Vib. **90**, 129 (1983).
2. S.W. Shaw, J. Sound Vib. **99**, 199 (1985).
3. F. Peterka and J. Vacík, J. Sound Vib. **154**, 95 (1992).
4. Q. Zhong, D. Inniss, K. Kjoller and V.B. Ellings, Surf. Sci. **290**, L688 (1993); J.P. Spatz, S. Sheiko, M. Möller, R.G. Winkler, P. Reineker and O. Marti, Nanotechnology **6**, 40 (1995). A better understanding of impact oscillators may lead to a significant improvement of the sensitivity of atomic force microscopy.
5. A.B. Nordmark, J. Sound Vib. **145**, 279 (1991).
6. W. Chin, E. Ott, H.E. Nusse, and C. Grebogi, Phys. Rev. E **50**, 4427 (1994).
7. The amplitude is controlled such that in amplitude scans the measured amplitude changes strictly monotonically. This is important for a correct assessment of the apparent hysteresis. Because the visco-elastic properties of the damping material are temperature dependent, the temperature needed to be kept constant to within $0.02°C$.
8. Singularities in this system in a more general context are discussed in G.S. Whiston, J. Sound Vib. **152**, 427 (1992).
9. H.E. Nusse and J.A. Yorke, Physica D **57**, 39, 1992; H.E. Nusse, E. Ott and J.A. Yorke, Phys. Rev. E **49**, 1073 (1994).
10. J. de Weger, D.J. Binks, J. Molenaar and W. van de Water, Phys. Rev. Lett. **76**, 3951 (1996).
11. J. de Weger, J. Molenaar, D.J. Binks and W. van de Water, to be published.
12. F.C. Moon and S.W. Shaw, Int. J. Non-Linear Mechanics **18**, 465 (1983).
13. An efficient numerical scheme for solving Eq. (1) in the presence of grazing impacts uses the analytical solutions for the motion between impacts. The (exact) positions $\xi(t)$ are computed in a small number of discrete points $t_1, \ldots, t_k$ in each drive period. It is crucial not to miss boundary crossings of $\xi(t)$. These crossings are detected both directly by checking $\xi(t_1), \ldots \xi(t_k)$ and by computing the position $\xi(t)$ at the turning points.

MODEL-BASED LOW-ORDER CONTROL DESIGN FOR MECHANICAL SERVO-SYSTEMS

Closed-loop Balanced Reduction & the use of Weighting Functions

P.M.R. WORTELBOER
Philips Center for Manufacturing Technology (CFT)
P.O.Box 218, 5600 MD Eindhoven, The Netherlands

Abstract. Detailed models of mechanical servo-systems can be used for low-order control design if we invoke closed-loop reduction in an iterative process of *model reduction – optimal control synthesis – controller reduction*. The twin feedback configuration is the key to formulate closed-loop reduction for both the mechanical model and the controller. Closed-loop balanced reduction is a straightforward extension of balanced reduction that can be applied in each step of the reduction process. Using a dedicated MATLAB toolbox, the change in performance after reduction can be monitored and by means of graphical manipulation of extra frequency weights the reduction process can be controlled. In many examples the final controller was close to the optimal full-order controller. The proposed approach is a step towards optimal fixed-order control of high-order systems.

1. Introduction

This paper concerns the integration of order reduction in control design. One of the remaining theoretical problems in linear optimal control is the limitation of the controller order. Theory is now quite mature for synthesizing a (full-order) linear controller for a given linear model and a performance measure in H_2 or H_∞ sense. The order of the optimal controller equals the order of the model. However, once the order of the resulting linear controller is bounded for implementation or cost reasons, there is no theory for synthesizing the optimal controller directly (Harn and Kosut, 1993). It is even possible that there is not a unique solution. The order problem is less serious in the control of systems that can be described by relatively low-order models (in process industry for instance), than in the control of mechanical servo-systems for which it is now well-known that the neglect

D. H. van Campen (ed.), IUTAM Symposium on Interaction between Dynamics and Control in Advanced Mechanical Systems, 425–432.

of vibration modes in the control-design model can have a disastrous effect on the performance, reaching as far as an unstable closed-loop system.

For instance, large flexible space structures equipped with feedback loops that have been designed on simplified models, may exhibit the 'spill-over' phenomenon (Balas, 1982), resulting in large energy levels in parasatic modes and bad control of the modelled modes.

All these observations have led to the awareness that modelling and control design are not two separate tasks but are to be tackled as one. In practice this means that cycles of modelling and control design are repeated untill the controlled system behaves satisfactorily. Satisfactory closed-loop behaviour is often realized by amazingly simple feedback. Experienced control engineers are capable of finding such a feedback, but the question remains whether there is a better one and by which general procedure the best controller can be found. It is rather disappointing that model-based optimal control theory as such provides no solution since it says nothing about the type of model that should be used. Moreover, even if we assume that perfect models of real systems exist, the optimal controllers cannot be computed and certainly not implemented due to their high complexity (high order). Over the years, many researchers have proposed order reduction techniques to solve this problem, but in our research for low-order optimal control of a compact-disc player many were found to be of academic use only, and there was still a need for a truly effective and general reduction strategy: this stimulated the work on closed-loop balanced reduction (Wortelboer, 1994; Wortelboer and Bosgra, 1994) and the design of the WOR-toolbox for Weighted Order Reduction based on MATLAB (Wortelboer, 1995). This paper tries to put this work in an up-to-date perspective. For technical details, notation and most references to pre-1994 work, the reader is referred to Wortelboer (1994).

2. Structural dynamics in control design

The modelling problem and controller design problem cannot be solved independently (Liu and Skelton, 1993). The behaviour of the structure highly depends on the interaction with the controller that is to be implemented. This amounts to the sensor-actuator location and the filtering properties of the controller. The same phenomenon can be seen in dynamic substructuring (Craig and Bampton, 1968): the most effective description of the substructure behaviour can only be found in conjunction with the main structure. Again this amounts to the connection points on the substructure and the filtering characteristics of the structure. Note that this similarity is perfect in case the feedback controller has colocated sensors and actuators.

In case the controller dynamics is not known yet, it is dangerous to strive to efficient, i.e. low-order, dynamic models of the structure. Then it is safer to model all realistic vibration modes. By means of the finite element method and possibly by some identification method we can make dynamic models of many vibration modes. This model can then be used as a general-purpose reference model. For many control design purposes it will be much too detailed (for instance for slow and stiff servo-mechanisms it is not necessary to model flexible modes), whereas for high-performance control it will hopefully be sufficiently detailed such that the danger of unexpected spill-over is small.

Accurate general-purpose modelling of mechanical systems implies high-order models. The allowable order of the controller however is often restricted and the challenge is to construct a high-performance servo-system using a low-order controller. The problem then is to find the dynamic phenomena that are essential in the interaction between the specific controller that is yet to be found, and the structure. At this point mechanical engineering and control engineering should meet. The gap between structural dynamics modelling and model-based full-order control design can be narrowed by order reduction techniques as will be shown in the sequel.

3. Standard feedback configurations

3.1. STANDARD FORM OF LINEAR STRUCTURAL DYNAMICS MODEL

We have observed that in structural dynamics, one is mainly concerned with the open-loop dynamics. Linear structural dynamics models are often expressed in vector second-order form:

$$M\ddot{x} + L\dot{x} + Kx = f$$

with M, L, K the inertia, linear viscous damping and stiffness matrix, and x, f the degree of freedom and external force vector.

3.2. STANDARD PLANT FORMULATION IN OPTIMAL CONTROL

The optimal control design problem is often stated as the minimization over all stabilizing controllers K of some system norm (mostly the H_2-norm or H_∞-norm) of a closed-loop system $\mathcal{F}_l(N, K)$.

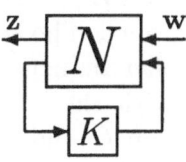

$\mathcal{F}_l(N, K)$ is the lower feedback configuration with N the so-called standard plant. N contains the true model (of the structure) and shape or weight functions such that $\mathbf{w}$ is a generalized disturbance and $\mathbf{z} = \mathcal{F}_l(N, K)\,\mathbf{w}$ is used to measure the generalized disturbance attenuation.

3.3. THE TWIN FEEDBACK CONFIGURATION

In Wortelboer (1994), the following standard configuration has been introduced:

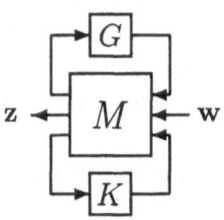

$$\mathbf{z} = \mathcal{I}(G, M, K)\, \mathbf{w} \qquad \text{(twin feedback configuration)}$$

G is the system model, M the master weight, and K the controller. The reason for isolating G (and M) from N is twofold: it makes the system model accessible for order reduction, and separates the design variables (the weighting functions in M) from the variables given by nature (G).

4. The need for different types of order reduction

We assume that G_h gives an accurate description of the system behaviour in closed-loop, no matter what specific controller is used in the feedback loop; this model has order h which is in general much larger than the allowable controller order r.

Today's control synthesis solutions provide full-order controllers. In order to find an rth-order controller, we can apply order reduction in two ways. Assume that M is of order zero. The first method reduces G_h to G_r and then finds the controller K_r that controls G_r optimally. K_r is not likely to yield satisfactory results on G_h due to the spill-over danger. The second method aims at finding K_h that optimally controls G_h and then reduces K_h to $\tilde{K}_r$. The latter approach often fails in practice due to numerical inaccuracies in determining the high-order K_h. The objective in controller reduction,

$$\min_{K_r \text{from } K_h} \|\mathcal{I}(G_h, M, K_r)\|$$

is closely related to the control objective itself, $\min_{K_r}\ \|\mathcal{I}(G_h, M, K_r)\|$. This explains the preference for controller reduction above model reduction. Figure 1 shows the order (radial coordinate) for models (vertical axis) and controllers (horizontal axis). The quarter circle arrow represents full-order control synthesis. Given the practical constraints mentioned above, it is advised to repeat the following cycle: $G_h \rightarrow G_n$, $G_n \nrightarrow K_n$, and $K_n \rightarrow K_r$, with n free. We did not yet explain how the reduction is performed. The key word is 'closed-loop'. Both in model reduction and in controller reduction, we want to retain that part that interacts most strongly with the other 'twin'. So, once again, closed-loop reduction of G depends on G and K. Note that after one complete cycle we do have a (preliminary) controller that helps in reducing the model in closed-loop sense.

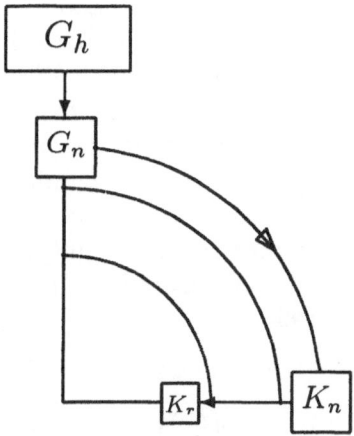

Fig. 1: The order of models and controllers

5. Closed-loop balanced reduction and its manipulation

The topic of this section is worked out in detail in Wortelboer (1994). Here the main points are summarized.

Closed-Loop Balanced Reduction (CLBR) (Ceton *et al.*, 1993) is a straightforward extension of standard balanced reduction (Moore, 1981). CLBR is equivalent to frequency-weighted balanced reduction (Enns, 1984) (see (Wortelboer, 1994) and (Schelfhout, 1996)). The advantages of CLBR are

- More direct computation [1]
- Allows unstable model and controller as long as closed-loop stability is guaranteed

Closed-loop balancing differs from standard balancing in the same way as frequency-weighted balancing differs from standard balancing: the actual balancing of controllability and observability is only performed on separate parts of the state vector associated with the model or controller. Consideration to choose the balancing strategy were:

1. Model and controller reduction can be defined in a unified framework
2. Major computational burden is the balancing, the reduction is a mere truncation and can be done very cheaply for each possible r
3. Extra frequency weights can be naturally incorporated; these can be used to further improve the reduction

The third consideration is very important. It is well-known that balanced reduction is not optimal in H_2 or H_∞ sense and examples are avail-

---

[1]see (Schelfhout, 1996) for a further improvement for observer-based controllers

able that can hardly be reduced by pure balanced reduction. It is less known that simple frequency interval weights can decrease the reduction error impressively (Gawronski and Juang, 1990; Wortelboer, 1994).

The WOR-toolbox for use with MATLAB exploits frequency interval weights in the following way: for each input and output a frequency magnitude function can be created in a graphics window. They are used as a sort of penalty function in the reduction steps. The effects of these frequency weights on the performance can be monitored. The user interface, piecewise quadratic frequency functions and frequency pulse functions are all explained in Wortelboer (1994).

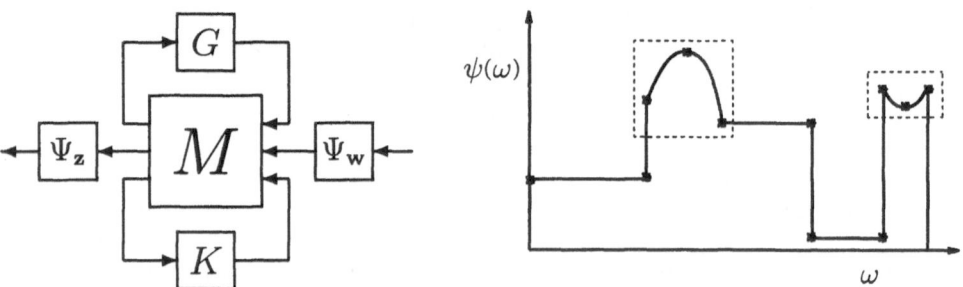

Extra freq.-weights $\Psi = \mathrm{diag}(\psi(\omega))$ Graphically defined piecewise quadratic ψ

6. Low-order control explained on a CD-player

The procedure for finding a low-order controller based on a high-order model, is illustrated by means of a CD-player control problem. In Wortelboer (1994) it is explained how the CD-player model has been derived. The reader is referred to Wortelboer (1994) and the references therein for details about the principles of CD-player control. Our starting point is a model G_{120} containing 60 vibration modes, a fifth-order master weight M_5 for H_∞ control design and a preliminary low-order controller of order two: K_2 stabilizes the CD-player but the 'bandwidth' is insufficient. A better controller of order 4 is sought. In this paper's notation: $h = 120, r = 4$. Choices to be made in the process are: the number of iterations, values for n and possibly the extra frequency weights. Figure 2 shows the first cycle. It is important to evaluate the model reduction by a relative measure such as

$$\frac{\|\mathcal{I}(G_{120}, M, K_2) - \mathcal{I}(G_{32}, M, K_2)\|_\infty}{\|\mathcal{I}(G_{120}, M, K_2)\|_\infty} \ll 1$$

The optimal H_∞-controller for G_{32} is K_{37}. Next, we investigate the possibility of controller order reduction. Using closed-loop balanced controller reduction, we found that for $r \geq 6$ the performance degradation is very

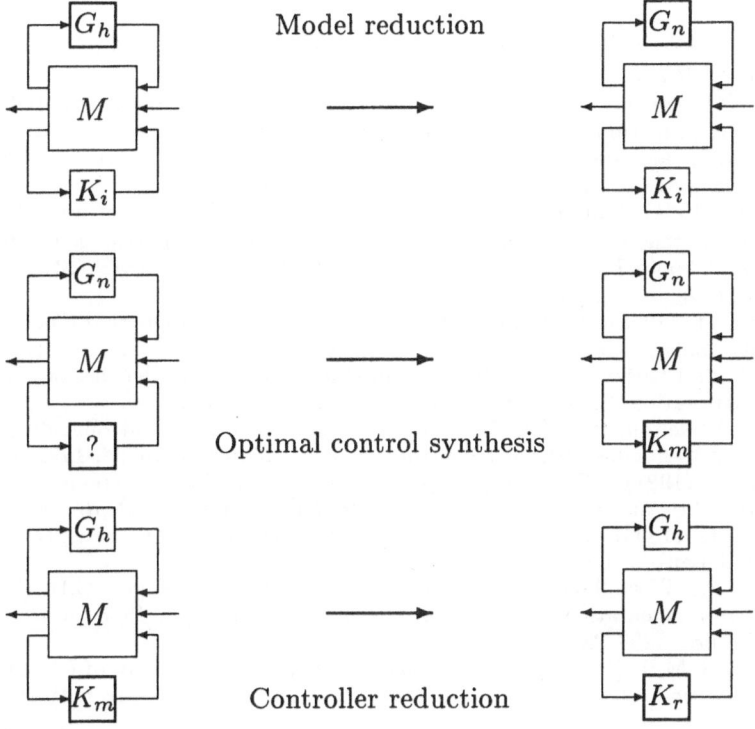

Fig. 2: Three-step cycle for constructing a low-order CD-player controller

small. For $r = 4$ the performance degradation can only be kept small by applying one extra frequency weight to penalize the appearance of a closed-loop resonance. This result cannot be improved much by doing more design and reduction cycles. Alternative approaches of controller reduction (Goddard, 1995) show similar results.

7. Conclusion

It has been shown that low-order control design of mechanical servo-systems starting from detailed structural dynamics models hinges on careful 'in-the-loop' reduction. The twin feedback configuration is the key to formulate closed-loop reduction for both the mechanical model and the controller. Closed-loop balanced reduction is a straightforward extension of balanced reduction that can be applied in each step of the proposed iterative reduction strategy. The choice of the order and the graphical manipulation of extra frequency weights enlarges the chance to find a low-order controller that is close to the optimal one.

References

Balas, M. (1982). "Trends in large space structure control theory: fondest hopes, wildest dreams", *IEEE Trans. Automat. Contr.*, AC-27, no.3, 522–535.

Ceton C., Wortelboer, P.M.R. and Bosgra, O.H. (1993). "Frequency weighted closed-loop balanced reduction", *Proc. 2nd-European Control Conference*, 697–701.

Craig Jr, R.R. and Bampton, M.C.C. (1968). "Coupling of substructures for dynamic analyses", *AIAA J.*, vol.6, no.7, 1313–1319.

Enns, D.F. (1984). "Model reduction with balanced realization: an error bound and a frequency weighted generalization", *Proc. 23rd IEEE Conf. Dec. & Contr.*, 127–132.

Gawronski, W. and Juang, J-N. (1990). "Model Reduction for Flexible Structures", *Control and Dynamic Systems*, vol.36, 143–222.

Goddard, P.J. (1995). *Performance-preserving Controller Approximation*, PhD Thesis, Trinity College, Cambridge.

Harn, Y-P. and Kosut, R.L. (1993). "Optimal low-order controller design via 'LQG-like' parametrization", *IFAC Automatica*, vol.29, 1377–1394.

Liu, K. and Skelton, R.E. (1993). "Integrated modeling and Controller design with application to flexible structure control", *IFAC Automatica*, vol.29, 1291–1314.

Moore, B.C. (1981). "Principal component analysis in linear systems: controllability, observability, and model reduction", *IEEE Trans. Automat. Contr.*, AC-26, 17–32.

Schelfhout, G. (1996). *Model Reduction for Control Design*, PhD Thesis, Katholieke Universiteit Leuven, Leuven.

Steinbuch, M., Wortelboer, P.M.R., van Groos, P.J.M. and Bosgra, O.H. (1994). "Limits of Implementation - A CD player control case study", *Proc. American Control Conference*, 3209–3213.

Wortelboer, P.M.R. (1994). *Frequency-weighted Balanced Reduction of Closed-loop Mechanical Servo-systems: Theory and Tools*, PhD Thesis, Delft University of Technology, Delft.

Wortelboer, P.M.R. and Bosgra, O.H. (1994). "Frequency Weighted Closed-Loop Order Reduction in the Control Design Configuration," *IEEE Conf. Decision & Control*, 2714–2719.

Wortelboer, P.M.R. (1995). *The WOR-toolbox for Weighted Order Reduction*, ©Philips Research, anonymous ftp: `ftp-mr.wbmt.tudelft.nl /pub/wortelboer`.

AUTHOR INDEX

434

AUTHOR INDEX (cont).

ADDRESSES OF PARTICIPANTS
(Session Chairmen are identified by an asterisk)

Prof. S. Arimoto
University of Tokyo, Fac. of Engineering
Dept. of Math. Eng. and Information Physics
Bunkyo-Ku, Tokyo 113
Japan
Tel: +81 3 3812 2111/6930   Fax:+81 3 3816 7805
E-mail: arimoto@sr3.t.u-tokyo.ac.jp

Prof. A.K. Bajaj*
Purdue University,
School of Mechanical Engineering,
West Lafayette,
IN 47907-1288, U.S.A.
Tel: +1-317-494-6896   Fax:+1-317-494-0539
E-mail: bajaj@ecn.purdue.edu

Prof. V.V. Beletsky
Russian Academy of Sciences
Keldysh Institute of Applied Mathematics
Miusskaya pl. 4,
Moscow, 125047, Russia
Tel: +7-095-250-5615   Fax:+7-095-972-0737
E-mail: beletsky@applmat.msk.su

Prof. F. Benedettini
Universita Dell'Aquila,
Dipartimento Ingegneria Delle Strutture
67040 Monteluco Di Roio,
L'Aquila, Italy
Tel: +39-862-434513/434531
Fax:+39-862-434548
E-mail: ben@vaxaq.cc.univaq.it

Prof. V.E. Berbyuk
Ukrainian National Academy of Sciences,
Inst. of Appl. Problems of Mech. & Math.
Naukova ul. 3-B
Lviv, 290601, Ukraine
Tel: 380-322-636000   Fax:380-322-654240
E-mail:
Kalyniak%IPPMM.Lviv.UA@LITech.Lviv.UA

Prof. D. Bestle
Brandenburg. Techn. University Cottbus
Institute of Machine Dynamics
Karl-Marx-Strasse 17, P.O. Box 101344
03013 Cottbus, Germany
Tel: +49-3-55-69-3024   Fax:+49-3-55-69-3028

Dr. S.R. Bishop
University College London
Centre for Nonlinear Dynamics and its Appl.
Civil Engineering Building, Gower Street
London WC1E 6BT, U.K.
Tel: +44-171-380-7729   Fax:+44-171-380-0986
E-mail: s.bishop@ucl.ac.uk

Prof. R. Bogacz
Institute of Fundamental
Technological Research
ul. Swietokrzyska 21,
00-049 Warsaw, Poland
Tel: +48-22-261281 ext. 263   Fax:+48-22-269815
E-mail: rbogacz@ippt.gov.pl

Prof. J. Brindley*
University of Leeds
Dept. of Applied Mathematical Studies
Leeds LS2 9JT
U.K.
Tel: +44 532 335134   Fax:+44 532 429925
E-mail: amtjb@gps.leeds.ac.uk

Prof. D.H. van Campen
Eindhoven University of Technology
Dept. of Mechanical Engng., WFW
P.O. Box 513
5600 MB Eindhoven, The Netherlands
Tel: +31 40 2472851   Fax:+31 40 2447355
E-mail: secr@wfw.wtb.tue.nl

Prof. F.L. Chernousko*
Russian Academy of Sciences
Institute for Problems in Mechanics
pr. Vernadskogo 101
Moscow, 117526, Russia
Tel: +7 095 434 0207   Fax:+7 095 938 2048
E-mail: chern@ipm.msk.su

Prof. T.H. Chin
Sophia University
Dept. of Electrical and Electronics Engng.
7-1 Kioicho, Chiyoda-ku
Tokyo 102, Japan
Tel: +81-3-3238-3335   Fax:+81-3-3238-3885

Dr. K. Czolczynski
Technical Univ. of Lodz
Div. of Machine Dynamics (K-13)
Stefanowskiego 1/15
90-924 Lodz, Poland
Tel: +48-42-312231   Fax:+48-42-365646

Mr. A. Darby
Cambridge University
Dept. of Engineering
Trumpingon St.
Cambridge CB2 1PZ, U.K.
Tel: +44-1223-332854   Fax: +44-1223-332662
E-mail: apd@eng.cam.ac.uk

436

Dr. M.E. Davies
University College London
Centre for Nonlinear Dynamics and its Appl.
Civil Engineering Building, Gower Street
London WC1E 6BT, U.K.
Tel: +44-171-380-7050/2727
Fax:+44-171-380-0986
E-mail: michael.davies@ucl.ac.uk

Prof. A. De Carli
University of Rome La Sapienza
Dept. of Computers and Systems Sciences
Via Eudossiana 18
I-00184 Roma, Italy
Tel: +39 6 4458 5357   Fax:+39 6 4458 5367

Mr. A. Dequidt
LAMIH-CSME,
Le Mont Houy BP311,
59304 Valenciennes
France
Tel: +33-27-141374   Fax:+33-27-141305
E-mail: adequidt@univ-valenciennes.fr

Prof. M. Ding
Florida Atlantic University
Center for Complex Systems
Boca Raton,
FL 33431, U.S.A.
Tel: +1-407-367-2324   Fax:+1-407-367-3634
E-mail: ding@daffy.ccs.fau.edu

Dr. U. Dressler
Daimler-Benz Research Institute
Goldsteinstr. 235
D-60528 Frankfurt
Germany
Tel: 049-69-6679326   Fax:049-69-6679210
E-mail: dressler@dbag.fra.daimlerbenz.com

Prof. W.L. Esmeijer
Vesaliuslaan 44
5644 HL Eindhoven
The Netherlands
Tel: +31-40-2113419   Fax:+31-40-2447355

Ir. L.F.P. Etman
Eindhoven University of Technology
Dept. of Mechanical Engng., WFW
P.O. Box 513, 5600 MB Eindhoven
The Netherlands
Tel: +31-40-2472807   Fax:+31-40-2447355
E-mail: pascal@wfw.wtb.tue.nl

Dr. R. Faglia
Universita di Brescia
Dipartimento di Ingegneria Meccanica
Via Branze 38, 25123 Brescia
Italia
Tel: +39-30-3715401   Fax:+39-30-3702448
E-mail: rfaglia@bsing.ing.unibs.it

Dr. R.H.B. Fey
TNO Building and Construction Research
Centre for Mechanical Engineering
P.O. Box 49, 2600 AA Delft
The Netherlands
Tel: +31-15-2608423   Fax:+31-15-2564102
E-mail: R.Fey@bouw.tno.nl

Dr. C. Gähler
ETH-Technopark, PFA F18
Internat. Center for Magnetic Bearings
Pfingstweidstr. 30
CH-8005 Zürich, Switzerland
Tel: +41-1-445-1335   Fax:+41-1-445-1365
E-mail: gaehler@ifr.mavt.ethz.ch

Dr. A. Gasparetto
DIEGM-Dipartimento di Ingegneria
University of Udine, Via delle Scienze 208
33100 Udine,Italy
Tel: +39-432-558254   Fax:+39-432-558251
E-mail: gasparetto@picolit.diegm.nnind.it

Prof. C. Grebogi*
University of Maryland at College Park
Lab. for Plasma Research,
College Park, Energy Research Building
Maryland 20742-3511, U.S.A.
Tel: +1 301 405 5021   Fax:+1 301 405 1678
E-mail: grebogi@chaos.umd.edu

Dr. J.G.A.M. v. Heck
DAF-Trucks N.V.
P.O. Box 90065
5600 PT Eindhoven
The Netherlands
Tel: +31 40-2143508   Fax:+31 40-2144331
E-mail: J.G.A.M.v.Heck@wtb.tue.nl

Dr. T.A.G. Heeren
Oce Nederland B.V.
P.O. Box 101
5900 MA Venlo
The Netherlands
Tel: +31 77-3594421   Fax:+31 77-3595472
E-mail: tah@oce.nl

Ir. M.F. Heertjes
Eindhoven University of Technology
Dept. of Mechanical Engng., WFW
P.O. Box 513, 5600 MB Eindhoven
The Netherlands
Tel: +31-40-2472811   Fax:+31-40-2447355
E-mail: marcelh@wfw.wtb.tue.nl

437

Dr. G.H.M. van der Heijden
University College London
Centre for Nonlinear Dynamics
Gower Street, London WC1E 6BT
U.K.
Tel: +44 171-3877050/2927
Fax:+44 171-3800986
E-mail: g.heijden@ucl.ac.uk

Prof. H.Y. Hu
Institute of Vibration Engineering Research
Nanjing University of Aeron. and Astron.
210016 Nanjing,
P.R. China
Tel: +86 25 4494933/3227   Fax:+86 25 4498069

Prof. R.A. Ibrahim*
Wayne State University
Department of Mechanical Engineering
Detroit,
MI 48202, U.S.A.
Tel: +1-313-577-3885   Fax:+1-313-577-8789
E-mail: raouf_ibrahim@eng.wayne.edu

Prof. H. Irschik
Johannes Kepler University of Linz
Institute of Technical Mechanics
Altenbergerstr. 69
A-4040, Linz, Austria
Tel: +43-732-2468-9760   Fax:+43-732-2468-9763
E-mail: Irschik@mechatronik.uni-linz.ac.at

Prof. J.B. Jonker
University of Twente,
Dept. of Mechanical Engrg.
P.O. Box 217, 7500 AE Enschede
The Netherlands
Tel: +31-53-4892591   Fax:+31-53-4893663
E-mail: jonker@wb.twente.nl

Prof. T. Kapitaniak
Technical University of Lodz
Division of Control and Dynamics
Stefanowskiego 1/15
90.924 Lodz, Poland
Tel: +48 42 312231   Fax:+48 42 365646
E-mail: Tomaszka@LODZ1.P.LODZ.PL

Prof. R. Kasper
Otto-von-Guericke-Universität Magdeburg
Facultät für Maschinenbau, Inst. für Maschi-
nen und Antriebstechnik, Universitätsplatz 2,
Postfach 4120, D-39106 Magdeburg, Germany
Tel: +49-391-6718607   Fax:+49-391-6712151
E-mail: IMAT@Uni-magdeburg.de

Dr. P. Kiriazov
Bulgarian Academy of Sciences
Institute of Mechanics
Acad. G. Bonchev Str., bl. 4BG-1113 Sofia,
Bulgaria
Tel: +359-2-7135213   Fax:+359-2-702056
E-mail: dors@bgcict.acad.Bg

Ir. L. Kodde
Eindhoven University of Technology
Dept. of Mechanical Engng., WFW
P.O. Box 513, 5600 MB Eindhoven
The Netherlands
Tel: +31-40-2473174   Fax:+31-40-2447355

Prof. J.J. Kok*
Eindhoven University of Technology
Dept. of Mechanical Engng., WFW
P.O. Box 513,
5600 MB Eindhoven, The Netherlands
Tel: +31 40 2472798   Fax:+31 40 2447355
E-mail: j.j.kok@ctrl.phys.tue.ln

Dr. W.P. Koppens
TNO Road Vehicles Res. Inst.
Crash-Safety Res. Centre
P.O. Box 6033, 2600 JA Delft
The Netherlands
Tel: +31-15-2696373   Fax:+31-15-2624321
E-mail: koppens@wt.tno.nl

Dr. K. Kostadinov
Lab. Mechatronic Systems, Inst. of Mechanics
Acad. G. Bonchev St., Block 4,
BG-1113 Sofia,
Bulgaria
Tel: +359 2706264   Fax:+359 2702056
E-mail: kkostad@bgcict.acad.bg

Prof. E.J. Kostelich
Arizona State University
Dept. of Mathematics
Tempe,
AZ 85287-1804, U.S.A.
Tel: +1 602-965-5006   Fax:+1 602-965-8119
E-mail: eric@saddle.la.asu.edu

Dr. A. de Kraker
Eindhoven University of Technology
Dept. of Mechanical Engng., WFW
P.O. Box 513
5600 MB Eindhoven, The Netherlands
Tel: +31 40 2472847   Fax:+31 40 2447355
E-mail: kraker@wfw.wtb.tue.nl

438

Dr. R. Krause
Daimler-Benz AG, Forschung u. Technik,
Goldsteinstr. 235,
60528 Frankfurt,
Germany
Tel: +49-69-6679-575   Fax:+49-69-6679-418
E-mail:

Prof. E. Kreuzer*
TU Hamburg-Harburg
Arbeitsbereich Meerestechnik II
Eißendorfer Str. 42
D-21071 Hamburg, Germany
Tel: +49-40-7718-3120   Fax:+49-40-7718-2028
E-mail: Kreuzer@tu-harburg.d400.de

Ir. A.H.W.M. Kuijpers
Eindhoven University of Technology
Dept. of Mechanical Engng., WFW
P.O. Box 513, 5600 MB Eindhoven
The Netherlands
Tel: +31-40-2472811   Fax:+31-40-2447355
E-mail: ard@wfw.wtb.tue.nl

Dipl.-Ing. O. Kust
TU Hamburg-Harburg
Arbeitsbereich Meerestechnik II
Eißendorfer Str. 42
D-21071 Hamburg-Harburg, Germany
Tel: +49-40-7718-2815   Fax:+49-40-7718-2028
E-mail: kust@tu-harburg.d400.de

Ir. M.H.L.H. Kusters
Océ Nederland b.v.
P.O. Box 101
Research and Development
5900 MA Venlo, The Netherlands
Tel: +31 77 3593765   Fax: +31 77 3595472
E-mail: mku@oce.nl

Prof. P. Maißer
Technische Universität Chemnitz-Zwickau
Institut für Mechatronik e.v.
Reichenhainer Straße 88
D-09126 Chemniz, Germany
Tel: +49 371 531 4670   Fax:+49 371 531 4669
E-mail: IfMmail@lfm.tu-chemnitz.de

Prof. H. Mann
Czech Technical University of Prague
Computing Centre CVUT
Zikova Street 4
CZ-166 35 Prague 6, Czech Republic
Tel: +42-2-311-2454   Fax:+42-2-2431-0271
E-mail: mann@vc.cvut.cz

Dr. J.P. Meijaard
Delft University of Technology
Fac. WbMT, vakgr. Technische Mechanica
Mekelweg 2, P.O. Box 5033
2600 GA Delft, The Netherlands
Tel: +31-15-2786509   Fax:+31-15-2782150
E-mail: J.P.Meijaard@wbmt.tudelft.nl

Prof. P. Meijers
Delft University of Technology
Fac. WbMT, vakgr. Technische Mechanica
Mekelweg 2, P.O. Box 5033
2600 GA Delft, The Netherlands
Tel: +31-15-2786524   Fax:+31-15-2782150
E-mail: P.Meijers@wbmt.tudelft.nl

Dipl. Phys. R. Mettin
TH Darmstadt
Inst. für Angewandte Physik
Schloßgartenstr. 7, D-64289 Darmstadt
Deutschland
Tel: +49-6151-16-2786   Fax:+49-6151-16-4534
E-mail: robert.mettin@physik.th-darmstadt.de

Dr. S.A. Mikhailov
Safety Control Engineering
University of Wuppertal
GaußStr. 20,
D-42097 Wuppertal, Germany
Tel: +49-202-439-2586   Fax:+49-202-439-2901
E-mail: mihialov@wrcs1.urz.uni-wuppertal.de

Dr. M.J.G. van de Molengraft
Eindhoven University of Technology
Dept. of Mechanical Engng., WFW
P.O. Box 513, 5600 MB Eindhoven
The Netherlands
Tel: +31-40-2472841   Fax:+31-40-2447355
E-mail: rmolen@wfw.wtb.tue.nl

Prof. P.C. Müller*
University of Wuppertal
Safety Control Engineering
GaußStr. 20
D-42097, Wuppertal, Germany
Tel: +49 202 439 2017/2018
Fax:+49 202 439 2901
E-mail: mueller@wrcs1.urz.uni-wuppertal.de

Prof. A.H. Nayfeh*
Virginia Polytechnic Inst.and State Univ.
Dept. of Engineering Science and Mechanics
Blacksburg, Virginia 24061-0219
U.S.A.
Tel: +1 540 231 5453   Fax:+1 540 231 4574
E-mail: anayfeh@vt.edu

Dr. K. Nederveen
Kluwer Academic Publishers
P.O. Box 17
3300 AA Dordrecht
The Netherlands
Tel: +31-78-6392306   Fax:+31-78-6392254
E-mail: nederveen@wkap.nl

Dr. H. Nijmeijer
Twente University, Dept. Applied Math.
Vakgroep Systeem- en Besturingstechnolgie
P.O. Box 217
7500 AE Enschede, The Netherlands
Tel: +31 53-4893442   Fax:+31 53-4340733
E-mail: h.nijmeijer@math.utwente.nl

Dr. Z. Papp
TNO-TPD
P.O. Box 155
2600 AD Delft
The Netherlands
Tel: +31-15-2692234   Fax:+31-15-2692111
E-mail: papp@tpd.tno.nl

Prof. A.D. de Pater
Delft Univ. of Technology, Fac. of Mech. and
Naval Engng., Lab. for Eng. Mechanics
Krammer 4
2641 TZ Pijnacker, The Netherlands
Tel: +31-15-3695346   Fax:+31-15-2782150
E-mail:

Prof. F. Peterka
Institute for Thermomechanics
Academy of Sciences of the Czech Republic
Dolejskova 5
CS 182 000 Prague 8, Czech Republic
Tel: +42 2 6605 3083   Fax:+42 2 858 4695
E-mail: peterka@bivoj.it.cas.cz

Dr. N.B.O.L. Pettit
UMIST, E.E. & E. Department
Control Systems Centre,
P.O. Box 88,
Manchester M60 1QD, U.K.
Tel: +44 161 200 4668   Fax:+44 161 200 4647
E-mail: pettit@csc.umist.ac.uk

Prof. F. Pfeiffer
TU-München,
Lehrstuhl B für Mechanik
Arcisstr. 21, Postfach 202420,
D-80290 München 2, Germany
Tel: +49-89-289-15200   Fax: +49-89-289-15213
E-mail: pfeiffer@lbm.mw.tu-muenchen.de

Dr. A. Pirrotta
Dipartimento di Ingegneria Strutturale
e Geotecnica, Viale delle Science,
90128 Palermo
Italy
Tel: +39-91-6568424   Fax:+39-91-6568451
E-mail: futura@cuc.unipa.it

Prof. D. Pogorelov
Bryansk Inst. Transport Engineering
Dept. Applied Mechanics
b. 50 let Orlovskaya 7, Bryansk, 241035,
Russia
Tel: +7/083-2-568637   Fax:+7/083-2-560533
E-mail: bitm@bitmcnit.bryansk.su

Prof. K. Popp*
Universität Hannover
Institut für Mechanik
Appelstraße 11
D-30167 Hannover, Germany
Tel: +49 511 762 4161   Fax:+49 511 762 4164
E-mail: popp@ifm.uni-hannover.de

Ir. K. Pottie
BWIP
Parallelweg 6
4878 AH Etten-Leur
The Netherlands
Tel: +31-76-5028200   Fax:+31-76-5028487

Dr. H.J. Pradlwarter
University of Innsbruck
Institute of Engineering Mechanics
Technikerstraße 13, A-6020 Innsbruck
Austria
Tel: +43-512-507-6841   Fax:+43-512-507-2905
E-mail: helmut.pradlwarter@uibk.ac.at

Mr. I. Pratt
Loughborough University of Technology
Dept. of Electronic & Electrical Engng.
Ashby Road, Loughborough, LE 11 3TU
U.K.
Tel: +44-1509-228106   Fax:+44-1509-222854
E-mail: elip@lut.ac.uk

Prof. M. Radeş*
Polytechnic University of Bucharest
Splaiul Independenţei 313
79590 Bucharest,
Romania
Tel: +40-1-4100387   Fax:+40-1-4100387
E-mail: rades@form.resist.pub.ro

440

Prof. G. Rega
Universita di Roma La Sapienza
Dip.Ingegneria Strutturale e Geotecnica
Via A. Gramsi 53,
00197 Roma, Italy
Tel: +39-6-44589195   Fax:+39-6-3221449
E-mail: rega@hp720.dsg.uniroma1.it

Dr. O.P. Salimova
Russian Academy of Sciences
Starobitsevskaya st. 19-1-73
Moscow, 113628,
Russia
Tel: +7-095-711-8030   Fax:+7-095-711-6636
E-mail: beletsky@applmat.msk.su

Prof. W.O. Schiehlen*
University of Stuttgart
Institute B of Mechanics
Pfaffenwaldring 9
D-70550 Stuttgart 80, Germany
Tel: +49 711 685 6388   Fax:+49 711 685 6400
E-mail: wos@mechb.uni-stuttgart.de

Prof. K. Schlacher
Johannes Keppler University of Linz
Institute of Automatic Control
Altenbergerstr. 69, A-4040 Linz, Austria
Tel: +43-732-2468-9730   Fax:+43-732-2468-1056
E-mail: kurt@regpro.mechatronic.uni-linz.ac.at

Dr. W.E. Seemann
TH Darmstadt,
Institut für Mechanik II,
Hochschulstr. 1, D-64289 Darmstadt
Germany
Tel: +49-6151-163385   Fax:+49-6151-164125
E-mail: dg6d@hrzpub.th-darmstadt.de

Prof. R.S. Sharp*
Cranfield University
School of Mechanical Engineering
Cranfield
Bedford MK43 OAL, U.K.
Tel: +44-1234-754708   Fax:+44-1234-750728
E-mail: R.S.SHARP@CRANFIELD.AC.UK

Dr. M. Steinbuch
Philips Research Labs.
Prof. Holstlaan 4
P.O. Box 80.000
5600 JA Eindhoven
Tel: +31 40-2743597   Fax:+31 40-2744810
E-mail: Steinbuc@natlab.research.philips.com

Prof. G. Stépán*
Technical University of Budapest,
Department of Applied Mechanics,
H-1521 Budapest,
Hungary
Tel: +36-1-463-1369   Fax:+36-1-463-3471
E-mail: stepan@mm.bme.hu

Dr. D.A. Tortorelli
University of Illinois
Mechanical and Industrial Engineering
1206 W. Green Street Urbana,
IL 61801 U.S.A.
Tel: +1 217-333-5991   Fax:+1 217-244-6534
E-mail: dtortore@uiuc.edu

Prof. H. True*
The Technical University of Denmark
Institute of Mathematical Modelling
Building 321
DK-2800 Lyngby, Denmark
Tel: +45-45-25-3016   Fax:+45-45-93-2373
E-mail: HT@IMM.DTU.DK

Prof. H. Ulbrich
Universität GH Essen
Lehrstuhl für Mechanik, FB12
Schutzenbahn 70
D-45117 Essen, Germany
Tel: +49-201-183-2913   Fax:+49-201-183-2871
E-mail:
Ulbrich/Bormann@hrz-ntas.hrz.uni-essen.de

Dr. F.E. Veldpaus
Eindhoven University of Technology
Dept. of Mechanical Engng., WFW
P.O. Box 513
5600 MB Eindhoven, The Netherlands
Tel: +31 40 2472796   Fax:+31 40 2447355
E-mail: frans@wfw.wtb.tue.nl

Dr. G. Verbeek
Eindhoven University of Technology
Dept. of Mechanical Engng., WFW
P.O. Box 513, 5600 MB Eindhoven
The Netherlands
Tel: +31-40-2474838   Fax:+31-40-2447355
E-mail: bertv@wfw.wtb.tue.nl

Dr. E.L.B. van de Vorst
TNO Road Vehicles Res. inst.
Crash Safety Res. Centre
P.O. Box 6033, 2600 JA Delft
The Netherlands
Tel: +31-15-2697436   Fax:+31-15-2624321
E-mail: vandeVorst@wt.tno.nl

Ir. O. de Vrij
DAF Special Products
Eindhovenseweg 120,
P.O. Box 436
5660 AK Geldrop, The Netherlands
Tel: +31 40 2809228   Fax:+31 40 2809123

Drs. J.G. de Weger
Eindhoven University of Technology
Dept. of Physics, Building W&S 0.46
P.O. Box 513,
5600 MB Eindhoven, The Netherlands
Tel: +31 40 2472143   Fax:+31 40 2464151
E-mail: weger@tnh.phys.tue.nl

Dr. P.M.R. Wortelboer
Philips CFT
Development Support
P.O. Box 218/SAQ-2730
2600 MD Eindhoven, The Netherlands
Tel: +31-40-2742917   Fax:+31-40-2744810
E-mail: wortel@prl.philips.nl

Ir. N. van de Wouw
Eindhoven University of Technology
Dept. of Mechanical Engng., WFW
P.O. Box 513, 5600 MB Eindhoven
The Netherlands
Tel: +31-40-2473358   Fax:+31-40-2447355
E-mail: nathan@wfw.wtb.tue.nl

LIST OF SPONSORS

Generous financial support contributed to the success of the Symposium. The support of the following sponsors is gratefully acknowledged

- International Union of Theoretical and Applied Mechanics

- Royal Dutch Academy of Sciences

- Division of Mechanics, Royal Dutch Institute of Engineers

- Division MRBT, Royal Dutch Institute of Engineers

- Eindhoven University of Technology

- Stevin Centre for Computational and Experimental Engineering Science, Eindhoven

- TNO Institute of Applied Physics, Delft

- TNO Road Vehicles Research Institute, Delft

- TNO Institute of Building and Construction Research, Delft

- Shell Netherlands B.V., Rotterdam

- Philips Medical Systems, Best

- Philips Research, Eindhoven

- Philips Centre for Manufacturing Technology, Eindhoven

- DAF Trucks N.V., Eindhoven

- Van Doorne's Transmissie B.V., Tilburg

- Océ Netherlands B.V., Venlo

- DAF Special Products, Eindhoven

- MARC Analysis Research Corp. Europe, Zoetermeer

- Silicon Graphics B.V. Computer Systems

- LMS International N.V., Leuven, Belgium

- Kluwer Academic Publishers, Dordrecht

- World Trade Center Eindhoven N.V.

Mechanics

SOLID MECHANICS AND ITS APPLICATIONS

Series Editor: G.M.L. Gladwell

Aims and Scope of the Series

The fundamental questions arising in mechanics are: *Why?*, *How?*, and *How much?* The aim of this series is to provide lucid accounts written by authoritative researchers giving vision and insight in answering these questions on the subject of mechanics as it relates to solids. The scope of the series covers the entire spectrum of solid mechanics. Thus it includes the foundation of mechanics; variational formulations; computational mechanics; statics, kinematics and dynamics of rigid and elastic bodies; vibrations of solids and structures; dynamical systems and chaos; the theories of elasticity, plasticity and viscoelasticity; composite materials; rods, beams, shells and membranes; structural control and stability; soils, rocks and geomechanics; fracture; tribology; experimental mechanics; biomechanics and machine design.

Kluwer Academic Publishers – Dordrecht / Boston / London

Mechanics

Kluwer Academic Publishers – Dordrecht / Boston / London

Mechanics

SOLID **MECHANICS AND ITS APPLICATIONS**
Series Editor: G.M.L. Gladwell

Kluwer Academic Publishers – Dordrecht / Boston / London

Mechanics

FLUID MECHANICS AND ITS APPLICATIONS
Series Editor: R. Moreau

Aims and Scope of the Series

The purpose of this series is to focus on subjects in which fluid mechanics plays a fundamental role. As well as the more traditional applications of aeronautics, hydraulics, heat and mass transfer etc., books will be published dealing with topics which are currently in a state of rapid development, such as turbulence, suspensions and multiphase fluids, super and hypersonic flows and numerical modelling techniques. It is a widely held view that it is the interdisciplinary subjects that will receive intense scientific attention, bringing them to the forefront of technological advancement. Fluids have the ability to transport matter and its properties as well as transmit force, therefore fluid mechanics is a subject that is particularly open to cross fertilisation with other sciences and disciplines of engineering. The subject of fluid mechanics will be highly relevant in domains such as chemical, metallurgical, biological and ecological engineering. This series is particularly open to such new multidisciplinary domains.

Kluwer Academic Publishers – Dordrecht / Boston / London

Mechanics

FLUID MECHANICS AND ITS APPLICATIONS
Series Editor: R. Moreau

Kluwer Academic Publishers – Dordrecht / Boston / London